Self-Cleaning of Surfaces and Water Droplet Mobility

Self-Cleaning of Surfaces and Water Droplet Mobility

Bekir Sami Yilbas
Mechanical Engineering Department and Center of Excellence in Renewable Energy, King Fahd University of Petroleum & Minerals, Dhahran, Saudi Arabia

Abdullah Al-Sharafi
Mechanical Engineering Department, King Fahd University of Petroleum & Minerals, Dhahran, Saudi Arabia

Haider Ali
Mechanical Engineering Department, King Fahd University of Petroleum & Minerals, Dhahran, Saudi Arabia

ELSEVIER

Elsevier
Radarweg 29, PO Box 211, 1000 AE Amsterdam, Netherlands
The Boulevard, Langford Lane, Kidlington, Oxford OX5 1GB, United Kingdom
50 Hampshire Street, 5th Floor, Cambridge, MA 02139, United States

Notices

Knowledge and best practice in this field are constantly changing. As new research and experience
broaden our understanding, changes in research methods, professional practices, or medical
treatment may become necessary.

Practitioners and researchers must always rely on their own experience and knowledge in evaluating
and using any information, methods, compounds, or experiments described herein. In using such
information or methods they should be mindful of their own safety and the safety of others, including
parties for whom they have a professional responsibility.

To the fullest extent of the law, neither the Publisher nor the authors, contributors, or editors, assume
any liability for any injury and/or damage to persons or property as a matter of products liability,
negligence or otherwise, or from any use or operation of any methods, products, instructions, or ideas
contained in the material herein.

British Library Cataloguing-in-Publication Data
A catalogue record for this book is available from the British Library

Library of Congress Cataloging-in-Publication Data
A catalog record for this book is available from the Library of Congress

ISBN: 978-0-12-814776-4

For information on all Elsevier publications
visit our website at https://www.elsevier.com/books-and-journals

Publisher: Matthew Deans
Acquisition Editor: Christina Gifford
Editorial Project Manager: Peter J. Ileweyn
Production Project Manager: R. Vijay Bharath
Cover Designer: Mark Rogers

Typeset by MPS Limited, Chennai, India

Working together
to grow libraries in
developing countries

www.elsevier.com • www.bookaid.org

Contents

Preface

Climate change results in regular dust storms around the world. This phenomenon is particularly noteworthy in desert areas such as in the Middle East, North Africa, Central Asia, and some regions in North America. Dust particles have various sizes and shapes. Small dust particles can suspend in air over many days after a storm and some of these particles can settle on exposed surfaces in the open environment. Although these particles are small in size, they cover large areas over time and modify optical, textural, and other characteristics of surfaces. Mineral dust particles have a tendency to absorb and scatter solar radiation, lowering the solar power reaching surfaces and thus altering surface temperature. While some of the solar power loss is recovered by long wave emission of radiation by the dust particles, it may not contribute considerably to solar power harvesting in terms of useful energy generation through concentrated solar heating, electricity generation by photovoltaics, etc. Minimization of dust particle settlement on the active surfaces of energy-harvesting devices is currently a challenge in the solar energy field. Efforts to generate cost-effective antifouling surfaces for energy-harvesting applications is also ongoing. Dust particles are comprised of various elements including alkaline and alkaline earth metals such as Ca, K, and Na. Some dust compounds can dissolve in water condensate in humid ambient conditions while forming a chemically active solution. This, in turn, causes a liquid layer to form on the solid surface under the gravitational potential energy. The film formed by the liquid solution has several effects on surfaces, and can result in erosion and corrosion while permanently damaging the solid surface. In addition, upon drying the liquid solution enhances dust particle adhesion on solid surfaces and thus requires great effort to remove.

Due to the above, self-cleaning of surfaces offers several advantages over conventional cleaning methods. Self-cleaning does not involve energy-intensive processes unlike conventional cleaning methods that use water, brush, or air jet for cleaning. Using nonconventional cleaning methods of surfaces is difficult in some areas such as rural regions where water scarcity is an issue and electricity supply is limited. In general, self-cleaning of surfaces requires hydrophobic wetting on surfaces, which reduces particle adhesion on surfaces significantly while enhancing particle mobility. The wetting state of surfaces depends on the interfacial energies of the solid and liquid, surface texture parameters, and the LaPlace pressure of the droplet fluid;

however, the surface hydrophobicity of solid substrates is mainly influenced by the surface texture parameters and the surface free energy. The low value of the surface free energy and the surface texture with micro/nano pillars are favorable for achieving the hydrophobic wetting state on surfaces. Mimicking the nature of hydrophobic surfaces, such as lotus leaves surfaces, enables creating surfaces with superhydrophobic characteristics, resulting in a substantial increase in hydrophobicity when the right combination of chemical modification and surface roughness of the substrates is obtained. Many techniques and processes have been proposed for enhancing the hydrophobicity of surfaces. Some of these techniques involve multistep procedures and harsh conditions or require specialized reagents and equipment. Deposition of functionalized nano-silica particles, sol-gel coating, and crystallization via immersion techniques are some of the promising methods available for improving surface hydrophobicity. Although these techniques are fast and generate hierarchical texture structures on surfaces with low surface energies, the optical characteristics of the resulting surfaces may degrade because of increased scattering and absorption of the incident optical radiation. Consequently, further research is needed to improve the optical transmittance of these hydrophobic surfaces.

In this book, assessment methods of the surface characteristics of self-cleaning applications are introduced. Model studies are explained for particle adhesion on surfaces, and the wetting states of surfaces involved with wetting state enhancement are presented. Characterization studies for environmental dust particles are also introduced and the mud formed from the dust particles on solid surfaces in humid ambient conditions is explored. Self-cleaning aspects of hydrophobic surfaces are discussed and the dynamics of the water droplets on inclined hydrophobic surfaces are presented. The limitations of liquid droplet rolling and sliding on hydrophobic surfaces are also covered, and the relation between droplet rolling and self-cleaning of surfaces is introduced. The dust particles and dry mud adhesion on hydrophobic surfaces are analyzed and tangential force measurement for removal of the dust particles and dried mud from the surfaces is presented. The effects of thermocapillary forces on droplet mobility are analyzed and simulated, incorporating appropriate conditions for self-cleaning applications. The findings are discussed in comparison with the experimental results. The influence of the inclination angle of hydrophobic surface together with droplet size on self-cleaning of surfaces is presented. The motion of dust particles in droplet fluid under thermocapillary forces is included and droplet fluidity due to flow forces is formulated and predicted accordingly. Practical applications of self-cleaning processes are introduced and the findings are presented in detail. Future treatments will review other issues and related processes not covered here.

Acknowledgments

I would like to acknowledge the role of King Fahd University of Petroleum & Minerals (KFUPM) for extending strong support from the beginning to the end of this project and facilitating its completion. The author would also like to thank the Deanship of Scientific Research at KFUPM for supporting the writing of this book through project #BW171003. Thanks also to the colleagues who contributed to the work presented here. In particular, thanks to Drs. Nasser Al-Aqeeli, Numan Abu-Dheir, Mazen Khaled, Fahd Al-Sulaiman, Hussain Al-Qahtani, Mr. Ghassan Hassan Hajhamed, Mr. Muhammad Rizwan Yousaf, Mr. Aditia Rifai, and all my other graduate and undergraduate students.

Chapter 1

Introduction

Chapter Outline

Climate change increases the frequency of regular dust storms around the globe particularly in the Middle East and the Sahara region. The performance of solar energy-harvesting devices is highly dependent on the amount of incident solar energy reaching the active surface of the energy-harvesting devices. Dust settlements on such surfaces degrade device performance in terms of efficiency and output power and require additional efforts to remove dust from surfaces. Several methods have been proposed for dust removal from surfaces and some of these include sonic and mechanical excitation of dust particles, mechanical brushing, air jet blowing, and water cleaning. Most of these methods involve sophisticated devices or external efforts such as energy to operate. Scarcity of the clean water limits the practical applications of water-jet and water-film cleaning of surfaces and, hence, minimization of water consumption during cleaning becomes necessary. Natural powers such as gravitation and wind power are not sufficient for the dust-removal process. The mechanical techniques used to clean surfaces include blowing, brushing, and ultrasonic driving and vibrating. However, there are many problems associated with these mechanical methods of cleaning. The brushing cleaning method is inefficient because of the strong adhesion and small size of the dust particles. Surfaces can also be damaged by the brush during the wiping process. Blowing is also ineffective due to difficulties in maintainability and high energy consumption. Thus the self-cleaning method is a promising technique for cleaning dusty surfaces. Adopting self-cleaning can also minimize the energy required for cleaning energy-harvesting device surfaces while providing an effective cleaning process for removing dust particles. In general, the self-cleaning process utilizes hydrophobic characteristics of surfaces with low free energy of surface and reduced particle contact area at the surface, due to air gap within surface texture, which provides weak adhesion of particles on surfaces. In addition, low-contact-angle hysteresis of hydrophobic surfaces enables water droplet rolling on the inclined hydrophobic surface. This, in turn, facilitates picking up of weakly adhered

Self-Cleaning of Surfaces and Water Droplet Mobility. DOI: https://doi.org/10.1016/B978-0-12-814776-4.00001-X

dust particles from the hydrophobic surface by rolling water droplets. Self-cleaning of optically transparent wafers is also of interest in many applications and has become critical for efficient harvesting of solar energy in harsh environments [1]. Generating self-cleaning characteristics of optically transparent wafer surfaces is one of the current challenges because the task of achieving durable surfaces in harsh environmental conditions is extremely difficult. Lotus leaves, red rose petals, fish scales, etc. inspire designs and fabricate artificial surfaces with self-cleaning characteristics [2]. These examples, as observed in nature, show the importance of surface hydrophobicity in achieving self-cleaning characteristics at the surface. Hydrophobic characteristics of surfaces mainly depend on surface texture and the surface free energy of the substrate materials. Surface texture composed of micro/nanopillars with low surface energy is required for achieving hydrophobic characteristics of surfaces [3]. Although several methods and strategies are introduced to create hydrophobic surfaces, some of these methods are involved with multistep procedures in harsh conditions, specialized reagents, and high cost. These include phase separation [4], electrochemical deposition [5], plasma treatment [6], sol—gel processing [7], electrospinning [8], laser texturing [9], and solution immersion [10]. Challenges are faced during surface texturing and chemical/physical modification of surfaces in reducing surface free energy because of high cost, long processing time, and equipment and skilled manpower requirements. Consequently, development of innovative technologies for cost-effective processing for improving surface hydrophobicity of optically transparent wafers is essential.

Solar energy harvesting plays a crucial role in renewable energy applications around the globe. Utilization of high-temperature-resistant ceramic components is unavoidable for solar thermal applications. Because of its excellent optical properties, in·terms of absorption and thermal emission, Zirconium nitride can serve as a selective surface for solar energy harvesting in thermal systems [11]. However, climate change has created dust storms around the Middle East [12]. Dust settlement on solar energy-harvesting surfaces reduces device performance in terms of output power and efficiency [13]. The environmental dust particles are composed of various elements and compounds including alkaline and alkaline earth metals [14]. In humid ambient air, water condensates on the dust particles and compounds of the dust particles dissolve into the condensate water. This forms a chemically active liquid solution, which accumulates at the interface of the device surface and the dust particles under gravity [14]. Chemically active liquid solution causes some asperities on the device surface, such as pin holes and pit sites, while causing permanent damages [15]. In addition, once the liquid solution dries at the device surface, adhesion between the dried liquid and the surface becomes strong and the efforts required to remove the dried solution and the mud from the surface increase significantly. One of the solutions to create self-cleaning characteristics at the surface is to improve surface hydrophobicity. In general, surface hydrophobicity is associated with the micro/nanoscale surface texture and low surface free energy. However, plane yttria-stabilized zirconia surface has high surface free energy and demonstrates hydrophilic characteristics. Consequently, altering the characteristics of zirconia surface through surface processing is essential to generate surface hydrophobicity. Although several methods have been suggested and many techniques have been reported for generating hydrophobic characteristics at the surfaces [4–10], laser gas-assisted texturing offers considerable advantages over the multistep processes for hydrophobizing surfaces. This is particularly true for ceramic surfaces [16]. Some of these advantages include high-speed processing, precision of operation, and low cost. In the laser texturing process, combination of melting and evaporation of ceramic surface takes place via ablation [16,17]. This in turn generates surface texture consisting of micro/nanopoles and cavities. The use of high-pressure assisting gas, such as nitrogen, generates nitride compounds on the laser textured surface [16]; in which case, nitride compounds have low surface energy than the oxides [16]. In addition, arrays of micro/nanosize pole/pillar formed at the surface can improve capacity of the laser textured surface. However, dust settlement on the laser textured surface changes surface characteristics such as surface hydrophobicity. Hence, investigation of environmental dust particle adhesion and influence of dried mud solution on the characteristics of laser textured zirconia surface is essential.

According to the World Meteorological Organization protocol, the events of dust have been classified into different categories based on visibility. The first group is the dust-in-suspension that reduces visibility up to 10 km. The second category is the blowing dust that raises sand and dust at the observation time and reduces the visibility from 1 to 10 km. The third group is the dust storm where strong winds lift soil and dust particles and visibility is reduced to 200–1000 m. Finally, there are severe dust storms with strong winds that lift large dust quantities and decrease the visibility to 200 m [18]. It is reported that dust emission varies with space and time and greatly depends on land-surface and atmospheric conditions. Different technologies and measurement methods have been established to track the flow of dust around the globe. Satellite has been used extensively to track dust migration and identify physical quantities. Estimation of the dust physical quantities is difficult due to the complicated signal structure (from the dust, clouds, land surface, and other aerosol) that are detected by the satellite. Therefore satellite data can be integrated with other radiation measurements to ensure accurate estimations [18]. Dust storms originate in semiarid and arid regions (e.g., Mongolia, the Middle East, the Sahara). The dust storms throughout these places add more than 5000 million tons of dust into the atmosphere per year. North Africa is considered as the world's largest source of dust to the atmosphere, and emits around 10^{12} g of dust annually. Of this, 250 Tg of dust is transported over the tropical North Atlantic Ocean [18]. The dust storms that originate from the Sahara (North Africa) migrate and influence the air quality and climate of the Middle East, Africa, Asia, Europe, the Americas, and the Caribbean. The Middle East is considered as the most affected region by dust because of the frequent dust events around the year [19]. The wind force passing over wide deserts motivates the loosely sand particles to vibrate and leap. Then, the small dust particles start to travel in suspension. The regional sand—dust storms have a wide extension that covers different countries. Removing of vegetation cover is a direct cause of dust storms that result in loosening the top soil cover that prevents the existence of loose sand particles [20].

On the other hand, the dust settlement onto the surfaces located in open environments causes irrecoverable damages on surfaces and lowers system performance, such as those associated with solar thermal and solar photovoltaic (PV) applications. There are many surface treatment methods being reported in the literature for minimizing the dust and mud effect on surface characteristics and surface performance [21]. However, self-cleaning or cost-effective removal of dust particles from such surfaces still remains challenging. As indicated earlier, the dust particles absorb water vapor in ambient humid air and form mud at the surface. Once the mud is dried at high-temperature conditions under the solar radiation, it becomes difficult to remove from the surfaces. This is because the adhesion force between the dry mud and surfaces is not only governed by the physical forces, such as van der Waals forces, but also chemically induced forces, such as forces due to ionic and covalent bonding. This soiling process on solar panels—because of airborne dust—happens due to dry dust deposition from gravitational settling and wet particle deposition. Such soiling absorbs and reflects the incident solar radiation leading to a reduction of light transmittance. However, the dissolved ions (Na^+, K^+, Ca^{+2}) from dust particles attract mud and clay mineral structures. These ions penetrate the mud layers and hold them together. The process of dust—water interaction and mud formation has the following steps: firstly, soil and dust particle surfaces are composed of hydroxyls and oxygen layers, so hydroxyls attract the negative corners and oxygen with positive corners, which easily generate hydrogen bonding between water molecules. Dust particle reactions take place with molecules of polarized water that separate the single ions (Na^+, K^+, Ca^{+2}) surrounded by hydration sheaths and dissolved in the water molecules while forming oxygen and hydroxyls (KOH, NaOH, etc.) stretch. Solid undissolved dust particles have pendant hydroxyl ($-OH$) groups that can form bonds or strong polar attractions to mud solution and inorganic surfaces. During drying, the dissolved ions (Na^+, K^+, Ca^{+2}, Cl^-, SiO^-) attract mud structure due to electrostatic and ionic bonding force. These ions dissolve in the mud solution and holding the dust particles together and form crystals in between, staying in the mud structure during water evaporation and increasing the adhesion force.

In reviewing the various dust-impact mitigation approaches, the focus is on the techniques that deal with the effect of dust as it is one of the most significant environmental factors affecting PV module performance in the regions where solar power may be a particularly viable alternative to fossil fuels as a result of the high level of solar radiation. To recover the performance of a PV module covered with dust, it is important to carry out periodic cleaning. However, the required frequency of cleaning depends on environmental conditions. There are a variety of methods that have been used or developed to mitigate the dust effect on PV module performance. Some also contribute to mitigating other climate factors as well, such as temperature and humidity. The reported dust mitigation methods in the literature can be divided into four categories: spontaneous, mechanical and electromechanical, electrostatic shields, and micro- and nanoscale surface functionalization. The purpose of micro- and nanoscale surface fabrication is to develop self-cleaning surfaces with optimal optical properties. This method enables the creation of a super-hydrophobic surface that has low wettability and high water droplet mobility. Such a surface can enhance the cleaning efficiency and thereby reduce the necessary cleaning frequency. Super-hydrophobic surfaces are comprised of a micro- or nanostructure surface coated by a thin film of low surface energy material or vice versa [22].

Most of today's self-cleaning technology has been derived from nature. Several surfaces in nature have self-cleaning features, for example, butterfly wings and plant leaves. During the 20th century, this technology received a great deal of attention due to the wide range of applications (PV cleaning, cements to textiles, etc.). In addition, this technology provides different solutions to reducing maintenance cost, reducing cleaning time, and eliminating manpower efforts [23]. Coating used in self-cleaning applications is classified based on the action of water into hydrophobic and hydrophilic categories. In the hydrophobic technique, the droplet rolls and slides thereby cleaning the surface, while in the hydrophilic technique the cleaning takes place by water spreading on the surface [23]. The self-cleaning property is mainly affected by surface contact angle (CA). The surface CA is the angle formed between the solid surface and liquid droplet at the three phase contact line. The surfaces are classified based on water CA as: hydrophilic surface for CA < 90 degrees, hydrophobic surface for CA > 90 degrees, super (ultra) hydrophobic for CA > 150 degrees, and super (ultra) hydrophilic for CA close to 0 degree. Super-hydrophobic surfaces with water-repelling characteristics are of great importance in different areas (e.g., water-repellent automotive parts, waterproofing of textiles, biofouling prevention, drag reduction in microchannels, and self-cleaning windows) [24]. The common method for studying super-hydrophobic materials is based on static and dynamic CA. Dynamic CA are measured by evaluating the receding and advancing CA for the sliding water drops [25].

The study of dust and dry mud on selective surfaces involves several experimental techniques (mechanically and chemically) using different characterization tools. The characterization tools used for dust particle examinations can be divided into three main groups: mechanical testing tools, microstructure characterization tools, and chemical investigation instruments. The mechanical equipment includes tensile testing machines and scratching testing machines used to measure the interfacial forces (adhesion and cohesion force) between the dry mud and substrate. In addition, the microstructure characterization methods include scanning electron microscopy (SEM), energy dispersive spectroscopy (EDS), atomic force microscopy (AFM), X-ray diffraction (XRD), and three dimensional (3D) imaging. The SEM is a multipurpose device available for microstructure identification, chemical composition, and morphology exploration of dust particles, while EDS is an elemental analysis technique that is used to determine the dust particle elements and the percentage of their quantities. The AFM provides a 3D profile of the dust particle and dry mud layer in nanoscale. The AFM measures the forces between the surface and sharp probe (usually made of silicon nitride) at short distance. The XRD known as the compound analysis technique is used for phase identification and unknown solids determination of dust particle contents. The 3D optical microscope is a light microscope based on white light interferometry that uses the light photons beam to form detailed 3D images from the specimen surface. The advantage of the image taken by the light electron microscope is its accuracy and high magnification that show more information about surface topography. The 3D optical microscopy can provide detailed information about the shape and size of dust particles as well as measure surface topography of the dry mud layer. Dust particles and dry mud solution undergo several chemical investigations using Fourier transform infrared spectroscopy (FTIR) and pH measurements. The FTIR technique can be used to obtain the emission or absorption infrared spectrum of dust solid particles. FTIR analysis aids in identifying unknown dust materials, determining the quality or consistency of a sample, and the amount of dust components in a mixture.

In general, the dust particles are composed of various sizes and shapes. Small dust particles attach to large particles, and are associated with the electrostatic charges of small particles. In this regard, it is important to evaluate the dust particles based on their size and shape. The shape of dust particles can be categorized by two key geometric parameters. However, the shape factor is related to the inverse of the particle circularity in relation to the complexity of the particle; in which case, the shape factor of unity corresponds to the perfect circle. The aspect ratio corresponds to the ratio of major to minor axes of the ellipsoid best fit to the particle, which is associated with the approximate particle roundness. Consequently, from the measured particle sizes, the equivalent circular area can be determined for the round shapes. In the case of noncircular shapes, an ellipse is considered by assuming that the longest projection as the major axis and preserving the particle cross-sectional area. The relation between the particle size and the aspect ratio or the shape factor is not simple. However, a simplified assessment can be introduced to classify dust particles in terms of their shapes. Since an inverse relation is observed between the particle size and the aspect ratio, increasing particle sizes gives rise to low aspect ratios. However, a direct relation is present in between the particle size and the shape factor. Consequently, the shape factor increases while aspect radio decreases with increasing particle size. The shape factor becomes almost unity for the small size particles (≤ 2 μm) and the median shape factor approaches almost three for the large size particles (≥ 10 μm) [14]. Although the particle size varies from submicrometer to tens of micrometer, the average particle size is in the order of 1.2 μm [26]. The EDS data gives (wt.%) of dust particles. Dust particles contain various elements including Si, Ca, Mg, Na, K, Cl, S, O, and Fe [26]. The existence of salt and oxide compounds in dust particles can be identified from X-ray diffractogram for dust particles. The peaks of Na and K are probably related to the salt, and iron is likely related to the clay-aggregated hematite (Fe_2O_3) [26]. The AFM image of the dust particles and dry mud removed from the surface demonstrate the various texture morphologies at the surface. Using the line scan mode, the texture height and the average surface roughness of the surface can be measured. The pH of the solution demonstrates the base nature and the high rate of increase pH is associated with the presence of OH^- ions in mud solution, which is related to the dissolution of alkaline (Na, K) and alkaline earth (Ca) metals in dust particles [26].

In this book, the importance, formulation, and processes of wetting states of surfaces pertinent to self-cleaning applications are presented. The influence of environmental dust particles on optically transparent and opaque surfaces is discussed in harsh environments with high humidity and high temperature. Droplet dynamics on surfaces are also considered for water droplet cleaning of soiled surfaces. The droplet heat transfer is explored to account for the thermocapillary effects on the droplet behavior on various surfaces.

REFERENCES

[1] J. Strauss, P. Soave, R. Ribeiro, F. Horowitz, Absorber and self-cleaning surfaces on modified polymer plates for solar harvesting in the humid (sub) tropics, Solar Energy 122 (2015) 579−586.

[2] B. Bhushan, Y.C. Jung, K. Koch, Self-cleaning efficiency of artificial superhydrophobic surfaces, Langmuir 25 (5) (2009) 3240−3248.

[3] G. Azimi, R. Dhiman, H.-M. Kwon, A.T. Paxson, K.K. Varanasi, Hydrophobicity of rare-earth oxide ceramics, Nat. Mater. 12 (4) (2013) 315.

[4] J.T. Han, X. Xu, K. Cho, Diverse access to artificial superhydrophobic surfaces using block copolymers, Langmuir 21 (15) (2005) 6662−6665.

[5] N.J. Shirtcliffe, G. McHale, M.I. Newton, G. Chabrol, C.C. Perry, Dual-scale roughness produces unusually water-repellent surfaces, Adv. Mater. 16 (21) (2004) 1929–1932.

[6] H. Kinoshita, A. Ogasahara, Y. Fukuda, N. Ohmae, Superhydrophobic/superhydrophilic micropatterning on a carbon nanotube film using a laser plasma-type hyperthermal atom beam facility, Carbon 48 (15) (2010) 4403–4408.

[7] S.S. Latthe, H. Imai, V. Ganesan, A.V. Rao, Superhydrophobic silica films by sol–gel co-precursor method, Appl. Surface Sci. 256 (1) (2009) 217–222.

[8] M. Ma, Y. Mao, M. Gupta, K.K. Gleason, G.C. Rutledge, Superhydrophobic fabrics produced by electrospinning and chemical vapor deposition, Macromolecules 38 (23) (2005) 9742–9748.

[9] A. Dunn, T.J. Wasley, J. Li, R.W. Kay, J. Stringer, P.J. Smith, et al., Laser textured superhydrophobic surfaces and their applications for homogeneous spot deposition, Appl. Surface Sci. 365 (2016) 153–159.

[10] X. Zhang, Y. Guo, P. Zhang, Z. Wu, Z. Zhang, Superhydrophobic CuO@ Cu2S nanoplate vertical arrays on copper surfaces, Mater. Lett. 64 (10) (2010) 1200–1203.

[11] J.-P. Meng, X.-P. Liu, Z.-Q. Fu, K. Zhang, Optical design of $Cu/Zr_{0.2}AlN_{0.8}/ZrN/AlN/ZrN/AlN/Al_{34}O_{62}N_4$ solar selective absorbing coatings, Solar Energy 146 (2017) 430–435.

[12] A.J. Parolari, D. Li, E. Bou-Zeid, G.G. Katul, S. Assouline, Climate, not conflict, explains extreme Middle East dust storm, Environ. Res. Lett. 11 (11) (2016) 114013.

[13] N.S. Beattie, R.S. Moir, C. Chacko, G. Buffoni, S.H. Roberts, N.M. Pearsall, Understanding the effects of sand and dust accumulation on photovoltaic modules, Renew. Energy 48 (2012) 448–452.

[14] B.S. Yilbas, H. Ali, N. Al-Aqeeli, M.M. Khaled, S. Said, N. Abu-Dheir, et al., Characterization of environmental dust in the Dammam area and mud after-effects on bisphenol-A polycarbonate sheets, Sci. Rep. 6 (2016) 24308.

[15] B.S. Yilbas, G. Hassan, H. Ali, N. Al-Aqeeli, Environmental dust effects on aluminum surfaces in humid air ambient, Sci. Rep. 7 (2017) 45999.

[16] B. Yilbas, H. Ali, N. Al-Aqeeli, M. Oubaha, M. Khaled, N. Abu-Dheir, Laser gas assisted nitriding and sol–gel coating of alumina surfaces: effect of environmental dust on surfaces, Surface Coatings Technol. 289 (2016) 11–22.

[17] B. Yilbas, Laser treatment of zirconia surface for improved surface hydrophobicity, J. Alloys Compounds 625 (2015) 208–215.

[18] Y. Shao, C. Dong, A review on East Asian dust storm climate, modelling and monitoring, Global Planet. Change 52 (1–4) (2006) 1–22.

[19] D. McGee, G. Winckler, J. Stuut, L. Bradtmiller, The magnitude, timing and abruptness of changes in North African dust deposition over the last 20,000 yr, Earth Planet. Sci. Lett. 371 (2013) 163–176.

[20] V. Sissakian, N. Al-Ansari, S. Knutsson, Sand and dust storm events in Iraq, J. Nat. Sci. 5 (10) (2013) 1084–1094.

[21] M.A. Ramli, E. Prasetyono, R.W. Wicaksana, N.A. Windarko, K. Sedraoui, Y.A. Al-Turki, On the investigation of photovoltaic output power reduction due to dust accumulation and weather conditions, Renew. Energy 99 (2016) 836–844.

[22] K. Yadav, B.R. Mehta, K.V. Lakshmi, S. Bhattacharya, J.P. Singh, Tuning the wettability of indium oxide nanowires from superhydrophobic to nearly superhydrophilic: effect of oxygen-related defects, J. Phys. Chem. C 119 (28) (2015) 16026–16032.

[23] V.A. Ganesh, H.K. Raut, A.S. Nair, S. Ramakrishna, A review on self-cleaning coatings, J. Mater. Chem. 21 (41) (2011) 16304–16322.

[24]   J. Zimmermann, F.A. Reifler, G. Fortunato, L.C. Gerhardt, S. Seeger, A simple, one-step approach to durable and robust superhydrophobic textiles, Adv. Funct. Mater. 18 (22) (2008) 3662−3669.

[25]   M. Miwa, A. Nakajima, A. Fujishima, K. Hashimoto, T. Watanabe, Effects of the surface roughness on sliding angles of water droplets on superhydrophobic surfaces, Langmuir 16 (13) (2000) 5754−5760.

[26]   B.S. Yilbas, H. Ali, M.M. Khaled, N. Al-Aqeeli, N. Abu-Dheir, K.K. Varanasi, Influence of dust and mud on the optical, chemical, and mechanical properties of a PV protective glass, Sci. Rep. 5 (2015) 15833.

Chapter 2

Wetting Characteristics of Surfaces

Chapter Outline

2.1 INTRODUCTION

The wetting state of surfaces plays a major role in self-cleaning applications of surfaces. Texturing the surfaces generating the micro/nanosize pillars on the surface reduces the contact area between the surface and the particle. In addition, reducing the surface free energy lowers the van der Waals forces between the surface and the particle. This arrangement remains true for liquids. In the case of wetting state of the surfaces, the liquid droplet contact is essential to assess the wetting state. Two main categories of wetting state were introduced previously based on Wenzel and Cassi and Baxter states. In any case, the surface possesses either hydrophilic or hydrophobic characteristics, which can be assessed by the droplet contact angle measurements. In addition, the contact angle hysteresis remains important for droplet mobility on surfaces. In this chapter, the wetting characteristics of surfaces are presented and formulation of adhesion force on surfaces is introduced. The assessment of methods associated with the surface wetting characteristics is also discussed.

Self-Cleaning of Surfaces and Water Droplet Mobility. DOI: https://doi.org/10.1016/B978-0-12-814776-4.00002-1

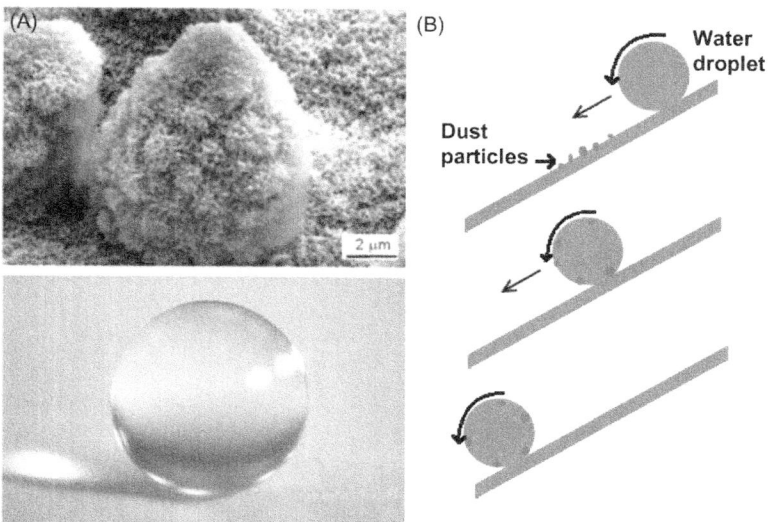

FIGURE 2.1 Water-repellent surfaces: (A) Lotus leaf [3] and (B) self-cleaning effect where the dust particles are carried away by the rolling droplet on a superhydrophobic surface.

The wetting states of surfaces can be determined by the behavior of water droplets on surfaces. Recently, superhydrophobic surfaces have gained a lot of interest for self-cleaning applications [1,2], where water droplets on such surfaces roll and slide easily while cleaning undesired contaminations including dust particles. Natural superhydrophobic surfaces like lotus plant leaves can remain clean since water droplets roll away at small inclination angles and take away dust particles along their path (Fig. 2.1). Due to this fact, self-cleaning is named as the lotus leaves effect. Self-cleaning applications extend to include windshields, eye glasses, windows, paints for buildings, solar energy harvesting surfaces, and antifouling.

2.2 WETTING STATES AND PROPERTIES

Wetting of surfaces is a complex phenomenon, influenced by the surface energies of the relevant phases and the surface texture of the solid substrate. Wetting of a surface shows its ability to be in contact with liquids resulting in intermolecular interactions. Wettability is influenced by a force balance between adhesive and cohesive forces [4] and deals with the three phases of materials: solid, liquid, and gas. Surface tension (γ) can be defined as the work per unit area that is required to create a free surface of a liquid, which is the reason for the elastic tendency of liquids to form the least possible surface area. Surface tension attains high values for the stronger bonding materials: $10-20 \text{ mJ/m}^2$ for fluorocarbons, $20-40 \text{ mJ/m}^2$ for hydrocarbons,

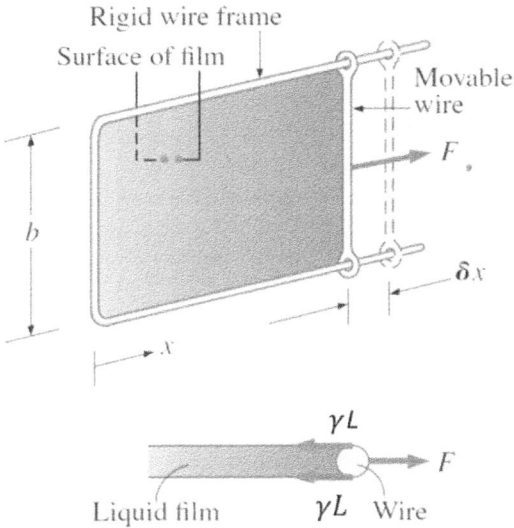

FIGURE 2.2 Stretching a liquid film with a U-shaped wire and the forces acting on the movable wire of length b [6].

73 mJ/m^2 for water, and 485 mJ/m^2 for mercury [5]. The surface tension of most liquids decreases with increasing temperature and reaches zero at the point of evaporation. If a liquid film is suspended on a wireframe as illustrated in Fig. 2.2, a specific force is needed to stretch this film by the movable part of the wireframe. The magnitude of this force should be equivalent to the microscopic forces between the liquid molecules at the liquid–air interfaces. These microscopic forces are perpendicular to any line in the surface, and the force generated by these forces per unit length is the surface tension. Therefore the work that is needed to stretch the film is the surface tension work, which is determined from $W = \int \gamma dA$, where A is the surface area.

Surfaces wettability is usually measured by the contact angle of a liquid droplet. The force balance of a spherical liquid droplet is secured by a force resulting from the difference between the internal and the atmospheric pressures against the surface tension force. This balance can be formulated as:

$$\Delta p = \frac{2\gamma}{R} \tag{2.1}$$

where γ is the liquid surface tension and R is the droplet radius. This pressure is called Laplace or capillary pressure [5]. A contact angle of 180 degrees represents no spreading and the whole droplet is located on the surface as a complete sphere, whereas a contact angle of 0 degree represents full spreading on the surface and the droplet liquid completely wets the surface. Surfaces with droplet contact angle greater than 90 degrees are called hydrophobic surfaces and those with water contact angle less than 90 degrees are called hydrophilic surfaces. Many factors influence the contact angle including surface energy, surface roughness, and its cleanliness [3]. Surfaces with high surface energy formed by polar molecules energy have hydrophilic characteristics, whereas surfaces with low surface energy and built of nonpolar molecules have a tendency for hydrophobic characteristics. Surfaces with a contact angle less than 10 degrees are called superhydrophilic while surfaces with a contact angle between 150 and 180 degrees are called superhydrophobic.

In the coming sections, the theoretical background for droplet behavior on smooth and rough surfaces is presented. For case of droplets on rough surfaces, two different states, Wenzel and Cassie−Baxter, are discussed and followed by a brief overview of lubricant impregnated surfaces.

2.2.1   Wetting on Smooth Surfaces

A surface's wettability is most often measured by the contact angle (θ). It is defined as the angle formed between the solid/liquid interface and the liquid/vapor interface when a droplet is placed on a surface or as the angle formed by a liquid at the three-phase boundary where a liquid, gas, and solid intersect. This is illustrated in Fig. 2.3.

The contact angle is influenced by the equilibrium of the interfacial tensions at the three-phase contact line, that is:

$$\gamma_{la}cos\theta + \gamma_{sl} = \gamma_{sa} \qquad (2.2)$$

where γ_{la}, γ_{sl}, and γ_{sa} are liquid−air, solid−liquid, and solid−air interfacial tensions, respectively.

Eq. (2.2) can also be expressed as:

$$cos\theta = \frac{\gamma_{sa} - \gamma_{sl}}{\gamma_{la}} \qquad (2.3)$$

The above equation is called Young's equation of contact angle and is applicable to smooth surfaces [7]. γ_{sl} in the above equation can be estimated from [8]:

$$\gamma_{sl} = \gamma_{sa} + \gamma_{la} - 2\sqrt{\gamma_{sa} \times \gamma_{la}} \qquad (2.4)$$

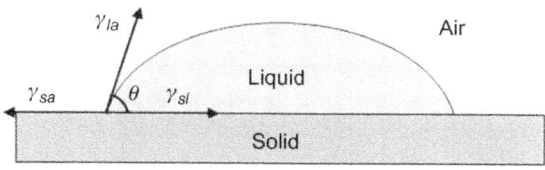

FIGURE 2.3   Force balance of interfacial tensions at three-phase contact line.

According to Eqs. (2.3) and (2.4), a surface can be made hydrophobic by decreasing the solid–air interfacial tension. The lowest reported solid–air interfacial tension is for surfaces with trifluoro methyl (−CF$_3$) groups (6 mN/m) [9] and the corresponding water contact angle observed on such surfaces is 120 degrees. A further increase in contact angle requires texturing the surfaces.

2.2.2 Wetting on Rough Surfaces

There are two primary wetting regimes on a rough surface: (1) Wenzel regime or homogeneous regime and (2) Cassie–Baxter regime or nonhomogeneous/composite regime.

Assuming that the liquid completely wets a surface, Wenzel developed a model that relates the contact angle of a droplet on a rough surface to that on a smooth surface [10]:

$$cos\theta_{rough} = rcos\theta_{flat} \tag{2.5}$$

where θ_{rough} and θ_{flat} are the contact angles of the droplet on rough and flat surfaces, respectively. r is the roughness ratio and is defined as the ratio of surface area to the flat projected area. Now if such surface has hydrophobic characteristics with, for example, θ_{flat} is greater than 90 degrees, then introducing roughness on the surface makes it more hydrophobic (since r is always greater than 1). The new contact angle (θ_{rough}) will be greater than θ_{flat}. If a surface has hydrophilic characteristics with, for example, θ_{flat} is less than 90 degrees, then introducing roughness on the surface will make it more hydrophilic. The new contact angle θ_{rough} will be less than θ_{flat} [11]. This is illustrated in Fig. 2.4.

The Wenzel state dictates that the liquid will completely fill the texture of a solid surface. However, complete submergence of texture with the liquid becomes less energetically favorable for a hydrophobic surface. In this case, the system is relatively at a higher energy when the liquid is fully wetting the texture. The Wenzel state for a droplet on a rough surface is shown in Fig. 2.5.

A more stable state is attained when air pockets are formed in between the texture. Cassie and Baxter extended Wenzel's work and incorporated the effect of trapped air pockets resulting in a composite solid–liquid–air interface as opposed to a homogeneous solid–liquid interface [12].

$$cos\theta_{rough} = r\phi_s cos\theta_{flat} + \phi_A cos\theta_{la} \tag{2.6}$$

where ϕ_s and ϕ_A are the fractional liquid–solid and liquid–air interfacial areas ($\phi_s + \phi_A = 1$). Since $\theta_{la} = 180$ degrees (contact angle of water with air), entrapment of air pockets will result in an increased contact angle. Putting ($cos\theta_{la} = -1$) in Eq. (2.6) gives:

$$cos\theta_{rough} = r\phi_s cos\theta_{flat} - (1 - \phi_S) \tag{2.7}$$

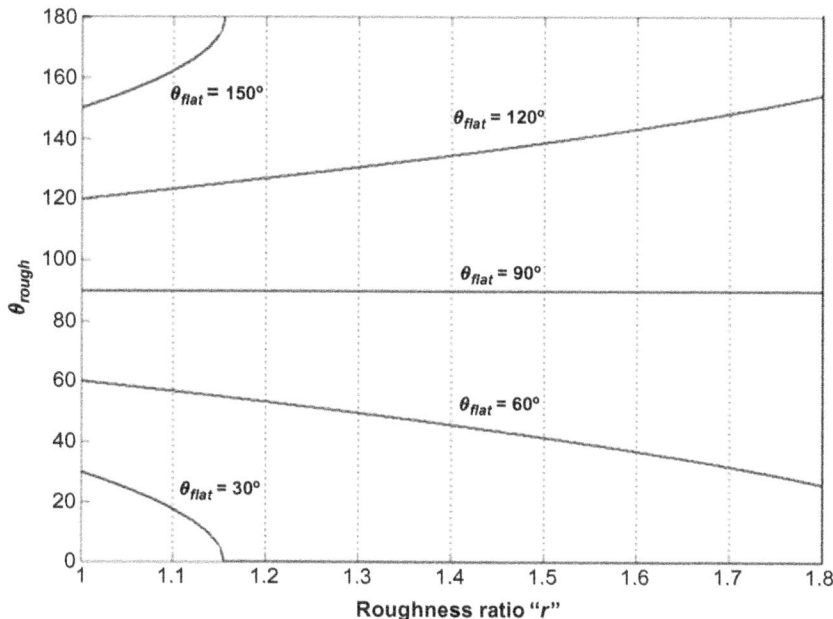

FIGURE 2.4 Evolution of contact angle with roughness ratio "r" for a surface. *Reprinted with permission from M. Nosonovsky, B. Bhushan, Roughness optimization for biomimetic superhydrophobic surfaces, Microsyst. Technol. 11 (2005) 535–549. © Springer.*

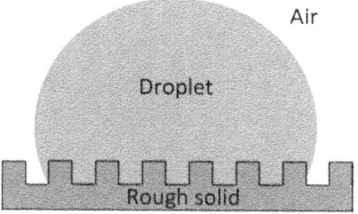

FIGURE 2.5 Wenzel state for a liquid droplet placed on a rough surface.

Based on Eq. (2.7), a decrease in the value of ϕ_s can increase the value of θ_{rough} and a surface can be made superhydrophobic (contact angle is greater than 150 degrees). The Cassie—Baxter state for a droplet placed on a rough surface is shown in Fig. 2.6.

Eq. (2.7) can also be expressed in terms of the fractional liquid—air interfacial area (ϕ_A):

$$cos\theta_{rough} = rcos\theta_{flat} - \phi_A(rcos\theta_{flat} + 1) \tag{2.8}$$

The evolution of contact angle with roughness ratio (r) and the liquid-air fraction area (ϕ_A) on surfaces for which θ_{flat} is 120 and 150 degrees, respectively, is shown in Fig. 2.7.

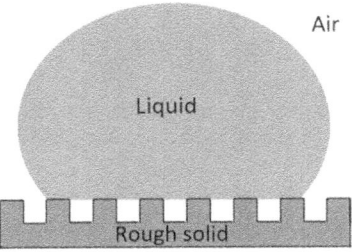

FIGURE 2.6   Cassie–Baxter state for a liquid droplet placed on a rough surface.

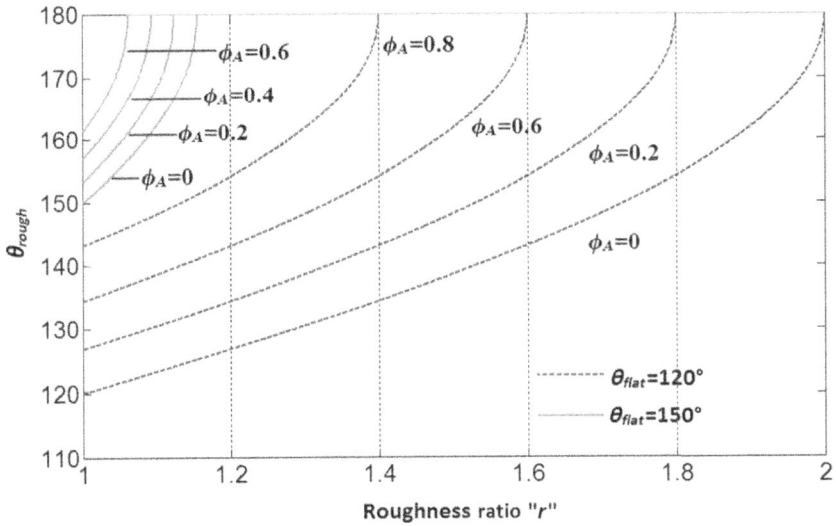

FIGURE 2.7   Evolution of contact angle with roughness ratio (r) and liquid–air fractional area (ϕ_A) for surfaces for which θ_{flat} is 120 and 150 degrees, respectively. *Reprinted with permission from M. Nosonovsky, B. Bhushan, Roughness optimization for biomimetic superhydrophobic surfaces, Microsyst. Technol. 11 (2005) 535–549. © Springer.*

For both the surfaces ($\theta_{flat} = 120$ degrees and $\theta_{flat} = 150$ degrees), increasing the roughness ratio (r) and liquid–air fractional area (ϕ_A) leads to increase contact angle.

In general, surface hydrophobicity is associated with low surface energy and surface texture composed of micro/nanopillars. Introducing roughness on the surfaces alters the wetting status and surface energy [13]. There are several methods to create hydrophobic characteristics on the surfaces including laser texturing [14,15], modified silica nanoparticle deposition [16,17], polydimethylsiloxane (PDMS) replica molding [18,19], etc. Glass is an excellent material in applications requiring high transmittance. Due to its hydrophilic nature, water droplets spread on the glass surface. In a dusty environment, water droplets catch the dust particles and form a mud solution at the surface, which greatly reduces their transmittance. This can be avoided by covering the glass surface with a superhydrophobic coating. Silica nanoparticle coating is an ideal option because of its matching refractive index with glass, which greatly improves the optical transmittance of the coating. Silica nanoparticle-based coatings also possess good scratch- and wear-resistant properties [20]. Another facile approach to obtain optically transparent superhydrophobic coatings is to replicate textured surfaces with superhydrophobic characteristics. PDMS, a silicon-based organic polymer, is an ideal choice because of its high transparency, hydrophobic nature, and the ability to copy submicron features during the replication process. Filling the texture of a superhydrophobic surface with some lubricant oil can overcome many limitations associated with conventional superhydrophobic surfaces. A lubricant oil provides a smooth and homogeneous interface that can repel a variety of fluids, give very low contact angle hysteresis, and improve the optical transparency of coatings [21,22].

Fig. 2.8 shows replicated PDMS surfaces made by copying the surface structure of silicon wafers with 1.5 mm thickness, which was used to texture micropost arrays at the surface using a lithographic technique with different sizes of micropost array spacing (*b*). Fig. 2.9 shows the optical images of the water droplet forming the Cassie−Baxter state (Fig. 2.9A) and Wenzel state on the replicated PDMS micropost arrays surface (Fig. 2.9B and C).

Polycarbonate surface can be textured via surface crystallization. The process starts by cleaning a bare polycarbonate wafer and immersing it in acetone for 4 minutes [23]. The hierarchical texture is obtained after solution crystallization of the polycarbonate surface. The crystallized surface composed of micro/nanospherules and fibrils. In this case, the droplet static contact angle can attain values in the range of 134−136 degrees [24]. Fig. 2.10 shows a scanning electron microscopy (SEM) of a solvent polycarbonate surface. Crystallization results in a hierarchical texture consisting of closely spaced spherules (Fig. 2.10A) and fibrils (Fig. 2.10B). The hierarchical texture is also evident from Fig. 2.11, which shows a 3D optical image (Fig. 2.11A) and landscape of surface texture height (Fig. 2.11B). The size of the spherules is in the order of a few micrometers and submicrometer size fibrils emanate from the spherule surface. In addition, no asperities such as microcracks or large-size cavities are observed on the crystalized surfaces. The crystallization takes place in three stages including initiation of crystallization, primary crystallization, and secondary crystallization. During crystallization, a fetus is formed as the polymer chains align in a parallel way and as the process progresses gradually, the chains are added to the fetus. As the fetus size becomes large enough, the growth of the nucleus initiates spontaneously. The nucleation results in the bundle-like or lamellar crystals formations. The types of crystallization depend on the length of the primary nucleus and the free energy of the surface normal to the chain direction per unit area. In the case of solution crystallization, the mixture of bundle-like

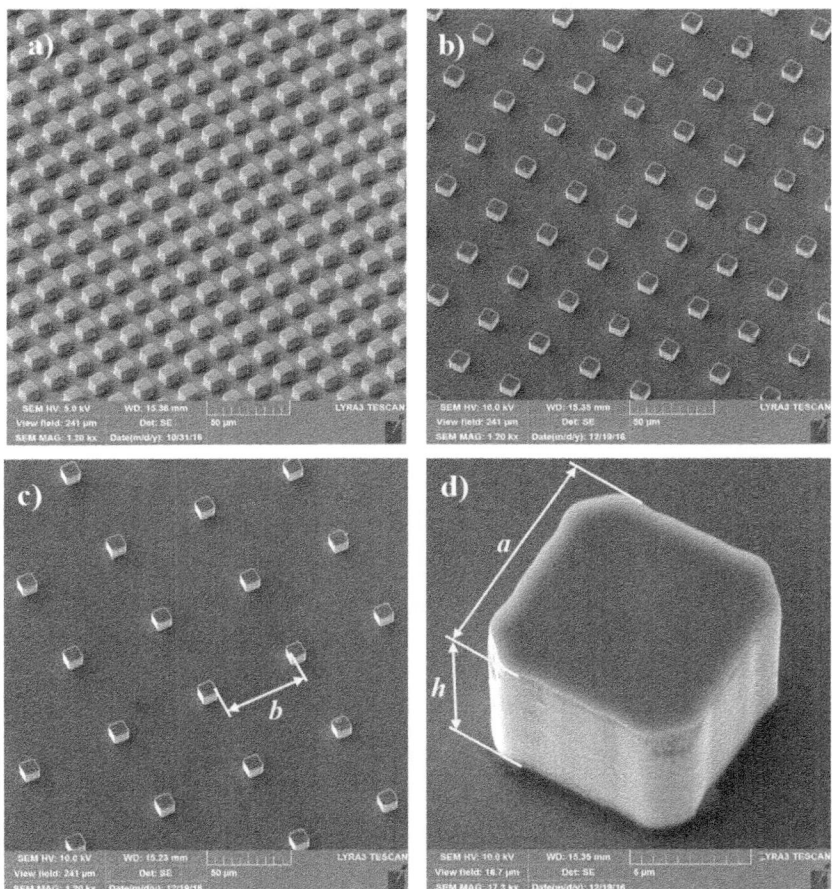

FIGURE 2.8 SEM micrographs of micropost arrays surface (a is the micropost width, b is micropost arrays spacing size, and h is the micropost height): (A) micropost arrays with $a = b = h = 10\,\mu m$; (B) micropost arrays with $a = h = 10\,\mu m$ and $b = 25\,\mu m$; (C) micropost arrays with $a = h = 10\,\mu m$ and $b = 50\,\mu m$; and (D) micropost [13].

and lamellar nucleus can be formed through the buildup by a series of additions of the repeating units during the solution crystallization process [24].

Octadecyltrichlorosilane (OTS) can be used to improve surface hydrophobicity of the textured polycarbonate wafers [25]. In this case, a deep coating technique is applied to form an OTS layer of 40 nm thickness. OTS-deposited surfaces were rinsed first with chloroform, and later ultrapure water. Inconel 718 alloy powders with a nominal size of 30 μm are cold sprayed onto the hydrophobic polycarbonate surface. The sprayed powders attach to the surface strongly and the presence of Inconel 718 alloy powders modifies the surface texture and further improves the surface hydrophobicity.

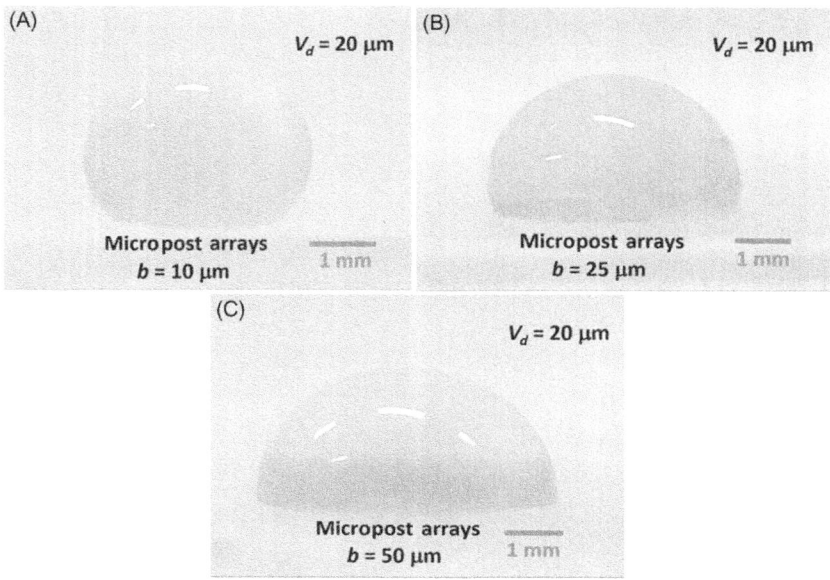

FIGURE 2.9   Optical images of the water droplet located on: (A) micropost arrays with $a = b = h = 10\ \mu m$; (B) micropost arrays with $a = h = 10\ \mu m$ and $b = 25\ \mu m$; and (C) micropost arrays with $a = h = 10\ \mu m$ and $b = 50\ \mu m$ [13].

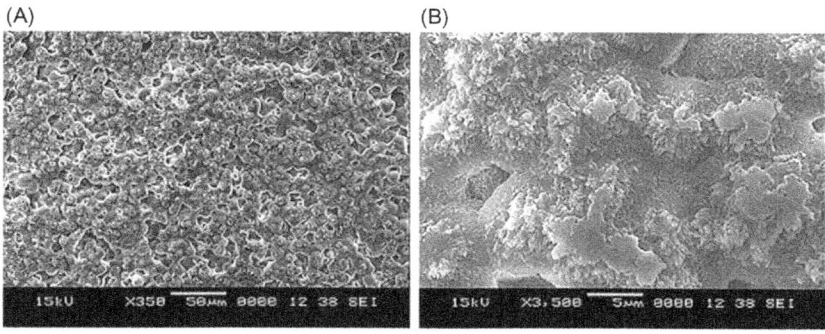

FIGURE 2.10   SEM micrograph of solvent crystallized polycarbonate surface: (A) texture with spherules and (B) fibrils on spherules surface [24].

Fig. 2.12 shows an SEM micrograph of powders (Fig. 2.12A) and optical images of the particle distribution (Fig. 2.12B) at the hydrophobic polycarbonate surface [25]. Since the powder size varied, the particle distribution resulted in nonregular patterns at the treated polycarbonate surface. However, the spacing in between the powder particles was in the range of $10-100\ \mu m$, which was much smaller than the water droplet diameter ($\approx 4 \times 10^{-3}$ m).

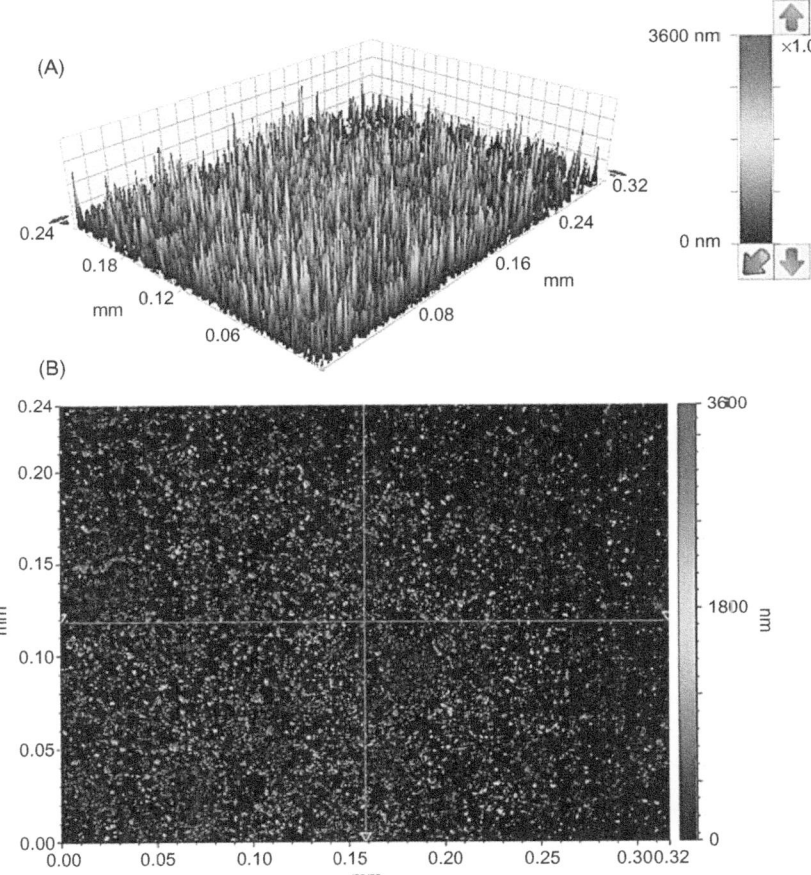

FIGURE 2.11   (A) Texture height landscape of textured polycarbonate surface and (B) 3D optical image of crystallized polycarbonate surface [24].

Fig. 2.13A shows the typical optical images of water droplets on the treated polycarbonate surface with the presence of metallic powders. Fig. 2.13B shows the images of droplets used for the contact angle measurements [25].

2.2.3   Contact Angle Hysteresis

In general, there are two types of contact angle measurements: static and dynamic. The static contact angle is measured by placing the droplet on a flat surface and then recording the value of the angle once an equilibrium is established. Dynamic contact angles are measured during the droplet growth and shrinkage. Dynamic contact angle measurement can also be performed by tilting the surface and noting the values of the contact angles at the front

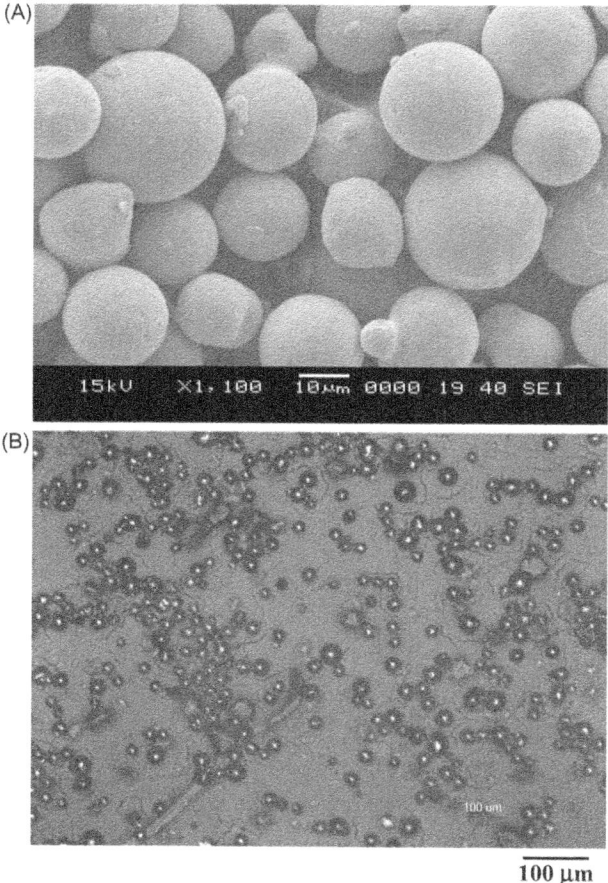

100 μm

FIGURE 2.12 SEM and optical micrographs of Inconel 718 particles: (A) SEM micrograph of loose particles and (B) optical image of particles sprayed on to the treated polycarbonate surface (crystallized and OTS coated prior to powder spraying) [25].

and the back of the droplet. These contact angles are called advancing (at the front of the droplet) and receding contact angles (at the back of the droplet), and the difference between their values is called contact angle hysteresis. The contact angle hysteresis is highest just before the droplet starts sliding/rolling on the surface. In addition, surface defects can lead to differences in the advancing and receding contact angles due to surface heterogeneity. Contact angle hysteresis is a measure of how much energy is dissipated during droplet motion [26]. For surfaces with low contact angle hysteresis, water rolls off at very small tilt angles and carries away contaminations such as dust particles along its path, giving rise to the so called self-cleaning effect where high contact angle and low contact angle hysteresis is required [26].

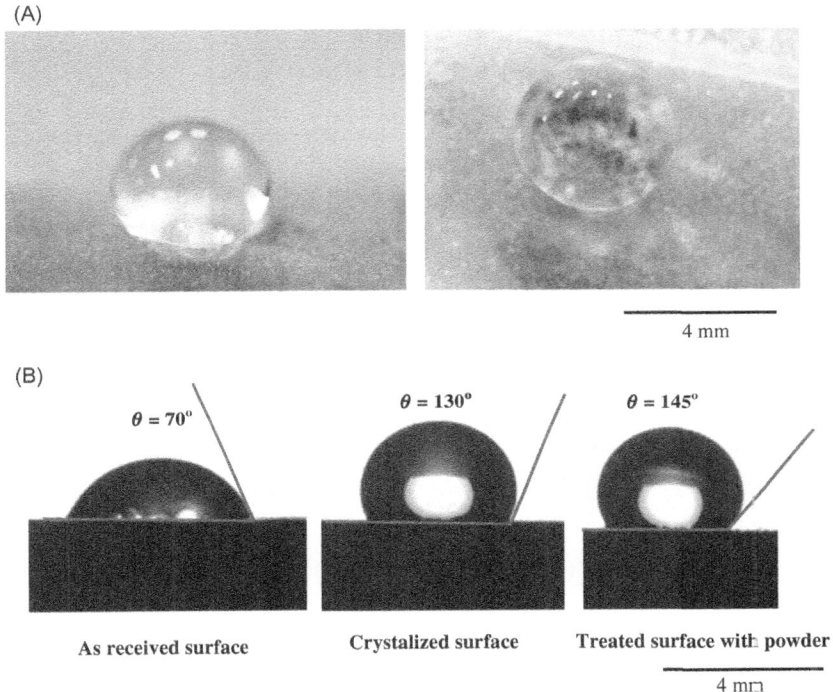

FIGURE 2.13 Water droplets on the workpiece surfaces: (A) Optical images of water droplets at the treated and powder sprayed surfaces and (B) water droplets onto the workpiece surfaces and the contact angles.

Using Eq. (2.8), a relation for the contact angle hysteresis for a rough surface can be derived. The resultant equation is:

$$cos\theta_{rough\ adv} - cos\theta_{rough\ rec} = r(1 - \phi_A)\left(cos\theta_{flat\ adv} - cos\theta_{flat\ rec}\right) + H_r \quad (2.9)$$

$\left(cos\theta_{flat\ adv} - cos\theta_{flat\ rec}\right)$ represents the contact angle hysteresis for the smooth surface and H_r represents the surface roughness effect.

The above equation can be simplified to obtain the following expression [27]:

$$\theta_{rough\ adv} - \theta_{rough\ rec} = r\sqrt{1 - \phi_A}\ \frac{cos\theta_{rec\ flat} - cos\theta_{adv\ flat}}{\sqrt{2\left(rcos\theta_{flat} + 1\right)}} \quad (2.10)$$

For a homogeneous or Wenzel interface, $\phi_A = 0$. Increasing roughness ratio (r) would increase the contact angle hysteresis. For a composite or Cassie−Baxter interface, contact angle hysteresis is directly proportional to $(1 - \phi_A)$, that is, a decrease in liquid−air fractional area will increase the contact angle hysteresis. This brings up the importance of microair pockets in the texture, which increase the contact angle and reduce the contact angle hysteresis for a surface. The presence of a nanoscale texture on top of a microscale texture can also significantly increase the liquid—air fractional area and reduce the contact angle hysteresis. Therefore a nanoscale texture prevents the pinning of the droplet and allows it to roll off at small inclination angles, thereby yielding the self-cleaning effect.

2.2.4 Lubricant Impregnated Surfaces

Conventional superhydrophobic surfaces rely on air pockets trapped between the texture to maximize the droplet contact angle with the surface. Such surfaces continue to exhibit nonwetting behavior as long as stable air pockets are maintained beneath the droplet [28]. These tiny air pockets, however, are unstable and collapse under conditions involving large wetting pressure, high temperature, or humidity [29,30], or when damage occurs to the surface texture [31]. A low surface tension liquid can also sometimes displace these air pockets and penetrate the texture [32]. In all of these cases, the droplet pins to the surface and hence the surface loses its self-cleaning ability.

To overcome the limitations associated with lotus leaf-inspired surfaces, a new type of pitcher-plant (*Nepenthes*)-inspired surfaces called SLIPS (slippery lubricant-infused porous surfaces) has been reported [21]. These surfaces do not rely on trapped air pockets inside the texture to repel liquids. The texture is instead filled with a lubricant and provides a surface with an overlying liquid interface that is ultrasmooth, chemically homogeneous, continuous, and provides an extremely low contact angle hysteresis for a broad range of liquids [21].

Substrate materials with high optical transmittance can be used in wide range of applications such as in the construction industry, automobile industry, optoelectronics, and energy conversion devices line PV panels. In all such applications, maintaining high optical transmittance is important for aesthetic and performance reasons. Surfaces are prone to lose their transmittance due to accumulation of airborne dust particles at the surface. Therefore maintaining optical clarity of the surfaces becomes challenging especially at large scales such as skyscraper windows and protective covers of PV cells in solar farms. All such cases bring about the importance of optically transparent surfaces that are cost-effective to clean.

2.2.5 Particle Adhesion on Surfaces

Dust settlement on surfaces gives rise to adhesion of dust particles that have detrimental effects on solar energy harvesting devices due to optical scattering and absorption of solar radiation at the surface. Several methods have been introduced to remove dust particles from the surface. A simple and traditional cleaning method is done by using a mop and water supply, a method that is labor intensive. Pressurized water can be used to avoid mopping the surfaces while minimizing mechanical scratches formed at the surface. This process increases process time and lowers cleaning efficiency; however, it enables cleaning spots that are difficult to be cleaned by a mop. The steam and compressed air can be used to remove the dust accumulation from the surfaces. In addition, there are methods where the need for the cost of labor is minimized. Some of these include robot devices, automatic brushes, and sprinklers. Introducing electric current to remove and repel dust particles from the surface is an alternative method to be implemented for surface cleaning. Utilizing mechanical vibration or centrifugal forces are examples of incorporating the dynamic methods for cleaning of surfaces from the contamination of environmental dust and small-size waste particles. In order to assess the dust particle adhesion, understanding of the contact mechanics associated with the particles is essential.

In general, the adhesion of dust particles is affected by many factors such as van der Waals force, static electric charge, relative humidity, size of the contact area, roughness of the surface, chemistry of particle agglomeration, duration of contact, local temperature, and other factors. It is assumed that van der Waals forces are the most common forces that contribute to the particle adhesion on the surface. The models describing adhesion of particles on surfaces are reviewed briefly in the following.

2.2.5.1 Johnson–Kandall–Roberts Model

A mechanical work, which is required to separate particles in contact, should be more than the adhesion forces. To separate the particles, energy is required to create a new surface area. This energy is called the solid surface energy. If two smooth spheres are brought into contact (Fig. 2.14), the equilibrium depends on the elastic forces in the contacting area. Under light

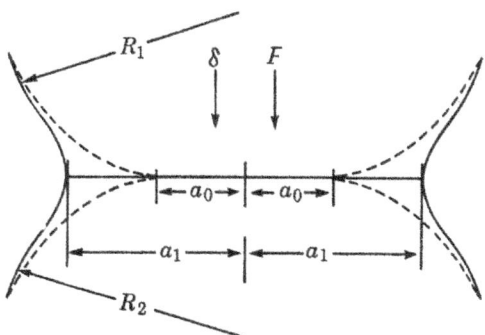

FIGURE 2.14 The contact between two convex bodies of radii, R_1 and R_2, under a normal load of F; δ is the elastic displacement both in the presence (contact radius a_1) and absence (contact radius a_0) of surface forces [33].

loading conditions between an elastic—solid surface, adhesion of spherical particles on the flat surfaces is understood using van der Waals forces, which is known as the Johnson—Kandall—Roberts (JKR) model for adhesion [33]. The total energy (U_T) of a system of two elastic spheres in contact equilibrium can be formulated in terms of contact radius (a). Equilibrium will be obtained if [33]:

$$\frac{dU_T}{da} = 0 \tag{2.11}$$

In the absence of the normal force F_0, the contact radius a_0 is given by the generalized Hertz equation:

$$a_0^3 = \frac{RF_0}{K} \tag{2.12}$$

where $R = R_1R_2/(R_1 + R_2)$ and $K = 4/3\pi(k_1 + k_2)$. Here k_1 and k_2 are the elastic constants of the material of each sphere, which can be estimated from $k_1 = (1 + v_1^2)/(\pi E_1)$ and $k_2 = (1 + v_2^2)/(\pi E_2)$, where v_1 and v_2 are the Poisson ratios and E_1 and E_2 are the Young's modulus of each material of the spheres.

The elastic displacement δ due to the applied load F is:

$$\delta = \frac{a^2}{R} \tag{2.13}$$

When adhesive forces reach equilibrium across the joining surfaces, the radii of the contact area leads to $a_1 > a_0$. At low applied forces the contact areas between these bodies are considerably larger than those predicted by Hertz [33]. Although the applied load remains at F, an apparent Hertz load F_1 corresponding to the contact radius a_1 may be defined such that:

$$a_1 = \sqrt{\frac{RF_1}{K}} \tag{2.14}$$

The total energy U_T of this system has three components: the stored elastic energy U_E, the mechanical energy in the applied load U_M, and the surface energy U_s.

Loading the system with a load F_1 that requires energy U_1 results in contact radius a_1. The final state of the system can be achieved by idealizing the load displacement while keeping the contact radius at a_1 and reducing the load F and resulting energy U_2. The stored elastic energy U_E is:

$$U_T = U_1 - U_2 \tag{2.15}$$

where $U_1 = \int_0^{F_1} \frac{2}{3}\frac{F^{\frac{2}{3}}}{K^{\frac{2}{3}}R^{\frac{1}{3}}}\,dp = \frac{2}{5}\frac{F_1^{\frac{5}{3}}}{K^{\frac{2}{3}}R^{\frac{1}{3}}}$. The load displacement can be written as [33]:

$$\delta = \frac{2}{3}\frac{F}{Ka_1} \tag{2.16}$$

Therefore $U_2 = \int_{F_0}^{F_1} \frac{2}{3} \frac{F}{Ka_1} dp = \frac{1}{3K^{\frac{2}{3}}R^{\frac{1}{3}}} \left(\frac{F_1^2 - F_0^2}{F_1^{\frac{1}{3}}} \right)$. This yields:

$$U_E = \frac{2F_1^{\frac{5}{3}}}{5K^{\frac{2}{3}}R^{\frac{1}{3}}} - \frac{(F_1^2 - F_0^2)}{3K^{\frac{2}{3}}R^{\frac{1}{3}}F_1^{\frac{1}{3}}} = \frac{1}{K^{\frac{2}{3}}R^{\frac{1}{3}}} \left(\frac{1}{15} F_1^{\frac{5}{3}} + \frac{1}{3} F_0^2 F_1^{-\frac{1}{3}} \right) \tag{2.17}$$

The mechanical potential energy U_M of the applied load F is [33]:

$$U_M = \frac{-F_0}{K^{\frac{2}{3}}R^{\frac{1}{3}}} \left(\frac{1}{3} F_1^{\frac{2}{3}} + \frac{2}{3} F_0 F_1^{-\frac{1}{3}} \right) \tag{2.18}$$

The surface energy U_s is given by [33]:

$$U_s = -\gamma \pi a_1^2 = -\gamma \pi \left(\frac{RF_1}{K} \right)^{\frac{2}{3}} \tag{2.19}$$

where γ is the energy of adhesion of both surfaces. The total energy is the summation of the stored elastic energy U_E, the mechanical energy in the applied load U_M, and the surface energy U_s, which is:

$$\begin{aligned} U_T = U_E + U_M + U_S = &\frac{1}{K^{\frac{2}{3}}R^{\frac{1}{3}}} \left(\frac{1}{15} F_1^{\frac{5}{3}} + \frac{1}{3} F_0^2 F_1^{-\frac{1}{3}} \right) \\ &- \frac{1}{K^{\frac{2}{3}}R^{\frac{1}{3}}} \left(\frac{F_0 F_1^{\frac{2}{3}}}{3} + \frac{2}{3} F_0^2 F_1^{-\frac{1}{3}} \right) - \gamma \pi \left(\frac{RF_1}{K} \right)^{\frac{2}{3}} \end{aligned} \tag{2.20}$$

Equilibrium is secured when $\frac{dU_T}{da_1} = 0$, which is $\frac{dU_T}{dF_1} = 0$:

$$\frac{dU_T}{dF_1} = \frac{F_1^{-\frac{4}{3}}}{K^{\frac{2}{3}}R^{\frac{1}{3}}} \left(F_1^2 - F_0^2 - 2F_0 F_1 + 2F_0^2 - 6\gamma \pi RF_1 \right) \tag{2.21}$$

Hence:

$$\begin{cases} F_1^2 - 2F_1(F_0 + 3\gamma \pi R) + F_0^2 = 0, \\ F_1 = F_0 + 3\gamma \pi R + \sqrt{\left[(F_0 + 3\gamma \pi R)^2 - F^2 \right]} \end{cases} \tag{2.22}$$

In the JKR model [33], the Hertz equation is modified to take into account the surface energy effect.

$$a^3 = \frac{R}{K} \left(F + 3\gamma \pi R + \sqrt{6\gamma \pi RF + (3\gamma \pi R)^2} \right) \tag{2.23}$$

At zero applied load the contact area is finite and given by:

$$a^3 = \frac{R(6\gamma \pi R)}{K} \tag{2.24}$$

For a real solution to be obtained for Eq. (2.23), $6\gamma \pi R \le (3\gamma \pi R)^2$ and $F = -\frac{3}{2}\gamma \pi R$. Separation of the spheres will occur when

$$F = -\frac{3}{2}\gamma \pi R \tag{2.25}$$

where R is the radius of the particle and γ is the surface energy between the two surfaces. However, the above equation is applicable for large-size soft bodies having the high surface energies.

2.2.5.2 Derjagin–Muller–Toropov Model

Derjaguin et al. [34] developed a model for a sphere with a high elastic modulus whose profile does not change outside the contact area. They found the relation of contact radius and the corresponding pull of force to be:

$$a_{DMT}^3 = \frac{d_P}{2K}[P + \pi d_P W_A]$$
(2.26)

and

$$F_{DMT}^{JKR} = \pi d_P W_A$$
(2.27)

For negligible elastic deformation of the sphere on a rigid surface in the absence of an externally applied force, the Derjagin–Muller–Toropov (DMT) model is given as:

$$E_s = 2\pi R \Delta \gamma$$
(2.28)

The DMT equation is valid for both hard and small particles, which has low surface energy. There are different assumptions in JKR and DMT models that lead to different predictions. The JKR model predicts the particle–surface contact area to be finite at the moment of separation while the DMT predicts a zero contact radius. Another difference is that the surface force according to JKR analysis is assumed to act inside the contact region of particle and substrate. However, the DMT analysis considers the force outside the contact area. From these differences in the assumptions, we can conclude that in terms of predicted pull-off force, the DMT model results in greater values than the JKR model.

Muller et al. [41] reported that the JKR model is appropriate for large particles with high surface energy and low Young's modulus. However, the DMT model is more applicable for hard and small-size particles with low surface energy and high Young's modulus. The transition between the two models can represented in terms of μ_T as:

$$\mu_T = \left(\frac{8d_P W_A^2}{9K^2 z_0^3}\right)^{\frac{1}{3}}$$
(2.29)

2.2.5.3 Hamaker's Model

Hamaker considered spherical particles despite the fact that the dust particle shape is irregular in nature [35]. The energy of interaction between two particles containing q atoms per cm^3 is:

$$E = - \int V_1 dv_1 \int V_2 dv_2 \frac{q^2 \lambda}{r^6}$$ (2.30)

where dv_1, dv_2, V_1, and V_2 designate volume elements and total volumes of the two particles, respectively; r denotes the distance between dv_1 and dv_2; and λ is the van der Waals constant.

In the Hamaker model, a sphere (of radius R_1 and center o and a point P outside at a distance $OP = R$) is considered. The schematic is shown in Fig. 2.15.

The first sphere (with radius R_1 and center o) will cut out from the second sphere (with radius r and center P) a surface ABC, which is:

$$Surface(ABC) = \int_0^{2\pi} d\varphi \int_0^{\theta_0} d\theta r^2 sin\theta$$ (2.31)

where θ_o can be estimated from $R_1^2 = R^2 + r^2 - 2rR cos\theta_o$. Integration of Eq. (2.31) gives:

$$Surface(ABC) = \pi \frac{r}{R} \left(R_1^2 - (R-r)^2 \right)$$ (2.32)

The potential energy of an atom at P is:

$$E_P = - \int_{R-R_1}^{R+R_1} \frac{\lambda q}{r^6} \pi \frac{r}{R} \left(R_1^2 - (R-r)^2 \right) dr$$ (2.33)

For a second sphere of radius R_2, which is a distance C from the center of the first sphere (Fig. 2.16), the total energy of interaction is expressed as [35]:

$$E = - \frac{\pi^2 q^2 \lambda}{6} \left\{ \frac{2R_1 R_2}{C^2 - (R_1 + R_2)^2} + \frac{2R_1 R_2}{C^2 - (R_1 - R_2)^2} + ln \left(\frac{C^2 - (R_1 + R_2)^2}{C^2 - (R_1 - R_2)^2} \right) \right\}$$ (2.34)

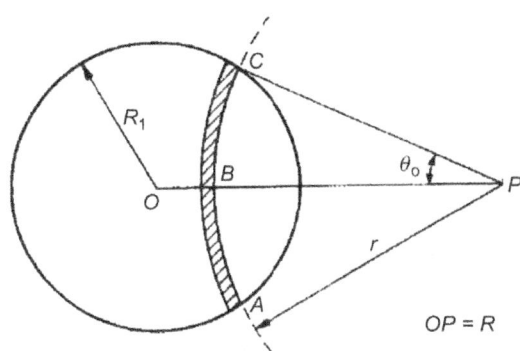

FIGURE 2.15   A hard spherical solid particle of radius R_1 and center o and a point P outside at a distance $OP = R$ [35].

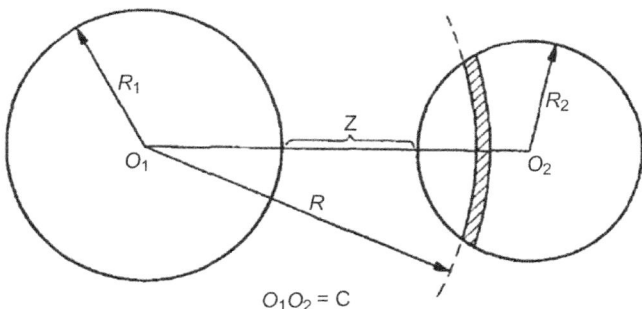

FIGURE 2.16 Two spherical solid particles of radius R_1 and R_2. The centers are a distance C apart [35].

This is an expression for energy in terms of the radii R_1 and R_2 and of the distance C between the center.

From Fig. 2.16, the distance between the two sphere centers can be written as $C = R_1 + R_2 + z$ where z is the shortest distance between the two particles. Then,

$$x = \frac{z}{2R_1} = \frac{z}{D_1} \text{ and } y = \frac{D_2}{D_1} = \frac{R_2}{R_1} \qquad (2.35)$$

In this case, x gives the ratio of the shortest distance z to the diameter of sphere 1 and y is the diameter D_2 of sphere 2 expressed in D_1.

Defining $a = \pi^2 q^2 \lambda$ and modifying Eq. (2.34) gives:

$$E = -a\frac{1}{12}\left\{\frac{y}{x^2 + xy + x} + \frac{y}{x^2 + xy + x + y} + 2ln\left(\frac{x^2 + xy + x}{x^2 + xy + x + y}\right)\right\} \qquad (2.36)$$

If $x \ll 1$, then Eq. (2.36) can be written as:

$$E = -a\frac{1}{12}\left(\frac{y}{x(y+1)}\right) \qquad (2.37)$$

Hamaker assumed values for D_1 and then predicted the diameter of the smallest spheres. Subsequently, x and y measure the distance and the diameter of the largest sphere in terms of the diameter of the smallest sphere where y will vary between 1 and ∞. In this case, the constant a will be different from case to case. Therefore and knowing the values of a, Hamaker introduced a table for the function $E_y(x)$ for different values of y and x [35]. For example, for two spheres of equal size, $y = 1$, then:

$$E_1(x) = \frac{1}{12}\left\{\frac{1}{x^2 + 2x} + \frac{1}{x^2 + x2x} + 2ln\left(\frac{x^2 + 2x}{x^2 + 2x + 1}\right)\right\} \qquad (2.38)$$

and when $x \ll 1$:

$$E_1(x) \approx \frac{1}{24x} \tag{2.39}$$

Similarly, when $y = \infty$, which represents the case of a sphere and an infinite mass bounded by a flat surface (Fig. 2.17):

$$E_\infty(x) = \frac{1}{12}\left\{\frac{1}{x} + \frac{1}{x+1} + 2ln\left(\frac{x}{x+1}\right)\right\} \tag{2.40}$$

and when $x \ll 1$:

$$E_\infty(x) \approx \frac{1}{12x} \tag{2.41}$$

The values of E_2, E_5, and E_{10} lie between those of E_1 and E_∞.

Differentiating the above equations give relations for the forces instead of energies:

$$F = \frac{\partial E}{\partial d} = \frac{a}{D_1}\frac{\partial E_y(x)}{\partial x} = \frac{a}{D_1}F_y(x) \tag{2.42}$$

The force calculation from Eq. (2.42) depends on the values of x and y. For two equal spheres ($y = 1$):

$$F_1(x) = \frac{1}{6}\left\{\frac{2(x+1)}{x^2+2x} - \frac{x+1}{(x^2+2x)^2} - \frac{2}{x+1} - \frac{1}{(x+1)^3}\right\} \tag{2.43}$$

and when $x \ll 1$:

$$F_1(x) = -\frac{1}{24}\frac{1}{x^2} \tag{2.44}$$

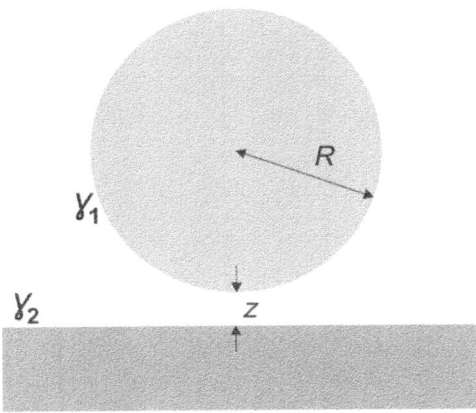

FIGURE 2.17 A sphere and an infinite mass bounded by a flat surface. Here, γ is the surface energy of mediums and z is the minimum spacing of contacting particle from the plat surface.

Similarly, when $y = \infty$ (Fig. 2.17):

$$F_\infty(x) = \frac{1}{12}\left\{\frac{2}{x} - \frac{1}{x^2} - \frac{2}{x+1} - \frac{1}{(x+1)^2}\right\} \tag{2.45}$$

and when $x \ll 1$:

$$F_\infty(x) = -\frac{1}{12}\frac{1}{x^2} \tag{2.46}$$

In general, for a spherical particle of radius R close to the surface and when the separation distance goes to zero:

$$F = \frac{aR}{12z^2} \tag{2.47}$$

where a is the Hamaker constant and z is the separation distance between the particle and the flat surface, which is in the order of 0.3 or 0.4 nm [36] and $x = z/D$. Hamaker's model did not consider the particle contact area to the surface.

2.2.5.4 Rumpf–Rabinovich Model

This model utilizes van der Waal's forces acting on a particle when located on the rough surface. Rumpf's model was modified by Rabinovich et al. [37]. Rumpf's introduced an expression to calculate the force of adhesion between a spherical particle and spherical surface asperity [38]:

$$F = \frac{a}{6z^2}\left(\frac{rR}{r+R} + \frac{R}{\left(1+\frac{r}{z}\right)^2}\right) \tag{2.48}$$

where a is the Hamaker constant, R and r are the radii of the adhering particle and asperity, and z is the separation distance between the particle and the flat surface as shown in Fig. 2.18.

Rabinovich et al. [37] developed the relationship between the radius of the asperity and the RMS (root mean square) of the surface roughness. In this case, the horizontal projection r_1 of the point on the asperity can be expressed as:

$$r_1 = r\sin\alpha \tag{2.49}$$

where r is any radius of the asperity and α is the angle between any radius and vertical axis. On the other hand, the vertical distance to the average surface plane (y) is:

$$y = r\cos\alpha \tag{2.50}$$

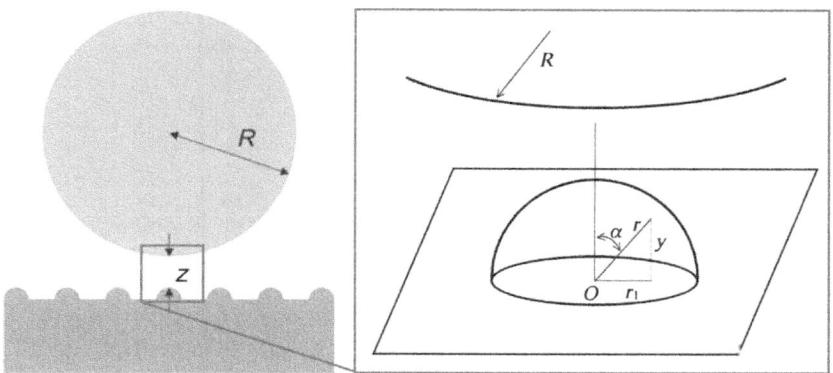

FIGURE 2.18 Schematic illustration of the geometry for the interaction of an adhering particle with a rough surface [38].

RMS can be defined as:

$$RMS = \sqrt{\frac{32 \int_0^r y^2 r_1 dr_1}{\lambda^2} k_p}$$ (2.51)

where $\lambda = 4r$ is the peak to peak distance and k_p is the surface packing density for close-packed spheres, 0.907 [37]. Putting Eqs. (2.49) and (2.50) into Eq. (2.51) gives:

$$RMS = 0.637r$$ (2.52)

and

$$r = 1.485RMS$$ (2.53)

In this case, statistical distribution of surface asperities and the effect of *RMS* on the rough surface on the adhesion force are considered. The resulting adhesion force takes the following form [37]:

$$F = \frac{aR}{6z^2} \left(\frac{1}{1 + \frac{R}{1.48RMS}} + \frac{1}{\left(1 + \frac{1.48RMS}{z^2}\right)^2} \right)$$ (2.54)

In general, the equations developed determining the adhesion force can be used according to the surface texture characteristics. However, equations based on JKR and DMT models are the only equations incorporating the free energy of the surfaces.

2.3 DUST PARTICLE REMOVAL AND SURFACE CLEANING

Particle removal mechanism requires an external force. This external force mechanism is generated from different sources such as substrate vibration and fluid flow (hydrodynamic forces) around small particles. One of the methods to remove the dust particles is to utilize the centrifugal acceleration at the surface via rotation of the surface. In this case, various forces should be considered for the dust removal from the surface by rotating the surface. These forces are listed as follows.

2.3.1 Gravitational Force

One of the forces contributing to particle adhesion onto the surface is gravitational force, which can be written for a spherical particle as:

$$F_g = mg = \frac{4}{3}\pi R^3 \rho g \qquad (2.55)$$

where m is the mass of the particle, R is the radius of the particle, ρ is the density of the particle, and g is the gravitational acceleration.

2.3.2 Lift Force (Inertial Force and Shear Stress)

Lift force can be generated by inertial force during the rotation of a solid body such as a dust particle. The inertial lift force can cause torque acting on the particle at the point contact between the particle and the center of rotation. In addition, another type of lift force is also generated because of the flow shear stress acting on the particle. The combination of these forces gives rise to lifting of the particle from the surface. The lift force can be derived from the Navier–Stokes equation for a spherical shape particle where the particle is almost touching the surface in the case where the Reynolds number remains below unity. Therefore the lift can be written as:

$$F_L = \frac{9.22\mu^2}{\rho}\left(Re^*\right)^3 \qquad (2.56)$$

where μ is the dynamic viscosity of fluid and Re^* is the shear Reynolds number, which is defined by:

$$Re^* = \frac{Ru^*}{v} \qquad (2.57)$$

where v is the kinetic viscosity and u^* is the friction velocity:

$$u^* = \sqrt{\frac{2\tau_o}{\rho}} \qquad (2.58)$$

where τ_o is the shear stress at the wall.

The shear stress of the rotating disk system incorporating the simplified form of the Navier–Stokes equation can be written as:

$$\tau_o = \rho r G_o' \sqrt{v\omega^3} \qquad (2.59)$$

where τ_o is the shear stress at the wall ($y = 0$), r is the position of the particle from the center of rotation, G'_o is the dimensionless constant at the wall ($G'_{z=0} = -0.61592$), and ω is the angular velocity of the disk.

2.3.3 Drag Force due to Pressure and Shear on the Particle Surface

In general, drag force can be categorized into three groups. The first group is pure pressure drag force when the area normal to the flow is relatively large; the second group is involved with the pure shear friction drag force, due to friction; and the last group is associated with the combination of pressure and shear drag forces. The drag force acting on the spherical particle can be considered as the third type, which involves with the combination of pressure and shear drag force.

The drag force for the spherical solid body can be written as:

$$F_D = 10.2\pi\mu Ru \qquad (2.60)$$

where u is the flow velocity, μ is the dynamic viscosity of fluid, and R is the radius of the particle.

2.3.4 Centrifugal Force

This is related to the rotational acceleration of the particle during rotation. The dust adhesion on the rotating disk system is highly affected by the centrifugal force that provides the force required to move the particle from the center to the edge of rotation. For the spherical particle, it takes the form:

$$F = m\omega^2 r_d = \frac{4}{3}\pi R^3 \rho \omega^2 r_d \qquad (2.61)$$

where m is the mass of the particle, ω is the angular velocity of the disk, r_d is the distance from the position of the particle to the center of rotation, and R is the particle radius.

2.3.5 Friction Force

This force is associated with the friction factor between the particle and the surface and it can be written as:

$$F_f = \mu_f N = \mu_f F_g = \mu mg = \frac{4\mu}{3}\pi R^3 \rho g \qquad (2.62)$$

where μ_f is the friction coefficient, m is the mass of the particle, R is the particle radius, ρ is the density of the particle, and g is the gravitational acceleration.

2.4 MEASUREMENT AND ASSESSMENT OF WETTING STATE

Various contact angle measurement techniques have been developed to evaluate the wetting state of a surface taking into account the principles of wettability discussed in the previous section. Contact angle measurement techniques can be categorized into two groups: on flat surfaces and on particles or rough surfaces [39]. Contact angle can be directly measured by side view optical microscopy. Due to technology progress and current development in optical recording devices, droplet images and free surface profile analysis are favorable for wetting state assessment and contact angle measurement [40].

2.4.1 Smooth Surfaces

Measurements of flat surface contact angle can be carried out using many techniques. Optical contact angle measurement is popular and is used to observe droplet images. Producing such surface and keeping it clean without contaminations in the lab environment are critical for precise contact angle measurements. Surface preparation is described in Ref. [41] where grinding and polishing of the surface in a precise and controlled manner give repeatable readings of the contact angle measurements while achieving homogenous properties of the surface; however, the uncertainty of the contact angle measurements is high for small contact angles.

2.4.1.1 Droplet Dimensions

Droplet dimensions can be used to determine the contact angle of the liquid droplet on the surface. In this case, some measuring parameters are used to determine the free surface profile of an axisymmetric droplet using the Laplace equation, which describes the profile of fluid interfaces [39,42]:

$$\Delta P \gamma \left(\frac{1}{R_1} + \frac{1}{R_2} \right) \tag{2.63}$$

where R_1 and R_2 are the two principal radii of the free surface curvature and ΔP is the pressure difference across the interface. This method uses the mathematical expression to predict small contact angles of a droplet knowing the radius of the droplet curvature from an image and the droplet volume.

2.4.1.2 Droplet Free Surface Profile

Analyzing the droplet profile is a direct method for contact angle measurements on smooth surfaces where a telescope goniometer is used to capture a droplet image and measure the profile surface [43]. A small quantity of the liquid is needed to form a droplet on the small area of the solid surface; however, to assess the wettability of the surface and ensure independent results of the surface heterogeneity, contact angle measurements should be repeated at different locations of the surface. High magnification images are recorded and the droplet free surface is traced. The contact angle can be measured from the images, which is the angle between the solid−liquid contact line and the tangent line drawn from the triple point. For relatively large contact angle droplets (>20 degrees), a measurement accuracy of ± 2 degrees can be used [44]. Advancing (θ_A) and receding (θ_A) contact angles can also be measured by the droplet profile method. In this case, images of the water droplet are taken and the advancing contact angle can be measured by increasing the droplet volume while the receding one by decreasing the droplet volume.

2.4.1.3 Axisymmetric Droplet Shape Analysis-Profile

In this method, the shape of axisymmetric menisci is used to predict the interfacial tension and contact angle [42,45]. This method predicts theoretically the drop profile that closely matches the corresponding one extracted from the image and calculates the surface tension and contact angle that satisfy the Laplace equation of capillarity. In this case, it is assumed that the droplet is axisymmetric and only gravity force is taken into consideration [39]. Three factors affect the accuracy of this method: (1) how many data points are needed to describe the profile of the droplet, (2) the number of significant figures of the profile data points, and (3) the randomness of the choice of data points.

2.4.1.4 Wilhelmy Method

This technique was introduced by Wilhelmy [46] based on the balance of the total force acting on the sample. In this method, a vertical palate (Wilhelmy plate) is immersed in a liquid. In this case, two situations were observed: the first one is when the contact angle is less than 90 degrees a downward force, W_F, is exerted on the plate, which is [39]:

$$W_F = p\gamma_{LV}cos\theta \tag{2.64}$$

where p is the perimeter of the contact line. The second situation is when the depth of immersion is not equal to zero; then, a volume of liquid will be displaced. Therefore the buoyancy affect cannot be neglected when carrying the force balance on the plate:

$$W_F = P\gamma_{LV}cos\theta - \forall g\Delta\rho \tag{2.65}$$

Therefore the contact angle can be determined if the surface tension is known. A dynamic Wilhelmy contact angle measurement involves the motion of a smooth solid plate, which is immersed in or pulled out of a known surface tension liquid. In this case advancing and receding angles can be determined.

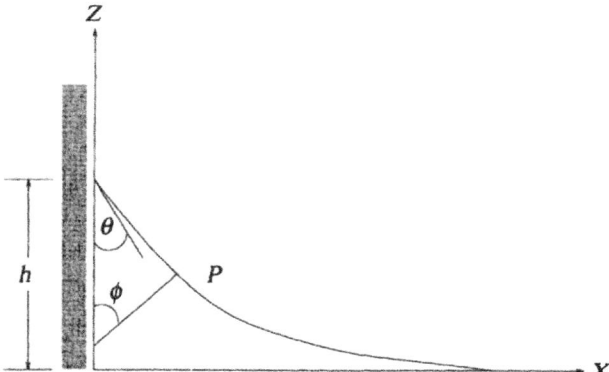

FIGURE 2.19 Schematic of capillary rise at a vertical plate. Here ϕ is the angle between the vertical axis and the normal at a point on the liquid–vapor surface, P, and θ is the contact angle [47].

2.4.1.5   Capillary Rise at a Vertical Plate

A modification of the Wilhelmy method is possible and has been used to determine the capillary rise of a vertical plate (h), as shown in Fig. 2.19. When the vertical plate is brought into contact with the liquid, the capillary effect will force the liquid to rise. Assuming that the vertical plate is very wide and measuring h from the experiment, the Laplace equation can be used to determine the contact angle as follows [47]:

$$sin\theta = 1 - \frac{\Delta\rho gh^2}{2\gamma_{la}} \qquad (2.66)$$

Here $\Delta\rho$ is the density difference between the liquid and the gas and γ_{la} is the liquid–air interfacial tension.

2.4.2   Rough Surfaces

The morphology of rough surfaces makes contact angle measurements difficult and hard to accomplish with high accuracy. This is due to the structure of the rough surfaces and imperfections. In this case, it is hard to find the contact line and draw a tangent from the three-phase (solid–liquid–gas) contact point and measure the contact angle using an image of the droplet. Therefore wettability assessment using capillary penetration techniques is needed.

2.4.2.1   Axisymmetric Droplet Shape Analysis Diameter Method

In the axisymmetric droplet shape analysis diameter (ADSA-D) method, the top view of the droplet is considered instead of the side view and the diameter of the contact area is measured. With known droplet volume, surface tension, and contact diameter, the contact angle can be computed using the Laplace equation of capillarity. The ADSA-D method is particularly powerful for low contact angles ($\theta < 20$ degrees) where other methods present difficulties [48].

2.4.2.2   Capillary Penetration Methods for particles

Contact angle measurement for porous media including powders and granules is difficult due to their nature. Therefore packing and compressing these powders in and applying a liquid droplet on the surface is one solution for wetting state assessment [49]. However, due to the original porous structure, the liquid may diffuse and penetrate through the material and the measurements will not be repeatable [4]. In the capillary penetration method, the capillary pressure is used to force liquid to move and penetrate through the packed and compressed powder in a capillary tube [50]. The liquid rises in the capillary tube and attains a specific height where the gravitational force balances the force due to the capillary pressure. The Laplace pressure, which drives liquid into a capillary bed, can be described by the Laplace equation as:

$$\Delta P = \frac{2\gamma_{la}cos\theta}{r_{eff}}$$

(2.67)

where r_{eff} is the effective capillary radius, which is [51]:

$$r_{eff} = \frac{2(1-\emptyset)}{\emptyset\rho A}$$

(2.68)

where $\emptyset$ is the volume fraction of the solid in the packed bed, ρ is the density of the solid material, and A is the specific surface area per gram of solid. Combining Eqs. (2.67) and (2.68) yields the Laplace–White equation, which is a strict thermodynamic expression for ΔP in porous media [51]:

$$\Delta P = \frac{\gamma_{LV}cos\theta\emptyset A\rho}{(1-\emptyset)}$$

(2.69)

2.5   FREE ENERGIES OF SURFACES AND INTERFACES

Surface energy is the energy per unit area that is required to form a surface in liquids and solids that is located in a gas or vacuum environment. The energy that is normally associated with bonding to other atoms is available at the surface. This energy required to create a new surface is referred to as free surface energy.

When a bond is formed between two materials with surface energies per unit area γ_1 and γ_2, interface with an interfacial energy per unit area, say,

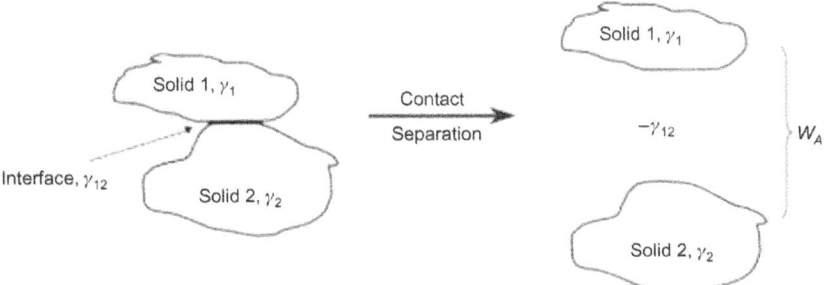

FIGURE 2.20 The work of adhesion during contact-interface separation [27].

γ_{12}, is formed. The work of adhesion is the energy required to separate the two surfaces per unit area, and it can be written as a Dupré equation [27]:

$$W_{ad} = \Delta\gamma = \gamma_1 + \gamma_2 - \gamma_{12} \tag{2.70}$$

and

$$\gamma_{12} = |\gamma_1 - \gamma_2| \tag{2.71}$$

where $\Delta\gamma$ represents the reduction in the surface energy of the system per unit area (a negative value in mJ/m^2, dynes/cm). For identical materials in contact, $\Delta\gamma$ represents the cohesion work that equals 2γ. It was observed from surface free energy theory that higher surface energy leads to stronger bonds between solid surfaces in contact. The total energy change during separation of two surfaces in contact is demonstrated in Fig. 2.20.

Surface energy can be defined also as the excess energy at the surface. The surface energy of solid–liquid in contact ranges from 0.05 to 0.2 J/m^2 while for solids–vapor in contact it is less than 2 J/m^2. The surface energy varies with the deformation of solid surfaces due to stretching of the interatomic bonds of the solid surface. This deformation can be written as:

$$GS = \gamma + \left(\frac{\partial G}{\partial A}\right) \tag{2.72}$$

A critical situation occurs when different phases meet together (gas, liquid, and solid). The material's surface energy can be derived from Young's equation (Eq. 2.3).

The adhesion work of solid-to-liquid (W_{ad}) can be expressed by combing Young's equation with the Dupré equation. This combination indicates that the strength of adhesion is directly related to the contact angle between solid and liquid. It can be observed that the adhesion energy can be measured using the contact angle:

$$W_{ad} = \gamma_{sa} + \gamma_{la} - \gamma_{sl} = \gamma_{la}(1 + cos\theta) \tag{2.73}$$

Rough surfaces have different effects on the contact angles of wetting liquids. Therefore a modified form of Young's equation is important to satisfy the practical working conditions. Cassie–Baxter and Wenzel are the two main models that attempt to describe the wetting of textured surfaces.

2.6   SPREADING COEFFICIENT OF LIQUIDS

Film spreading of a liquid on surfaces is involved with an important phenomenon. It is difficult for oil to penetrate the surface wetted by water and vice versa. The spreading coefficient is a criterion used to assess the behavior of liquids in contact with such surfaces (spreading or nonspreading) [52]. The spreading can be from the thermodynamics of surfaces. When a liquid droplet is deposited on the surface of another liquid spreading may occur. If this happens, the surface of the underlying liquid disappears, while its place is taken by substantially an equal area of the surface of the droplet liquid, provided that the surface of the underlying liquid and the interface of both liquids do not lose their identity. If they do, then only one composite surface takes the place of the surface droplet fluid. Since only large-scale motion is of importance in spreading, it is only the free surface energies that are involved. The spreading can be written as [52]:

$$S = \gamma_1 - (\gamma_2 + \gamma_{12}) \tag{2.74}$$

where γ_1 and γ_2 are the surface energies per unit area of liquid-1 and liquid-2, respectively, and γ_{12} is the interfacial energy per unit area.

The work of adhesion W_{ad}, or the energy required to separate the two surfaces per unit area, is given by the Dupré equation (Eq. 2.45), whereas the cohesion work equals $W_{co} = 2\gamma$. The spreading coefficient is a criterion used to measure the predisposition of a fluid to spread on a surface, which can also be defined by combining Eqs. (2.45) and (2.49) as:

$$S = W_{ad} - W_{co} \tag{2.75}$$

If the adhesion between the two liquids is greater than the cohesion in the liquid, spreading takes place. If the cohesion is greater than the adhesion, spreading does not occur. It is obvious that a positive value of the spreading coefficient corresponds to spreading and a negative value corresponds to nonspreading. It is also evident that because the liquid-2 spreads upon liquid-1, it is not at all a necessary conclusion that liquid-1 spreads upon liquid-2. Thus the spreading coefficient is given above for the case where liquid-1 is the liquid whose surface is already formed. The spreading coefficient for liquid-1 to spread upon liquid-2 is $S = \gamma_2 - (\gamma_1 + \gamma_{12})$ so a high surface energy for the liquid-1 acts in favor of spreading when liquid-1 is at the bottom liquid, and it does not spread when the liquid-2 is at the bottom [52]. Fig. 2.20 shows the silicon oil cloaking of a dust particle. In the initial state the silicon oil wets the dust particle and forms a ring-like

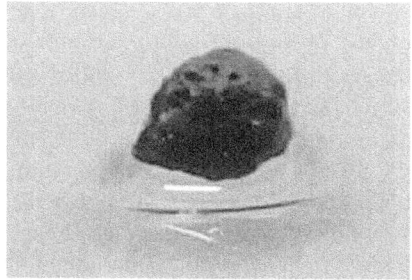

20 µm

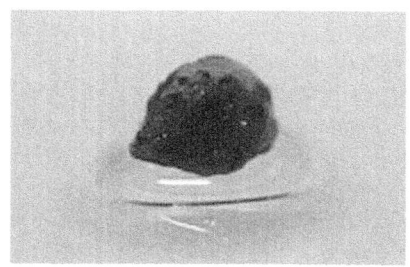

FIGURE 2.21 Silicon oil cloaking of dust particle surface over time.

ridge around the dust particle. As time progresses the wetting of the dust particles enhances while the silicon oil covers almost the whole surface of the dust particle (Fig. 2.21).

REFERENCES

[1] R. Fürstner, W. Barthlott, C. Neinhuis, P. Walzel, Wetting and self-cleaning properties of artificial superhydrophobic surfaces, Langmuir 21 (3) (2005) 956–961.

[2] B. Bhushan, Y.C. Jung, K. Koch, Self-cleaning efficiency of artificial superhydrophobic surfaces, Langmuir 25 (5) (2009) 3240–3248.

[3] B. Bhushan, Y.C. Jung, Natural and biomimetic artificial surfaces for superhydrophobicity, self-cleaning, low adhesion, and drag reduction, Prog. Mater. Sci. 56 (1) (2011) 1–108.

[4] Y. Yuan, T.R. Lee, Contact angle and wetting properties, Surface Science Techniques, Springer, 2013, pp. 3–34.

[5] Y.-W. Chung, Micro-and Nanoscale Phenomena in Tribology, CRC Press, 2011.

[6] Y.A. Cengel, M.A. Boles, Thermodynamics: An Engineering Approach, -PDF, McGraw-Hill, 2008.

[7] T. Young III, An essay on the cohesion of fluids, Philosop. Trans. Royal Soc. London 95 (1805) 65–87.

[8] J.N. Israelachvili, Intermolecular and Surface Forces, Academic Press, 2011.

[9] S. Shibuichi, T. Onda, N. Satoh, K. Tsujii, Super water-repellent surfaces resulting from fractal structure, J. Phys. Chem. 100 (50) (1996) 19512–19517.

[10]   R.N. Wenzel, Resistance of solid surfaces to wetting by water, Ind. Eng. Chem. 28 (8) (1936) 988–994.

[11]   M. Nosonovsky, B. Bhushan, Roughness optimization for biomimetic superhydrophobic surfaces, Microsyst. Technol. 11 (2005) 535–549.

[12]   A. Cassie, S. Baxter, Wettability of porous surfaces, Trans. Faraday Soc. 40 (1944) 546–551.

[13]   A. Al-Sharafi, B.S. Yilbas, H. Ali, Droplet heat transfer on micropost arrays with hydrophobic and hydrophilic characteristics, J. Heat Transfer 140 (7) (2018) 072402.

[14]   B. Yilbas, A. Matthews, C. Karatas, A. Leyland, M. Khaled, N. Abu-Dheir, et al., Laser texturing of plasma electrolytically oxidized aluminum 6061 surfaces for improved hydrophobicity, J. Manuf. Sci. Eng. 136 (5) (2014) 054501.

[15]   B. Yilbas, M. Khaled, N. Abu-Dheir, N. Al-Aqeeli, S. Said, A. Ahmed, et al , Wetting and other physical characteristics of polycarbonate surface textured using laser ablation, Appl. Surf. Sci. 320 (2014) 21–29.

[16]   L. Pan, G. Yu, D. Zhai, H.R. Lee, W. Zhao, N. Liu, et al., Hierarchical nanostructured conducting polymer hydrogel with high electrochemical activity, Proc. Natl. Acad. Sci. 109 (24) (2012) 9287–9292.

[17]   M. Jin, X. Feng, J. Xi, J. Zhai, K. Cho, L. Feng, et al., Super-hydrophobic PDMS surface with ultra-low adhesive force, Macromol. Rapid Commun. 26 (22) (2005) 1805–1809.

[18]   G. Shao, J. Wu, Z. Cai, W. Wang, Fabrication of elastomeric high-aspect-ratio microstructures using polydimethylsiloxane (PDMS) double casting technique, Sensors Actuat. A: Phys. 178 (2012) 230–236.

[19]   L. Yang, X. Hao, C. Wang, B. Zhang, W. Wang, Rapid and low cost replication of complex microfluidic structures with PDMS double casting technology, Microsyst. Technol. 20 (10–11) (2014) 1933–1940.

[20]   D. Ebert, B. Bhushan, Transparent, superhydrophobic, and wear-resistant coatings on glass and polymer substrates using SiO$_2$, ZnO, and ITO nanoparticles, Langmuir 28 (31) (2012) 11391–11399.

[21]   T.-S. Wong, S.H. Kang, S.K. Tang, E.J. Smythe, B.D. Hatton, A. Grinthal, et al., Bioinspired self-repairing slippery surfaces with pressure-stable omniphobicity, Nature 477 (7365) (2011) 443.

[22]   X. Yao, Y. Hu, A. Grinthal, T.-S. Wong, L. Mahadevan, J. Aizenberg, Adaptive fluid-infused porous films with tunable transparency and wettability, Nat. Mater. 12 (6) (2013) 529.

[23]   B. Yilbas, H. Ali, N. Al-Aqeeli, M. Khaled, N. Abu-Dheir, K. Varanasi, Solvent-induced crystallization of a polycarbonate surface and texture copying by polydimethylsiloxane for improved surface hydrophobicity, J. Appl. Polym. Sci. 133 (22) (2016).

[24]   A. Al-Sharafi, B.S. Yilbas, H. Ali, N. AlAqeeli, A. Water, Droplet pinning and heat transfer characteristics on an inclined hydrophobic surface, Sci. Rep. 8 (1) (2018) 3061.

[25]   A. Al-Sharafi, H. Ali, B.S. Yilbas, A.Z. Sahin, N. Al-Aqeeli, F. Al-Sulaiman, et al., Internal flow and heat transfer in a droplet located on a superhydrophobic surface, Int. J. Therm. Sci. 121 (2017) 213–227.

[26]   B. Bhushan, Y.C. Jung, K. Koch, Micro-, nano- and hierarchical structures for superhydrophobicity, self-cleaning and low adhesion, Philosop. Trans. Royal Soc. London A: Math. Phys. Eng. Sci. 367 (1894) (2009) 1631–1672.

[27]   N. Michael, B. Bhushan, Hierarchical roughness makes superhydrophobic states stable, Microelect. Eng. 84 (3) (2007) 382–386.

[28]   A. Lafuma, D. Quéré, Superhydrophobic states, Nat. Mater. 2 (7) (2003) 457.

[29]   T. Deng, K.K. Varanasi, M. Hsu, N. Bhate, C. Keimel, J. Stein, et al., Nonwetting of impinging droplets on textured surfaces, Appl. Phys. Lett. 94 (13) (2009) 133109.

[30] M. Reyssat, J. Yeomans, D. Quéré, Impalement of fakir drops, Europhys. Lett. 81 (2) (2007) 26006.

[31] L. Bocquet, E. Lauga, A smooth future? Nat. Mater. 10 (5) (2011) 334.

[32] A. Tuteja, W. Choi, M. Ma, J.M. Mabry, S.A. Mazzella, G.C. Rutledge, et al., Designing superoleophobic surfaces, Science 318 (5856) (2007) 1618–1622.

[33] K.L. Johnson, K. Kendall, A. Roberts, Surface energy and the contact of elastic solids, Proc. R. Soc. Lond. A 324 (1558) (1971) 301–313.

[34] B.V. Derjaguin, V.M. Muller, Y.P. Toporov, Effect of contact deformations on the adhesion of particles, J. Colloid. Interface. Sci. 53 (2) (1975) 314–326.

[35] H. Hamaker, The London—van der Waals attraction between spherical particles, Physica 4 (10) (1937) 1058–1072.

[36] Q. Li, V. Rudolph, W. Peukert, London-van der Waals adhesiveness of rough particles, Powder Technol. 161 (3) (2006) 248–255.

[37] Y.I. Rabinovich, J.J. Adler, A. Ata, R.K. Singh, B.M. Moudgil, Adhesion between nanoscale rough surfaces: I. Role of asperity geometry, J. Colloid. Interface. Sci. 232 (1) (2000) 10–16.

[38] H. Rumpf, Particle Technology, Springer Science & Business Media, 2012.

[39] T. Chau, A review of techniques for measurement of contact angles and their applicability on mineral surfaces, Minerals Eng. 22 (3) (2009) 213–219.

[40] T. Zhao, L. Jiang, Contact angle measurement of natural materials, Colloids Surfaces B: Biointerfaces 161 (2018) 324–330.

[41] I. Wark, A. Cook, An experimental study of the effect of xanthates on contact angles at mineral surfaces, Trans. Am. Inst. Min. Eng. 112 (1934) 189–244.

[42] P. Cheng, A. Neumann, Computational evaluation of axisymmetric drop shape analysis-profile (ADSA-P), Colloids Surfaces 62 (4) (1992) 297–305.

[43] W. Bigelow, D. Pickett, W. Zisman, Oleophobic monolayers: I. Films adsorbed from solution in nonpolar liquids, J. Colloid Sci. 1 (6) (1946) 513–538.

[44] R.J. Hunter, Foundations of Colloid Science, Oxford University Press, 2001.

[45] J. Spelt, Y. Rotenberg, D. Absolom, A. Neumann, Sessile-drop contact angle measurements using axisymmetric drop shape analysis, Colloids Surfaces 24 (2–3) (1987) 127–137.

[46] L. Wilhelmy, Concerning the dependence of the Capillaritäts-Constante the alcohol of substance and shape of the wetted solid body, Annalen Der Physik und Chemie 119 (1863) 177.

[47] D. Kwok, D. Li, A. Neumann, Capillary rise at a vertical plate as a contact angle technique, Surf. Sci. Ser. (1996) 413–440.

[48] O. Del Rıo, A. Neumann, Axisymmetric drop shape analysis: computational methods for the measurement of interfacial properties from the shape and dimensions of pendant and sessile drops, J. Colloid. Interface. Sci. 196 (2) (1997) 136–147.

[49] G. Zografi, S.S. Tam, Wettability of pharmaceutical solids: estimates of solid surface polarity, J. Pharm. Sci. 65 (8) (1976) 1145–1149.

[50] E.W. Washburn, The dynamics of capillary flow, Phys. Rev. 17 (3) (1921) 273.

[51] L.R. White, Capillary rise in powders, J. Colloid. Interface. Sci. 90 (2) (1982) 536–538.

[52] W.D. Harkins, A. Feldman, Films. The spreading of liquids and the spreading coefficient, J. Am. Chem. Soc. 44 (12) (1922) 2665–2685.

Chapter 3

Surfaces for Self-Cleaning

Chapter Outline

3.1 INTRODUCTION

Surfaces with the ability to remove any debris or avoid the deposition of dirt on their surfaces are known as self-cleaning surfaces. This chapter introduces the basic terminology related to the topic, then describes the surface characteristics of self-cleaning surfaces. In the latter half of the chapter the development of self-cleaning surfaces and their analysis and assessment will be presented.

3.2 SURFACE CHARACTERISTICS

In order to understand the surface characteristics of self-cleaning surfaces, it is necessary to introduce some basic terminology related to wetting of surfaces. Wettability is one of the important properties of a solid surface which is influenced by the surface energies of the relevant phases and surface texture of the solid substrate. The contact angle (θ) measurement is normally used to characterize the wettability of the surfaces. Contact angle is defined by the angle formed between the solid/liquid interface and the liquid/vapor interface when a droplet is placed on a surface as shown in Fig. 3.1.

Self-Cleaning of Surfaces and Water Droplet Mobility. DOI: https://doi.org/10.1016/B978-0-12-814776-4.00003-3
45

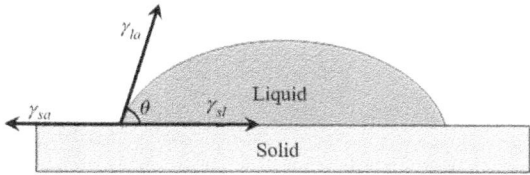

FIGURE 3.1 Force balance of interfacial tensions at three-phase contact line.

The contact angle is dictated by the equilibrium of the interfacial tensions at the three-phase contact line, that is, $\gamma_{la}cos\theta + \gamma_{sl} = \gamma_{sa}$, where γ_{la}, γ_{sl}, and γ_{sa} are the liquid–air, solid–liquid, and solid–air interfacial tensions, respectively. The relation is known as the Young's equation, which is valid for smooth surfaces [1]. Surfaces with a water contact angle (WCA) greater than 90 degrees are called hydrophobic and those having a WCA less than 90 degrees are called hydrophilic. For rough surfaces, theoretical work regarding the contact angle measurement was carried out by Wenzel and Cassie–Bexter. Assuming that the liquid completely wets a surface, Wenzel developed a model predicting the relation between the contact angle of a droplet on a rough surface to that on the smooth surface as $cos\theta_{rough} = rcos\theta_{flat}$, where θ_{rough} and θ_{flat} are the contact angles of the droplet on rough and flat surfaces, respectively. The roughness ratio (r) is defined as the ratio of surface area to the flat projected area. Cassie–Baxter extended Wenzel's work while incorporating the trapped air pocket on the rough surface. Hence, the solid surface consists of solid–vapor–liquid phase and the contact angle equation takes the form $cos\theta_{rough} = r\phi_s cos\theta_{flat} + \phi_A cos\theta_{LA}$, where ϕ_s and ϕ_A are the fractional liquid–solid and liquid–air interfacial areas ($\phi_s + \phi_A = 1$). Since $cos\,\theta_{LA} = 180$ degrees (contact angle of water with air), entrapment of air pockets will result in an increased contact angle. Using $cos\,\theta_{LA} = 180$ degrees, the contact angle equation yields $cos\theta_{rough} = r\phi_s cos\theta_{flat} - (1 - \phi_s)$. Before moving further, it is important to introduce the concept of contact angle hysteresis. In general, two types of contact angle measurements are performed: static and dynamic.

For the static contact angle measurement droplet placed on flat surface and measurements taken once the equilibrium is established, whereas dynamic contact angles are measured during the droplet growth and shrinkage. One another method to measure the dynamic contact angle is tilting the surface and noting the front and back contact angle of the droplet. Front and back angles are referred as advancing and receding contact angles respectively. Contact angle hysteresis is defined as the difference between the advancing and receding. For surfaces with low contact angle hysteresis, water rolls-off at very small tilt angles carry away dust particles, giving rise to the so-called self-cleaning effect [2]. Apart from the hydrophilic (contact angle <90 degrees) and hydrophobic (contact angle >90 degrees), there are two special case of wettability. One is a superhydrophilic surface where the

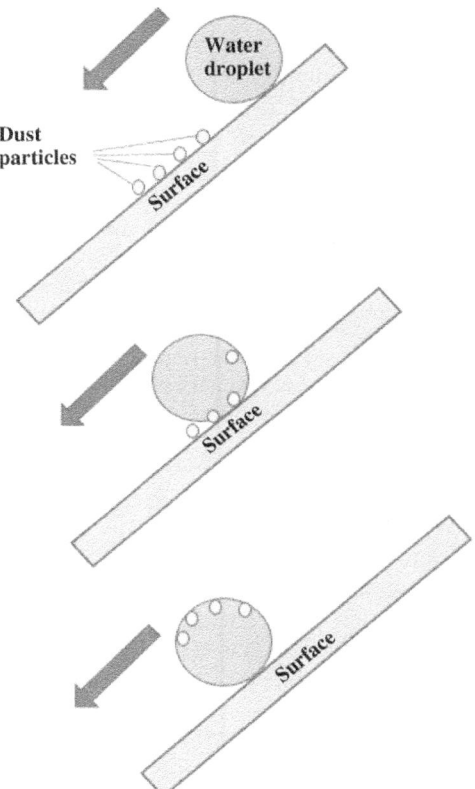

FIGURE 3.2 Lotus or self-cleaning effect whereby the dust particles are carried away by the rolling droplet on a superhydrophobic surface.

WCA is less than 5 degrees. The other is a superhydrophobic surface where the WCA is greater than 150 degrees with low contact angle hysteresis [3].

Natural superhydrophobic surfaces like lotus plant leaves can remain clean since water droplets roll away at small inclination angles and take away dust particles along its path. This gives rise to the so-called lotus effect or self-cleaning effect as shown in Fig. 3.2. A superhydrophobic surface thus can clean itself by using very little water in contrast to conventional wetting surfaces.

Looking to nature for surfaces possessing self-cleaning characteristics, Barthlott and Neinhuis introduced the "lotus effect" [4], described a surface possessing superhydrophobic properties with self-cleaning attributes. As described, self-cleaning and superhydrophobic surfaces have high contact angle with low contact angle hysteresis due to the combination of hierarchical surface structure and low surface energy material [5].

3.3 DEVELOPMENT, SYNTHESIZING, AND FABRICATION PROCESSES

There are several techniques available for the development of superhydrophobic surfaces. These techniques include lithography, templating, etching, sol—gel method, and layer-by-layer.

3.3.1 Lithography

Lithography is one technique used in the development of superhydrophobic surfaces. There are various versions of lithography which provide good control over the topography of a surface. Many researchers [6—14] have used lithography for the development of superhydrophobic surfaces. There are two types of lithography: conventional and nonconventional.

3.3.1.1 Conventional Lithography

Conventional lithography is commercially available and widely implemented in manufacturing. The cost of conventional lithography is high. There are two well-known methods for conventional lithography: photolithography and particle beam lithography such as electron and ion beam lithography.

3.3.1.1.1 Photolithography

Photolithography is an optical means of transferring a pattern on a substrate. All the photolithography methods follow this principle. First, the photoresist is placed on the substrate. Then, the substrate is exposed to electromagnetic radiation which modifies the molecular structure followed by a change in the solubility of the material [15], while placing the mask of pattern. After exposure, etching is carried out. Then this substrate is immersed in a developer solution. Developer solutions are typically aqueous and dissolve away areas of the photoresist exposed to light. A schematic representation of photolithography is shown in Fig. 3.3. Zhou et al. [16] developed the superhydrophobic micropillar arrays using the photolithography.

3.3.1.1.2 Particle Beam Lithography

Particle beam lithography, such as electron or ion beam lithography, is a costly technique for small-size fabrication. These techniques are popular for academic research because of the flexibility in feature design. The flexibility and high precision of electron beam lithography provides the advantage toward the development of hierarchical structures, which leads to the development of high WCA surface. The flexibility for the development of the well-defined microstructures along with secondary nanostructures help in understanding the superhydrophobicity. Feng et al. [17] developed a

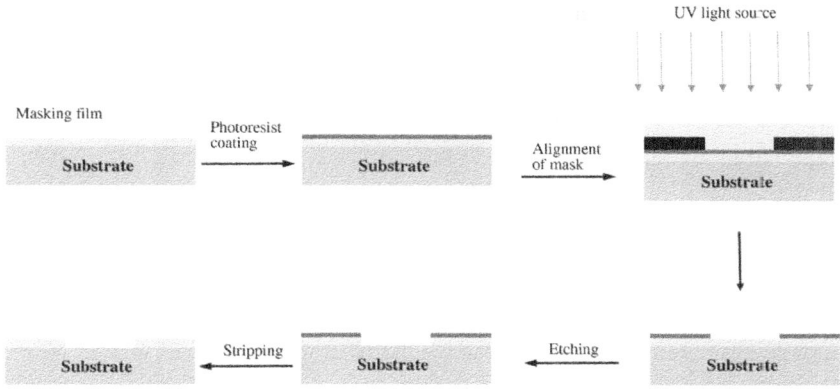

FIGURE 3.3   Schematic representation of the photolithographic process sequence.

superhydrophobic surface while using electron beam lithography, while fabricating a microscale and secondary nanoscale structure.

Ion beam lithography is similar to electron beam lithography. In ion beam lithography ion beams are used instead of electron beams. High-energy ions are able to penetrate resistant material with well-defined paths. The penetration depth depends on the ion energy. Since the interaction of ions with electrons does not provide significant deviation in the trajectory of the ion, the straight vertical wall with a high aspect ratio can be developed from ion beam lithography.

3.3.1.2 Unconventional Lithography

Unconventional lithography techniques appear to meet ever-increasing demand for nanofabrication with smaller features, lower cost, and more complicated geometries. In order to develop superhydrophobic patterns on polymers soft lithography is commonly used. Highly adhesive and superhydrophobic structures can be generated using soft lithography [18]. Pozzato et al. [6] fabricated superhydrophobic surfaces on silicon using soft lithography (nanoimprint lithography). Shiu et al. [7] used nanosphere lithography along with oxygen-plasma treatment for well-ordered superhydrophobic surfaces to get the WCA up to 170 degrees. The drawback of this method is the thermal instability. Many of the unconventional lithographic techniques cannot stand alone. They require the assistance of conventional lithographic techniques such as photolithography to design and make masks or masters. To address this challenge, an increasing amount of attention is being given to the use of self-assembly of molecules and colloidal particles for development of ingenious, cheap, bottom-up ways of masking. Colloidal lithography is one of the fast-developing unconventional lithographic techniques which is capable of fabricating superhydrophobic surfaces that can stand alone. Kothary et al. [11] developed a superhydrophobic structure, with a more than 150 degrees WCA and low water hysteresis angle on fluorosilane-modified polymer by developing hierarchical nanovoid arrays with large fractions of entrapped air.

3.3.2 Templating

Templating is a technique in which replication of a pattern take place while producing the inverse of the original pattern. For preparation of a polymeric superhydrophobic surface, templating is considered as the most cost and time effective procedure [19]. PDMS (polydimethylsiloxane), a silicon-based organic polymer, is an ideal choice for the development of transparent super-hydrophobic surfaces because of its high transparency, hydrophobic nature, and the ability to copy submicron features during the replication process.

Many researchers have fabricated superhydrophobic surfaces [20–22] using templating. Yilbas et al. [22] performed the crystallization of polycarbonate (PC) and then used PDMS to develop a replica of the crystallized PC structure. For templating, liquid PDMS was deposited and left on the crystallized PC surface for over 18 hours for curing purposes. The solidified PDMS was then removed from the crystallized PC surface after the curing period, resulting in a contact angle measurement of around 154 degrees.

3.3.3 Etching

Micropatterns can be created by etching process. Depending on the nature of the material superhydrophobicity can be achieved either by straightaway or a post treatment with a hydrophobizing agent. Chemical and plasma etching are the two main etching techniques used for the development of superhydrophobic surfaces.

Chemical etching is used on low surface energy materials for modulation of the geometrical microstructure and chemical composition [23]. Hierarchical surfaces fabricated using such chemical etching processes exhibit a high degree of hydrophobicity, with resultant water contact angles nearing 150 degrees [24].

Plasma etching is also used for the fabrication of superhydrophobic surface on metals, polymer, and elastomer [25]. Depending on the material crystal planes, etching rates are varied for the fabrication of desired surface structure [26]. Ion etching and laser ablation are also used for the fabrication of superhydrophobicity in polymers, whereas wet etching is used for metals [27].

3.3.4 Coating

3.3.4.1 Spin Coating

Spin coating is used for the fabrication of thin films to deposit uniform coating of organic materials on flat surfaces [28]. Fu et al. developed a transparent superhydrophobic transparent coating via spin coating [29]. Spin coating is performed in four steps, deposition, spin up, spin off, and evaporation, as

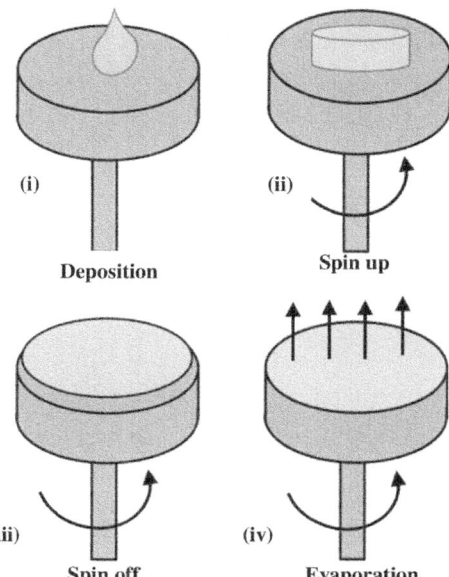

FIGURE 3.4 Stages of spin coating on substrate.

(i)   (ii)

Deposition   **Spin up**

(iii)   (iv)

Spin off   **Evaporation**

shown in Fig. 3.4. In the first stage the material is deposited on the turntable and then spin up and spin off occur in sequence while the evaporation stage occurs throughout the process. The applied solution on the turntable is distributed via centrifugal force. High spinning speed results in thinning of the layer. This stage is followed by drying of the applied layer. Uniform evaporation of the solvent is possible because of rapid rotation. High volatile components are removed from the substrate because of the evaporation or simply drying and the low volatile components of the solution remain on the surface of the substrate. Thickness of the deposited layer is controlled by the viscosity of the coating solution and the speed of rotation [30].

One of the main disadvantages of spin coating is the size of the substrate. As the size increases, the high-speed spinning becomes difficult, because film thinning becomes difficult. The material efficiency of spin coating is very low. In general, 95%−98% of material is flung off and disposed of during the process and only 2%−5% of material is dispensed onto the substrate [31].

3.3.4.2   Spray Coating

In industry, for the coating of complex-shaped polymeric substrates spray coating is used [32]. Atomizers or nebulizers, as shown in Fig. 3.5, are used for the formation of the fine droplets required for the spray coating. Fine droplets, from the automizer, reach the coating chamber with the help of carrier gas and are deposited on the substrate by gravity or with an electrostatic field. The size of the droplets determines the quality of the coating. The size

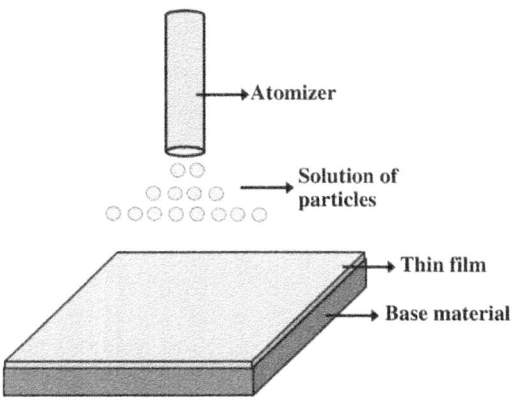

FIGURE 3.5 Spray coating on substrate.

of the droplets can be reduced by decreasing the viscosity of the solution and by increasing the atomizing pressure.

The quality of the coating is determined by the size of the droplet, which can be reduced by decreasing the viscosity of the solution, increasing the atomizing pressure, or by using a Venturi nozzle. Although spray coating is a fast coating process that wastes less coating sols, the spray-coating method offers limited control of the uniformity of thickness [33]. Using spray coating, Ogihara et al. [34] formed a hydrophobic silica nanoparticle coating on a glass surface. Functionalized silica nanoparticles suspended in propanol were used for coating purposes. The obtained coating consisted of very fine micro- and nanostructures and yielded high optical transparency. The effect of surface treatment of the glass on the hydrophobic characteristics of the resultant surfaces was also investigated. For untreated glass surfaces, the silica nanoparticle films did not show superhydrophobic characteristics. This is due to the high surface energy of the glass surfaces, which caused the water droplets to stick. For glass surfaces treated with fluorine groups, formation of a uniform silica nanoparticle film is difficult because of the poor wettability of the glass. Glass surfaces treated with dodecyl groups are most suited for forming a transparent superhydrophobic silica nanoparticle film.

3.3.4.3 Dip Coating

In the dip-coating process, the substrate is coated by immersing it in a liquid. A typical schematic view of the dip-coating process is shown in Fig. 3.6. In controlled atmosphere the substrate was withdrawn from the liquid with a specific speed. The thickness of the coating can be altered through the withdrawal rate and the liquid viscosity [30]. Through this technique the flat plane but also the cylinders and complex geometry with large surface can be coated.

Ling et al. [35] employed a simple dip-coating technique to obtain a silica nanoparticle coating on glass. Prior to dip coating, an amine-terminated self-assembled monolayer was grown on the surface of the glass. It enabled

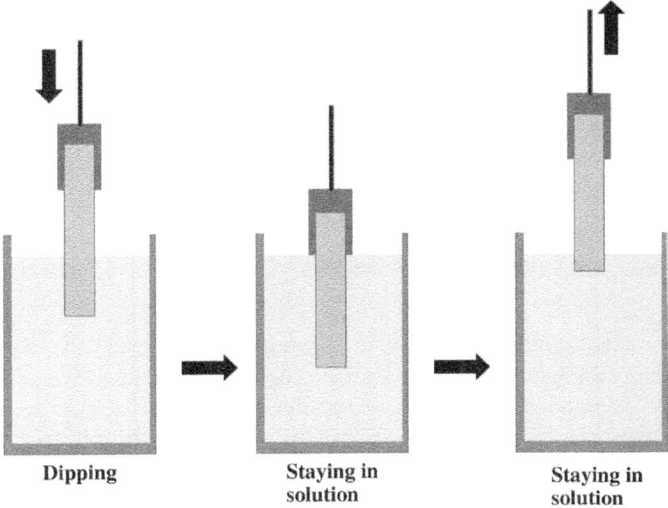

Dipping   Staying in   Staying in
solution   solution

FIGURE 3.6 A schematic view of the steps in dip coating.

the silica nanoparticles to leave large gaps in between while forming a coating at the glass surface. Static WCA of 152 degrees was observed after treatment with perfluorooctyltrichlorosilane (PFOTS) (trichloro ($1H,1H,2H,2H$-perfluorooctyl). The contact angle hysteresis, however, was relatively higher, averaging around 25 degrees. The UV/visible spectrum showed a slight decrease in transmittance at smaller wavelengths.

3.3.5 Sol–Gel Method

The sol–gel process is carried out by the synthesis of an inorganic network by a chemical reaction at low temperature [30]. Colloidal suspension of solid particles in a liquid is known as sol. In the colloidal suspension, the dispersed phase is so small and interaction is dominated by short-range forces such as van der Waals attraction as opposed to gravitational force [30]. The continuous solid skeleton made of colloidal particles is called a gel. Through the sol–gel processing, it is possible to create three-dimensional (3D) structures of the underlying surface. This method can create superhydrophobic surfaces with a WCA up to 160 degrees [36].

3.3.6 Layer-by-Layer Fabrication

In 1966 a novel method for the development of multilayers of inorganic colloidal particles was introduced by Iler [37]. This novel method is known as layer-by-layer fabrication. It facilitates the fabrication of coatings with varied compositions and structures. For the fabrication of hydrophobic surfaces the layer-by-layer method is one of the low cost and simple processes. This method can be used on flat as well as rough surfaces with large area. Cao and Gao [38] use the layer-by-layer method for the fabrication of transparent superhydrophobic of silica nanoparticles and sacrificial polystyrene nanoparticles. Superhydrophobicity was also generated on commercial polyurethane sponge surface using layer-by-layer coating of polydopamine films and silver nanoparticles onto the surface [39].

3.4   ANALYSIS AND ASSESSMENT

Analysis and assessment of self-cleaning surfaces is done by experimentation. The contact angle measurement is the main test for the confirmation of superhydrophobicity. High static contact angle along with low contact angle hysteresis are required for superhydrophobicity. In order to examine the surface structure scanning electron microscopy (SEM) and atomic force microscopy (AFM) are used. Study of the surface structure helps in the assessment and analysis of self-cleaning surfaces, and different case studies are presented here.

3.4.1   Oil-Impregnated Hydrophobic Glass Surfaces in Relation to Self-Cleaning

Hydrophobicity improvement of glass surfaces was carried out using colloidal silica particles by Rifai et al. [40]. Resulting surfaces were characterized by analytical tools including scanning electron and atomic force microscopes, water droplet contact angle measurements, and UV−visible transmittance. The dip-coating technique was adopted to deposit silica particles onto the glass surfaces. In order to generate the lotus effect through hierarchical structures at the deposited surfaces, various sizes of colloidal silica particles (30, 75, and 220 nm diameters) were used. In order to improve the optical transmittance and contact angle hysteresis of deposited surfaces, oil impregnation was introduced at the treated surface. The self-cleaning ability of silica particles deposited on glass surfaces was tested by incorporating water droplet cleaning of environmental dusts with a sample tilt angle of 10 degrees.

Piranha solution, which is a mixture of sulfuric acid (H_2SO_4) and hydrogen peroxide (H_2O_2), is widely used to modify the substrate surface prior to coating [41−43]. Surface preparation of glass was performed using two different methods: by piranha treatment and by simple direct cleaning treatment by acetone and deionized (DI) water. Two cleaned glass samples from each method were characterized by SEM to assess the effectiveness of the cleaning process. The SEM images in Fig. 3.7 show that some debris remained on the acetone/water cleaned glass. In contrast, the piranha-treated glass was completely cleaned and showed a very smooth surface without any

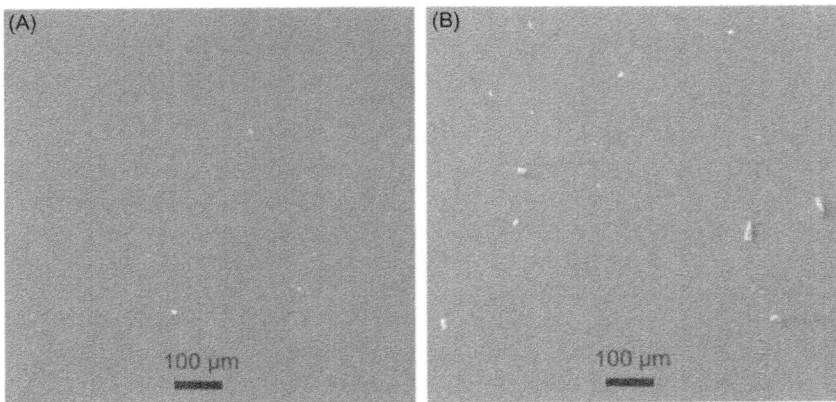

FIGURE 3.7 SEM micrographs of glass (A) with and (B) without piranha treatment [40].

contaminants left. Piranha treatment improve the surface cleanliness by forming −OH functional group on the surface [42]. This functional group improves the deposition process by increasing adherence of the surface to the coating material solution [44]. Piranha solution was prepared by mixing 1:7 volume ratio of hydrogen peroxide to sulfuric acid. Hydrogen peroxide was poured in the beaker glass, and then sulfuric acid by drop was added. The solution was left for 1 hour to complete exothermic reactions. The glass specimens 23 mm × 23 mm × 1 mm (width × length × thickness) in size were washed with soap, distilled water, and acetone successively. The cleaned glass was dried at room temperature and later soaked in piranha solution for 1 hour to create −OH functional groups. The treated glass specimens were rinsed with DI water several times and dried in open atmosphere stream for 3 hours. The used piranha solution was neutralized by adding sodium bisulfite and then left for 1 hour to complete reactions.

Dip coating was used due to its ability to produce very regular structure of nanoparticle coating [45−49]. First, colloidal silica nanoparticles (30, 75, and 220 nm) were diluted by ethanol with three different volume ratios of colloidal silica-to-ethanol solvent: 1:10, 1:15, and 1:20. The diluted colloidal silica was mixed together and dispersed in ultrasonic bath for 15 minutes. The treated glass was dipped in mixed-size colloidal-silica solution for 5 seconds and 1 mm/s withdrawal rate. The coated glass was kept at room temperature for 1 hour to let the silica nanoparticle arrange its structure. The silica-coated glass was sintered in the furnace for 2 hours at 200°C, and then immediately cooled to room temperature. Trimethylchlorosilane (TMCS) was used as a surface modifier to enhance the hydrophobicity of various materials [47]. TMCS was diluted in hexane to obtain 5% volume ratio. Dilution process was done by stirring the mixed solution for 15 minutes and then dispersing it in ultrasonic bath for 15 minutes.

The coated glass was soaked in diluted silane solution for 2 hours at room temperature. The silane-soaked sample was washed by hexane solution to remove unreacted silane. The sample was kept in room temperature to let the silylating agent arrange a self-assembly monolayer structure. The silated sample was heat treated in the furnace for 1 hour at 100°C. Later, it was taken out immediately into room temperature. A water droplet of 15 µL volume was used to test the droplet rolling on the silica/silane surface. Moreover, krytox-fluorinated oil was used to impregnate the interparticle spaces on the silica/silane surface to enhance the water-rolling effect by filling air pockets with the oil [50,51]. The surface-modified samples were dipped in fluorinated lubricant for 5 minutes and retracted with 1 mm/s rate. The lubricant bonded strongly to the carbon groups of monolayer silylating agent, making it adhere on the surface for prolonged durations, even for many years. The complete procedure is illustrated in Fig. 3.8.

Mixed-size silica nanoparticles were deposited on the piranha-treated glass surface by dip coating. The uniformity of thickness and transmittance of the mixed-size silica nanoparticles layer were assessed by using optical microscope (Fig. 3.9) and UV−visible spectrophotometer (Fig. 3.10).

The optical microscope images show that the 1:10 dilution ratio gives rise to highly agglomerated particles. This, in turn, results in reduced transmittance as compared to the 1:15 and 1:20 dilution ratios. The 1:15 dilution ratio is found to give improved uniformity and good optical transmittance;

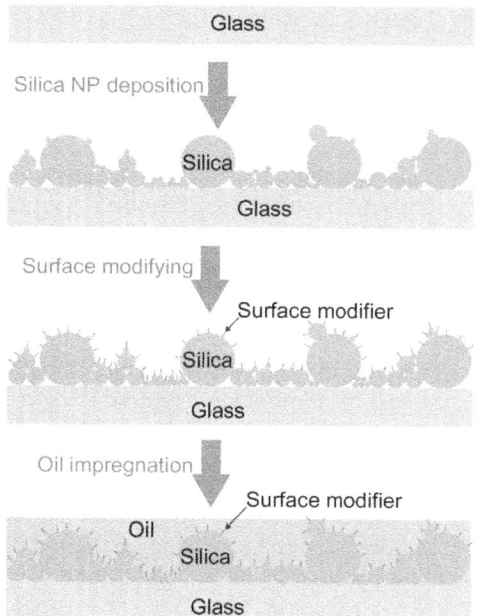

FIGURE 3.8 Procedure of oil-impregnated silica nanoparticle surface [40].

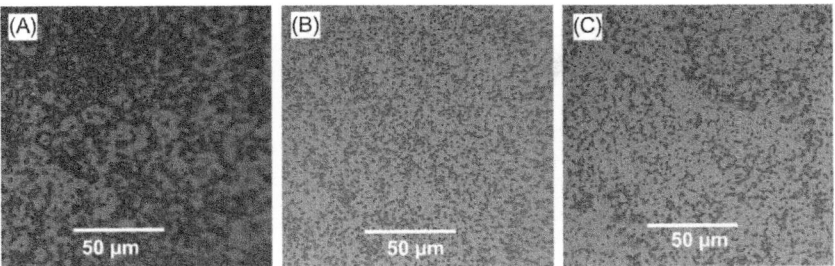

FIGURE 3.9 Optical microscope images of mixed-size silica nanoparticle deposition on glass with (A) 1:10, (B) 1:15, and (C) 1:20 volume ratio of colloidal silica nanoparticles to ethanol [40].

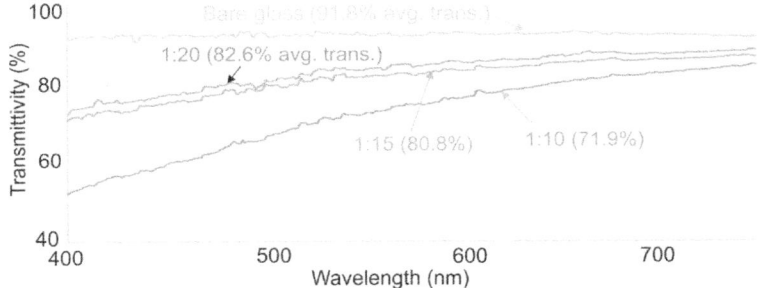

FIGURE 3.10 Optical transmittance of mixed-size silica nanoparticle deposition with different volume ratios of colloidal silica nanoparticle to ethanol [40].

therefore it is considered to be the optimum coating parameter used for the deposition process. The mixed-size silica nanoparticle deposition is performed because of the consideration that the mixed size enhances the hydrophobicity of the surface since it generates surface texture. This can be observed from the SEM micrographs as shown in Fig. 3.11A−D. Consequently, it reduces the liquid−solid contact fraction in the Cassie−Baxter state. The solid−liquid contact reduction decreases the free surface energy of the surface and thus the water droplet adhesion force is less than that of the smooth surface.

The main parameters influencing the water droplet contact angle are the surface free energy of the substrate material and the surface texture. The contact angle of a liquid droplet on chemically homogenous plain surfaces can be determined by Young's equation [51]; however, Young's equation is limited to extremely smooth and chemically homogenous surfaces. Moreover, the Wenzel and Cassie−Baxter equations for apparent contact angle, that includes surface roughness, provide more realistic data [48]. On the other hand, when dealing with heterogeneous surfaces, the Wenzel model is not sufficient for the assessment of surface hydrophobicity; in which case, a model incorporating surface

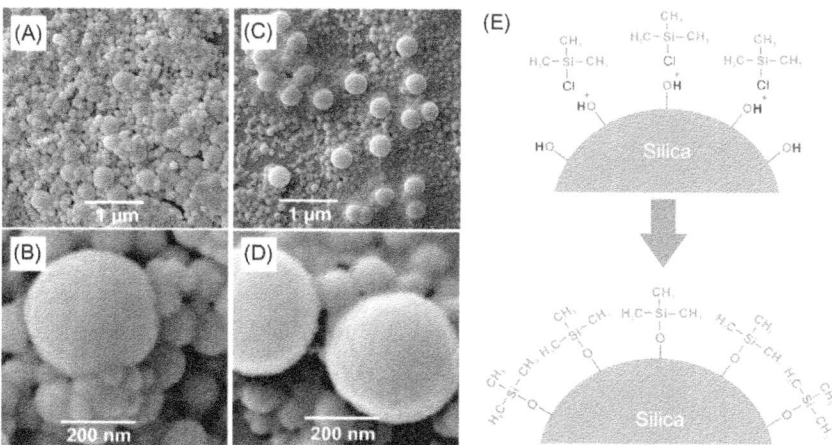

FIGURE 3.11   SEM micrographs of (A and B) sample O: silica mixed-size and (C and D) sample A11: TMCS 5 vol.% on silica mixed-size nanoparticle; left side is the illustration (E) of sylating process of the silica nanoparticle surface [40].

asperities such as roughness is needed [48]. The most common equation used to estimate the water droplet contact angle of rough surfaces is [48]:

$cos\theta_c = f_1 cos\theta_1 + f_2 cos\theta_2$, where θ_c is the apparent contact angle, f_1 is the surface fraction of the liquid−solid interface, f_2 is the surface fraction of the liquid−vapor interface, θ_1 is the contact angle for the liquid−solid interface, and θ_2 is the contact angle for the liquid−vapor interface. For the air−liquid interface, f_1 can be represented as f, which is the solid fraction, and air fraction (f_2) becomes $(1 - f)$. The parameter f ranges from 0 to 1; in which case, $f = 0$ is the case where the liquid droplet is not in contact with the surface and $f = 1$ is the case where the surface is completely wetted. In addition, the mode of surface hydrophobic state can change from the Cassie−Baxter state to the Wenzel state [49] when the surface texture becomes sparse or when the droplets impact on to the surface with high velocity [52].

The water droplet contacts measured for various surface conditions are shown in Fig. 3.12. In the case of mixed-size silica nanoparticles, the contact angle increases from 15.5 to 118.1 degrees. Consequently, the mixed-size silica nanoparticle deposition enhances the hydrophobicity of the glass surface significantly. However, the sliding angle parameter, which shows the mobility of the water droplet on the surface, shows some variations in the contact angle hysteresis. The nonfunctionalized silica particles, in which silica nanoparticles do not have functional group at the particle surfaces, cannot create the lotus effect to repel water droplets from the surface. Nevertheless, the presence of silica nanoparticles on glass surface improves hydrophobicity through forming nanoscale textures on the surface.

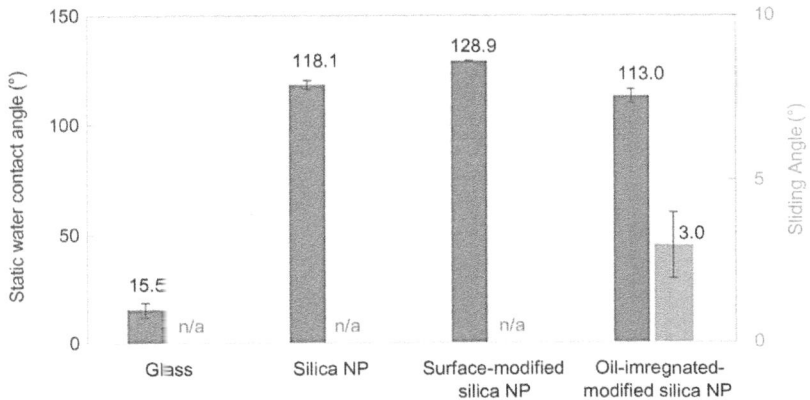

FIGURE 3.12 The water droplet contact angle with surface conditions [40].

The silicon oil impregnation results in fully wetting of the modified surface of the substrate. It should be noted that for a lubricant, which infuses the surface, fully wets the surface. In this case, for the oil-impregnated surface to shed the droplets at small tilt angles with little or no hysteresis, a thin lubricating film must be present underneath as well as outside the droplet. This situation can occur if the coefficient of spreading $(S_{os,a})$ for the oil on the surface is equal to or greater than zero [53]. In another words, $S_{os,a} = \left(\gamma_{s,a} - \gamma_{s,o} - \gamma_{o,a} \right) \geq 0$ and $S_{os,w} = \left(\gamma_{s,w} - \gamma_{s,o} - \gamma_{o,w} \right) \geq 0$. γ is the interfacial surface energy and the subscripts s, o, a, and w represent the surface, oil, air, and water phases, respectively. If for a surface $\theta_{os}(a) = 0$ and $\theta_{os}(w) = 0$, then by Young's equation of contact angle, both conditions $(S_{os}(a) \geq 0$ and $S_{os}(w) \geq 0)$ are satisfied. In the present case, both contact angles are found to be zero and a thin film totally encapsulates the surface texture, thereby allowing the water droplet to slide at the surface with close to zero tilt angle and negligible contact angle hysteresis. However, the water droplet can be cloaked by the impregnated oil when the spreading coefficient is greater than zero [54]. In the present case, the spreading coefficient remains almost zero at various locations on the surface. Therefore no cloaking takes place around the water droplet, which is also observed from the images of the sessile droplets.

Surface modifying by using 5 vol.% TMCS−Hexane mixture was performed to improve the hydrophobicity and oliophilicity of the samples. TMCS function coating is used is to create an interface layer between the silica and the impregnated oil layer. This enables oils to wet the silica particles, since silica nanoparticles do not have functional groups to attract oil. On the other hand, TMCS has $-CH_3$ functional groups, which attract oil for impregnation, as is shown schematically in Fig. 3.11E. Fig. 3.12 shows the improvement of hydrophobicity of the TMCS-modified sample from 118.1

to 128.9 degrees for WCA. The TMCS layer function is used to create an interface between the silica particles and the oil layer since silica does not have a good functional surface to adhere oil completely. It should be noted that the TMCS-modified surface does not result in any sliding angle when a 15 μL water droplet was used in the tests. The transition metal chalcogenides (TMCS) layer is a very thin coating layer and does not change the surface morphology and roughness considerably as can be seen in Fig. 3.13C and D when compared to the silica coating in Fig. 3.11A and B.

The very thin TMCS layer is used to avoid the surface morphology change since a thicker layer of TMCS can fill almost all the grooves in the silica layer and may reduce oil impregnation. AFM images in Fig. 3.13 also show the surface morphology from a 3D perspective, which agrees with the finding from the SEM images, showing no significant change in surface morphology of mixed-size silica nanoparticle- and TMCS-modified layer. AFM can measure the exact value of average surface roughness; the silica and TMCS layer have a 79.15 and 109.1 nm root mean square (rms) which is significantly rougher than the glass surface (1.36 nm rms).

Another drawback of TMCS coating is the transmittance decrease from 80.0% to 63.9% as shown in Fig. 3.14. In this case, the TMCS layer has many −CH functional groups that generate small roughness on the surface while decreasing the transmittance due to the light-scattering effect. The reduction of optical transmittance results in almost 35% power loss of the photovoltaic cells, which is considered to be significantly high. However, this drawback can be resolved by incorporating the oil-impregnated layer.

Oil impregnated on the TMCS-modified silica surface is very stable and results in increased water mobility. Samples with oil impregnation show very low sliding angle, which is in the order of 3 degrees; however, the static WCA decreases to 113 degrees, which is still in the range of the hydrophobic surface category as shown in Fig. 3.15. The static WCA reduces since the surface morphology becomes a smooth-like texture as a result of the oil filling the grooves around the silica nanoparticles.

Since oil possesses hydrophobic characteristics, the static WCA remains above 90 degrees. In order to examine the droplet mobility of the cleaning effect, the dusted surface is prepared by using environmental dust particles collected from desert in the Dhahran area. The dust particles collected were analyzed in terms of size and chemical composition, and consisted of various shapes and fine-size dust particles, in the order of 100−300 nm, that were attached at the surface of the large particles. In this case, the dust particles constituted the dissolved chlorine but not the salt crystals. It was, therefore, ensured that the dust particles, when in contact with water droplets, do not dissolve in water, but rather stay as solid particles at the treated sample surfaces. The treated sample was placed on a tripod and the tilting angle set at 10 degrees, because of easy rolling of water droplets under the gravitational force at the treated surface. It is also worth mentioning that the available tripod did not allow an accurate repeatable measurement of a lower angle of 10 degrees.

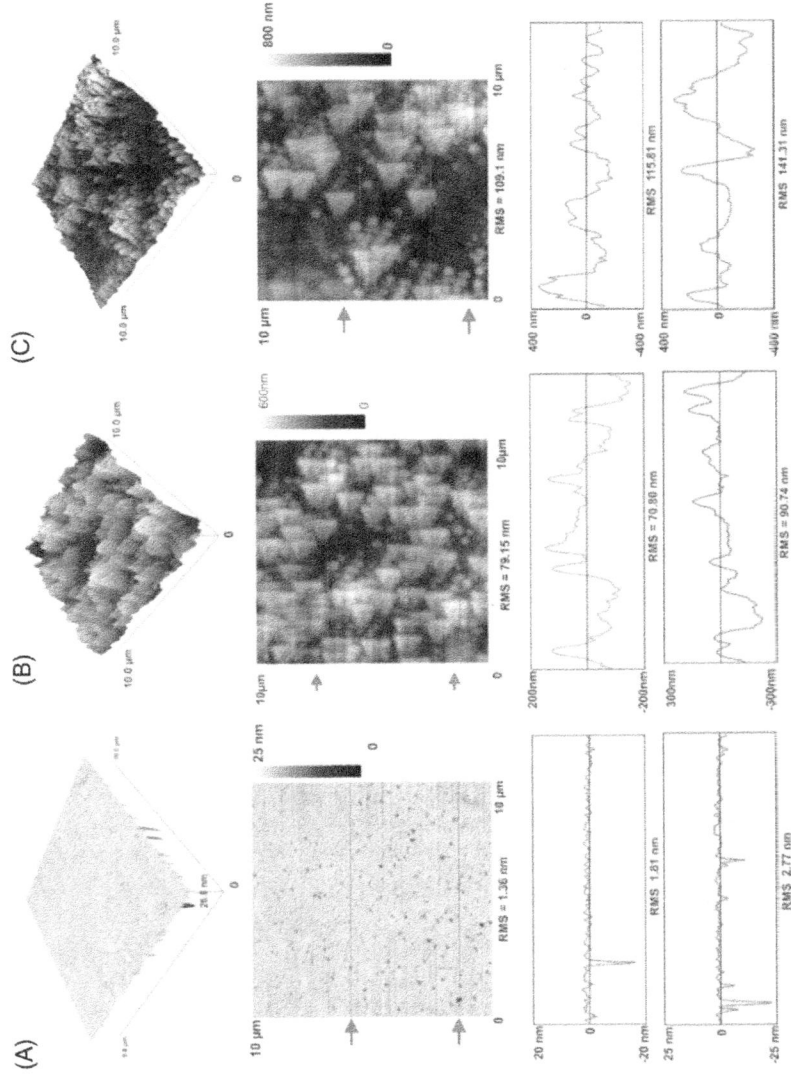

FIGURE 3.13 AFM scanning image of (A) glass substrate, (B) mixed-size silica nanoparticle, and (C) TMCS 5 vol.% coated mixed-size silica nanoparticle [40].

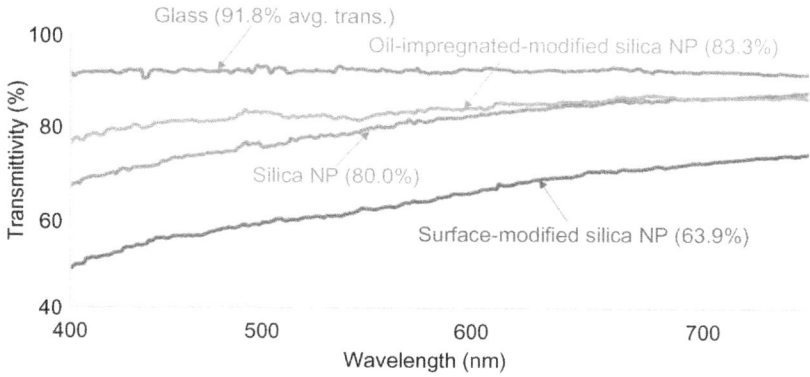

FIGURE 3.14 Transmittance curve of oil-impregnated silica nanoparticle (NP) samples from 300 to 900 nm wavelength [40].

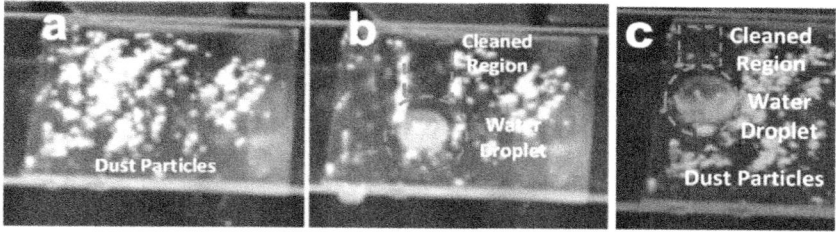

FIGURE 3.15 Dust-cleaning test on oil-impregnated silica nanoparticle sample: (A) dust-deposited surface; (B) self-cleaning process by a water droplet (cleaned region and water droplet are marked by dotted lines); and (C) self-cleaning process by a water droplet without oil impregnation on silica nanoparticle-deposited sample (cleaned region and water droplet are marked by dotted lines) [40].

Fig. 3.15 shows the cleaning test on a dusted surface that demonstrates that the water droplet slides smoothly and cleans the dusted surface along the way. The TMCS-coated surface gives rise to partial removal of the dust particles (Fig. 3.15C) as compared to the oil-impregnated surface (Fig. 3.15B). Therefore oil impregnation not only improves water droplet mobility, but results in high transmittance of the samples.

These findings reveal that surface-functionalized silica nanoparticles of different sizes can successfully create a combination of micro/nanohierarchical structures on treated glass surfaces [40]. This arrangement gives rise to the lotus effect on the deposited surface; in this case, the water droplet contact angle improved significantly. Deposited functionalized silica particles lower the optical transmittance significantly, which is mainly scattering of light due to the presence of particles at the treated surface. On the other hand, the silica oil impregnation on the silica-deposited surface improves the transmittance significantly; in this case, transmittance reached 83% on average. In addition, oil-impregnated and treated surfaces in the presence of functionalized silica nanoparticle results in water droplet contact angle in the order of 113 degrees. This in turn improved the self-cleaning of surfaces from dust particles via sliding of water droplets on impregnated surfaces with a tilted angle of 10 degrees under gravity.

3.4.2   Solvent-Induced Crystallization

A solvent-induced crystallization technique was introduced by Owais et al. [55] where acetone was used in its two most common physical states, liquid and vapor, to create a hierarchically structured surface. They used PC sheets of thickness 1.6 mm and produced by Bayer Company. Liquid acetone, from Sigma Aldrich (bp: 56°C and MW: 58.08), was used for the crystallization process. For the texturing process, PC sheets were immersed in liquid acetone for 10 minutes at room temperature (18°C). In addition, the PC sheet was crystallized and textured by exposure to the acetone vapor at room temperature (18°C) for an average of 8 hours. A 1-cm distance between the liquid phase (acetone) surface and the exposed PC surface was maintained during the process. They assessed the morphological and hydrophobic characteristics of the surface via analytical tools, and found that the development of specific and unique features under different conditions in different areas of the textured PC glass surfaces. Fig. 3.16 shows an AFM micrograph illustrating the appearance of a new surface texture after the immersion process. Immersion of the PC surface in liquid acetone resulted in large spherules appearing over the surface after immersing the sample for 10 minutes, Fig. 3.16A. The average width of the spherules observed is in the range of 12 μm. At lower scans, the surface of the spherule becomes more apparent and the texture on top of it can be clearly observed. Note that the surface of the spherule does not have uniform texture; in other words, the spherule's surface becomes hilly, and this hilly surface is full of grass-like textures. This structure can also be inferred from the 3D micrograph, Fig. 3.16B.

However, for the textured surface with acetone vapor, Fig. 3.17A, with a scanning scale of 40 μm, well-detailed spherules appear over the surface after the exposure of the smooth PC sample to the acetone vapor. The average width of the spherules observed, in the figure, is in the range of 8 μm. From Fig. 3.17B, it can be inferred that the hills are absent, and only the grass-like textures appear at the spherule's surface.

They also studied the topology of the textured surfaces by using surface profile micrographs. For the textured PC surface by immersion in acetone liquid, Fig. 3.18A, the height of the spherules was measured as 1.8 μm. The width of the middle spherule was about 13 μm. In Fig. 3.18B, a distinctive profile micrograph of the single spherule appears. The spherule contains three small elevations over its surface, and each elevation was textured. This promotes the observation of the hilly surface and the grass-like hills, which

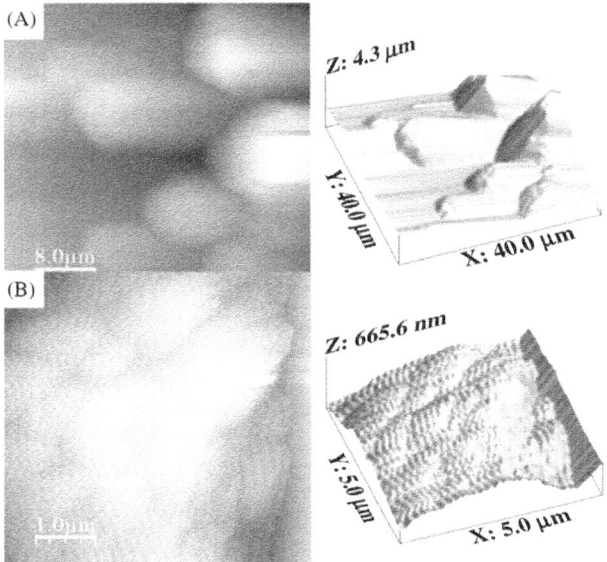

FIGURE 3.16 The 2D and 3D AFM micrographs for a textured polycarbonate surface by immersion in pure liquid acetone for 10 min at: (A): 40 μm scale and (B): 5 μm scale [55].

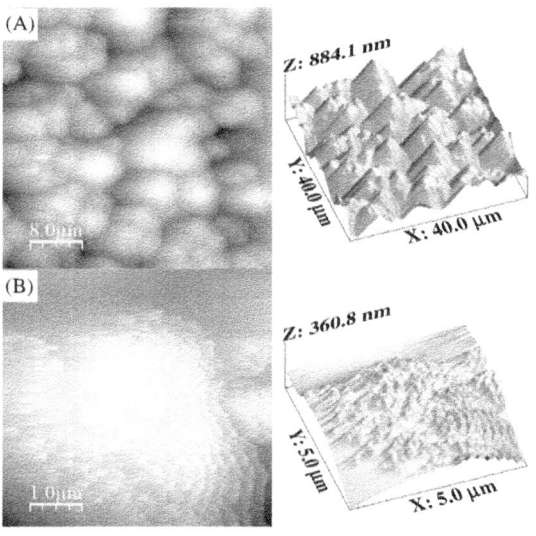

FIGURE 3.17 The 2D and 3D AFM micrographs for a textured polycarbonate surface by exposure to pure acetone vapor for 24 h at: (A) 40 μm scale and (B) 5 μm scale [55].

was noted in the description of Fig. 3.16A. In Fig. 3.18C, the height of the spherules can be easily measured, which is about 375 nm. The width of the left spherule is about 13.5 μm. In Fig. 3.18D, a distinctive profile micrograph of a single spherule's surface appears. The spherule contains a grass-like structure over its surface, and this grass resembles the surface texture.

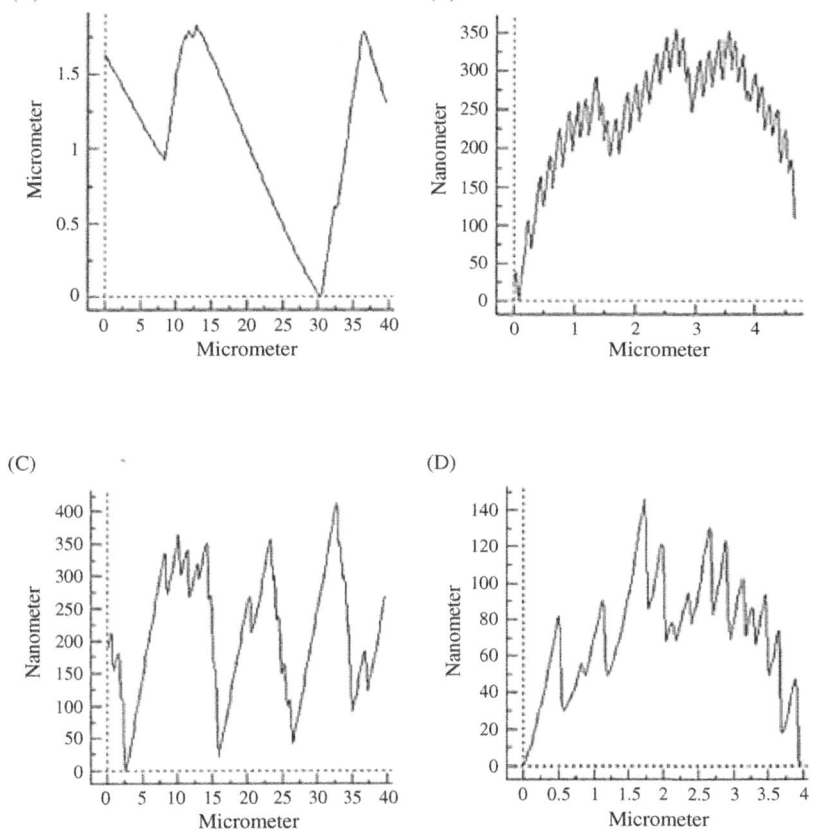

FIGURE 3.18 The (A) 40 μm scale and (B) 5 μm scale AFM texture profile micrographs for a textured polycarbonate surface by immersion in pure liquid acetone for 10 min. The (C) 40 μm scale and (D) 5 μm scale AFM texture profile micrographs for a textured polycarbonate surface by exposure to pure acetone vapor for 24 h [55].

The liquid-induced crystallization process of the PC surface process, by immersion in liquid acetone, leads to a molecular rearrangement on the surface of the PC sheet and to several microns in depth, leading to generation of spherule-like crystalline structures and creating a hierarchal textured surface of high roughness. This can be related to the hydrostatic pressure that is applied on the immersed PC surface in the liquid acetone. However, the textured PC sheet surfaces due to vapor-induced crystallization have small crystals due to the high nucleation density. Despite the enlargement of the crystal grain sizes due to increasing the exposure time of the PC sheet surface to the acetone vapor, reaching large crystals with perfectness still remains a challenge [56].

The size of the generated spherules and the polymer surface degree of crystallinity is directly proportional to the depth to which acetone diffuses. Furthermore, pore formation takes place if the layer depth is larger than the width of the spherule; however, incomplete spherule coverage results in less layer depth than the spherule dimensions. For the solid−vapor method of polymer crystallization, the acetone vapor condenses over the PC surface. Therefore the PC comes in contact with only 1 mL of acetone on average. As a consequence, the mass transfer in this case becomes considerably lower than that in the case of immersing the surface in the liquid acetone. Mass transfer, which is a driving-force dependent, affects the diffusion process significantly. The mass transfer equation between two phases is [57]: $N_A = ka\Delta C_A$, where N_A is the mass transfer rate of component A, k is the mass transfer coefficient, a is the transfer area, and ΔC_A is the concentration driving force. The driving force is the result of the difference between the concentration of the liquid in the bulk and the concentration of the liquid in the formed boundary−film interface. Therefore the concentration of the liquid in the bulk is larger when the bulk is immersed in liquid acetone, than that of exposing acetone vapor.

The contact angle of the patterned PC surface was measured using a DI water droplet over different spots on the textured surface and measured by using a Goniometer's camera. For the PC sheet immersed for 10 minutes, the contact angles of the three different areas over the surface are 138.3, 137.3, and 130.7 degrees, Fig. 3.19A−C, respectively. For the PC sheet surface exposed to acetone vapor for 30 minutes, the contact angles of the three different areas over the surface are 95.5, 85.3 and 86.9 degrees, Fig. 3.19D−F, respectively.

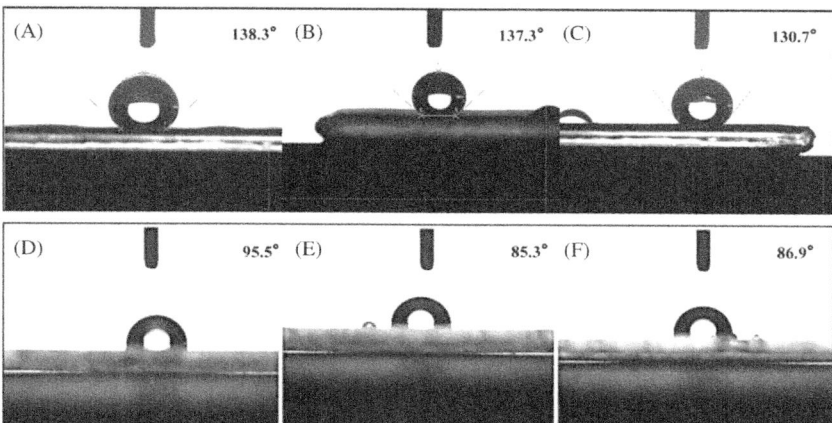

FIGURE 3.19 (A−C) Contact angle of a textured PC sample by immersion in pure liquid acetone for 10 min. (D−F) Contact angle of a textured PC sample by exposure to pure acetone vapor for 30 min [55].

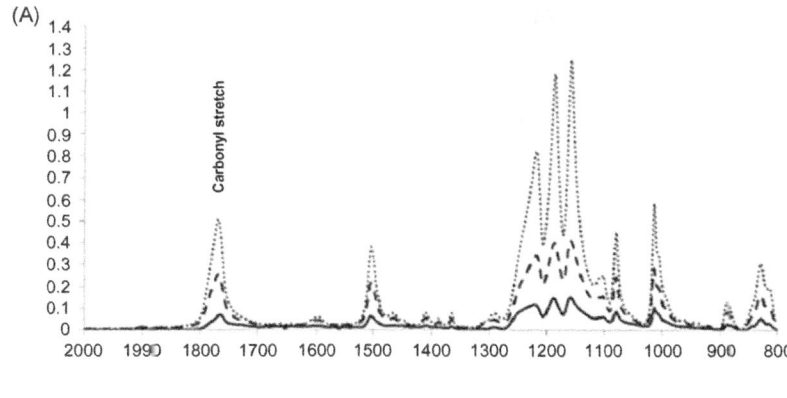

Carbonyl stretch

———— Liquid-acetone treated PC   – – – Acetone-vapor treated PC   ········· Untreated PC

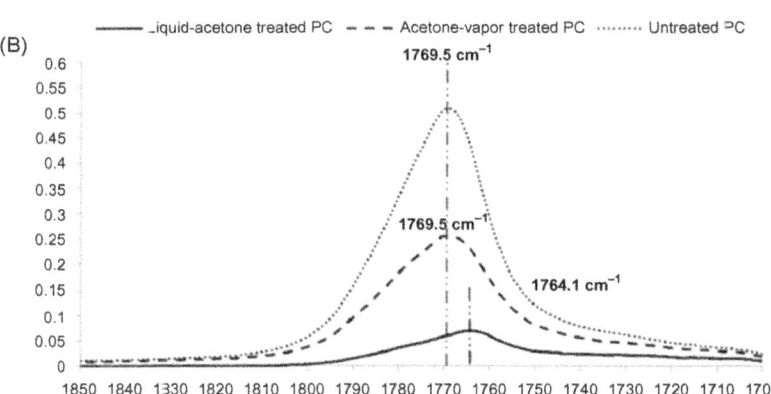

FIGURE 3.20 ATR FTIR spectra of smooth, liquid acetone-textured and vapor-acetone-textured PC sheet surfaces: (A) Whole spectra and (B) carbonyl stretching mode peak [55].

The attenuated total reflection Fourier transform infrared (ATR-FTIR) technique was used for detecting chemical, functional, side, and terminal groups, which are present in a chemical compound through the detection of the different stretching and bending modes of the bonds present in these groups. Fig. 3.20A shows the ATR-FTIR spectra of untreated smooth, vapor-acetone-textured, and liquid acetone-textured PC glass surfaces. The acetone treatment effect is observable at the wavelength at which the peak of carbonyl is C=O. Prior to exposure to acetone, the peak of the carbonyl stretching mode is observed at 1769.5 cm^{-1}, Fig. 3.20B. After exposure to acetone, an increase in crystallization is noted by a shift in the wavenumber of the carbonyl group to 1764 cm^{-1} [58]. The amorphous form of the polymer gives a relatively higher degree of freedom to the polymer's tail motion [59]. The dipole moment, due to asymmetric chemical structure, causes strong molecular vibrations. This also allows chain motion and bond vibrations in the crystalline phase, in which the molecules are tightly packed and consequently, the vibrations and motions are restricted [59]. This behavior indicates that the acetone interaction with the PC is a physical process because there is no new peak after the treatment process [58]. XRD was used to verify the crystallization of the PC surface, Fig. 3.21. However, CuKa ray penetrates a surface layer in the order of 600−900 mm. Although penetration depth is greater than the expected thickness of the crystalline spherulitic layer, the sharpening of the XRD peaks for samples with more aggressive crystallization where liquid acetone is used indicates an increase in the crystallinity of the total irradiated volume. Further, narrowing of the peak width shows an increase in the thickness of the crystalline layer.

3.4.3 Solvent-Induced Crystallization and Texture Copying by Polydimethylsiloxane

The surface texturing of PC wafers through acetone-induced crystallization and the replication of textured surfaces with PDMS were investigated by Yilbas et al. [22]. The influence of the crystallization periods on the surface texture and topology of the replicated surfaces is presented in light of the previous study [22]. PC wafers with 3-mm thicknesses were used as workpieces. The PC wafer was derived from a 4-phenyl-3-butenoic acid (*p*-hydroxyphenyl) with excellent optical clarity and high toughness. After ultrasonic cleaning, the PC wafers were immersed in liquid acetone for 2, 4, 6, 8, and 10 minutes. To select the immersion durations for the crystallization of PC wafers, several tests were conducted. The immersion durations resulting in crystal structures toward the formation of a surface texture for hydrophobic behavior were selected. Liquid PDMS, which belongs to a group of polymeric organosilicon compounds, was used to replicate the crystallized PC wafer surface. Liquid PDMS was deposited and left on the crystallized PC surface for over 18 hours for curing purposes. The solidified PDMS was then removed from the crystallized PC surface after curing. Fig. 3.22 shows a schematic view of the crystallization of the PC wafers and the copying of the resulting textured surface by PDMS.

Fig. 3.23 shows the SEM images of the crystallized PC surfaces with various immersion periods in liquid acetone. In general, the surfaces are composed of microstructures/nanostructures consisting of spherulites, pores, cavities, and nanosize fibrils (Fig. 3.23A and B). The spherulite size remains small for the case of a short immersion duration (2 minutes; Fig. 3.23A), and the spherulites cover a large area at the surface for 6 minutes of immersion (Fig. 3.23B). In addition, the scattering of a small number of spherulites is observed at the surface (Fig. 3.23A). As the immersion time progressed (2 minutes), the size of the spherulites increased, and nanosize fibrils were initiated from the spherulite surfaces (Fig. 3.23C). As the immersion time

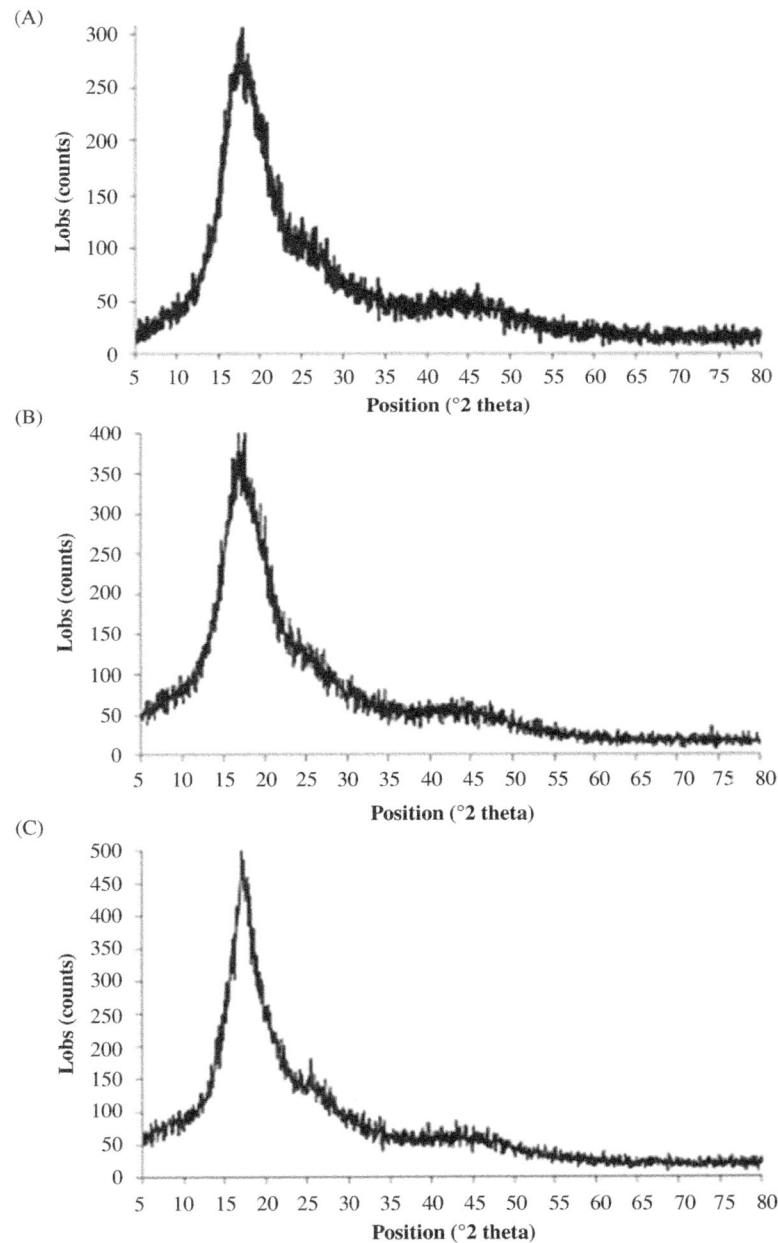

FIGURE 3.21   XRD data for (A) untreated PC sheet. (B) A textured PC sheet by immersion in pure liquid acetone for 10 min. (C) A textured PC sheet by exposure to pure acetone vapor for 30 min [55].

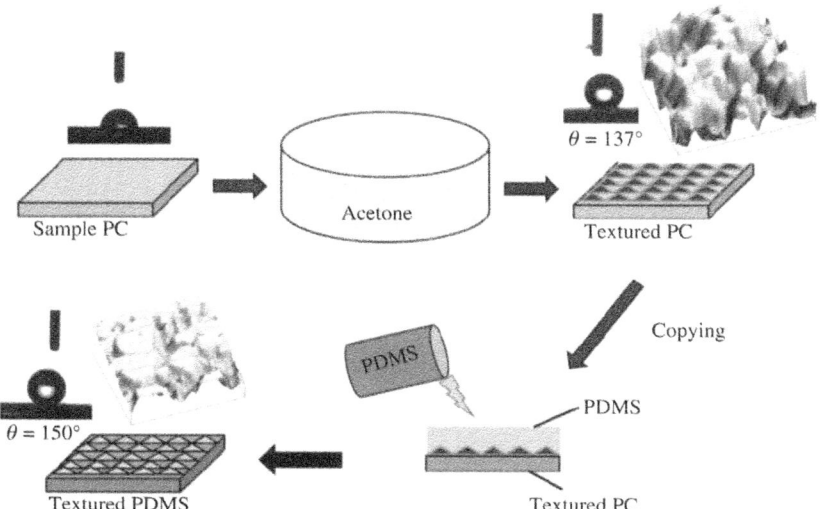

FIGURE 3.22 Schematic view of the crystallization of PC and copying of the crystallized surface by PDMS [22].

progressed further, the spherulites aggregated, and the fibrils covered almost the entire surface (Fig. 3.23D) and extended over the spherulite surface (Fig. 3.23E) while forming nanotextures. Moreover, the increase in the immersion duration enhanced the coverage area of the fibrils on the spherulite surface. In general, crystals grow radially from potential nucleation sites, and a few branches from the nucleation sites occur during radial growth (Fig. 3.23F). Intermittent branching results in further growth of crystals to form large spherules at the surface, particularly for long immersion durations (6 minutes). The formation of microtextures/nanotextures at the surface is associated with the process of acetone (solvent)-induced crystallization [60,61]. Because acetone has Hildebrand solubility parameter [62] it possesses miscible characteristics. Therefore acetone diffuses into the polymeric structure and forms a swollen film (gelated layer) at the surface during the dissolution. As the diffusion progresses, the glass-transition temperature of the polymeric film behind the diffusion front decreases while causing the plasticization of the swollen polymer [62]. Because the diffusion of acetone into PC is governed by a non-Fickian mechanism [63] the diffusion front between the swollen film and the solid PC penetrates at almost a constant velocity into the solid phase of the amorphous PC [63]. Once the PC is removed from the immersion bath, acetone residues at the surface evaporated, and spherulites are initiated to form at PC surface; this is similar to the literature [64]. The glass-transition temperature decreases during acetone evaporation from the surface; this, in turn, gives rise to the supercooling of

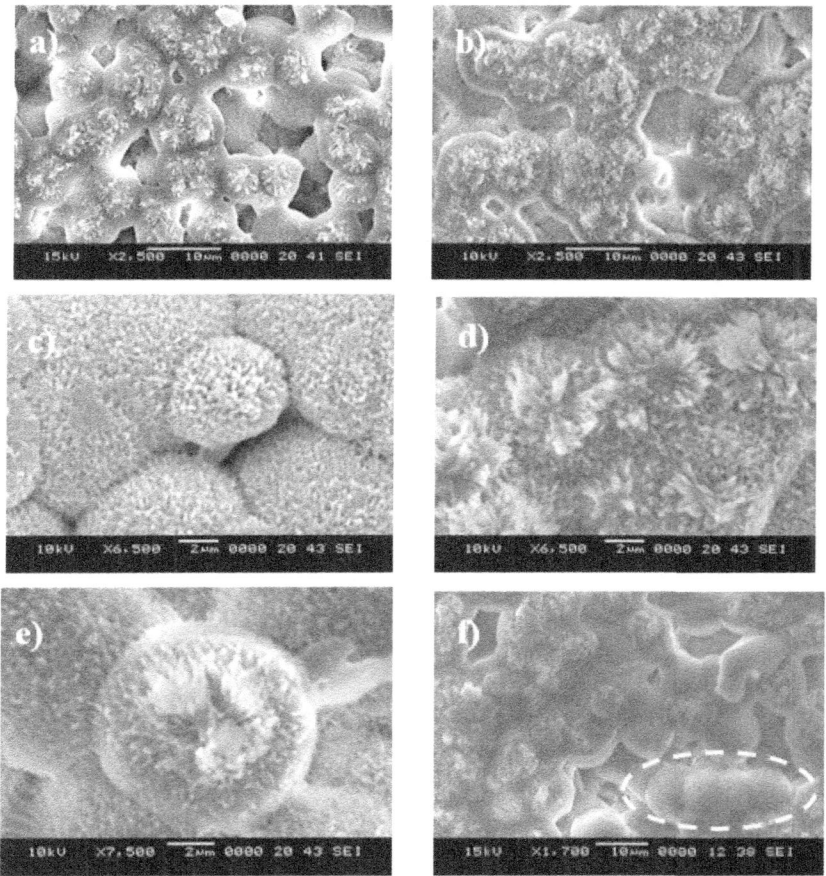

FIGURE 3.23 SEM micrographs of the crystallized PC surfaces: (A) PC surface after 2 min of immersion, (B) PC surface after 6 min of immersion, (C) initiation of fibrils after 2 min of immersion, (D) fibrils covering the almost spherulite surface, (E) fibrils formed at the spherulite surface, and (F) radial growth of the spherulites [22].

the swollen film and the formation of a surface texture consisting of micro-size/nanosize spherulites, cavities, pores, and fibrils. However, the spherulites does not form a well-defined hierarchical structure but instead scattered patterns at the surface (Fig. 3.23D). This behavior can be related to the non-uniform evaporation of acetone across the entire workpiece surface, giving rise to various nucleation densities across the surface. However, crystallization occurs in three consecutive phases or categories; these are the crystallization initiation, primary formation of crystals, and secondary crystal growth [65]. A nucleus emerges when the polymer chains gradually align in a parallel way, and the chains are added to the nucleus during the initiation of crystallization. Crystal growth becomes spontaneous after the nucleus size reaches the critical size [65]. The nucleation results in bundle-like or lamellar crystallization. The difference between the types of crystallization is associated with size of the primary nucleus and the free energy of the surface normal to the chain direction per unit area [66]. The crystallization mainly starts from the molten state of the swollen film. Therefore the mixture of bundle-like and lamellar nuclei can form because of the series of additions of repeating units during crystallization. Moreover, the presence of small-size cavities and pores results in microlevel/nanolevel waviness at the surface and contributes to the overall texture of the crystallized surface.

Fig. 3.24 shows the SEM micrographs of the PDMS surfaces after removal from the textured PC surface. Because liquid PDMS has excellent rheological properties, it wets the PC texture feature before solidification. Therefore it copies almost exactly the texture of the PC surface when it is removed in the solid phase from the PC surface (Fig. 3.24A). These result in microsize/nanosize voids and cavities formed at the PDMS surface (Fig. 3.24B). Some fibrils are not clearly observed at the PDMS surfaces (Fig. 3.24C). However, some fibrils on the spherulites are partially copied (Fig. 3.24D). This is attributed to the adhesion between the fibrils and

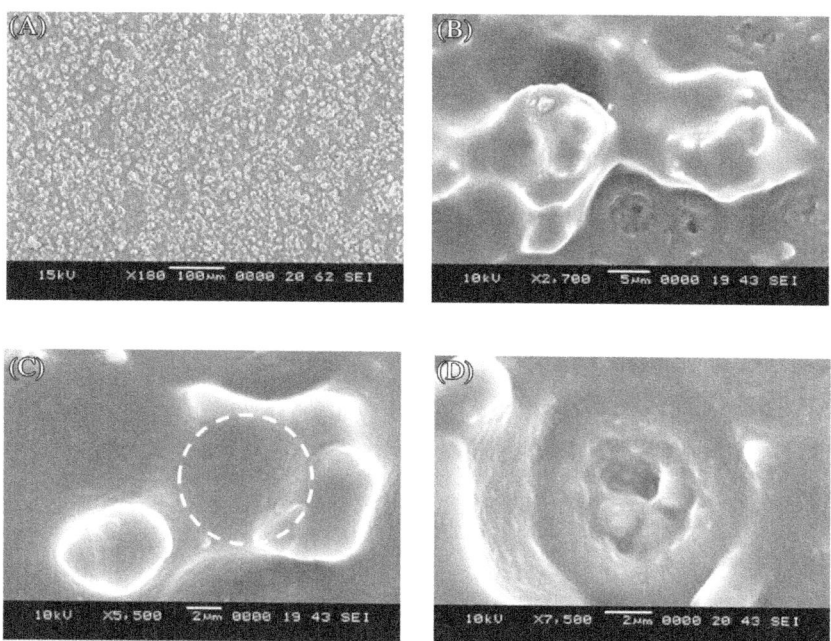

FIGURE 3.24   SEM micrographs of the PDMS surfaces after the crystallized PC surfaces were copied after 6 min of immersion: (A) surface copies, (B) copied local texture, (C) copied spherulite without fibrils, and (D) copied spherulite and fibrils [22].

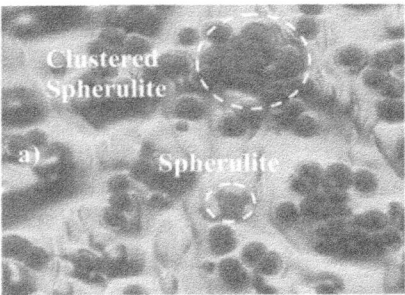

50 μm

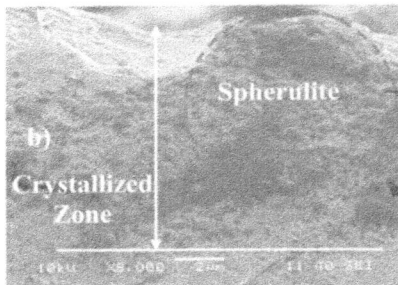

FIGURE 3.25 Optical and SEM images of the crystallized PC after 6 min of immersion: (A) optical image of the crystallized surface and spherulites, (B) SEM micrograph of the cross sections of the crystallized layer and spherulite, and (C) SEM micrograph of the cross section of the crystallized layer when no spherulite was formed [22].

PDMS, which caused broken fibrils during the removal of PDMS. The texture of the spherulite appearance is observed at the PDMS surface. This behavior was attributed to the low surface energy of the solid phase of PDMS, which acted like an elastic body during the removal from the textured PC surface. In the case of complex geometric shapes, such as the odd-shaped texture peaks on the PC surface, some residues of the solid-phase PDMS remained at the PC surface because of the strong adhesion between the solidified PDMS and the PC surface. Because the PDMS residues left at the PC surface are small, no significant rapturing of the PDMS surface is observed.

The typical optical image resulting after 6 minutes of immersion is shown in Fig. 3.25A. The findings reveal that the coverage area by the spherulites on the crystallized PC surface is almost 32% of the total surface area. In addition, Fig. 3.25B and C show the SEM micrographs of the cross section of the crystallized surface. It is evident that the depth of crystallization in the surface region is in the order of a few micrometers.

To assess the texture characteristics of the crystallized PC surfaces, AFM is used. Fig. 3.26 shows the AFM images of the crystallized PC surfaces and

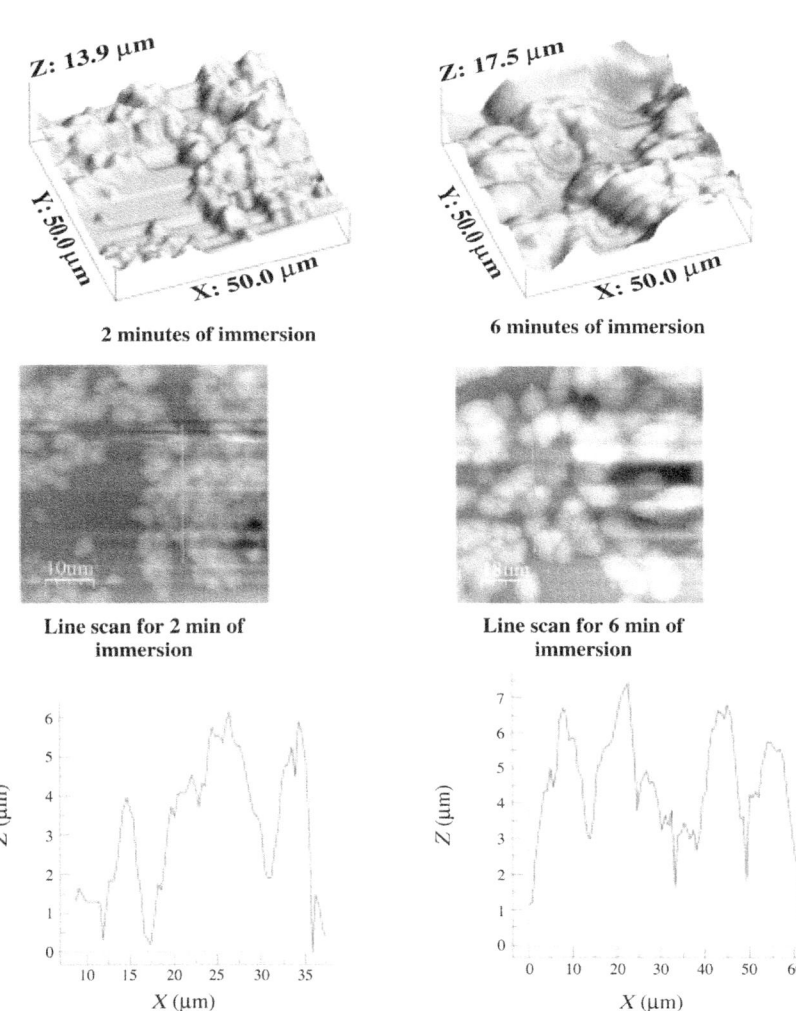

FIGURE 3.26   AFM images and texture profile of the crystallized PC surfaces for two immersion durations. Z is the surface texture profile height and X is the distance along the rakes [22].

line scans corresponding to two different immersion durations. It is evident that those formed after crystallization (Fig. 3.27) did not have exact spherical morphologies at the PC surface. Hence, the AFM images show nonspherical structures at the surface (Fig. 3.26). However, the height of the nonspherical structures increases slightly, and the coverage area of these structures increases at the surface with increasing immersion time. This behavior is

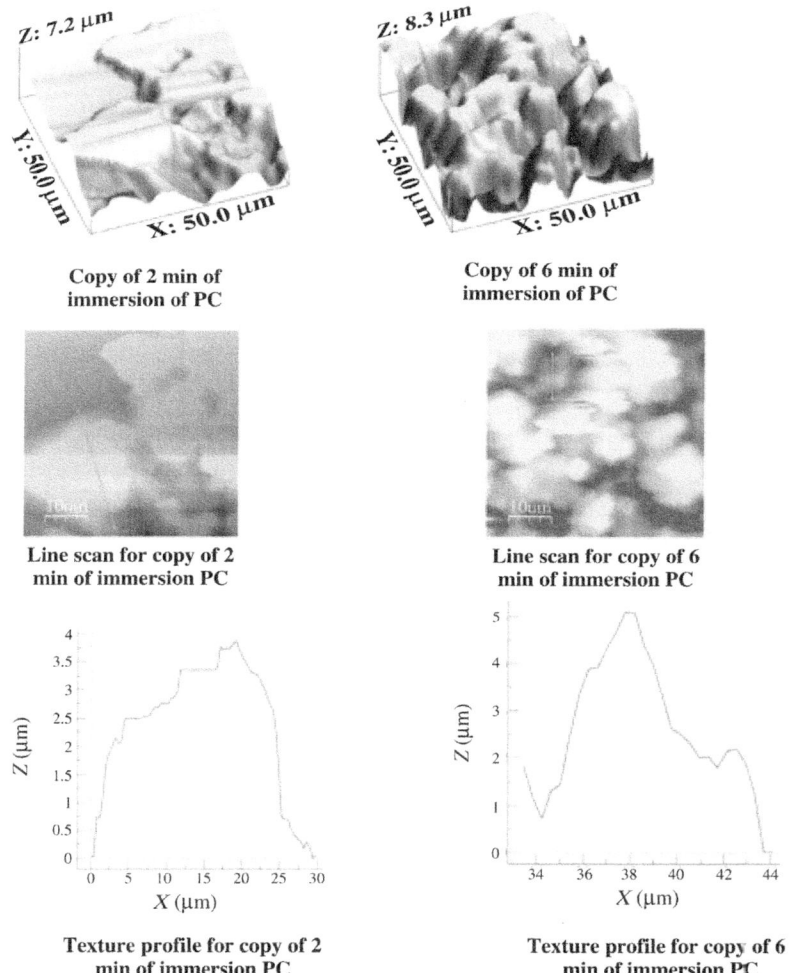

Copy of 2 min of immersion of PC

Copy of 6 min of immersion of PC

Line scan for copy of 2 min of immersion PC

Line scan for copy of 6 min of immersion PC

Texture profile for copy of 2 min of immersion PC

Texture profile for copy of 6 min of immersion PC

FIGURE 3.27 AFM images and texture profile of the EDMS surfaces copied from the PC crystallized surfaces for two immersion durations [22].

attributed to the rate of formation of nonspherical spherulites at the PC surface with immersion time. In this case, increasing the immersion time enhances the number of spherulites formed at the PC surface. The surface texture possesses microfibrils/nanofibrils, which form ripples/waviness in the texture profile, as shown in Fig. 3.26. Because crystal growth at the PC surface involves multidimensional features, no regular or standard pattern is observed along the texture profiles. The roughness of the crystallized PC surface varies within 3.6–4.3 μm; in this case, an increase in the immersion time increases the roughness of the surface. As shown in Fig. 3.27, in which

AFM images of the PDMS surface are shown, the presence of copied non-spherical spherulite-like structures are evident (Fig. 3.27). Although the crystallized PC is copied, some nanoscale and subnanoscale features are not copied properly from the crystallized PC surface. In this case, the surface texture of PDMS was rather smooth compared to that of the crystallized PC. This situation can also be seen from the texture profile; in this case, ripples/waviness disappear from the texture profiles of the PDMS surface (Fig. 3.27). Nevertheless, the crystallized PC surface is copied almost exactly at the PDMS surface, except randomly distributed nanosize fibrils and a few voids are not observed in the AFM images of PDMS because of the copied spherulites. But this does not notably alter the roughness of the PDMS surface.

Figs. 3.28 and 3.29 show the distribution of the contact angle and the corresponding images of droplets captured from the sessile contact angle tests for the crystallized PC and copied PDMS surfaces. The main parameters influencing the liquid drop contact angle are the surface free energy of the substrate material and the surface texture. Because the surface free energy is lower for PDMS (19.8 mN/m) [67] than for PC (34.2 mN/m) [67], the contact angle remains slightly higher for the copied PDMS surface. In addition, small differences between the contact angles, because of the crystallized PC and replica PDMS surface, are associated with the surface texture differences between the two surfaces. In this case, nanosize and subnanosize fibrils and voids are not copied exactly by PDMS from the crystallized PC surface. Therefore despite the large difference in the surface free energy of both

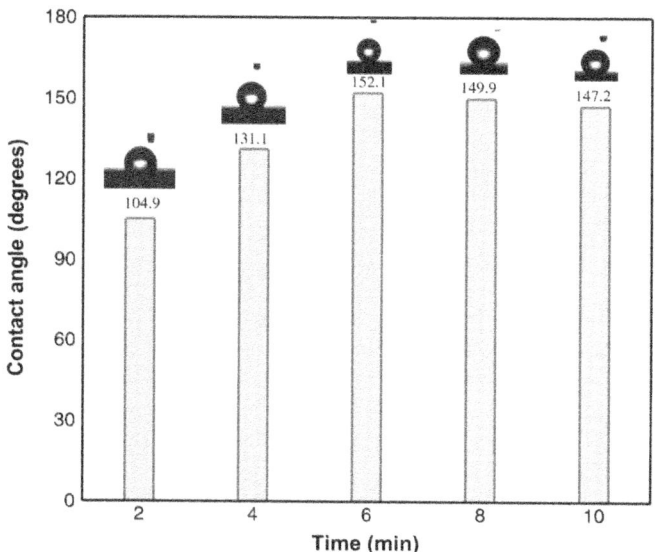

FIGURE 3.28 Contact angle variation for crystallized polycarbonate with immersion durations [22].

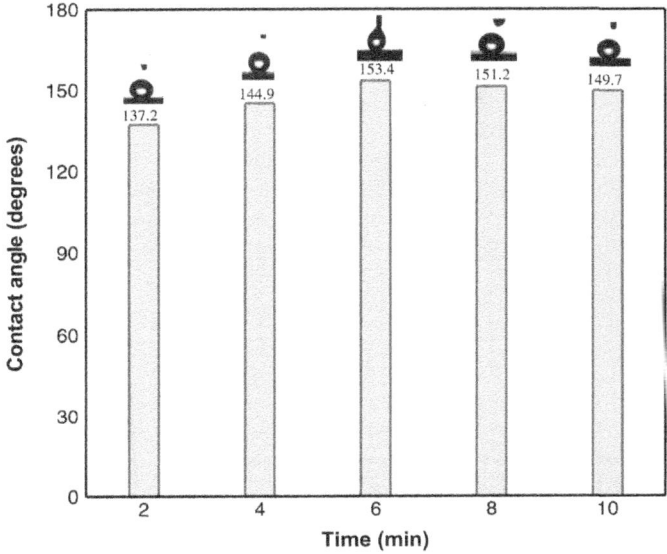

FIGURE 3.29 Contact angle variation with immersion durations for PDMS after copying crystallized polycarbonate surface [22].

TABLE 3.1 Contact Angles Measurement for Crystallized Polycarbonate Surface and Copied PMDS Surface

			Duration (min)				
		As Received	2	4	6	8	10
PC	Advancing	82	104	151	152	150	147
	Receding	56	98	147	148	145	143
PDMS	Advancing	104	137	145	153	151	149
	Receding	97	131	137	146	145	142

The measurement errors lie in between +4 and −4 degrees [22].

samples, the surface texture difference in detail suppresses the contact angle difference corresponding to both surfaces. Table 3.1 gives the hysteresis angles ($\theta_{hysteresis} = \theta_{Advancing} - \theta_{Receding}$, where $\theta_{Advancing}$ is the advancing angle and $\theta_{Receding}$ is the receding angle) of the crystallized and copied PDMS surfaces. The hysteresis corresponding to the copied PDMS surface is slightly higher than that of the hysteresis corresponding to the crystallized PC surfaces. This is attributed to the nonpresence of nanosize and subnanosize fibrils, which were responsible for the increased hysteresis of the copied PDMS surface.

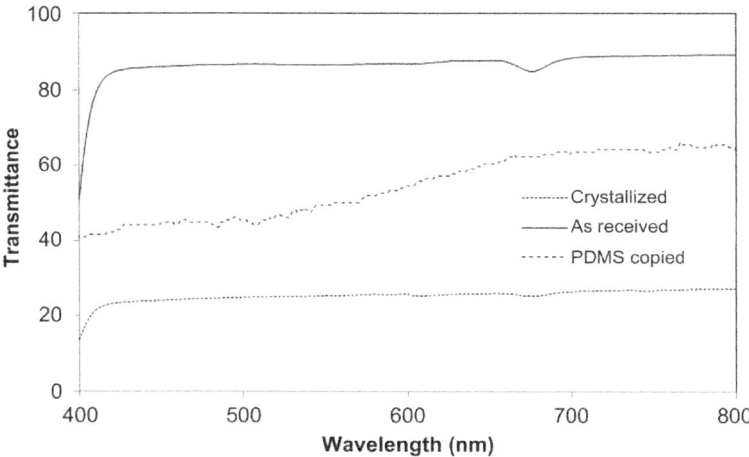

FIGURE 3.30 Transmittance data for as-received and crystallized polycarbonate after 6 min of immersion, and PDMS sample after copying polycarbonate surface immersed for 6 min in acetone [22].

Fig. 3.30 shows the optical transmittance of the crystallized PC wafers and their copies of PDMS with 6 minutes of immersion. The transmittance data for the as-received PC and PDMS are also included for comparison. The crystallization of the PC surface significantly reduces the transmittance; in this case, the increase in the immersion time lowers the transmittance of the crystallized PC wafer. In the case of the copied PDMS, the transmittance remains higher than that of the crystallized PC; however, the transmittance decreases almost 40% compared to the original PDMS transmittance. This is attributed to the scattering of incident radiation from the sample surfaces because of the textures developed after copying of the crystallized PC surface.

The surface hydrophobicity significantly improved for the crystallized PC, and the maximum contact angle achieved is in the order of 152 degrees, which corresponds to 6 minutes of immersion. The PDMS copying of the crystallized surface texture results in a higher contact angle than that of the crystallized PC surfaces. This is attributed to the low surface energy of PDMS, even though nanosize and subnanosize fibrils are not copied by PDMS. The optical transmittance of the crystallized PC wafer is reduced significantly because of the absorption and scattering of incident radiation by the surface texture; however, the optical transmittance remained reasonably high for the PDMS samples.

3.4.4 Polydimethylsiloxane Replication of Laser-Textured Surfaces

Yilbas et al. [21] textured the alumina tiles by a laser beam to obtain improved hydrophobic characteristics at the surfaces. Since textured surfaces are optically opaque, the resulting surfaces are copied and replicated by PDMS. Some of the texture features such as whisker-like structures with complex shapes cannot be copied by PDMS and synthesized silica particles are deposited onto PDMS-copied and PDMS-replicated surfaces to create a lotus effect. Introducing synthesized silica particles improves the lotus effect on the PDMS-copied and PDMS-replicated surfaces; in which case, the hysteresis of droplet contact angle remains significantly low. The process and findings are discussed in the light of the previous study [21].

Alumina (Al_2O_3) tiles (Ceram Tec-ETEC, 2010) with 3 mm thickness were used as workpieces. The CO_2 laser (LC-ALPHAIII) was used to irradiate the alumina tile surfaces. The nominal output power of the laser was 2 kW and the irradiated spot diameter at the workpiece surface was about 200 μm. High-pressure nitrogen gas jet emerging from the conical nozzle was utilized during the laser heating of the surfaces. The laser pulse frequency was set at 1500 Hz, which in turn resulted in an about 68% overlapping ratio for the irradiated spots at the surface. The initial tests were conducted to select the laser treatment parameters so that the laser parameters resulting in crack-free surfaces were selected. The results of the initial tests reveal that increasing laser power by 10% while keeping laser scanning speed constant gives rise to large cavity formation at the surface. On the other hand, reducing laser scanning speed by 10% while keeping the laser output power same results in cracks at the surface. Consequently, the laser-treated surface properties are highly dependent on the proper selection of the laser processing parameters. Therefore through controlling the laser power settings, beam intensity distribution, pulse repetition rate, spot size, and the scanning speed, crack-free surface texture can be realized. The laser treatment conditions are given in Table 3.2.

Liquid PDMS, which belongs to the group of polymeric organosilicon compounds, is used to replicate laser-textured surface. PDMS (Sylgard 184, Dow Corning) was prepared by mixing the elastomer base with a hardening agent in 10:1 wt.%. The mixture was deposited onto the laser-textured surface and degasified in a vacuum chamber at 0.1 bar for 30 minutes. The deposited and degasified PDMS was left in the oven at 150°C for 30 minutes for curing purposes. Solidified PDMS was then removed from the laser-textured surface after curing. PDMS was prepared again in 10:1 wt.%

TABLE 3.2 Laser Treatment Conditions [21]

Feed Rate (m/s)	Power (W)	Frequency (Hz)	Nozzle Gap (mm)	Nozzle Diameter (mm)	Focus Diameter (mm)	N_2 Pressure (kPa)
0.1	1800	1500	1.5	1.5	0.2	600

(elastomer base to hardening agent ratio), poured over the copied PDMS, degassed, and then cured in an oven using same parameters.

The procedure adopted for synthesizing of silica particles is similar to that reported in the previous study [68]. The process is shortly described here. Tetraethyl orthosilicate (TEOS) and isobuthytrimethoxysilane, ethanol, and ammonium hydroxide were used in the synthesizing process. In this case, 14.4 mL of ethanol, 1 mL of ultrapure water, and 25 mL of ammonium hydroxide were mixed and stirred for 12 minutes. Later, 1 mL TEOS diluted in 4 mL ethanol was added in the mixture. Thirty minutes after this process, 0.5 mL of TEOS diluted in 4 mL ethanol was added. After 5 minutes, a modifier silane molecule was added in a molar ratio of 3:4 with respect to the second edition of TEOS. The final mixture was stirred for 20 hours at room temperature, and later centrifuged and washed with ethanol for complete removal of reactants. The solvent casting was applied to deposit the solution onto PDMS-copied and PDMS-replicated surfaces. Upon vacuum drying until all solvent is evaporated, characterization of resulting surfaces is carried out.

Fig. 3.31 shows the SEM micrographs of a laser-textured alumina surface while Fig. 3.32 shows the AFM images of a 3D textured surface (Fig. 3.32A) and texture profile along the surface line scan (Fig. 3.32B). Laser-treated surface consists of regular laser scanning tracks (Fig. 3.31A), which are formed by the irradiated spots at the surface during the laser repetitive pulse heating. Since the laser pulsing frequency was 1500 Hz, the overlapping ratio of the consecutive irradiated spots was in the order of 70%. Although laser-controlled melting and ablation at the surface results in high-temperature gradients in the surface region [69], no thermally induced crack is observed at the surface. The crack-free laser-treated layer is also seen from the SEM micrographs of the cross section of the laser-treated section (Fig. 3.31D). In this case, a dense layer is formed at the surface because of the high cooling rates. The crack-free layer is associated with the closely spaced laser scanning tracks at the surface. In this case, heat conduction from recently formed laser scanning tracks toward the previously formed tracks modifies the cooling rate below the surface. This, in turn, creates a self-annealing effect in the treated region while lowering the thermal stresses formed in the treated layer. Moreover, laser power intensity across the irradiated spot is Gaussian, which gives rise to the occurrence of the laser peak intensity at the irradiated spot center. This results in surface evaporation at the irradiated spot center and melting toward the irradiated spot edges. Therefore a small cavity forms at the irradiated spot center (Fig. 3.31B). However, the melt flow from the neighboring irradiated spot modifies the cavity size during the consecutive pulses. This causes the formation of textures with micro/nanosize features at the surface (Fig. 3.31C). This situation is also observed in Fig. 3.32A in which a 3D image of the surface is shown. Since the regular laser scanning tracks take place at the surface, the

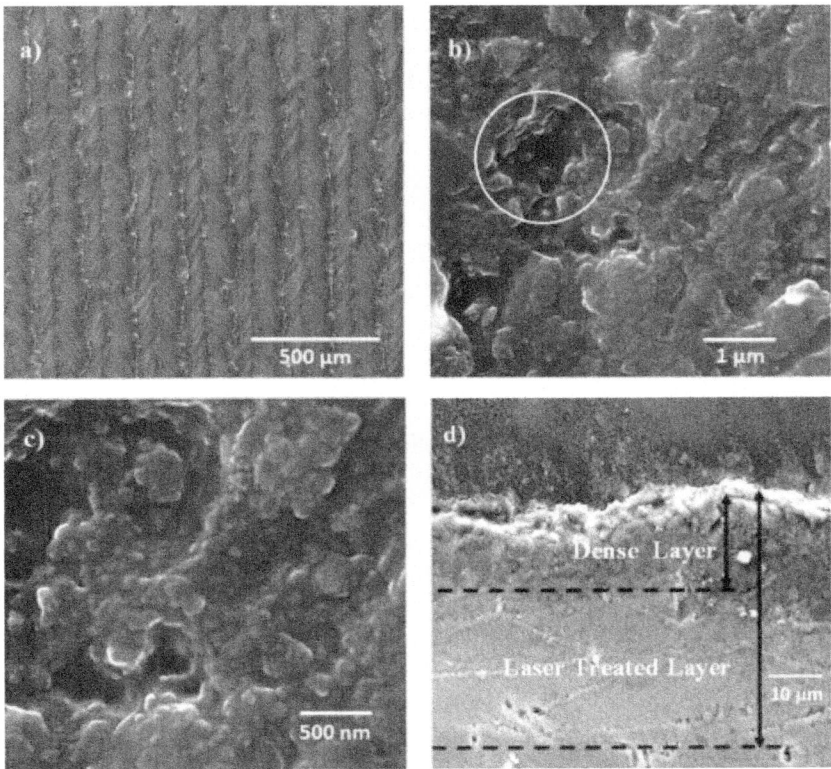

FIGURE 3.31 SEM micrographs of laser-textured surfaces: (A) regular laser scanning tracks, (B) small cavity formed at the surface (marked in a white color circle), (C) fine-size textures, and (D) cross section of laser-treated region [21].

hierarchical texture consisting of micro/nanopoles is formed on the laser-treated surface. The texture height varies, which can be seen from Fig. 3.32B, and the average surface roughness is in the order of 0.82 μm.

Fig. 3.33 shows the SEM micrographs of PDMS-replicated surfaces. PDMS-replicated and laser-textured surfaces are almost identical in terms of surface morphology (Figs. 3.31A and 3.33A); in this case, the presence of laser scanning tracks are evident on the PDMS-copied and PDMS-replicated surfaces. However, close examination of the replicated surfaces reveals that some of the texture details are not replicated properly (Fig. 3.33B and C). In this case, some nanosize whisker-like texture is not copied onto the PDMS surface. This is associated with the irregular alignment of these whisker-like textures on the laser-treated surface; therefore during peeling of PDMS from the laser-textured surfaces, these textures are broken and remained as residues in the PDMS-copied surface. This situation is also seen from Fig. 3.33D, in which the presence of alumina residue on the PDMS-replicated surface is evident.

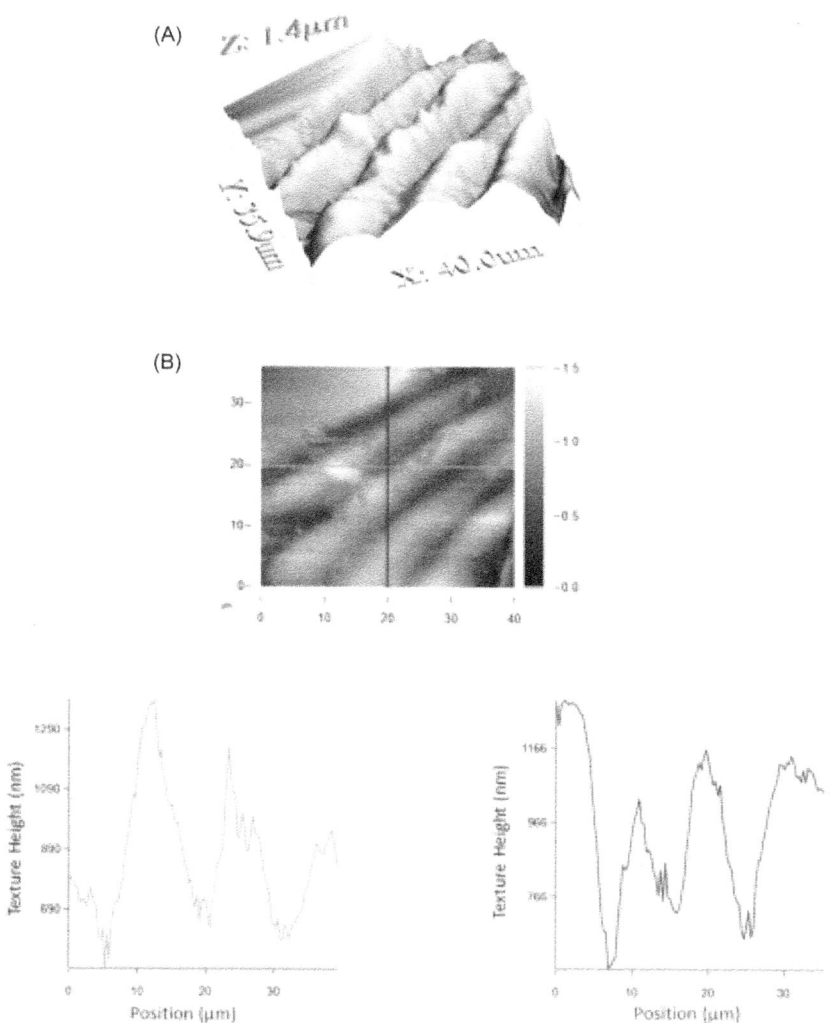

FIGURE 3.32   AFM images of laser-treated surface: (A) 3D view and (B) line scan showing texture heights at treated surface [21].

Fig. 3.34 shows AFM images of PDMS-copied and PDMS-replicated surfaces. The 3D image (Fig. 3.34A) shows that PDMS-copied surface is composed of micro/nanotextures with hierarchical structures. This is also true for the PDMS-replicated surface (Fig. 3.34B). However, the line scan at the PDMS-copied surface shows that some of the nanosize sharp waviness, as observed for the laser-textured surface (Fig. 3.32B), is replaced with rounded waviness. The same observation is made for the PDMS-replicated surface.

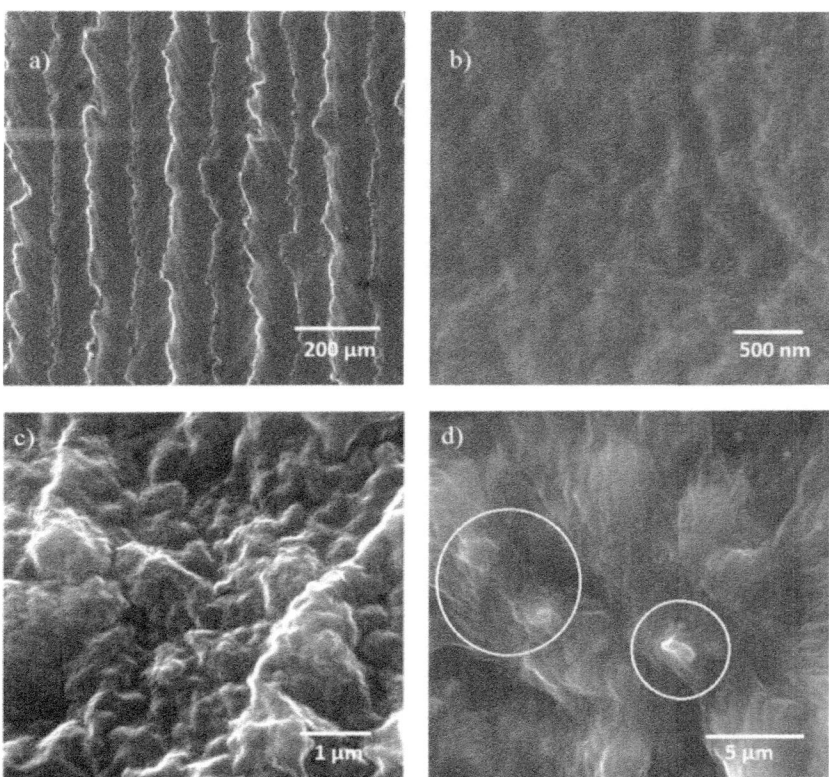

FIGURE 3.33 SEM micrographs of PDMS-replicated laser-textured surfaces: (A) regular laser scanning tracks, (B) textured surface, (C) fine-size textured morphology, and (D) some aluminum nitride (AlN) residues on the surface [21].

This is due to the broken pieces of nanosize whisker-like structures during the removal of PDMS from the laser-treated surface. In this case, the broken parts remain as residues on the PDMS-copied and PDMS-replicated surfaces. The average surface roughness of the PDMS-copied and PDMS-replicated surfaces is in the order of 0.72 and 0.87 mm, respectively. The small deviation of the average surface roughness is associated with the geometric feature of the surface texture. In this case, valleys on the PDMS-copied surface become hills on the PDMS-replicated surface since the PDMS-copied surface was used as a mold for the replicated PDMS surface. Since nanosize whiskers could not be copied and replicated, synthesized nanosize silica particles were introduced at PDMS-copied and PDMS-replicated surfaces to generate the lotus effect. It should be noted that the nanosize texture is crucial to creating a lotus effect on the surfaces, which lowers the contact angle hysteresis on the treated surface. Fig. 3.35 shows the SEM micrographs of

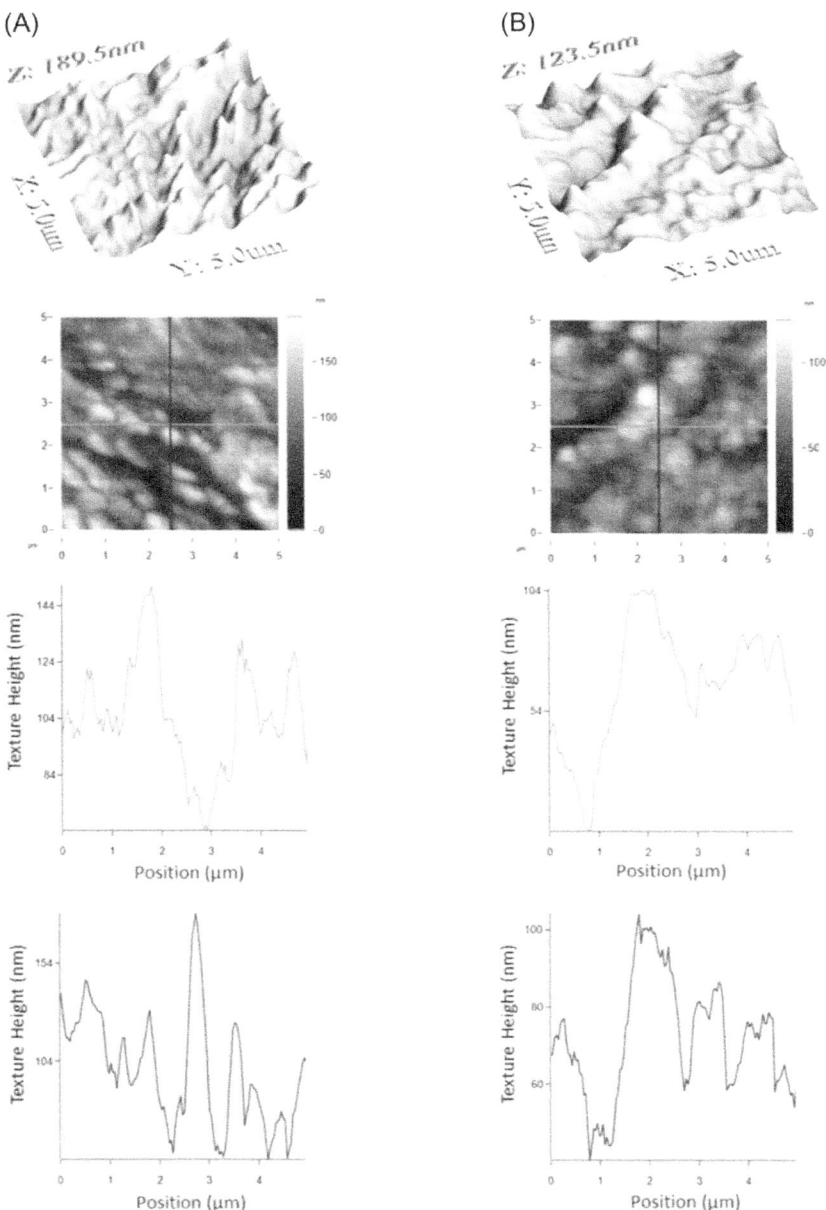

FIGURE 3.34   AFM images of PDMS-copied and PDMS-replicated surfaces, and line scans showing the texture heights at the surface: (A) PDMS-copied surface and (B) PDMS-replicated surface [21].

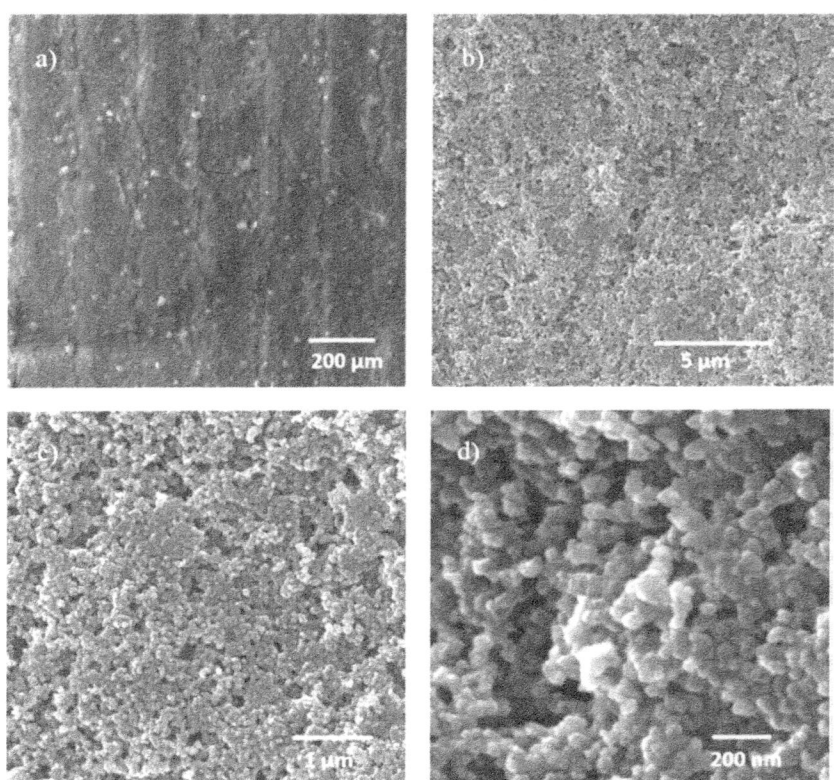

FIGURE 3.35 SEM micrographs of PDMS-replicated and functionalized silica particle-deposited surface: (A) functionalized particles on replicated surface, (B) large coverage area at PDMS-copied surface, (C) aggregated functionalized silica particles, and (D) spherical shape of functionalized and aggregated particles [21].

functionalized silica particles deposited onto PDMS-replicated surfaces. The deposited silica particles cover extensively PDMS-replicated surfaces and are closely spaced on the surface (Fig. 3.35B). Since TEOS is used during the synthesizing of silica particles, the functionalized shell distorts the surface roughness of the particles; in which case, a small increase in the cell size occurs [70]. This may be related to the condensing monomer units, which are growing at a faster rate than the nucleation rate [70]. Since diluted TEOS concentration with ethanol is incorporated during synthesizing of silica particles, the rate of formation of new nuclei is suppressed. This, in turn, resulted in aggregation and adhesion of the particles (Fig. 3.34C and D). It should be noted that the hydroxyl groups on the functionalized silica particle surfaces have different moieties and can have different reactivity toward the modifier molecules. Hence, modifier silane results in side reactions and condensation on the silica surface [71].

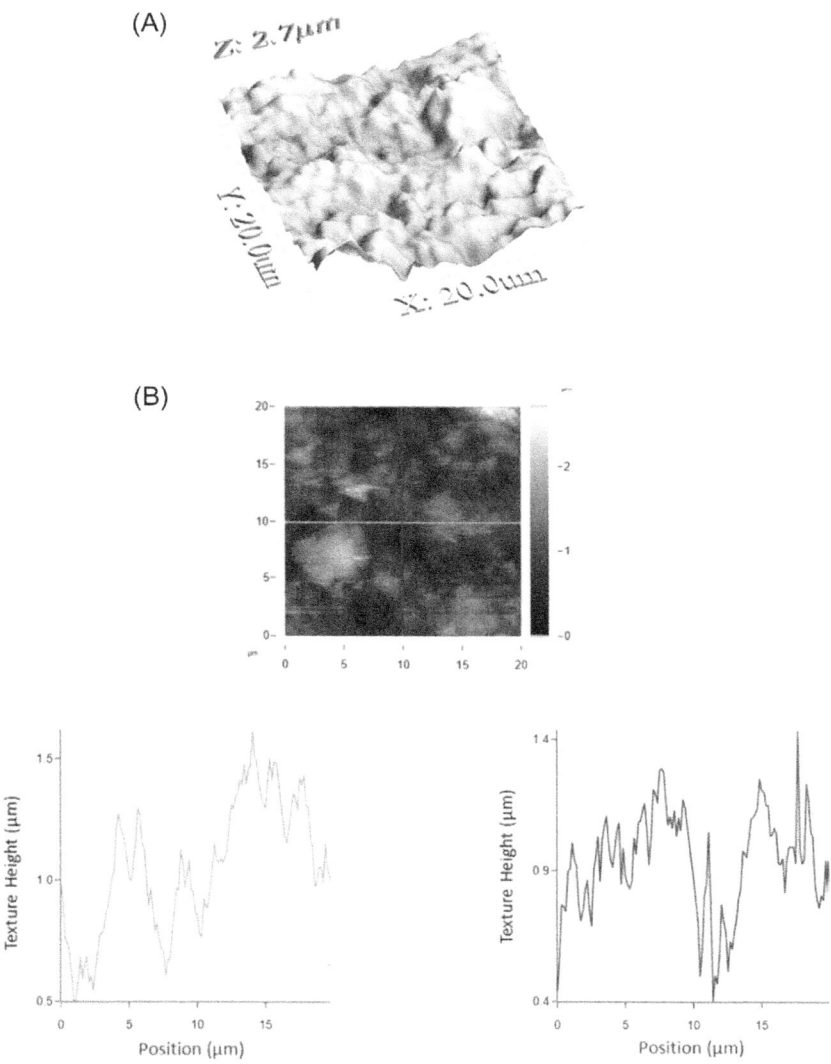

FIGURE 3.36 AFM image of PDMS-replicated and functionalized silica particle-deposited surface: (A) 3D surface texture and (B) line scan showing the texture height at the surface [21].

Fig. 3.36 shows AFM images of PDMS-replicated surface after deposition of functionalized silica particles. The surface texture does not change considerably after the deposition of silica particles on PDMS-replicated surfaces, which can be observed from the 3D image of the surface (Fig. 3.36A). The presence of silica particles is evident from the line scan (Fig. 3.36B). Since no loose silica particles are observed from SEM micrographs

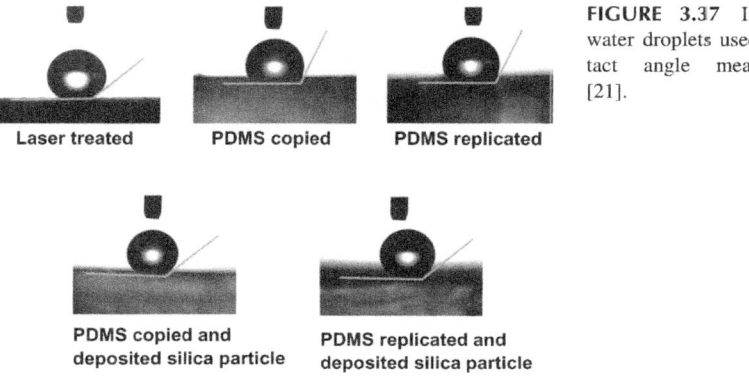

FIGURE 3.37 Images of water droplets used for contact angle measurements [21].

Laser treated   PDMS copied   PDMS replicated

PDMS copied and deposited silica particle

PDMS replicated and deposited silica particle

(Fig. 3.35) and AFM images (Fig. 3.36), functionalizing silica particles prior to deposition onto PDMS-replicated surface improves the adhesion between the particles and the surface.

Fig. 3.37 shows images of water droplets on laser-textured, PDMS-copied, PDMS-replicated, and colloidal particle-deposited surfaces. Water droplet contact angle remains high for the laser-textured surface, and then follows colloidal particle-deposited surfaces, PDMS-replicated, and PDMS-copied surfaces. In this case, low surface energy, due to AlN compounds formed at the surface, and texture of laser-treated alumina surface resulted in high water droplet contact angle. The contact angle of the droplets varies within 10% over the laser-treated surface. This is associated with one or all of the following: (1) surface free energy of laser-treated sample (51.6 mJ/m^2) varies across the surface because of nonuniform distribution of AlN compounds at the surface, and (2) nonhomogeneous texture morphology of laser-treated surface. Nevertheless, the area covered for high contact angle is considerably larger than that of the low contact angle region; in which case, the high contact angle region is in the order of 93% of the total area of the laser-treated surface. Since whisker-like geometric features could not be copied by PDMS, the lotus effect is suppressed. This, in turn, results in relatively lower water droplet contact angle for PDMS-copied and PDMS-replicated surfaces than that of laser-treated surfaces. This situation can be seen from Table 3.3, in which droplet contact angle and hysteresis are given. Deposition of functionalized silica particles on PDMS-copied and replicated surfaces increases water droplet contact angle and reduces significantly contact angle hysteresis. Consequently, nanosize silica particles create the lotus effect on the deposited surfaces.

3.4.5 Laser Treatment of Zirconia Surfaces and Surface Hydrophobicity

The surface energy of zirconia can be modified by laser heating at the surface [72], which may further improve surface hydrophobicity. In the present

TABLE 3.3 Contact Angle Data and Hysteresis for As-Received and Laser-Treated Surfaces, PDMS-Copied/Replicated Surfaces, and PDMS-Copied/Replicated and Functionalized Silica Particle-Deposited Surfaces [21]

	Contact Angle (degrees)	Hysteresis (degrees)
Untreated surface	64.3 (+5/ − 5)	42(+3/ − 3)
Laser-treated surface	150.3 (+5/ − 5)	18(+5/ − 5)
PDMS copied	122.62 (+0.2/ − 0.2)	17(+1.8/ − 1.8)
PDMS replicated	123.7 (+1.1/ − 1.1)	20.9(+1.9/ − 1.9)
PDMS copied and silica deposited	157 (+1.5/ − 1.5)	4.0(+3.3/ − 3.3)
PDMS replicated and silica deposited	155.5 (+0.8/ − 0.8)	2.6(+1.9/ − 1.9)

study, laser-controlled ablation of zirconia surface under high-pressure nitrogen-assisting gas is considered and the surface characteristics including microhardness, residual stress, fracture toughness, surface texture, and hydrophobicity are presented in light of the previous study [73].

Fig. 3.38 shows the SEM micrographs of the top surface of the treated layer. The laser scanning forms regular ablated/melted tracks at the surface. Due to the repetition of the laser pulses (1500 Hz) during the processing, laser-irradiated spots are overlapped at the surface during the scanning. The overlapping ratio of the spots is in the order of 72%, which provides continuous ablated/melted sites along the tracks. The laser power intensity and scanning speed are set such that high evaporation rate from the surface is avoided. This setting was established after carrying out several tests. The treated surface was free from large-scale asperities such as large-scale cracks or crack networks, pores, and cavities. In addition, no melt flow was observed across the laser scanning tracks. The treated surface was composed of micron/nanosized grooves consistent with the previous work [74].

The roughness of the treated surface is in the order of 0.3 μm. The surface roughness varies along the laser scanning tracks, which is associated with the ablation of the surface during the scanning. In this case, the laser beam partially impinges onto the previously formed cavity and partially onto the neighboring untreated surface along the scanning direction. Since the overlapping ratio is high, the area where the laser intensity incident onto the untreated neighboring surface becomes larger than that corresponding to the initially formed cavity surface. Therefore laser intensity remains higher on the untreated neighboring surface than that of the initially formed cavity surface because of the Gaussian intensity distribution of the incident laser

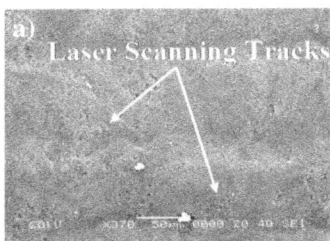

FIGURE 3.38 SEM micrographs of laser-treated surface: (A) laser-treated surface and presence of scanning tracks, (B) laser-treated smooth surface, (C) mixture of smooth and rough structures at laser-treated surface, and (D) laser-treated rough surface [73].

beam. Low intensity causes partial melting of the initially formed cavity surface while modifying surface texture and altering the surface roughness along the scanning tracks. The microsized cavities are few and are randomly distributed at the surface. The formation of microsized cavities can be explained in terms of the thermal agitation of the surface plasma at the irradiated surface. In this case, evaporated front absorbs the incident laser energy and forms small surface plasma, which becomes transiently hot [75]. The surface plasma acts like an additional heat source at the surface while increasing the cavity size during the ablation.

Fig. 3.39 shows SEM micrographs of the cross section of the laser-treated layer. Laser treatment results in a dense layer at the surface because of the high cooling rates in this region. The dense layer consists of fine grains and closely spaced fine dendrites forming a feathery-like structure in this region. The surface layer has higher density than that of the untreated substrate below the dense layer, which causes volume shrinkage in the surface region. This, in turn, causes some small-sized voids to be formed below the dense layer. However, the voids are small and few in numbers in the treated layer. As the depth below the surface increases, slightly longer and larger columnar structures are formed. This is attributed to the relatively lower cooling rates taking place in this region as compared to that in the surface region. Microcracks are not observed in the dense layer and in the large columnar region despite the occurrence of high cooling rates at the workpiece surface. This is because of the self-annealing effect of the later formed laser scanning tracks. In this case, later formed tracks act as heat source to

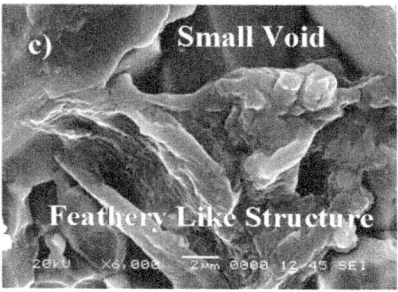

FIGURE 3.39 SEM micrographs of cross section of laser-treated layer: (A) dense layer formed at the surface vicinity, (B) Columnar structure formed below the surface, and (C) feathery-like structure in the treated layer [73].

initially formed tracks while generating a self-annealing effect on the previously formed tracks. The heat-affected zone is not observable at the interface of the laser-treated layer and the base material, which is associated with the low thermal conductivity of zirconia.

Fig. 3.40 shows XRD diffractograms of the laser-treated and as-received surfaces. The as-received material is comprised of tetragonal ZrO_2 (t-ZrO_2). The presence of ZrN peaks are observed from the diffractogram of the laser-treated surface. The use of nitrogen at pressure is responsible for the formation of ZrN compound at the surface. However, the ZrN compound is formed in a two-step process. First, is the transformation of the tetragonal structure of the zirconia (t-ZrO_2) into cubic zirconia (c-ZrO_2), which takes place at high temperatures at the surface. Secondly, is the oxygen release through the dissociation process, which results in formation of ZrN. Therefore the chemical process can be outlined as: t-$ZrO_2 \rightarrow$ c-ZrO_2 and $2ZrO_2 + N_2 \rightarrow ZrN + O_2$ However, the reactions taking place in the surface vicinity result in formation of vacancies in the zirconia [76], which alters the surface energy. Elemental composition across the treated surface remains almost uniform, which can be observed from Table 3.4, in which the energy dispersive spectroscopy (EDS) data obtained from the laser-treated surface are given. Although the quantification of light elements, such as nitrogen, from the EDS data involves error,

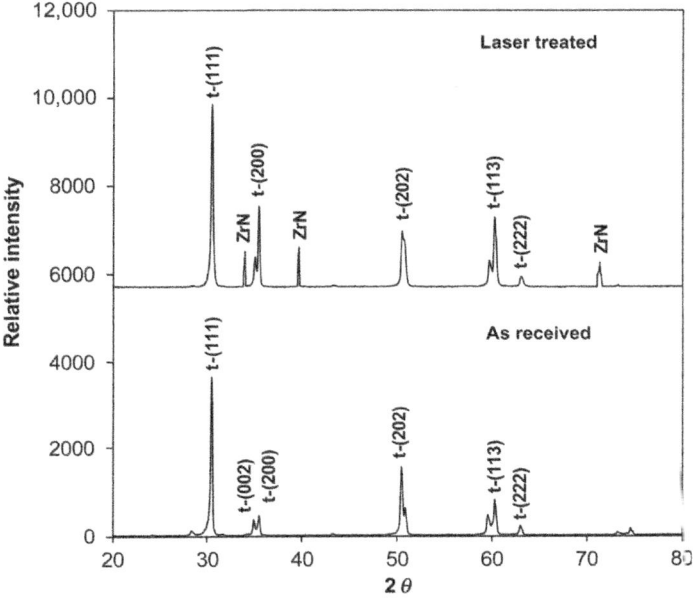

FIGURE 3.40 X-ray diffractogram of laser-treated and as-received surfaces [73].

TABLE 3.4 EDS Data for Elemental Composition of Laser-Treated Workpiece Surface (wt.%) [73]

Spectrum	Y	N	O	Zr
Spectrum 1	5	4	46	Balance
Spectrum 2	6	5	44	Balance
Spectrum 3	6	5	45	Balance

the presence of nitrogen is evident from the EDS data. This agrees with the presence of ZrN peaks in the XRD diffractogram. However, due to the unavailability of the WDS (wavelength dispersive X-ray spectroscopy) facility in our laboratory, accurate elemental analysis is not presented and left for the future study. Table 3.5 gives the microhardness and fracture toughness of

TABLE 3.5 Microhardness and Fracture Toughness and Data Used for Fracture Toughness Calculations [73]

	Hardness HV (GPa)	Fracture Toughness (MPa√m)	P (N)	a (μm)	c (μm)
As-received surface	15.7 ± 0.06	9.5 ± 0.4	5	20	50
Laser-treated surface	19.2 ± 0.06	7.2 ± 0.4	5	25	50

the laser-treated and as-received surfaces. It is evident that the laser treatment improves the microhardness at the surface; however, fracture toughness reduces because of the microhardness enhancement at the surface. The formation of a dense layer, feathery-like structures, and presence of ZrN compound in the surface vicinity are responsible for the microhardness enhancement at the surface. The residual stress predicted is compressive and is in the order of -1.6 ± 0.05 GPa.

Table 3.6 gives the contact angles measured at different locations on the surface. The surface texture and surface energy influence the wetting properties of the surface [77]. Since the surface texture is composed of micro/nanogrooves, the wetting state varies across the surface. In addition, formation of ZrN at the surface alters the wetting state due to modification of the surface energy; in which case the surface energy of laser-treated reduces due to formation of ZrN [78,79]. The surface energy of laser-treated workpieces is of the order of 50 mJ/m^2, which agrees with the previous study (52.6 mJ/m^2) for ZrN [79].

The combined wetting states can be explained through the Laplace pressure. In the case of nanosized texturing, having a heterogeneous interface, confining the water at the air/water interface takes place. The Laplace pressure can be expressed as [80]: $\Delta P = - \frac{2\gamma cos(\theta - \alpha)}{d_o + htan\alpha}$, where γ is the surface tension of water, θ is the contact angle, α is the inclination angle, h is the height of the groove, d_o is the groove width, and $\Delta P = P - P_o$ (where P is the pressure in the liquid of the meniscus and P_o is the ambient pressure [81]). Increasing the Laplace pressure allows more air to be trapped in the groove, which in turn prevents the droplet meniscus from touching the bottom surface of the groove while causing increased contact angle. The Laplace pressure estimated varies at the laser-treated surface because of the combined micro/nanotextures. The Laplace pressure estimated is in the range of $0.9 \times 10^4 - 0.1 \times 10^4$ Pa. This indicates the presence of Cassie state, which can also be seen from the images of contact angle measurements. However, at some locations, where the microtexturing is dominant, the Laplace pressure cannot be calculated correctly and water meniscus touches the treated

TABLE 3.6 Contact Angle Measurement Results Prior to and After the Laser Treatment Using Water, Glycerol, and Diiodomethane [73]

	Contact Angle (degrees)					
	Water		Glycerol		Diiodomethane	
	Low Roughness	High Roughness	Low Roughness	High Roughness	Low Roughness	High Roughness
Untreated	62.9 ± 5		50.8 ± 5		39.2 ± 5	
Laser-treated	121.4 ± 5	98.8 ± 5	114.6	93.4	41.1	40.2

surface. The Wenzel state dominates in this region and the contact angle reduces. Moreover, the contact angle due to a rough surface can be associated with that of the smooth surface through an equation [48]: $cos\theta = f_1(R_1 cos\theta + 1) - 1$, where f_1 is the fraction of the solid–liquid interface under the droplet and $f_1 = 1$ for the homogeneous interface (without the presence of air gab), R_1 is the roughness factor of the surface, which is equal to the ratio of the total surface area to its flat projection, and θ is the contact angle. Since the laser-treated surface contains nano- and microsized textures, the solid–liquid interface factor (f_1) should be within $0 \le f_1 \le 1$. Consequently, the composite surface texture with the combination of micro/nanogrooves results in two states of hydrophobicity at the treated surface. The droplet angle measurements repeated at nine locations at the treated surface, seven locations prevails the presence of Cassie state and two locations are in Wenzel state; therefore the Cassie state dominates over the Wenzel state. It can also be observed from the photographs that in the first case (rough surface), the Wenzel state takes place; therefore the water droplet penetrates into the grooves and touches the surface without the presence of the air gap.

REFERENCES

[1]   T. Young III, An essay on the cohesion of fluids, Philos. Trans. Royal Soc. London 95 (1805) 65–87.

[2]   B. Bhushan, Y.C. Jung, K. Koch, Micro-, nano-and hierarchical structures for superhydrophobicity, self-cleaning and low adhesion, Philos. Trans. Royal Soc. London A: Math. Phys. Eng. Sci. 367 (1894) (2009) 1631–1672.

[3]   J. He, Self-Cleaning Coatings: Structure, Fabrication and Application, Royal Society of Chemistry, 2016.

[4]   W. Barthlott, C. Neinhuis, Purity of the sacred lotus, or escape from contamination in biological surfaces, Planta 202 (1) (1997) 1–8.

[5]   Y.C. Jung, B. Bhushan, Contact angle, adhesion and friction properties of micro- and nanopatterned polymers for superhydrophobicity, Nanotechnology 17 (19) (2006) 4970–4980.

[6]   A. Pozzato, S. Dal Zilio, G. Fois, D. Vendramin, G. Mistura, M. Belotti, et al., Superhydrophobic surfaces fabricated by nanoimprint lithography, Microelectr. Eng. 83 (4–9) (2006) 884–888.

[7]   J.-Y. Shiu, C.-W. Kuo, P. Chen, C.-Y. Mou, Fabrication of tunable superhydrophobic surfaces by nanosphere lithography, Chem. Mater. 16 (4) (2004) 561–564.

[8]   N. Blondiaux, E. Scolan, G. Franc, R. Pugin, Manufacturing of superhydrophobic surfaces combining nanosphere lithography with replication techniques, in: 12th IEEE Conference on Nanotechnology (IEEE-NANO), IEEE, 2012, pp. 1–6.

[9]   H. Yang, X. Dou, Y. Fang, P. Jiang, Self-assembled biomimetic superhydrophobic hierarchical arrays, J. Colloid. Interface. Sci. 405 (2013) 51–57.

[10] Y. Lai, C. Lin, H. Wang, J. Huang, H. Zhuang, L. Sun, Superhydrophilic–superhydrophobic micropattern on TiO$_2$ nanotube films by photocatalytic lithography, Electrochem. Commun. 10 (3) (2008) 387–391.

[11] P. Kothary, X. Dou, Y. Fang, Z. Gu, S.-Y. Leo, P. Jiang, Superhydrophobic hierarchical arrays fabricated by a scalable colloidal lithography approach, J. Colloid. Interface. Sci. 487 (2017) 484–492.

[12] F. Wang, L. Wang, H. Wu, J. Pang, D. Gu, S. Li, A lotus-leaf-like SiO$_2$ superhydrophobic bamboo surface based on soft lithography, Colloids Surfaces A: Physicochem. Eng. Aspects 520 (2017) 834–840.

[13] Y.H. Sung, Y.D. Kim, H.-J. Choi, R. Shin, S. Kang, H. Lee, Fabrication of superhydrophobic surfaces with nano-in-micro structures using UV-nanoimprint lithography and thermal shrinkage films, Appl. Surf. Sci. 349 (2015) 169–173.

[14] C.W. Berendsen, M. Škereň, D. Najdek, F. Černý, Superhydrophobic surface structures in thermoplastic polymers by interference lithography and thermal imprinting, Appl. Surf. Sci. 255 (23) (2009) 9305–9310.

[15] C. Acikgoz, M.A. Hempenius, J. Huskens, G.J. Vancso, Polymers in conventional and alternative lithography for the fabrication of nanostructures, Eur. Polym J. 47 (11) (2011) 2033–2052.

[16] S. Zhou, M. Hu, Q. Guo, X. Cai, X. Xu, J. Yang, Solvent-transfer assisted photolithography of high-density and high-aspect-ratio superhydrophobic micropillar arrays, J. Micromech. Microeng. 25 (2) (2015) 025005.

[17] J. Feng, M.T. Tuominen, J.P. Rothstein, Hierarchical superhydrophobic surfaces fabricated by dual-scale electron-beam-lithography with well-ordered secondary nanostructures, Adv. Funct. Mater. 21 (19) (2011) 3715–3722.

[18] Y. Miyamura, C. Park, K. Kinbara, F.A. Leibfarth, C.J. Hawker, T. Aida, Controlling volume shrinkage in soft lithography through heat-induced cross-linking of patterned nanofibers, J. Am. Chem. Soc. 133 (9) (2011) 2840–2843.

[19] P. Peng, Q. Ke, G. Zhou, T. Tang, Fabrication of microcavity-array superhydrophobic surfaces using an improved template method, J. Colloid. Interface. Sci. 395 (2013) 326–328.

[20] C. Sun, L.-Q. Ge, Z.-Z. Gu, Fabrication of super-hydrophobic film with dual-size roughness by silica sphere assembly, Thin. Solid. Films. 515 (11) (2007) 4686–4690.

[21] B.S. Yilbas, M.R. Yousaf, H. Ali, N. Al-Aqeeli, Replication of laser-textured alumina surfaces by polydimethylsiloxane: improvement of surface hydrophobicity, J. Appl. Polym. Sci. 133 (41) (2016) 1–13.

[22] B. Yilbas, H. Ali, N. Al-Aqeeli, M. Khaled, N. Abu-Dheir, K. Varanasi, Solvent-induced crystallization of a polycarbonate surface and texture copying by polydimethylsiloxane for improved surface hydrophobicity, J. Appl. Polym. Sci. 133 (22) (2016) 1–12.

[23] Y.W. Hu, S. Liu, S.Y. Huang, W. Pan, Fabrication of superhydrophobic surfaces of titanium dioxide and nickel through electrochemical deposition on stainless steel substrate, Key Engineering Materials, Trans Tech Publ, 2010, pp. 496–498.

[24] E. Celia, T. Darmanin, E.T. de Givenchy, S. Amigoni, F. Guittard, Recent advances in designing superhydrophobic surfaces, J. Colloid. Interface. Sci. 402 (2013) 1–18.

[25] J.P. Fernández-Blázquez, D. Fell, E. Bonaccurso, A. Del Campo, Superhydrophilic and superhydrophobic nanostructured surfaces via plasma treatment, J. Colloid. Interface. Sci. 357 (1) (2011) 234–238.

[26] J. Gao, Y. Li, Y. Li, H. Liu, W. Yang, Fabrication of superhydrophobic surface of stearic acid grafted zinc by using an aqueous plasma etching technique, Central Europ. J. Chem. 10 (6) (2012) 1766–1772.

[27] P. Roach, N.J. Shirtcliffe, M.I. Newton, Progress in superhydrophobic surface development, Soft Matter 4 (2) (2008) 224–240.

[28] C. Lawrence, The mechanics of spin coating of polymer films, Phys. Fluids 31 (10) (1988) 2786–2795.

[29] Z. He, M. Ma, X. Lan, F. Chen, K. Wang, H. Deng, et al., Fabrication of a transparent superamphiphobic coating with improved stability, Soft Matter 7 (14) (2011) 6435–6443.

[30] C.J. Brinker, G.W. Scherer, Sol-Gel Science: The Physics and Chemistry of Sol-Gel Processing, Academic Press, 2013.

[31] N. Sahu, B. Parija, S. Panigrahi, Fundamental understanding and modeling of spin coating process: a review, Indian J. Phys. 83 (4) (2009) 493–502.

[32] A. Steele, I. Bayer, E. Loth, Inherently superoleophobic nanocomposite coatings by spray atomization, Nano. Lett. 9 (1) (2008) 501–505.

[33] B. Faure, G. Salazar-Alvarez, A. Ahniyaz, I. Villaluenga, G. Berriozabal, Y.R. De Miguel, et al., Dispersion and surface functionalization of oxide nanoparticles for transparent photocatalytic and UV-protecting coatings and sunscreens, Sci. Technol. Adv. Mater. 14 (2) (2013) 023001.

[34] H. Ogihara, J. Xie, T. Saji, Controlling surface energy of glass substrates to prepare superhydrophobic and transparent films from silica nanoparticle suspensions, J. Colloid. Interface. Sci. 437 (2015) 24–27.

[35] X.Y. Ling, I.Y. Phang, G.J. Vancso, J. Huskens, D.N. Reinhoudt, Stable and transparent superhydrophobic nanoparticle films, Langmuir 25 (5) (2009) 3260–3263.

[36] H. Xiang, L. Zhang, Z. Wang, X. Yu, Y. Long, X. Zhang, et al., Multifunctional poly-methylsilsesquioxane (PMSQ) surfaces prepared by electrospinning at the sol–gel transition: superhydrophobicity, excellent solvent resistance, thermal stability and enhanced sound absorption property, J. Colloid. Interface. Sci. 359 (1) (2011) 296–303.

[37] R. Iler, Multilayers of colloidal particles, J. Colloid. Interface. Sci. 21 (6) (1966) 569–594.

[38] L. Cao, D. Gao, Transparent superhydrophobic and highly oleophobic coatings, Faraday Discuss 146 (2010) 57–65.

[39] F. Liu, F. Sun, Q. Pan, Highly compressible and stretchable superhydrophobic coating inspired by bio-adhesion of marine mussels, J. Mater. Chem. A 2 (29) (2014) 11365–11371.

[40] A. Rifai, N. Abu-Dheir, M. Khaled, N. Al-Aqeeli, B.S. Yilbas, Characteristics of oil impregnated hydrophobic glass surfaces in relation to self-cleaning of environmental dust particles, Solar Energy Mater. Solar Cells 171 (2017) 8–15.

[41] M. Köthe, M. Müller, F. Simon, H. Komber, H.-J. Jacobasch, H.-J. Adler, Examination of poly (butadiene epoxide)-coatings on inorganic surfaces, Colloids Surfaces A: Physicochem. Eng. Aspects 154 (1–2) (1999) 75–85.

[42] D. Maji, S. Lahiri, S. Das, Study of hydrophilicity and stability of chemically modified PDMS surface using piranha and KOH solution, Surface Interface Analysis 44 (1) (2012) 62–69.

[43] N.R. Armstrong, C. Carter, C. Donley, A. Simmonds, P. Lee, M. Brumbach, et al., Interface modification of ITO thin films: organic photovoltaic cells, Thin. Solid. Films. 445 (2) (2003) 342–352.

[44] L. Hallmann, A. Mehl, N. Sereno, C.H. Hämmerle, The improvement of adhesive properties of PEEK through different pre-treatments, Appl. Surf. Sci. 258 (18) (2012) 7213–7218.

[45] D. Tarn, C.E. Ashley, M. Xue, E.C. Carnes, J.I. Zink, C.J. Brinker, Mesoporous silica nanoparticle nanocarriers: biofunctionality and biocompatibility, Acc. Chem. Res. 46 (3) (2013) 792–801.

[46]   X. Liu, J. He, Hierarchically structured superhydrophilic coatings fabricated by self-assembling raspberry-like silica nanospheres, J. Colloid. Interface. Sci. 314 (1) (2007) 341−345.

[47]   S.A. Mahadik, D. Mahadik, M. Kavale, V. Parale, P. Wagh, H.C. Barshilia, et al., Thermally stable and transparent superhydrophobic sol−gel coatings by spray method, J. Sol-Gel Sci. Technol. 63 (3) (2012) 580−586.

[48]   Y.C. Jung, B. Bhushan, Wetting transition of water droplets on superhydrophobic patterned surfaces, Scripta Materialia 57 (12) (2007) 1057−1060.

[49]   R.N. Wenzel, Resistance of solid surfaces to wetting by water, Ind. Eng. Chem. 28 (8) (1936) 988−994.

[50]   H. Ogihara, J. Xie, J. Okagaki, T. Saji, Simple method for preparing superhydrophobic paper: spray-deposited hydrophobic silica nanoparticle coatings exhibit high water-repellency and transparency, Langmuir 28 (10) (2012) 4605−4608.

[51]   J. Bravo, L. Zhai, Z. Wu, R.E. Cohen, M.F. Rubner, Transparent superhydrophobic films based on silica nanoparticles, Langmuir 23 (13) (2007) 7293−7298.

[52]   T. Deng, K.K. Varanasi, M. Hsu, N. Bhate, C. Keimel, J. Stein, et al., Nonwetting of impinging droplets on textured surfaces, Appl. Phys. Lett. 94 (13) (2009) 133109.

[53]   J.D. Smith, R. Dhiman, S. Anand, E. Reza-Garduno, R.E. Cohen, G.H. McKinley, et al., Droplet mobility on lubricant-impregnated surfaces, Soft Matter 9 (6) (2013) 1772−1780.

[54]   B. Yilbas, A. Matthews, C. Karatas, A. Leyland, M. Khaled, N. Abu-Dheir, et al., Laser texturing of plasma electrolytically oxidized aluminum 6061 surfaces for improved hydrophobicity, J. Manuf. Sci. Eng. 136 (5) (2014) 054501.

[55]   A. Owais, M.M. Khaled, B.S. Yilbas, N. Abu-Dheir, K.K. Varanasi, K.Y. Toumi, Surface and wetting characteristics of textured bisphenol-A based polycarbonate surfaces: acetone-induced crystallization texturing methods, J. Appl. Polym. Sci. 133 (14) (2016) 43074.

[56]   Z. Fan, C. Shu, Y. Yu, V. Zaporojtchenko, F. Faupel, Vapor-induced crystallization behavior of bisphenol-A polycarbonate, Polym. Eng. Sci. 46 (6) (2006) 729−734.

[57]   A. Sinha, P. De, Mass Transfer: Principles and Operations, PHI Learning Pvt. Ltd, 2012.

[58]   C.K. Liu, C.T. Hu, S. Lee, Effect of compression and thickness on acetone transport in polycarbonate, Polym. Eng. Sci. 45 (5) (2005) 687−693.

[59]   F.L.O. de Oliveira, M.C.M. Leite, L.O. Couto, T.R. Correia, Study on bisphenol-A polycarbonates samples crystallized by acetone vapor induction, Polym. Bull. 67 (6) (2011) 1045−1057.

[60]   Y. Cui, A.T. Paxson, K.M. Smyth, K.K. Varanasi, Hierarchical polymeric textures via solvent-induced phase transformation: a single-step production of large-area superhydrophobic surfaces, Colloids Surfaces A: Physicochem. Eng. Aspects 394 (2012) 8−13.

[61]   E. Turska, W. Benecki, Studies of liquid-induced crystallization of bisphenol a polycarbonate, J. Appl. Polym. Sci. 23 (12) (1979) 3489−3500.

[62]   J. Brandrup, E. Immergut, E.A. Grulke, Polymer Handbook, fourth ed., A Wiley-Interscience Publication, New York, 1999.

[63]   D. Van Krevelen, K.T. Nijenhuis, in Properties of Polymers: Their Correlation With Chemical Structure: Their Numerical Estimation and Prediction From Additive Group Contributions, fouth completely rev./ed., Elsevier, Amsterdam, Boston, MA, 2009.

[64]   P. Dayal, A.J. Guenthner, T. Kyu, Morphology development of main-chain liquid crystalline polymer fibers during solvent evaporation, J. Polym. Sci. Part B: Polym. Phys. 45 (4) (2007) 429−435.

[65]   J.I. Lauritzen Jr, J.D. Hoffman, Formation of polymer crystals with folded chains from dilute solution, J. Chem. Phys. 31 (6) (1959) 1680−1681.

[66] F.C. Frank, M. Tosi, On the theory of polymer crystallization, Proc. R. Soc. Lond. A 263 (1314) (1961) 323–339.

[67] S. Tension, Solid surface energy data (SFE) for common polymers, 2017. Available from: <http://www.surface-tension.de/solid-surface-energy.htm> (accessed 14.01.13).

[68] W.Y.D. Yong, Z. Zhang, G. Cristobal, W.S. Chin, One-pot synthesis of surface functionalized spherical silica particles, Colloids Surfaces A: Physicochem. Eng. Aspects 460 (2014) 151–157.

[69] B. Yilbas, S. Akhtar, C. Karatas, Laser gas assisted melting of preprepared alumina surface including TiC particles at surface, Surface Eng. 27 (6) (2011) 470–476.

[70] M.S. Hellsing, H.M. Kwaambwa, F.M. Nermark, B.B. Nkoane, A.J. Jackson, M.J. Wasbrough, et al., Structure of flocs of latex particles formed by addition of protein from Moringa seeds, Colloids Surfaces A: Physicochem. Eng. Aspects 460 (2014) 460–467.

[71] J. Lin, H. Chen, Y. Ji, Y. Zhang, Functionally modified monodisperse core–shell silica nanoparticles: silane coupling agent as capping and size tuning agent, Colloids Surfaces A: Physicochem. Eng. Aspects 411 (2012) 111–121.

[72] S. Norouzian, M.M. Larijani, R. Afzalzadeh, Effect of nitrogen flow ratio on structure and properties of zirconium nitride films on Si (100) prepared by ion beam sputtering, Bull. Mater. Sci. 35 (5) (2012) 885–887.

[73] B. Yilbas, Laser treatment of zirconia surface for improved surface hydrophobicity, J. Alloys Compounds 625 (2015) 208–215.

[74] B. Luo, P.W. Shum, Z. Zhou, K. Li, Preparation of hydrophobic surface on steel by patterning using laser ablation process, Surface Coatings Technol. 204 (8) (2010) 1180–1185.

[75] B. Yilbas, R. Davies, A. Gorur, Z. Yilbas, F. Begh, N. Akcakoyun, et al., Investigation into development of liquid layer and formation of surface plasma during CO_2 laser cutting process, Proc. Inst. Mech. Eng. Part B: J. Eng. Manuf. 206 (4) (1992) 287–298.

[76] X. Huting, F. Yongqing, M. Chandrasekaran, A. Batchelor, X-ray imaging of laser remelted plasma sprayed zirconia coating, J. Mater. Sci. Lett. 17 (2) (1998) 163–165.

[77] S. He, M. Zheng, L. Yao, X. Yuan, M. Li, L. Ma, et al., Preparation and properties of ZnO nanostructures by electrochemical anodization method, Appl. Surf. Sci. 256 (8) (2010) 2557–2562.

[78] A. Noro, M. Kaneko, I. Murata, M. Yoshinari, Influence of surface topography and surface physicochemistry on wettability of zirconia (tetragonal zirconia polycrystal), J. Biomed. Mater. Res. Part B Appl. Biomater. 101 (2) (2013) 355–363.

[79] C.-C. Sun, S.-C. Lee, W.-C. Hwang, J.-S. Hwang, I.-T. Tang, Y.-S. Fu, Surface free energy of alloy nitride coatings deposited using closed field unbalanced magnetron sputter ion plating, Mater. Trans. 47 (10) (2006) 2533–2539.

[80] Y. Liu, W. Lin, Z. Lin, Y. Xiu, C. Wong, A combined etching process toward robust superhydrophobic SiC surfaces, Nanotechnology. 23 (25) (2012) 255703–255710.

[81] A.W. Adamson, A.P. Gast, Physical Chemistry of Surfaces, Interscience, New York, 1967.

Chapter 4

Environmental Dust on Surfaces

Chapter Outline

4.1 INTRODUCTION

Environmental dust particles have detrimental effects on surfaces, particularly for those incorporated in optically sensitive devices such as reflectors, photovoltaic active surfaces, and solar thermal power troughs. Recent climate changes have resulted in extreme weather conditions including frequent dust storms around the globe, especially in regions of desert areas such as Central America, North Africa, India, Central Asia, Middle East, and other similar areas. Dust particles settled on surfaces have multifold effects. Some of these effects include diffusion and absorption of the incident optical radiation, creating thermal resistance on the surface, and becoming chemically active when dissolved in water. Environmental dust particles are mobile and easily carried by the wind from one region to another. Dust mitigation also takes place from one continent to another. Dust particles have various sizes, shapes, and elemental compositions. The morphology of dust particles is usually regionally dependent and depends on the geologic structures of the landscape. In general, dust particles vary from within a few nanometers to tens of micrometers in size and do not conform to any regular shape. Dust particles are formed from alkaline and alkaline earth metals together with silica, oxygen, sulfur, chlorine, and other elements. The weight percentage of these elements mainly depends on the geographic position and geologic structure of the region. On the other hand, in humid ambient air, when water condensates on the dust particles, some of alkaline and alkaline earth metals dissolves forming a chemically active liquid solution. The liquid solution dispenses on the solid surface under the influence of gravitational potential energy while forming a film. Depending on the material properties, local erosion and corrosion can result while damaging the solid surface permanently. In addition, it forms a mud-like mixture on the surface via mixing of the dust particles and the liquid solution. Upon drying of the mud-like mixture on the surface, the adhesion between the solid surface and the mud-like mixture becomes extremely high and efforts to remove it more difficult. In this chapter, analyses of dust particles are introduced and the chemo-mechanics of mud formed from the dust particles presented.

4.2 CLASSIFICATION OF DUST PARTICLES

In general, the characteristics of surfaces exposed to the environment are degraded by dust accumulation, which in turn degrades the optical, tribological, and thermal properties of the surface. Dust removal requires additional energy and cost to restore the surface to its original state, and is increasingly complicated and challenging in high-humidity environments. As dust accumulates, mud forms because of the accumulation of dust particles and the condensation of water vapor on the surface of the dust particles. Dust is composed of soluble particles, such as alkaline metals (e.g., Na, K), and nonsoluble compounds such as silica and calcite (e.g., $CaCO_3$). The soluble compounds alter the base of the solution and increase adhesion of mud onto the surface by forming covalent bonds between the mud and the solid surface. Although adhesion between dry dust particles and the substrate surface is governed by van der Waals forces, the cohesive effect due to the crystallized solution at the interface increases mud adhesion at the surface. Consequently, mud residues that remain at the substrate surface modify the chemical and physical characteristics of the surface including its surface texture, optical and tribological properties, stress levels, and surface hydrophobicity. Such changes can consequently reduce the performance of the solid substrate in a given application.

Considerable studies have been carried out to examine dust particles and their effects on surfaces. Airborne dust and its characteristics have been thoroughly examined [1−4]; however, studies regarding the effects of environmental dust on surfaces have not been extensively reported [5]. Characterization of atmospheric airborne dust during the wet seasons in East Africa was presented by Mkoma et al. [6]. They demonstrated that common crystal and sea-salt elements, including Na, Mg, Al, Si, Cl, Ca, Ti, Mn, Fe, Sr, NO_3^-, and P (and to a lesser extent Cu and Zn) tended to be coarse particles. In addition, the aerosol chemical mass content of the aerosol was determined to consist of 48% organic matter, 44% crustal matter, 4% sea salt, and

2% elemental carbon. A characterization of atmospheric aerosols was also carried out by Maenhaut et al. [7]. They showed that most of the Ca was water soluble; the mineral dust Ca was presumably mostly present as calcite, and perhaps also in part as gypsum. In contrast, only half of the K content was water soluble, indicating that it was to a large extent associated with insoluble mineral dust. Patterns of dust retention on urban trees in oasis cities were examined by Baidourela and Zhayimuj [8]. The findings revealed that dust that had accumulated on tree leaves was mainly of local urban origin, and the heavy metal concentrations at different sites varied significantly. The morphology of atmospheric particles in a semiarid region of India was studied by Mishra et al. [9]. They demonstrated that the influence of the dust aspect ratio on dust scattering was significant for dust with high hematite content. Petruk and Skinner [10] characterized particles in airborne dust and showed that small particles had low aspect ratios. The effects of sand and dust accumulation on photovoltaic modules were studied by Beattie et al. [11]. They demonstrated that the reduction in the active area of the modules was mainly due to the formation of clusters of particles on the surface, which reduced the available area for light capture to a much smaller area compared to particles resting directly on the glass surface. The adhesion of dust particles to common indoor surfaces in an air-conditioned environment was examined by Tan et al. [12]. They found that dust and activated carbon adhesion were highly sensitive to surface roughness with an inverse relationship between adhesion force and roughness due to the reduction in contact area between the particle and a rougher material surface. The effect of drought on dust production in the Sudano—Sahelian zone of the Sahara Desert was investigated by Middleton [13]. The data showed that dust-storm activity in the west and east of the Sudano—Sahelian belt had dramatically increased during the drought years; by a factor of 6 in Mauritania and up to a factor of 5 in Sudan. The transport of Asian dust around the globe was reported by Uno et al. [14]. The findings revealed that Asian dust could influence the global radiation budget by stimulating cirrus cloud formation and marine ecosystems by supplying nutrients to the open ocean. The eolian dust deposition in the western United States due to human activity was studied by Neff et al. [15]. They indicated that the larger dust flux, which persisted into the early 21st century, resulted in a more than five-fold increase in inputs of K, Mg, Ca, N, and P to the alpine ecosystems, with implications for surface-water alkalinity, aquatic productivity, and terrestrial nutrient cycling. Some research studies were carried out to examine the properties of mud formed in various environmental conditions. Mechanical interfacial properties of epoxy/red mud layers were investigated by Park et al. [16]. They showed that the mechanical interfacial properties of epoxy/treated red mud nanocomposites were intimately correlated with the improvement of interfacial adhesion between the red mud surface and epoxy matrix. Euganean thermal mud properties were studied by Rossi et al. [17]. They demonstrated that a chemico-mineralogical marker could identify the typical chemico-mineralogic characteristics of thermal mud, which were interconnected with the mud surface energy. Electrochemical analysis of hot-dip galvanized steel due alkaline mud adhesion was studied by Zhang et al. [18]. They demonstrated that the higher contents of dissolved oxygen and Cl⁻ ions in the mud played an important role in accelerating the corrosion. A study on mud-crack patterns during repeated drying cycles was carried out by Goehring et al. [19]. They indicated that stress field developed in the dry mud gave rise to crack formations either approximately rectilinear or hexagonal tiling.

In order to proceed with the dust particle analysis dust particles were collected over a period of 12 months from the energy laboratory of King Fahd University of Petroleum and Minerals, which is located close to the city of Dammam in Saudi Arabia. Dust particles accumulated on the surface of the protective glass of photovoltaic panels were removed by soft brushes and stored in an airtight container. The collected dust particles were first analyzed in terms of weight, size, shape, and elemental composition using the above-mentioned analytical tools. The findings revealed that the dust particles collected over a 1-week period within 12 months had similar characteristics in terms of elemental composition, size distribution, and shape. The amount of dust particles accumulated on the surface of the photovoltaic protective layer within 1 month was in the order of 20 g/m²; however, this number varied within 16% (by weight) over 6 months, which was attributed to the wind speed and direction. Although the wind speed and direction changed over time, the average wind speed was approximately 4 m/s over the year. The dust particles were characterized using SEM (JEOL 6460) and energy dispersive spectroscopy (EDS), which consisted of an INCA Mics microscope image capture system, including an INCA X-stream, X-ray acquisition, and detector control unit. X-ray photoelectron spectrometer (XPS) was performed incorporating ESCALAB 220 XL spectrometer. A monochromatic Al K_α X-ray source (1486.6 eV) was operated in the constant analyzer energy mode (CAE = 100 eV for survey spectra and CAE = 40 eV for high resolution spectra). X-ray diffraction (XRD, Model: D8 Advanced diffractometer, Manufacture: Bruker, USA) analysis was performed with CuKα radiation at typical settings of 40 kV and 30 mA. Water cloaking of the dust particles was monitored, and the cloaking velocity was measured using a high-speed camera (Model: SpeedSense 9040, Manufacture: Dantec Dynamic, Denmark).

4.2.1 Dust Particle Size Distribution, Shape, and Analysis

To assess the size and shape of the dust particles, a soft brush was used to collect the particles from photovoltaic active surfaces, which are located in the local region of Dammam in the Kingdom of Saudi Arabia. They were later secured in a sealed container prior to analyzing them. The analysis and the findings are presented here in light of the previous studies [20–24].

Fig. 4.1 shows SEM images of the dust particles accumulated on a PV protective glass test surface in Dammam, Saudi Arabia.

The size of the dust particles varied from the nanometer range to 20 μm while exhibiting an average size of 1.2 μm. Fig. 4.2 shows the distribution of the dust particle sizes. However, some small size dust particles in the sub-micrometer range attach to the surfaces of large dust particles (Fig. 4.1A and B). The bright areas in the SEM micrographs correspond to the small parti-

(A)   (B)

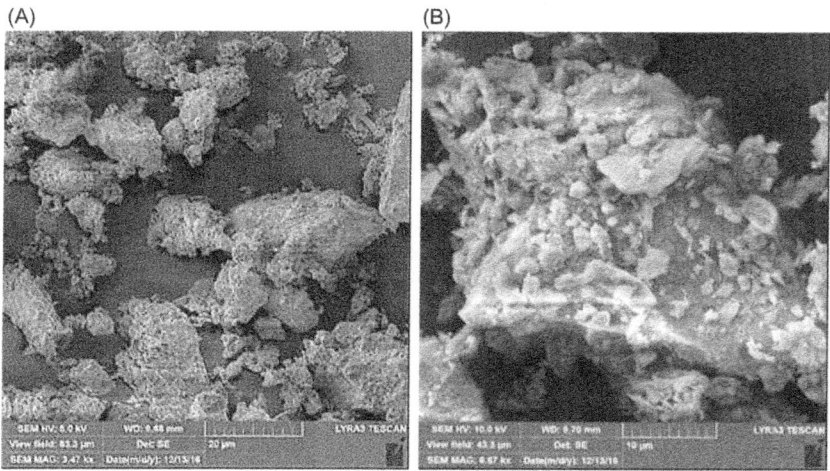

FIGURE 4.1 SEM micrographs of dust particles: (A) various sizes and shapes of dust particles, and (B) small dust particles attached to large dust particles.

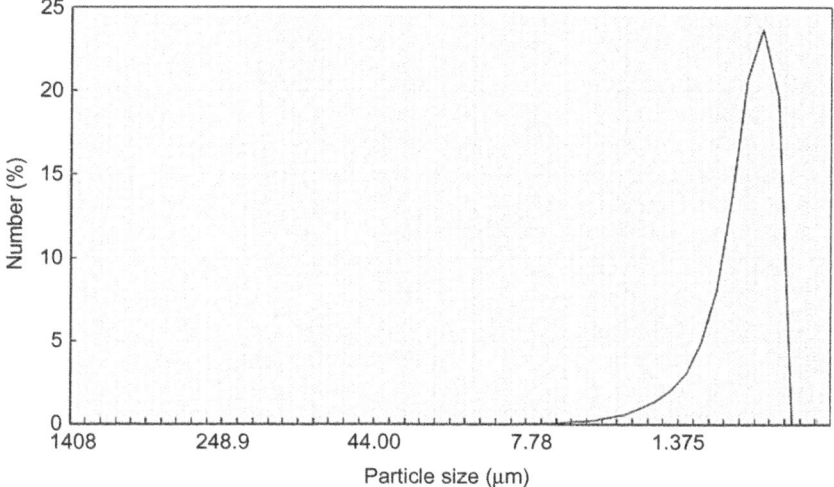

FIGURE 4.2 Size distribution of dust particles [25].

cles. The bright appearance of the small particles indicates the occurrence of electron charging in the SEM chamber during the micrographing. Consequently, this dictates that the small particles are charged, which create forces for attachment to the surfaces of the large particles. The small-size particles (average particle diameter) reside in the atmosphere for prolonged times and interact with solar radiation for longer durations than the large particles. Therefore, prolonged exposure to the atmosphere in regions closer to the sea causes the attachment of ionic compounds. This contributes to the charging of the small dust particles during SEM micrographing.

In general, the dust particles exhibited various shapes with round corners, sharp edges, or flake-like structures. The geometric features of dust particles can be classified by the shape factor, $R_{Shape} = \frac{P^2}{4\pi A}$, where P is the perimeter of the dust particle; the aspect ratio, $A_{Aspect} = \frac{\pi(L_{proj})^2}{4A}$, where A is the cross-sectional area; and L_{proj} is the longest projection length of the dust particle. The aspect ratio is related to the approximate particle roundness and represents the ratio of the major-to-minor axes of an ellipsoid that is best fit to the particle. The shape factor is the inverse of the particle circularity, which is associated with the complexity of the particle (i.e., a shape factor of unity corresponds to a perfect circle). The particle diameter and area can be obtained from the measurements, where the diameter of a circle with an equivalent area is considered for circular dust particles, and an ellipse model is used when the longest projection is assumed to be the major axis. The particle cross-sectional area is preserved for noncircular dust particles. The relationship between the particle size and the aspect ratio or the shape factor is not simple. An inverse relationship is observed between the particle size and the aspect ratio, whereas a direct relationship is observed between the particle size and the shape factor. In this case, the particle aspect radio reduces as the shape factor increases with increasing particle sizes. The cross-sectional area of a typical dust particle 1.8 μm in size is in the order of 2.5 μm², which results in a shape factor of 1.05. However, the cross-sectional area of a typical dust particle 15 μm in size is in the order of 175 μm², and the corresponding shape factor is approximately 3.18. The shape factor of the small particles (<1.2 μm) approaches unity, while the median shape factor of the large particles (≥10 μm) approaches 3. The elemental composition of the dust particles is given in Table 4.1 (wt.%). The most common elements in the dust particles are Si, Ca, Mg, Na, K, S, O, and Fe, irrespective of the size and shape of the dust particles; moreover, the concentrations of Na, K, Ca, and O are found to increase, and chlorine is also present in the small particles (<1.2 μm). The changes in the elemental composition of the small particles (<1.2 μm) can be attributed to their prolonged residence time in the atmosphere, during which long periods of interaction occur between solar radiation and the small dust particles. Hence, the prolonged residence of small-size particles in the atmosphere allows for the attachment of ionic

TABLE 4.1 Elemental Composition of Dust (wt.%) Determined by Energy Dispersive Spectroscopy (EDS)

	Si	Ca	Na	S	Mg	K	Fe	Cl	O
Size ≥ 1.2 μm	11.6	8.4	2.2	1.2	1.6	0.9	1.2	0.6	Balance
Size <1.2 μm	10.1	7.8	2.9	1.4	1.2	1.1	1.1	1.1	Balance
Dust residues	12.1	8.2	2.2	1.2	2.1	0.8	1.1	0.7	Balance

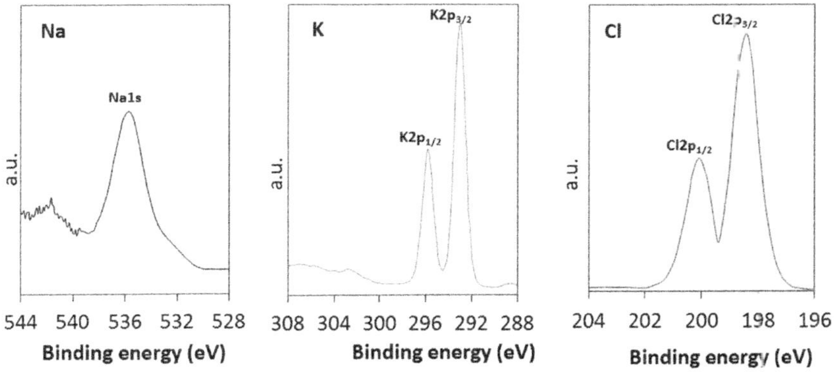

FIGURE 4.3 XPS peaks for binding energy of sodium, potassium, and chlorine [24].

compounds in regions near the Gulf Sea. The concentration of chlorine varied among the different small dust particles, and the EDS data are not consistent with the molar ratio of NaCl, as given in Table 4.1. The chlorine concentration in the small dust particles suggests that the dust particles do not contain salt crystals but rather chlorine dissolved in compounds. Since the concentrations of K, Cl, and Na are low in the dust particles (Table 4.1), XPS was carried out determine the concentrations in the dust particles. Fig. 4.3 shows the XPS data for the binding energy of K, Cl, and Na. The potassium, sodium, and chlorine on the outermost layers of the dust particle is evident from XPS data and chlorine composed of inorganic chloride; in which case, the Cl $2p_{3/2}$ peak intensity occurs at 198.9 eV, which is in good agreement with the previous study [26,27]. XPS peak for potassium K $2p_{3/2}$ occurs at 293.2 eV. The binding energy data for potassium K $2p_{3/2}$ and Cl $2p_{3/2}$ peaks demonstrates the presence of KCl in the dust. The concentration analysis via using CasaXPS indicates that the concentration ratio (after transforming in mass percentage) for K over Cl is in the order of 1.62, which is similar to that obtained from Table 4.1. Similar measurements were carried out for sodium and XPS binding data for Na 1s peak occurring at 1072.8 eV

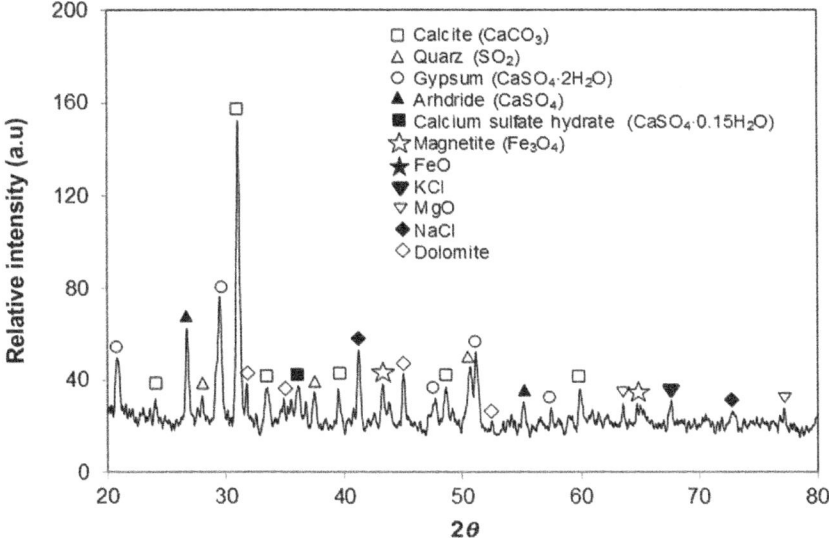

FIGURE 4.4 X-ray diffracogram of dust particles [22].

(Fig. 4.3), which agrees with the previous data reported [26,27]. The elemental concentration of the dust particles varies at different locations in the dust. In addition, it also changes with the dust particle shape and size. This indicates that the dust is composed of nonuniformly distributed elements and compounds. The quadrangular particle, which appears to be deformed from a cubic particle, is rich in sodium and chlorine; however, the aggregated particles are rich in calcium and oxygen. Flake-like particles are also observed; these particles are rich in calcium and silicon.

Fig. 4.4 shows an X-ray diffractogram of the dust particles. Potassium, sodium, calcium, sulfur, chlorine, and iron peaks are clearly visible. The iron peak is coincident with the aluminum and silicon peaks. The presence of sodium and potassium peaks is associated with sea salt because the region where the dust was collected is close to the Arabian Gulf. The concentration of chlorine changes for different dust particles, and the EDS data do not satisfy the molar ratio for NaCl (Table 4.1); therefore, the dust particles do not contain salt crystals and instead NaCl is dissolved in the compound form. The sulfur may form a monomer layer during the aging process in the atmosphere. However, the sulfur can be correlated with the calcium in the dust, such as the anhydrite or gypsum component ($CaSO_4$). The iron is most likely related to clay-aggregated hematite (Fe_2O_3).

4.2.2 Dust Particle Adhesion on Surfaces

Surface energy and assessment of cloaking of the dust particles in water or oil environments remains critical for dust removal from surface via water or light oils.

To assess the adhesion of dust particles to the hydrophobic surface, tangential force measurements for dust particle displacement on the hydrophobic surface were carried out by AFM [25]. When small-size dust particles attach on the surface of large-size particles, they agglomerate to form small dust clusters. The small-size dust particles have a charge field due to the prolonged duration of exposure to the atmosphere near the Gulf Sea, and the charges enhance the adhesion of these particles to the hydrophobic surface. The dust particles possess alkaline (Na, K) and alkaline earth (Ca) compounds (Table 4.1). These compounds contribute to ionic bonding at the surface under the influence of humidity while enhancing adhesion between the dust particles and the hydrophobic surface. Therefore, the presence of ionic bonding and electrostatic charge forces between the dust particles and the hydrophobic surface modifies the retention force. AFM was used to measure the retention force due to adhesion of a dust particle on the hydrophobic surface, where the AFM cantilever tip is proportional to the slope of the deflection of the tip when the tip is in contact with the surface. In this case, from the deflection relation, the retention force can be written as $F = k\sigma\Delta V$ [25], where k is the spring constant of the cantilever tip (N/m), σ is the slope of the displacement over the recorded probe voltage ($\Delta z/\Delta V$, m/V), and ΔV is the voltage recorded during surface scanning by the AFM tip in the contact mode. In the measurements, the following values were adopted: $k = 0.12$ N/m and $\Delta z/\Delta V = 1.481 \times 10^{-6}$ m/V. The maximum value of the retention force obtained from AFM analysis is in the order of 30 nN for a single dust particle with an average size of 1.2 μm; however, the retention force for a single dust particle with an average size of 5.2 μm on the hydrophobic surface is in the order of 65 nN.

4.2.3   Surface Energy and Cloaking of Dust Particles

This demonstrates that the wetting state of the dust particles is critical for the removal of the dust particles from the hydrophobic surface by the water droplet. Therefore, an experiment was carried out to assess the water cloaking of the dust particles. The water-cloaking velocity of the dust particles and wetting height were obtained from the high-speed camera data.

Fig. 4.5A shows the stages of the water cloaking of the dust particles and the cloaking velocity [23]. The cloaking velocity first increases rapidly and then reduces as time progresses. Because water-film cloaking occurs opposite to gravity, as the weight of the water film cloaking the dust particles increases, the net driving force opposing gravity for cloaking decreases. Therefore, the water cloaking experiment was extended to the functionalized dust particles for comparison. Water does not cloak the functionalized dust particles, as seen in Fig. 4.5B. This behavior is associated with the spreading rate of the water film at the dust particle–air interface. The spreading coefficient ($S_{op(a)} = \gamma_{pa} - \gamma_{pw} - \gamma_{wa}$, where γ_{pa} is the interfacial energy at the dust

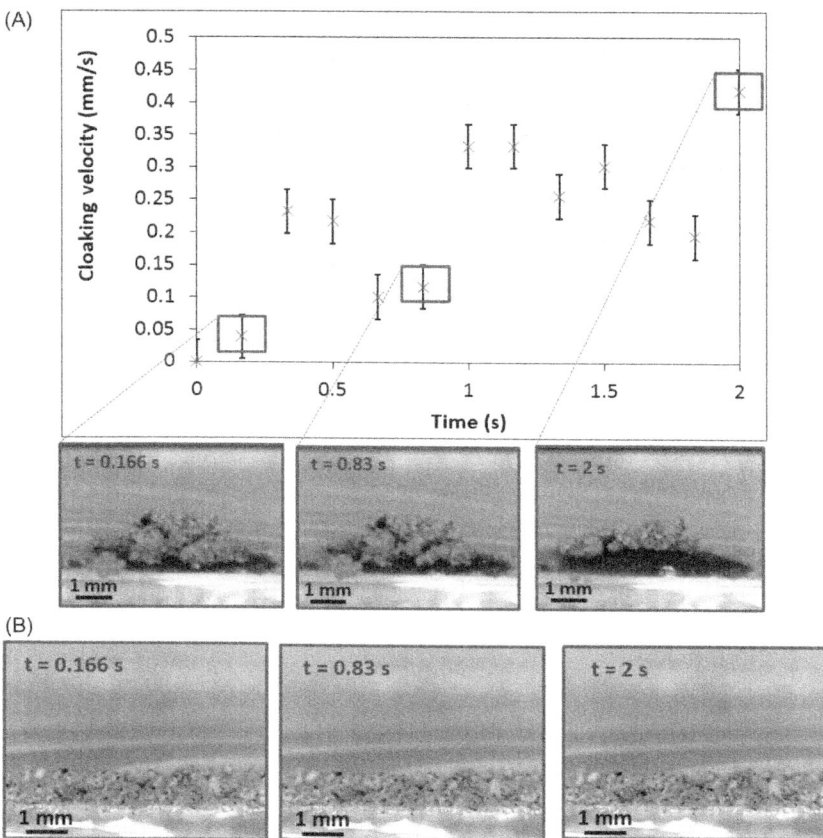

FIGURE 4.5   Water cloaking of normal and functionalized dust particles: (A) cloaking velocity and cloaking images of normal dust particles at different duration and (B) noncloaking images of functionalized dust particles at different durations. Water does not cloak the functionalized dust particles for extended periods [23].

particle−air interface, γ_{pw} is the interfacial energy at dust particle−water interface, and γ_{wa} is the interfacial energy at the water−air interface) must be greater than zero for water to cloak the outer surface of the dust particles. Although the interfacial energy between the dust particle and air is unknown, the condition $\gamma_{pa} > (\gamma_{pw} + \gamma_{wa})$ should be satisfied for water cloaking. Note that the dust particles were compacted into a pellet, and the surface energy of the dust pellet can be determined through experiments incorporating the Owens−Wendt (OW) method [28]. The surface energy of the pellet was determined to be in the order of 750 mJ/m^2, which is between the surface energy of calcite (347 mJ/m^2 [29] and silica (1500 mJ/m^2 [30]). The actual surface energy of the dust particles may slightly differ from that of the measured value; however, it should remain within a similar order of magnitude because of the major constituting elements, silica, and calcite. Because $\gamma_{wa} = 72\text{mJ/m}^2$, in any case, γ_{pw} should be less than 678 mJ/m^2. On the other hand, water spreading on the dust particles occurs into two stages. In the first stage, the balance between the surface tension gradient and the shear stress at the water–dust interface results in a monolayer of water spread on the dust particle. In the second stage, the location of water spreading follows Joos' law [31], and the spreading velocity can be related to $V_s \propto (3S_{ow(a)}/4\sqrt{\mu_o\rho_o})^{1/2}t^{-1/4}$), where μ_o is the dynamic viscosity of water, ρ_o is the density of water, and $S_{ow(a)}$ is the spreading coefficient of water on the dust particles [31].

The dissipating force during water spreading around a dust particle can be approximated by the Ohnesorge number ($Oh = \mu_o/\sqrt{\rho_o a \gamma_{oa}}$), where a is the characteristic size of the dust particle [31], which can be considered to be the equivalent diameter [32]. For an average dust particle size of 1.2 μm, Oh well exceeds unity ($Oh > 1$), which implies a large dissipation force for water cloaking of the dust particle. The cloaking rate is associated with cloaking time in the form of $\sim k_m t^{1/4}$, where k_m is the cloaking factor [32], and cloaking is not possible if $k_m t^{1/4} < 1$. In the present case, $k_m t^{1/4}$ was determined to be greater than unity for normal dust particles. Moreover, the cloaking velocity was determined from high-speed camera data, which is shown Fig. 4.5A, and its average value is in the order of 0.3×10^{-3} m/s. From the average cloaking velocity, the duration of complete cloaking of the dust particle within the range of 1.2–10 μm is in the order of 0.0315 s. Note that the cloaking velocity is inversely related and the cloaking time; in which case, the relation for the cloaking velocity can be approximated in the form of $\sim Ct^{-0.5}$, where C is a constant that varies with the shape of the dust and t is the cloaking time. Moreover, the distance corresponding to the cloaking time and traveled by the droplet on the hydrophobic surface is in the order of 9 μm, which is much less than the contact length of the 40 μL liquid droplet on the solid surface ($l \cong 0.002$ m). Therefore, the dust particles are picked up from the hydrophobic surface by the water droplet cloaking the dust particles. The dust particles picked up by the droplet remain in the droplet fluid and mix with the droplet liquid (Fig. 4.6B). On the other hand, the functionalized dust particles do not penetrate into the droplet liquid and instead remain at the droplet surface (Fig. 4.6B). Consequently, the droplet fluid and the functionalized dust particles do not mix. The adhesion of the functionalized dust particles on the water droplet surface can be attributed to the electrostatic attraction developed within the deposited surface of the functionalized dust particles [33,34].

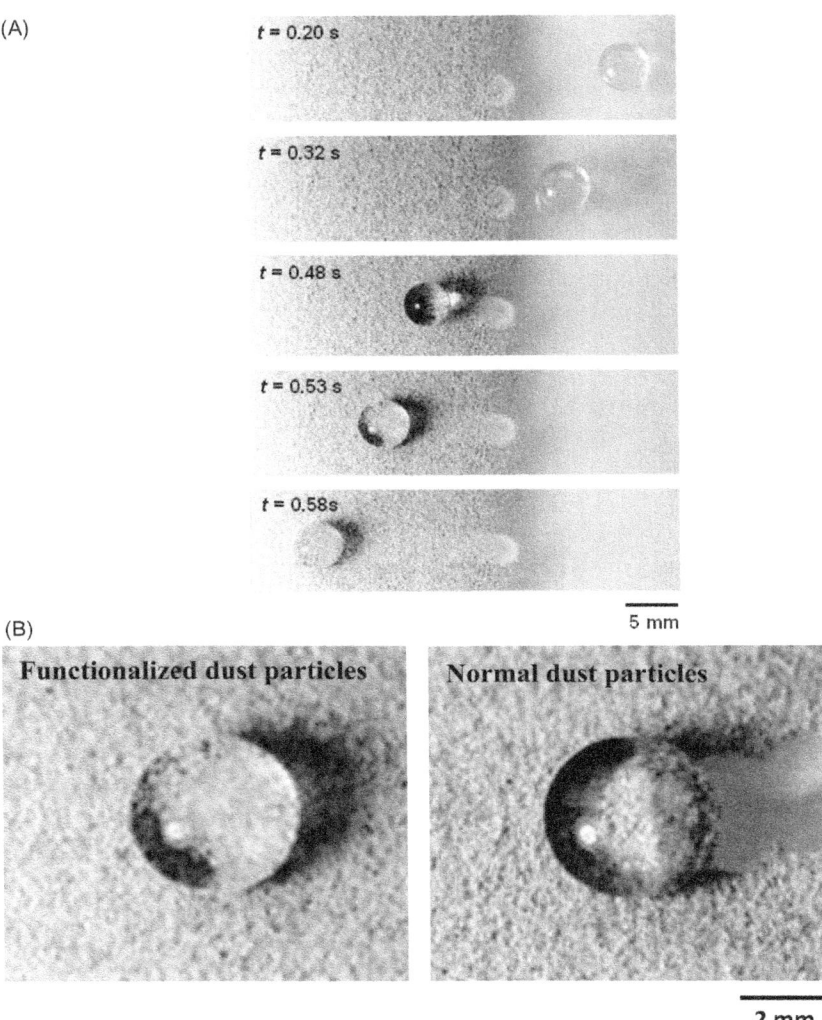

FIGURE 4.6 Optical images of a droplet on the hydrophobic surface with functionalized dust particles: (A) droplet on the hydrophobic surface with functionalized dust particles at different time duration and (B) top image of the droplet on the surface with functionalized and normal dust particles [24].

4.3   CHEMO-MECHANICS OF MUD FORMED FROM DUST PARTICLES

In humid air environments, water vapor condensates on the surface of dust particles and is then absorbed by particles forming mud on the solid surface. The formed mud dries as a result of solar radiation and adheres to the solid surface. Dry mud removal from surfaces becomes difficult due to the strong adhesion between the dry mud and surface. The chemo-mechanics of the mud formed from the dust particles are presented in the light of the previous studies [20,23].

Mekhilef et al. [35] reported that adhesion of dust particles to surfaces was affected by the humidity in the atmosphere. As the relative humidity decreased, the solar panel efficiency increased because less dust adhered to the surface. In addition, Adinoyi and Said [36] demonstrated that dust particles adhered to the surface of the PV panel cover glass due to humidity, which required external efforts to carefully clean the surfaces to restore the initial power output of the panels. Brown et al. [37] reported that applying an antisoiling hydrophilic coating to the glass cover reduced the amount of dust soiling on the surface. Conversely, capillary bridges formed on the solid surfaces because of the interaction between the dust particles and condensed vapor in the gaps between the particles and surface. This effect generated meniscus forces that increased the dust layer and the adhesion force between the dust particles and solid surfaces [38,39]. Corn [40] studied the adhesion force of solid particles and demonstrated that the adhesion force increased with the particle size. Furthermore, the contact area between a rough surface and particle was found to have a major role in the adhesion between the particles and surface. In addition, the relative humidity of the ambient air affected the adhesion force. McLean [41] presented the cohesive forces related to the sediment layers of dust particles. He demonstrated that an electrostatic precipitator had a significant cohesive force that influenced the sediment layers because of the electric field charging of the particles. Podczeck et al. [42] studied the effect of the relative humidity on particle adhesion and found that at high relative ambient humidity, the adhesion increased slightly, whereas the van der Waals forces became nearly 10 times greater than the electrostatic forces. Somasundaran et al. [43] examined the adhesion force between solutions on glass surfaces and found that cohesive forces on the glass surface increased when the pH of the solution decreased; this effect was associated with the amount of salt in the solution. In addition, the cohesion between particles and surfaces was reduced due to the interaction of an anionic surfactant within the polyethylene oxide layer. Fukunishi and Mori [44] investigated the adhesion force between particles in humid environments and demonstrated that the adhesive force between hydrophobic glass and the particles remained nearly constant for different humidity conditions. Kumar et al. [45] studied the influence of particle size on the adhesion of particles to smooth surfaces and used the Johnson—Kendall—Roberts (JRK) adhesion model to characterize the adhesion force for smooth surfaces. Jarząbek et al. [46] presented a measurement method to determine the particle adhesion of ceramic and the adhesion in ceramic-reinforced composite structures. They demonstrated that the presence of ceramics improved the adhesion of particles in the composite matrix. Knoll et al. [47] characterized the adhesion force generated by magnetic particles on protein surfaces. Their findings revealed that magnetic particles strongly adhered to protein surfaces and that an additional force was required for the separation of magnetic particles. Petean and Aguiar [48] determined the adhesive force between a particle and rough surface and compared experimental data with the simulation results of various models. Their findings indicated that among the models considered, the JKR model yielded results that were closest to the experimental values.

Yilbas et al. [22] reported that dust particles consist of ionic and neutral compounds. The alkaline and alkaline earth metallic compounds of dust particles dissolve in the water condensate on surfaces in humid environments, which gives rise to the formation of a chemically active mud solution that flows around dust particles under the effect of gravity and reaches the solid surface where the dust particles have settled. This, in turn, gives rise to the formation of a liquid film between the dust particles and solid surface. The liquid film dries together with the mud, thus forming a dry mud on the solid surface. However, the bonding force is a combination of the ionic and adhesion forces that depends on the wetting area of the solid surface and the dust particles.

4.3.1 Characteristics of Mud Formed From Dust Particles

The mud solution (liquid) was extracted from the mixture of dust particles and water. The pH of the mud liquid was measured and analyzed using inductively coupled plasma mass spectroscopy. The temporal variation of the mud liquid pH is shown in Fig. 4.7, and the data obtained from the inductively coupled plasma mass analysis are shown in Table 4.2.

The pH of the mud liquid increased significantly with time and achieved an approximate steady state after 10 days. The pH of the solution is basic in nature, and the high rate of increase in the pH is associated with the presence of OH^- ions in the mud solution, which is related to the dissolution of alkaline (Na, K) and alkaline earth (Ca) metals in the dust particles. This trend can also be observed from Table 4.1, where alkaline and alkaline earth metals are present. To assess the characteristics of the dry mud solution, mud liquid was placed on a glass surface and dried in a controlled environment (at 20°C and a pressure of 1 atm). Fig. 4.8 shows the SEM micrographs of the dried mud solution on the glass surface. Various crystal structures sizes were formed on the glass surface (Fig. 4.8A and B). Although drying took place in a controlled environment, the local heat transfer modifies the crystal size on the glass surface. In this case, the crystal structures become small when the local cooling rates are high at the surface (Fig. 4.8A), whereas the crystal structures are large under low cooling rates (Fig. 4.8B) [49]. The elemental composition of the crystallized structures on the glass surfaces was analyzed, and Table 4.3 provides the relevant EDS data. The crystallized layer shows that alkaline (K, Na), alkaline earth metals (Ca),

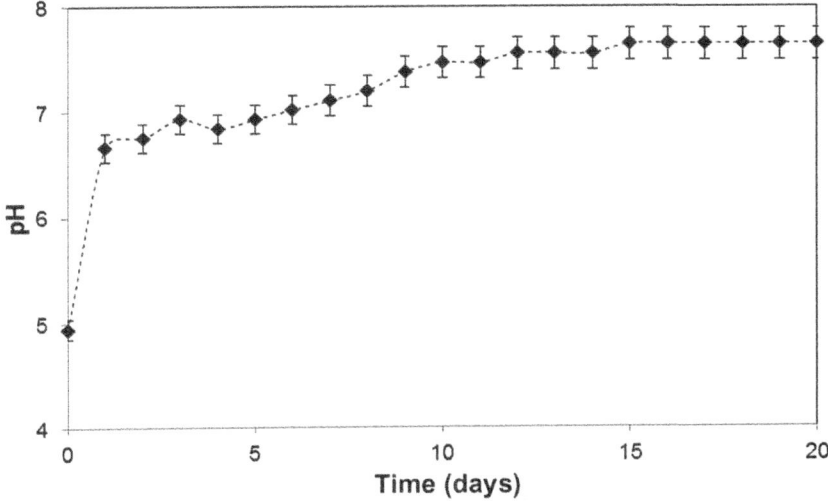

FIGURE 4.7   Temporal variation of the pH of the mud solution [20].

TABLE 4.2 Inductively Coupled Plasma Spectroscopy (ICP) Data (ppb) for the Mud Solution After the Dust Particles Were Dissoluted in Desalinated Water for 8 h

Ca	Na	Mg	K	Fe	Cl
309800	44600	69950	33400	1830	37600

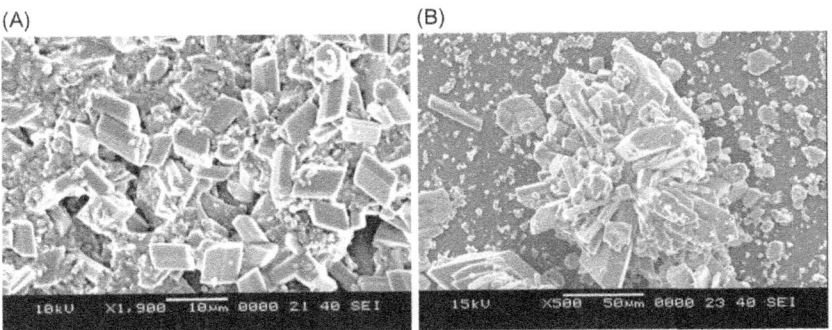

FIGURE 4.8   SEM micrographs of crystalized structures formed on the glass surface after the mud solution dried: (A) small-size crystals, and (B) clustered crystals [20].

TABLE 4.3 Elemental Composition of the Crystals Formed After Drying the Mud Solution on a Glass Surface (wt.%)

	Si	Ca	Na	S	Mg	K	Fe	Cl	O
40°C	0.6	20	1.4	9.1	0.4	0.4	0.6	1.7	Balance

(A)

(B)

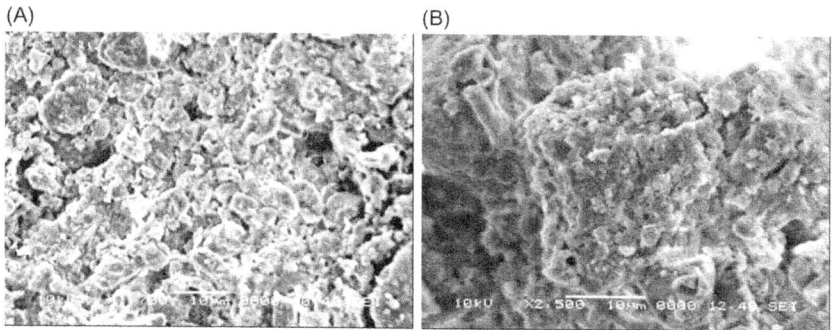

FIGURE 4.9 SEM micrographs of the mud surface: (A) voids formed around large dust particles, and (B) dense structures formed around small dust particles due to the mud solution [20].

oxygen, and chlorine were present in the crystal structures. Therefore, the alkaline and alkaline earth metal compounds dissolve in water and form a chemically active liquid at the glass surface. The mud solution has chemically active characteristics [18]. However, assessing the chemical potential of the mud solution is not within the scope of this study and will be studied in the future.

Figs. 4.9 and 4.10 show SEM micrographs of the dry mud surface and dry mud cross-section, respectively. Large particles together with closely spaced small particles were observed on the surface of the dry mud (Fig. 4.9A) whereas voids were observed near certain large particles (Fig. 4.9B). In addition, some small voids occurred on the dry mud surface; these voids are mainly associated with the evaporation of water from the mud surface during drying, which causes the porous-like morphology at the surface. However, mud contains alkali and alkali earth metal compounds (Table 4.3), which dissolve in water during the formation of the mud solution and flow across the mud cross-section toward the glass surface as a result of gravity. The mud solution accumulates on the glass surface and forms a thin liquid layer. Upon drying, a crystallized dry mud solution is formed between the dry mud and glass surface. This situation can be observed in Fig. 4.10A, in which the SEM micrographs of the dry mud cross-section are shown. The dry mud cross-section consists of porous-like

(A)

(B)

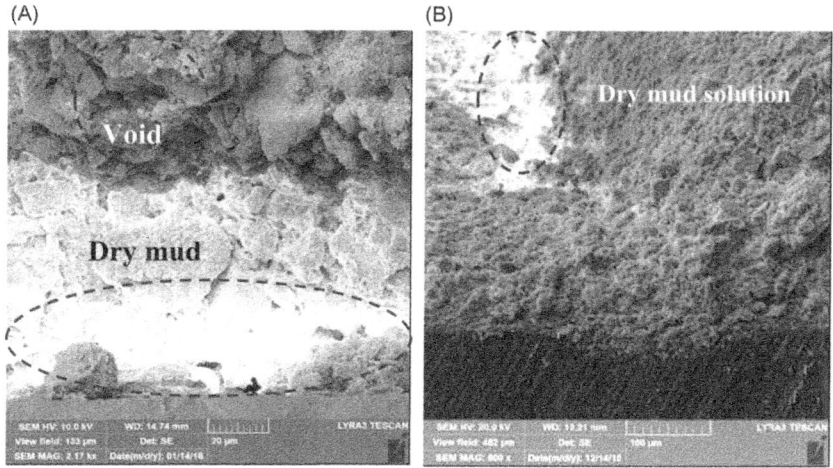

FIGURE 4.10   SEM micrographs of the dry mud cross-section: (A) dry mud solution at the dry mud−glass surface interface and (B) dry mud solution in a void [20].

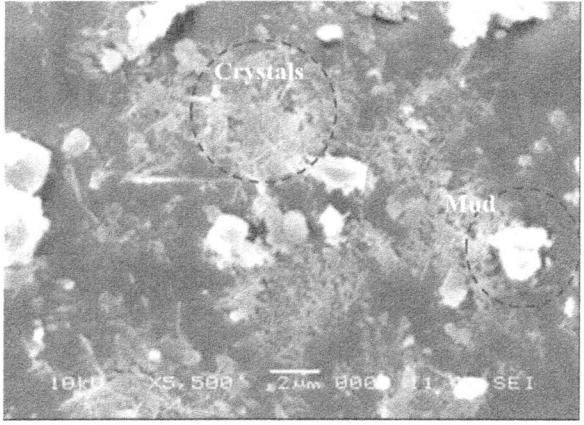

FIGURE 4.11   SEM micrographs of the glass surface in which the dry mud was removed using a pressurized desalinated waterjet [20].

structures, including some small cavities across the cross-section (Fig. 4.10B) because of the wide range of dust particle sizes (0.001−20 μm). This arrangement enables the liquid mud solution to flow in between these structures. However, a portion of the liquid mud solution sediments in the cavities across the cross-section (Fig. 4.10A). This sedimentation appears as a bright color in the SEM image (Fig. 4.10B). To assess the effect of dry mud and its solution on the glass surface, the dry mud was removed from the glass surface using a pressurized distilled waterjet. Fig. 4.11 shows the

TABLE 4.4 Elemental Composition of the Dry Mud Residues on a Glass Surface (wt.%)

	Si	Ca	Na	S	Mg	K	Fe	Cl	O
40°C	1.6	24	1.2	2.2	0.4	0.5	0.8	0.9	Balance

SEM micrographs of the dust particle residue on the glass surface. The residue of the dry mud was related to the dry mud solution in between dust particles and the glass surface, which increased the adhesion at the interface. The EDS analysis of the dry mud residue revealed that the dry mud residue was composed of Ca and Si, and Na, K, and Cl were also present (Table 4.4). These results show that the residue of the dry mud solution together with dry mud particles were present on the glass surface (Fig. 4.11).

In addition, close examination of the SEM micrographs (Fig. 4.11) reveals that some crystallized structures were formed on the surface; in this case, dissolved alkaline and earth alkaline metals are responsible for the crystalline morphology on the dry mud that was removed from the glass surface. The dry mud residue formed various texture morphologies on the glass surface. The maximum height varies within a range of 130 nm; therefore, the average surface roughness of the glass surface increases to 80 nm. In addition, some small cavities formed at the glass surface after the dry mud was removed. The formation of small cavities is associated with hydroxyl attached to the surface because of the high pH of the liquid mud solution prior to drying [22]. Consequently, dry mud residues modify the surface texture and surface structure of the glass through hydroxyl attacks on the surface.

4.3.2 Mechanical Properties of Dry Mud

The mechanical properties of dry mud were assessed through friction and tensile tests. The tangential force was monitored during the friction tests to determine the adhesion work required to remove dry mud from the glass surface. Tensile tests provide a combination of the adhesion and cohesion forces (binding forces) in the dry mud; therefore, the cohesion forces in dry mud caused by drying were also assessed by comparing the tangential force and tensile test data. Fig. 4.12 shows the friction coefficients of the as-received surface, dry mud solution, and dry mud removed using a pressurized distilled waterjet.

Fig. 4.13 shows the tangential forces obtained when the dry mud was removed from the glass surface, the frictional force of the glass surface without mud deposition, and the tangential force corresponding to the dry mud solution being removed from the glass surface. The area under the force

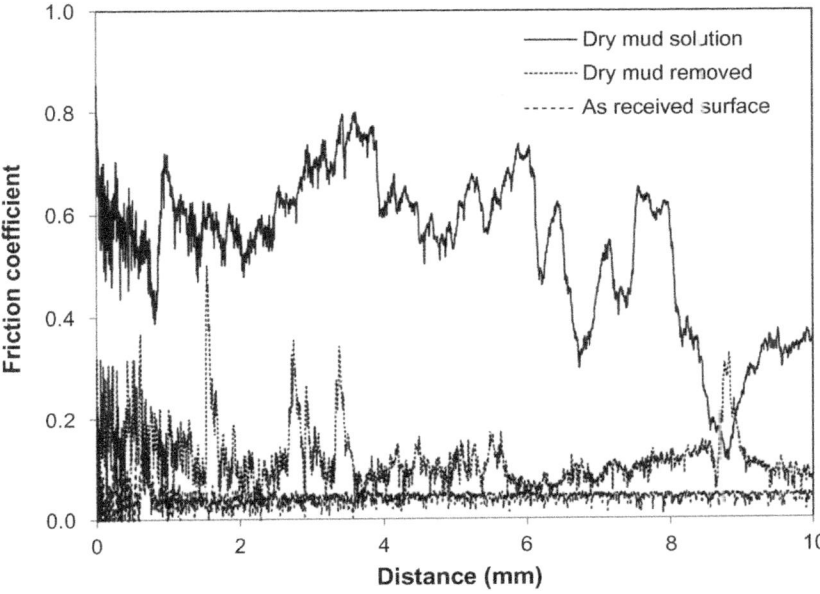

FIGURE 4.12   Friction coefficient for the dry mud solution, as-received glass surface, and surface for which the dry mud was removed using a pressurized desalinated waterjet [20].

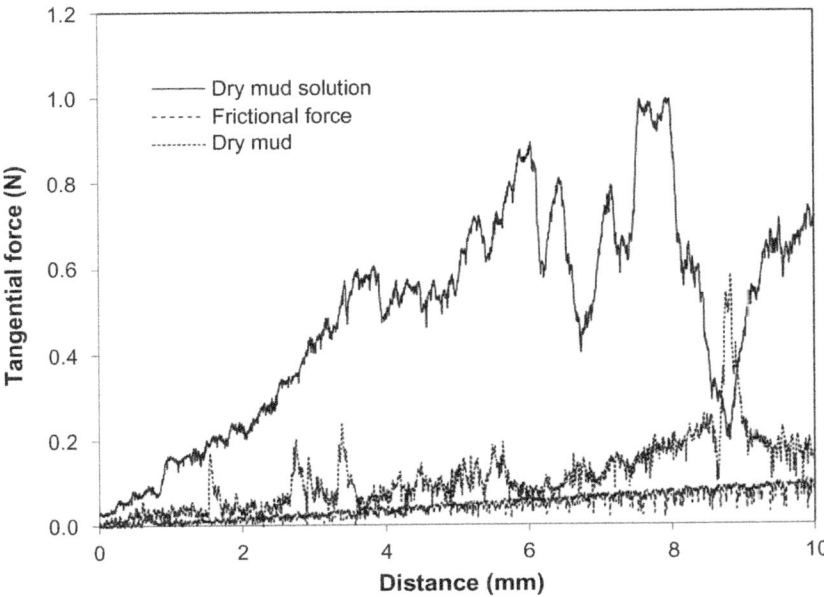

FIGURE 4.13   Tangential force obtained from the scratch tests for the dry mud solution and surface for which the dry mud was removed using a pressurized desalinated waterjet. The frictional force for the as-received glass surface is provided for comparison [20].

curves provides the frictional and adhesion work. The friction coefficient is largest for the dry mud solution, followed by the surface after the dry mud was removed using the pressurized distilled waterjet and then as-received surface.

The friction coefficient increases due to the strong adhesion between the indenter tip and dry mud solution surface, as shown in Fig. 4.13. Moreover, the increased friction coefficient for the surface for which the dry mud has been removed is associated with an increased surface roughness because of the mud residues on the surface. Increasing the surface roughness has been reported to increase the friction coefficient of the surface [50]. In addition, some small peaks in the friction curve are observed, which is related to the surface roughness due to the mud residue on the surface. In addition, a locally increasing friction coefficient is also related to the surface modification because of the hydroxyl attacks on the surface as the mud solution dries on the glass surface. Consequently, the increase in surface roughness due to the mud residue and the cavities formed because of the hydroxyl attacks contribute to the enhancement of the friction coefficient of the surface. The scratch marks left on the as-received glass surface and the surface for which the dry mud has been removed by a pressurized distilled waterjet extend nearly uniformly on the surface. However, a portion of the mud residue on the surface modifies the scratch marks due to the strong adhesion force between the mud and glass surface because of the thin film formed from the dried mud solution at the interface. Moreover, no microcracks were observed around the scar marks in the as-received glass and in the glass in which the mud was removed. This result demonstrates that the reduction in the fracture toughness due to surface modification by hydroxyl attack is not significant. In the case of the tangential force (Fig. 4.13), strong adhesion between the dry mud and glass surface resulted in a sharp increase in the tangential force. This appears locally on the force curve (Fig. 4.13). Consequently, the dry mud solution between the dry mud and glass surface is responsible for the increase in tangential force. Table 4.5 provides the adhesion work determined from the tangential force measured using the scratch tester during dry mud removal. The adhesion work is obtained by integrating the tangential force over the scratch distance. However, the adhesion work determined is

TABLE 4.5 Adhesion Work Obtained From the Tangential Force Measurements [20]

	Adhesion Work (mJ)
Dry mud	0.119 ± 0.008
Mud solution	0.536 ± 0.01

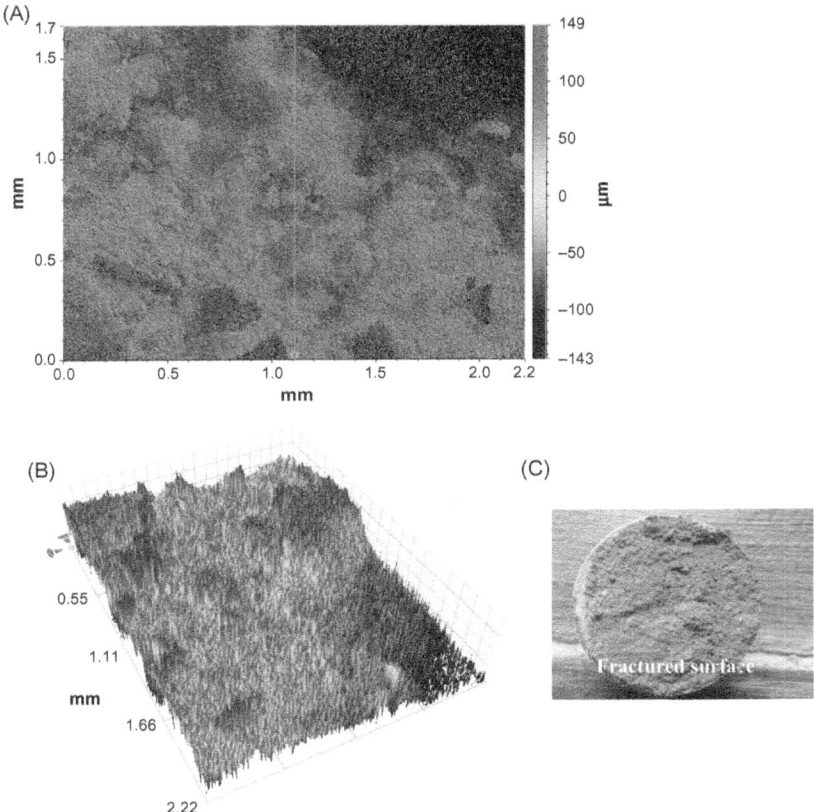

FIGURE 4.14   Optical image of the fractured pellet surface after tensile tests: (A) 2D image showing the texture height distribution, (B) 3D view of the texture of a fractured surface, and (C) optical image of the fractured surface (pellet diameter is 40 mm) [20].

corrected by subtracting the frictional work, which was obtained by integrating the frictional force over the scratch distance for the as-received glass surface. The adhesion work determined was in the order of 0.119 mJ. The experiments for the tangential force variation were repeated five times to assess any experimental error. The experimental error was in the order of 17% based on the repeatability tests. Fig. 4.14 shows the optical images of the dry mud pellet after the tensile tests. Fig. 4.14A is a zoomed-in 2D image of the fractured surface, Fig. 4.14B is the 3D image of the fractured surface, and Fig. 4.14C presents the fractured pellet.

Fig. 4.15 shows the pulling force with the displacement obtained from the tensile tests. In Fig. 4.15, the pulling force increases to reach a maximum without going through a yielding point, as opposed to the force observed for dense materials, such as metals. In addition, no necking was observed in the

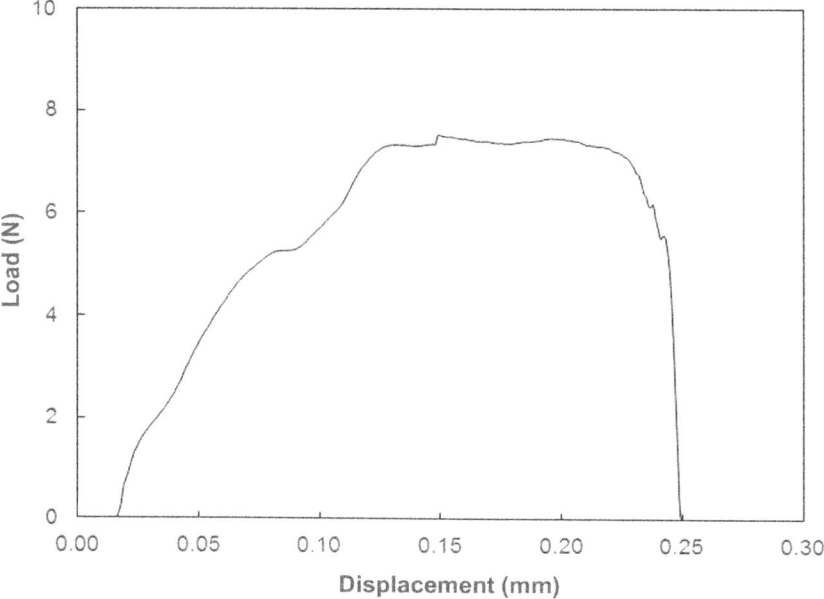

FIGURE 4.15 Load—displacement curve for the tensile test of a dry mud pellet [20].

tensile curve (Fig. 4.15). Necking is a mode of tensile deformation in which relatively large amounts of strain localize disproportionately in a small region of the material; in this case; the resulting decrease in the local cross-sectional area provides a form of plastic deformation where the necking occurs. The fractured facets are examined in detail to determine the closely spaced mud structure, where the dry mud solution covers the dust particle surfaces. Because mud liquid, when dried, causes volume shrinkage due to the evaporation of water, high strains are formed in the region where the mud solution dried. Consequently, a rather closely packed mixture of a dry mud solution and fine size dust particles is observed in certain regions, which are marked as "region A" in Fig. 4.14A. This phenomenon is also seen in Fig. 4.14B. However, in some other regions (marked as "region B" in Fig. 4.14A and B), the dust particles are large and are not covered completely by the mud solution because of voids formed in between the large dust particles. In this case, the mud solution flows toward the glass surface due to gravity, and the amount of mud solution present around the large dust particles remains low. Consequently, facets appear to have a porous-like texture (Fig. 4.14A). Moreover, the coverage area of the porous-like fractured surface over the total area of the fractured surface is estimated from the 3D images of the fractured surface (as shown in Fig. 4.14B). The porous-like structures appear as blue lines with a negative textured height (Fig. 4.14B), and the closely spaced dense layer appears in a brownish/

reddish color with a positive texture height (Fig. 4.14B). The area ratio of the porous-like face to the total area of the face after the tensile test is approximately 78%. This indicates that a closely spaced and dry-mud-solution-covered face surface is typically 22% of the total surface area of the dry mud pellet. The binding force in the closely spaced and mud-solution-covered region is mainly a combination of cohesive and adhesive forces. However, the porous-like textured region of the fractured surface is assumed to be dominated by the adhesion force. Based on this consideration, the ratio of the adhesion force to the combination of the adhesion and cohesion forces is in the order of 3.55, which indicates that the binding force in the dry mud pellet is mainly dominated by the adhesion force. The total work performed during a tensile test is determined by integrating the pulling curve along the displacement, which corresponds to the area under the curve in Fig. 4.15. Conversely, the tangential force is measured using a micro/nano tribometer; hence, the cohesion force can be determined from the tensile data shown in Fig. 4.15. In this context, the contributions of the cohesion and adhesion forces to the binding work, which is required to separate the dry mud pellet, are determined using the data in Table 4.5. In this case, the total work due to adhesion and cohesion is in the order of 1.408 mJ, which corresponds to the area under the curve in Fig. 4.15. Because the adhesion work covers 78% of the total area and the adhesion work estimated from the tangential force is in the order of 0.119 mJ, the cohesion work due to the dry mud solution around the small dust particles becomes 1.289 mJ. Although the percentage of the cohesion force based on the coverage area obtained from Fig. 4.14 is small, its effect on the overall binding work is significant, i.e., the ratio of cohesion work, which obtained from tangential force, over the total cohesion and adhesion work obtained from tensile tests is in the order of 90%. Therefore, the formation of a film from the dry mud solution at the interface of the dry mud and the solid surface plays a major role in the removal of the dry mud from surfaces.

4.3.3   UV—Visible Transmittance of Dusty and Dust-Removed Surfaces

In order to assess the optical transmittance of the surface, several tests were conducted. The findings are in line with the previous study [23,24]. In this case, dust removal from an inclined hydrophobic surface was analyzed. Firstly, solution crystallization of a polycarbonate surface was carried out to create a hierarchical, textured surface composed of micro/nanosize spheroids and fibrils. Synthesized silica nanoparticles were then deposited onto the textured polycarbonate surface to create hydrophobic characteristics, reduce the contact angle hysteresis, and generate the lotus effect on the surface. The optical performance of the prepared surfaces were assessed through UV—visible transmittance tests.

Fig. 4.16 shows the optical transmittance of the hydrophobic, dusty, and self-cleaned surfaces by the water droplet. In addition, a graph of the relative ratio of the optical transmittance of the surfaces before and after dust removal from the hydrophobic surface by the water droplet is inserted in Fig. 4.16. The ratio is determined from the difference between the optical transmittance of the water droplet-cleaned hydrophobic surface and transmittance of the dusty hydrophobic surface over the difference between transmittance of the hydrophobic surface and transmittance of the dusty hydrophobic surface ($= \frac{(T_{Cleaned} - T_{Dusty})}{(T_{Hydrophobic} - T_{Dusty})}$, where $T_{Cleaned}$ is the transmittance after dust removed by the water droplets, T_{Dusty} is the transmittance of the dusty surface, and $T_{Hydrophobic}$ is the transmittance of the hydrophobic surface). Dust accumulation on the surface reduces the optical transmittance of the hydrophobic surface significantly. A water droplet-cleaned surface improves the optical transmittance considerably. The transmittance improvement ratio (inset graph) shows a transmittance improvement almost over 60% on average over the incorporated wavelengths of the optical radiation. Consequently, the transmittance difference between the dusty and hydrophobic surfaces significantly increases when cleaned by water droplets. This improvement in transmittance demonstrates the feasibility of the self-cleaning of surfaces by water droplets, which can provide a possible cleaning process for solar energy harvesting applications.

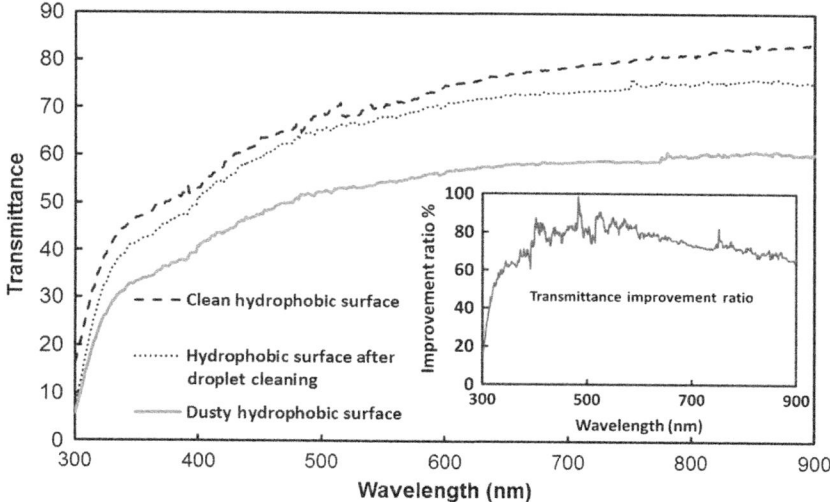

FIGURE 4.16 UV−visible transmittance of the hydrophobic surface before and after dust removal by water droplets. The inset figure depicts the improvement ratio of the transmittance. The ratio was determined from the difference between the optical transmittance before and after dust removal by water droplets over the difference between the optical transmittance of the hydrophobic surface before dust deposition and after removal of dust from the surface by water droplets [24].

4.4   DUST ON OIL-IMPREGNATED SURFACES

Crystallized polycarbonate and functionalized silica particle-deposited surfaces suffer from low optical transmittance. In this case, a refractive index correction liquid, such as silicon oil, is introduced to improve the resulting surface. However, the optical transmittance of functionalized silica particles deposited glass remains low as compared to as-received glass. Consequently, here the findings of oil-impregnated surfaces are discussed in line with the previous study [23] herein. This situation can be seen from Fig. 4.17, in which UV–visible transmittance of the substrate materials is shown. The low optical transmittance of functionalized silica-deposited glass surface is associated with scattering of incident optical radiation at the treated surface. This is related to the locally agglomerated functionalized silica particles and porous-like structures formed at the surface. However, incorporating corrected fluid (silicone oil) with a similar refractive index of glass improves optical transmittance greatly. This is true for thick and airjet-thinned silicone oil films. Although water droplet contact angle reduces after oil impregnation at the surface, optical transmittance improves considerably. It is reported that the particle mobility remained high on the silicone oil impregnated surface; consequently, reduced water droplet contact angle does not have a notable adverse effect on the self-cleaning ability of silicone oil-impregnated surface. In order to assess the optical transmittance of the silicone oil-impregnated surface in an open environment, oil-impregnated surfaces were

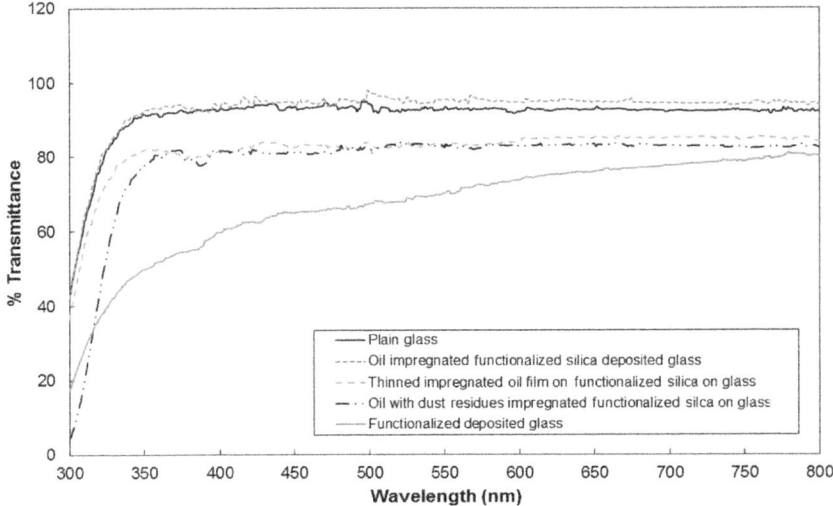

FIGURE 4.17   UV–visible transmittance of functionalized silica particle-deposited surface prior and after silicon oil-film deposition. Dissolved dust residues in silicon oil-impregnated functionalized silica particle-deposited surface is also shown for comparison [23].

left in an outdoor environment for 1 week; in which case, the environmental dust fell onto the surface of silicone oil-impregnated samples. The percentage of change of the average optical transmittance due to functionalized silica particle-deposited glass surface and silicone oil-impregnated and treated surface after the outdoor tests are shown in Fig. 4.18. The average optical transmittance was obtained after integrating transmittance over the wavelengths and dividing it by the average wavelength. The percentage of transmittance was determined from the ratio of optical transmittance difference prior and after the outdoor tests over optical transmittance prior to start of the tests ($= (T_{AT} - T_o)/T_o$, where T_o is the average transmittance prior to the outdoor tests and T_{AT} is the average transmittance after the outdoor tests). The optical transmittance of the glass reduces considerably after the tests. This is associated with the sedimentation of dust particles on the impregnated glass surface. In order to assess the behavior of the immersing and sedimentation of dust particles into the silicone oil and on the glass surface, tests for oil cloaking of dust particles were carried out.

Fig. 4.19 shows images of two different dust particles subjected to oil cloaking, together with oil-cloaking velocity and cloaking-film height. Slightly large size dust particles are selected to monitor oil cloaking with high accuracy. The error estimated for cloaking height and velocity measurement is in the order of 4%, which is based on the repeatability of experiments. Cloaking

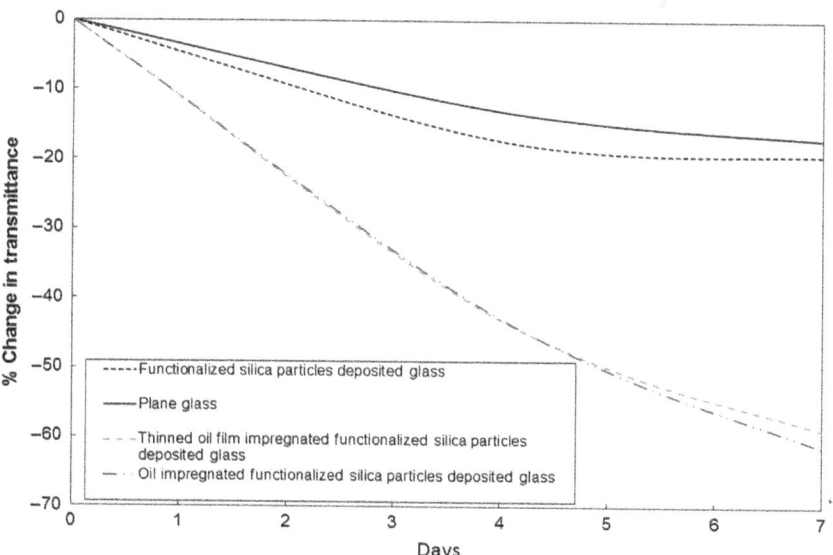

FIGURE 4.18 Percentage of change of UV−visible transmittance of functionalized silica-deposited surface after outdoor tests. (Percentage of transmittance change $= (T_{AT} - T_o)/T_o$, where T_o is the average transmittance prior to the outdoor tests and T_{AT} is the average transmittance after the outdoor tests [23].)

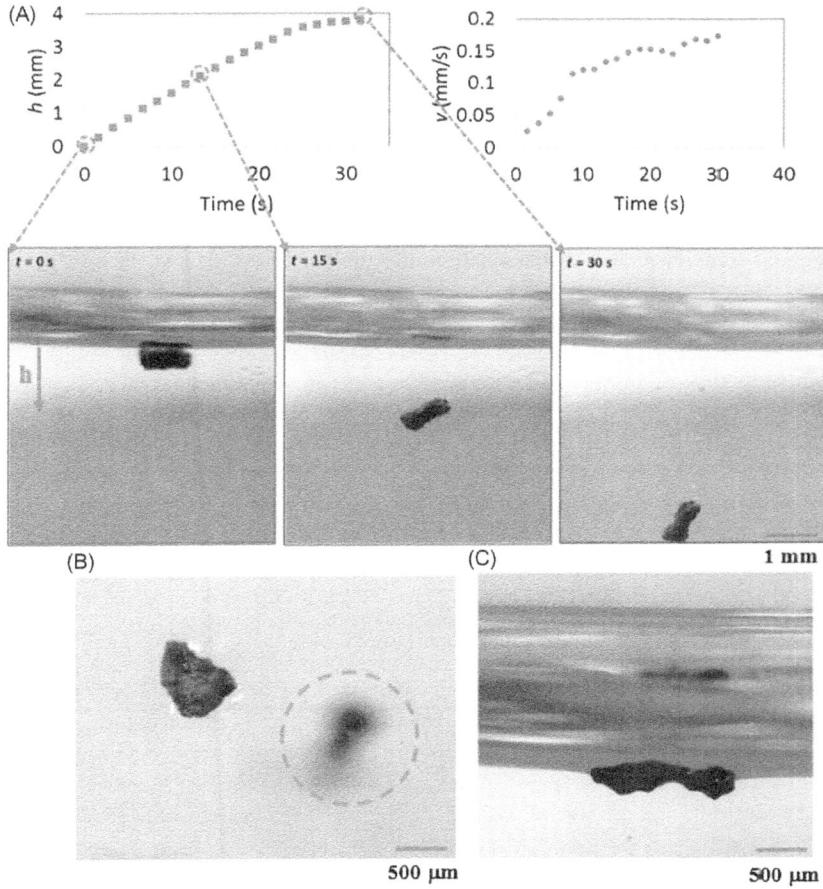

FIGURE 4.19 Immersion of dust particles into silicone oil due to gravitational force after oil cloaking surface: (A) height and speed of immersion and location of dust particle in silicone oil, and (B) top view of functionalized dust particle (floating on oil surface) and immersed dust particle (in blue circle), (C) side view of factionalized dust particle (floating on oil surface) [23].

velocity and height are obtained from high-speed camera records. Cloaking-film height and cloaking speed increase rapidly and the rate of increase reduces as the time progresses. Since oil-film cloaking occurs opposing to the gravity, weight of the liquid film, which is cloaking around dust particle, cloaking height increases while lowering the net driving force opposing to gravity for cloaking. For further investigation of oil cloaking, some dust particles were functionalized to reduce surface energy to overcome the cloaking. In this case, dust particles were functionalized via covering the surface of dust particles by Trichloro(1*H*,1*H*,2*H*,2*H*-perfluorooctyl) (PFOTS) via chemical vapor deposition. In general, silicone oil cloaking of the surface of dust particles resulted, and the height of oil cloaking over the dust surface increased by time. As the time progressed, cloaking covered the dust particles completely, and caused immersion of dust particles into silicone oil (Fig. 4.19A). The speed of immersion of dust particles into silicone oil increased with time under the influence of gravitational force (Fig. 4.19A). However, silicone oil cloaking does not take place for the functionalized dust particles and these particles float on oil surface (Fig. 4.19A). This is because oil cloaking on the surface of dust particles is attributed to the spreading rate of oil film at the air—dust particle interface. The spreading coefficient ($S_{op(a)} = \gamma_{pa} - \gamma_{po} - \gamma_{oa}$, where γ_{pa} is the interfacial energy at dust particle—air interface, γ_{po} is the interfacial energy at the dust particle—oil interface, and γ_{oa} is the interfacial energy at the oil—air interface) should remain greater than zero for possible cloaking of the outer surface of dust particle by silicone oil. Although interfacial energy between the dust particles and air, and the dust particles and silicone oil are unknown, the condition $\gamma_{pa} > (\gamma_{po} + \gamma_{oa})$ should be satisfied for oil cloaking. While $\gamma_{oa} = 21.2 \mathrm{mN/m}$, in any case, γ_{pa} should be greater than 21.2 mN/m.

On the other hand, vertical force balance for dust particles partially immersed in the silicone oil yield: $F_v = F_\gamma sin\theta + F_B - W_p$, where F_v is the net resulting force, F_γ is the surface tension force, θ is the dust contact angle on the oil film surface, F_B is the buoyancy force, and W_P is the weight of the dust particles. The net resulting force remains zero for the partially immersed and steadily floating dust particles on the oil film. The net resulting force becomes negative for continuously falling immersed dust particles in silicone oil film. However, once the oil cloaking initiates around dust particles due to spreading rate, the circumference of the air—oil interface around dust particle reduces, which in turn lowers surface tension force. However, the buoyancy force increases because of dust immersion into silicone oil. Consequently, the amount of dust particles floats on the oil surface (Fig. 4.19A). This is because oil cloaking on the surface of dust particles is attributed to the spreading rate of oil film at the air—dust particle interface.

Fig. 4.20 shows the surface tension force (F_γ) and difference between buoyancy and weight force with the height of immersed spherical dust particle in silicone oil. Surface tension force dominates over other forces when oil cloaking covers a small region of dust particle. As oil cloaking covers the surface of dust particle, the size of the rim around the dust particle at the interface between dust particle and oil film reduces. This in turn lowers surface tension force and causes dust particle to further immerse into silicone oil under gravitational force. The difference between buoyancy force and weight of dust particle remains negative as shown in Fig. 4.20, while demonstrating that buoyancy force remains less than the weight of dust particles. Consequently, environmental dust particles gradually immerse into silicone oil and settle on the treated glass surface.

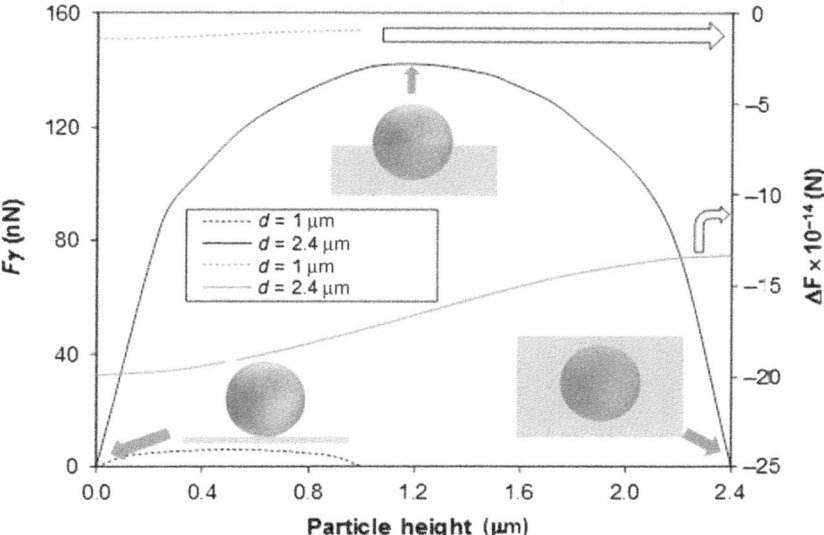

FIGURE 4.20   Surface tension force (F_γ) and the difference between weight and the buoyancy forces ($\Delta F = W - F_B$, where W is the weight of spherical particle and F_B is the buoyancy force) with the height of immersed spherical dust particles. For particle diameter of 2.4 μm, $h = 0$ represents dust particles on silicon oil surface, $h = 1.2$ μm corresponds to the halfway immersed particle, and $h = 2.4$ μm corresponds to totally immersed particles [23].

Fig. 4.21 shows the optical images of dust particles settled on glass and oil-impregnated glass surface after exposing oil surface in an outdoor environment for 1 week. The coverage area of dust particles over the total area of the surface exposed is in the order of 20% for thick (700 nm), and airjet-thinned silicone oil film (56.2 nm); however, dust particles coverage area reduces to 18% for as-received glass surface. Since all samples were located close to each other in the outdoor environment and starting and ending durations of the outdoor tests were set the same for all samples, the reduction in the coverage area of dust particles could attribute to the wind effect. In this case, some of the dust particles that settled at as-received glass surface could be removed by the local wind. However, dust particles immersed in thin and thick silicone oil film remained at the functionalized glass surface under the local wind conditions. The area covered by dust particles lowers optical transmittance of silicone oil impregnated glass (Fig. 4.18). The removal of sediment dust particles from functionalized silica particle-deposited glass surface is almost impossible unless impregnated silicone oil is removed from the glass surface. Since compounds of alkaline and earth alkaline earth metals in dust particles can dissolve in a fluid, these dust compounds in silicone oil can be dissolved. This situation was examined using inductively coupled plasma atomic emission spectroscopy. The findings are given in Table 4.6. The concentration of Fe,

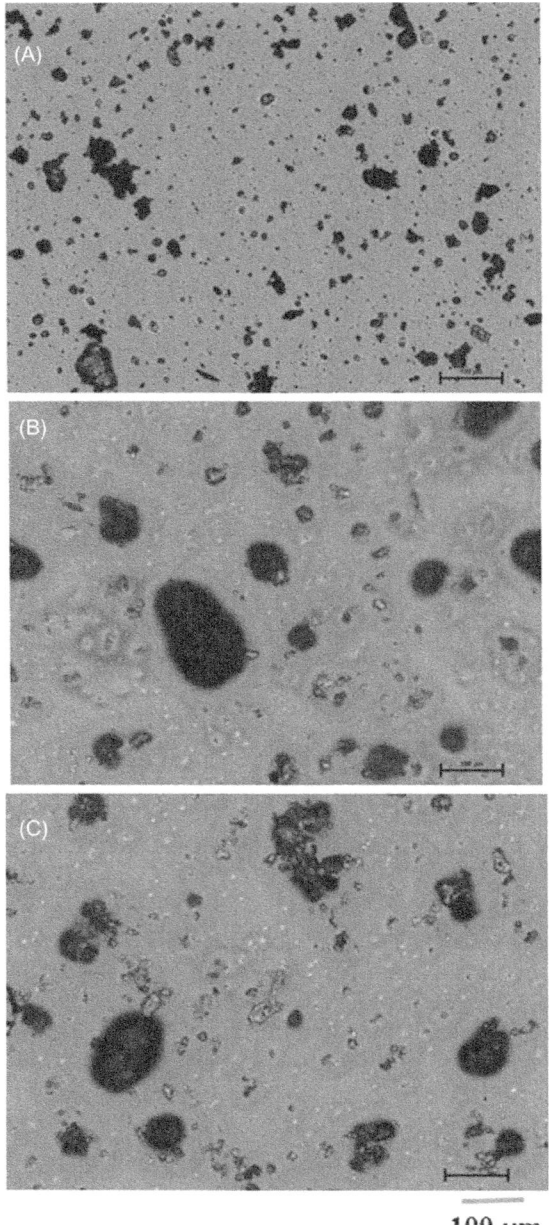

100 μm

FIGURE 4.21   Optical images of dust particles settled on glass and oil impregnation of functionalized silica particle-deposited glass surface: (A) glass surface, (B) thick layer (700 nm) of oil-impregnated functionalized silica particle-deposited glass surface, and (C) thin layer (56.2 nm) of oil-impregnated functionalized silica particle-deposited glass surface [23].

TABLE 4.6 ICP Data for Solution Extracted From Silicone Oil and Dust Mixtures for 1 Week

	Al (ppb)	Ca (ppb)	Fe (ppb)	K (ppb)	Mg (ppb)	Na (ppb)
Silicone oil	1782	2191	243	2159	5610	6730
Dissolved dust Compounds in oil	4687	4587	55860	61450	61050	49730

Na, Mg, K, and Cl were observed in silicone oil when dust particles are settled on the glass surface. This indicates dissolution of some dust compounds in silicone oil during the duration of dust settlement in silicone oil. To assess the effect of dissolved compounds on optical characteristics, UV—visible transmittance of silicone oil was measured after the removal of dust particles from silicone oil. Consequently, the presence of dissolved dust compounds in the silicone oil reduced optical transmittance of the oil, and this reduction was in the order of 12% (Fig. 4.18).

REFERENCES

[1] B.A. Maher, P. Dennis, Evidence against dust-mediated control of glacial—interglacial changes in atmospheric CO_2, Nature 411 (6834) (2001) 176—180.

[2] K.A. Farley, D. Vokrouhlický, W.F. Bottke, D. Nesvorný, A late Miocene dust shower from the break-up of an asteroid in the main belt, Nature 439 (7074) (2006) 295—297.

[3] I. Song, B. Zuckerman, A.J. Weinberger, E. Becklin, Extreme collisions between planetesimals as the origin of warm dust around a Sun-like star, Nature 436 (7049) (2005) 363—365.

[4] K. Tomeoka, K. Kiriyama, K. Nakamura, Y. Yamahana, T. Sekine, Interplanetary dust from the explosive dispersal of hydrated asteroids by impacts, Nature 423 (6935) (2003) 60—62.

[5] J. Wang, Y. Li, X. Liang, Y. Liu, Research of adhesion force between dust particles and insulator surface using atomic force microscope, High Voltage Eng. 39 (6) (2013) 1352—1359.

[6] S.L. Mkoma, W. Maenhaut, X. Chi, W. Wang, N. Raes, Characterisation of PM10 atmospheric aerosols for the wet season 2005 at two sites in East Africa, Atmos. Environ. 43 (3) (2009) 631—639.

[7] W. Maenhaut, N. Raes, W. Wang, Analysis of atmospheric aerosols by particle-induced X-ray emission, instrumental neutron activation analysis, and ion chromatography, Nucl. Instr. Methods Phys. Res. Sect. B: Beam Interact. Mater. Atoms 269 (22) (2011) 2693—2698.

[8] A. Baidourela, K. Zhayimu, Patterns of dust retention by urban trees in Oasis Cities, Nat. Environ. Pollut. Technol. 14 (1) (2015) 53—57.

[9] S.K. Mishra, R. Agnihotri, P.K. Yadav, S. Singh, M. Prasad, P.S. Praveen, et al., Morphology of atmospheric particles over Semi-Arid region (Jaipur, Rajasthan) of India: implications for optical properties, Aerosol Air Qual. Res. 15 (3) (2015) 974—984.

[10] W. Petruk, H. Skinner, Characterizing particles in airborne dust by image analysis, JOM 49 (4) (1997) 58−61.

[11] N.S. Beattie, R.S. Moir, C. Chacko, G. Buffoni, S.H. Roberts, N.M. Pearsall, Understanding the effects of sand and dust accumulation on photovoltaic modules, Renew. Energy 48 (2012) 448−452.

[12] C.L.C. Tan, S. Gao, B.S. Wee, A. Asa-Awuku, B.J.R. Thio, Adhesion of dust particles to common indoor surfaces in an air-conditioned environment, Aerosol. Sci. Technol. 48 (5) (2014) 541−551.

[13] N. Middleton, Effect of drought on dust production in the Sahel, Nature 316 (6027) (1985) 431−434.

[14] I. Uno, K. Eguchi, K. Yumimoto, T. Takemura, A. Shimizu, M. Uematsu, et al., Asian dust transported one full circuit around the globe, Nat. Geosci. 2 (8) (2009) 557−560.

[15] J. Neff, A. Ballantyne, G. Farmer, N. Mahowald, J. Conroy, C. Landry, et al., Increasing eolian dust deposition in the western United States linked to human activity, Nat. Geosci. 1 (3) (2008) 189−195.

[16] S.-J. Park, D.-I. Seo, C. Nah, Effect of acidic surface treatment of red mud on mechanical interfacial properties of epoxy/red mud nanocomposites, J. Colloid Interface Sci. 251 (1) (2002) 225−229.

[17] D. Rossi, P.G. Jobstraibizer, C.D. Bosco, A. Bettero, A combined chemico-mineralogical and tensiometric approach for evaluation of Euganean Thermal Mud (ETM) quality, J. Adhesion Sci. Technol. 27 (1) (2013) 30−45.

[18] H. Zhang, X.-G. Li, C.-W. Du, H.-B. Qi, Corrosion behavior and mechanism of the automotive hot-dip galvanized steel with alkaline mud adhesion, Int. J. Miner. Metall. Mater. 16 (4) (2009) 414−421.

[19] L. Goehring, R. Conroy, A. Akhter, W.J. Clegg, A.F. Routh, Evolution of mud-crack patterns during repeated drying cycles, Soft Matter 6 (15) (2010) 3562−3567.

[20] G. Hassan, B. Yilbas, S.A. Said, N. Al-Aqeeli, A. Matin, Chemo-mechanical characteristics of mud formed from environmental dust particles in humid ambient air, Sci. Rep. 6 (2016) 30253.

[21] B.S. Yilbas, H. Ali, N. Al-Aqeeli, M.M. Khaled, S. Said, N. Abu-Dheir, et al., Characterization of environmental dust in the Dammam area and mud after-effects on bisphenol-A polycarbonate sheets, Sci. Rep. 6 (2016) 24308.

[22] B.S. Yilbas, H. Ali, M.M. Khaled, N. Al-Aqeeli, N. Abu-Dheir, K.K. Varanasi, Influence of dust and mud on the optical, chemical, and mechanical properties of a PV protective glass, Sci. Rep. 5 (2015) 15833.

[23] B.S. Yilbas, M.R. Yousaf, A. Al-Sharafi, H. Ali, F. Al-Sulaiman, N. Abu-Dheir, et al., Silicone oil impregnated nano silica modified glass surface and influence of environmental dust particles on optical transmittance, RSC Adv. 7 (47) (2017) 29762−29771.

[24] B.S. Yilbas, G. Hassan, A. Al-Sharafi, H. Ali, N. Al-Aqeeli, A. Al-Sarkhi, Water droplet dynamics on a hydrophobic surface in relation to the self-cleaning of environmental dust, Sci. Rep. 8 (1) (2018) 2984.

[25] A. Rifai, N. Abu-Dheir, M. Khaled, N. Al-Aqeeli, B.S. Yilbas, Characteristics of oil impregnated hydrophobic glass surfaces in relation to self-cleaning of environmental dust particles, Solar Energy Mater. Solar Cells 171 (2017) 8−15.

[26] J. Stoch, M. Ladecka, An XPS study of the KCl surface oxidation in oxygen glow discharge, Appl. Surf. Sci. 31 (4) (1988) 426−436.

[27] N. Tsubouchi, H. Hashimoto, N. Ohtaka, Y. Ohtsuka, Chemical characterization of dust particles recovered from bag filters of electric arc furnaces for steelmaking: some factors

influencing the formation of hexachlorobenzene, J. Hazard. Mater. 183 (1−3) (2010) 116−124.

[28] S. Varagnolo, D. Ferraro, P. Fantinel, M. Pierno, G. Mistura, G. Amati, et al., Stick-slip sliding of water drops on chemically heterogeneous surfaces, Phys. Rev. Lett. 111 (6) (2013) 066101.

[29] A.T. Santhanam, Y. Gupta, Cleavage surface energy of calcite, International Journal of Rock Mechanics and Mining Sciences & Geomechanics Abstracts, Elsevier, 1968, pp. 253−259.

[30] Y.K. Shchipalov, Surface energy of crystalline and vitreous silica, Glass Ceram. 57 (11−12) (2000) 374−377.

[31] V. Bergeron, D. Langevin, Monolayer spreading of polydimethylsiloxane oil on surfactant solutions, Phys. Rev. Lett. 76 (17) (1996) 3152.

[32] S. Anand, K. Rykaczewski, S.B. Subramanyam, D. Beysens, K.K. Varanasi, How droplets nucleate and grow on liquids and liquid impregnated surfaces, Soft Matter 11 (1) (2015) 69−80.

[33] D. Ingber, D.C. Leslie, M. Super, A.L. Watters, A. Waterhouse, Modification of surfaces for fluid and solid repellency, in, Google Patents, 2016.

[34] A. Laukkanen, J.-E. Teirfolk, O. Ikkala, R. Ras, H. Mertaniemi, Hydrophobic coating and a method for producing hydrophobic surface, in, Google Patents, 2014.

[35] S. Mekhilef, R. Saidur, M. Kamalisarvestani, Effect of dust, humidity and air velocity on efficiency of photovoltaic cells, Renew. Sustain. Energy Rev. 16 (5) (2012) 2920−2925.

[36] M.J. Adinoyi, S.A. Said, Effect of dust accumulation on the power outputs of solar photo-voltaic modules, Renew. Energy 60 (2013) 633−636.

[37] K. Brown, T. Narum, N. Jing, Soiling test methods and their use in predicting perfor-mance of photovoltaic modules in soiling environments, in: Photovoltaic Specialists Conference (PVSC), 2012 38th IEEE, IEEE, 2012, pp. 001881−001885.

[38] L. Jing, Z. Zhi-Jun, Y. Ji-Lin, B. Yi-Long, A thin liquid film and its effects in an atomic force microscopy measurement, Chin. Phys. Lett. 26 (8) (2009) 086802.

[39] M.D. Kempe, Modeling of rates of moisture ingress into photovoltaic modules, Solar Energy Mater. Solar Cells 90 (16) (2006) 2720−2738.

[40] M. Corn, The adhesion of solid particles to solid surfaces, I. A review, J. Air Pollut. Control Assoc. 11 (11) (1961) 523−528.

[41] K.J. Mclean, Cohesion of precipitated dust layer in electrostatic precipitators, J. Air Pollut. Control Assoc. 27 (11) (1977) 1100−1103.

[42] F. Podczeck, J.M. Newton, M.B. James, The influence of constant and changing relative humidity of the air on the autoadhesion force between pharmaceutical powder particles, Int. J. Pharm. 145 (1−2) (1996) 221−229.

[43] P. Somasundaran, H. Lee, E. Shchukin, J. Wang, Cohesive force apparatus for interactions between particles in surfactant and polymer solutions, Colloids Surf. A: Physicochem. Eng. Aspects 266 (1−3) (2005) 32−37.

[44] A. Fukunishi, Y. Mori, Adhesion force between particles and substrate in a humid atmo-sphere studied by atomic force microscopy, Adv. Powder Technol. 17 (5) (2006) 567−580.

[45] A. Kumar, T. Staedler, X. Jiang, Role of relative size of asperities and adhering particles on the adhesion force, J. Colloid. Interface. Sci. 409 (2013) 211−218.

[46] D.M. Jarząbek, M. Chmielewski, T. Wojciechowski, The measurement of the adhesion force between ceramic particles and metal matrix in ceramic reinforced-metal matrix com-posites, Compos. Part A: Appl. Sci. Manuf. 76 (2015) 124−130.

[47] J. Knoll, S. Knott, H. Nirschl, Characterization of the adhesion force between magnetic microscale particles and the influence of surface-bound protein, Powder Technol. 283 (2015) 163–170.

[48] P. Petean, M. Aguiar, Determining the adhesion force between particles and rough surfaces, Powder Technol. 274 (2015) 67–76.

[49] A. Bejan, S. Lorente, B. Yilbas, A. Sahin, Why solidification has an S-shaped history, Sci. Rep. 3 (2013) 1711.

[50] M. Stoudt, J. Hubbard, S. Mates, D. Green, Evaluating the relationships between surface roughness and friction behavior during metal forming, 0148-7191, SAE Technical Paper, 2005.

Chapter 5

Water-Droplet Dynamics and Heat Transfer

Chapter Outline

5.1 INTRODUCTION

One of the methods for self-cleaning surfaces is to use water droplets on hydrophobic surfaces. In general, water droplets roll and/or slide on the hydrophobic surfaces when contact angle hysteresis is low. This in turn enables the water droplet to pick up the dust particles from the surface. However, the size, shape, and surface energy of the dust particles may influence the removal process. In this case, cloaking of the dust particles by the

Self-Cleaning of Surfaces and Water Droplet Mobility. DOI: https://doi.org/10.1016/B978-0-12-814776-4.00005-7
133

droplet fluid is critical for the dust removal from the hydrophobic surface via droplet rolling. When the droplet rolls/slides over the dust particles, the transition time for the particle cloaking is also important for the complete removal of the dust particles. On the other hand, the water-droplet mobility on surfaces plays a vital role in an effective self-cleaning process. Droplet motion consists of rolling and sliding and both are needed to remove dust particles from surfaces. In general, such dynamic motion of droplet depends on the characteristics of the solid surface and surface tension of the droplet liquid. Mimicking the lotus effect that occurs in nature is needed to lower the adhesion between the water droplet and the solid surface. This means that surface hydrophobicity and pinning of droplet—created by large contact angle hysteresis—is important. The reversible exchange of the wetting state of surfaces has applications in advanced multifunctional systems and is important for extraction, separation, surface chemistry, life science, and organic solvents. One of the methods for achieving a reversible exchange of wetting state is applying a thin coating of phase-change material on a hydrophobic surface. The surface remains hydrophobic in the solid phase of the phase-change material and becomes hydrophilic in the liquid phase of the phase-change material. The wettability in terms of spreading rate (S) of the phase-change material in a liquid phase plays a crucial role in the formation of continuous film on the hydrophobic surface. Since the surface remains hydrophilic in the liquid phase of the phase-change material, the interfacial and surface free energy of the liquid phase and the solid surface should satisfy the condition for the spreading rate $S > 0$. Therefore, the selection of phase material is important to give rise to a positive spreading rate on the hydrophobic surface in the liquid phase. One of the candidates of the phase-change material fulfilling the requirements of positive spreading rate is n-octadecane, which has low solidus and liquidus temperatures and positive spreading rate for most hydrophobic surfaces in the liquid phase. In this chapter, the droplet attachment and mobility, thermal effects on droplet internal fluidity, and reversible exchange of surface wetting states and droplet mobility are presented in light of previous studies [1−8].

5.2 DROPLET ADHESION ON HYDROPHOBIC SURFACE AND THERMAL EFFECTS

Water-droplet behavior on hydrophobic surfaces is of current research interest. Water-droplet adhesion on surfaces using hydrophobic and hydrophilic nanoparticles was studied by Ebert and Bhushan [9]. They indicated that the microstructure pitch value and nanostructure density were important in achieving the adhesion of the droplets on surfaces possessing rose petal textures. The droplet behavior on hydrophilic/hydrophobic surfaces was investigated by Kwon and Myong [10]. They showed that the strong nonequilibrium state inside the droplet, due to the synergistic effect of cohesion and adhesion forces of two surfaces, was the key mechanism for droplet movement. The droplet behavior on a combination of hydrophobic and hydrophilic surfaces were studied by Yang et al. [11]. They demonstrated that hydrophilic microsize dots demonstrated great adhesion toward water droplets without changing the contact angles, whereas the prewetted line grooves exhibited strong anisotropic water adhesion. The resistance force that restricted the droplet from detaching in directions parallel and perpendicular to the grooves was significantly different due to the different stress state of the droplet on the grooves. The liquid-droplet adhesion on slit and plain fins was examined by Shi et al. [12]. The findings revealed that the height of the surface conical posts, the spacing between the consecutive conical posts, the number of the droplets, the radius of the droplet, and the wettability property of the complex textured surface had an important effect on the spontaneous jumping of the coalesced droplet. For the liquid condensate adhesion on slit and plain fins, both the hydrophilic coating with small contact angle and the hydrophobic coating with large contact angle were effective at avoiding liquid bridge between fins. Sessile droplet freezing and ice adhesion on aluminum surface with different wettability characteristics were studied by Ou et al. [13]. They showed that as surface temperature decreased, some interfacial air gaps were squeezed out and such freezing retarding and adhesion lowering effect for superhydrophobic surfaces was reduced greatly. At low surface temperatures, ice adhesion on superhydrophobic surfaces was even greater than that on hydrophobic surfaces. In this case, the stability of interfacial air pockets played a major role at low surface temperatures. Wetting, adhesion, and friction of superhydrophobic and hydrophilic leaves and fabricated micro/nanopatterned surfaces were studied by Bhushan and Jung [14]. They presented the variation of the contact angles with the texture pitch and the importance of air pockets on droplet evaporation. A study of the hydrophilic/hydrophobic surface pattern design for oil-repellent function was carried out by Moronuki et al. [15]. They showed that a star-like hydrophobic pattern propelled oil droplet better than dot or rectangular patterns with the help of a directional force component. Reversible switching of surface wettability and water adhesion on a polymer nanocomposite coating was studied by Zhang et al. [16]. They showed that when a superhydrophobic coating was directly illuminated by UV light, the surface was fully covered with the hydrophilic groups, which created a state of superhydrophilic wettability on the surface. However, when UV light illuminated the surface through a mask, the surface maintained its superhydrophobic property, but the adhesion of the water droplet changed from sliding easily to a highly sticky movement. Analysis of single-droplet dynamics on striped surface domains using the lattice Boltzmann method was carried out by Chang and Alexander [17]. They validated the model study by demonstrating the consistency of the simulation results with an exact solution of capillary rise of liquid, and through qualitative comparison of computed dynamic contact line behavior with experimentally measured surface properties. They reported that the decrease in contact angle of the liquid droplets on a hydrophilic surface might lead to break up of the droplets for certain patterns.

On the other hand, the liquid-droplet adhesion on surfaces finds application in various engineering and medical fields [18–21]. A study of the adhesion of moving liquid droplets in microchannels was carried out by Chen et al. [18]. They demonstrated that there existed a critical capillary number for the droplet adhesion; in which case, the droplet adhesion could be utilized as a promising technology for recycling of emulsions in droplet microfluidics. The droplet mobility for the biodetection via surface reaction was investigated by Liu et al. [19]. They indicated that the droplet sorting could be realized by smart stimuli-responsive surfaces due to the different pHs of the droplets. Based on a similar mechanism, protein detection between antibodies and antigens could be achieved by immobilizing biotin ligands on a hydrophobic surfaces with or without streptavidin. Consequently, these surfaces possess potential applications in microfluidics, microreactors, and biodetection. The importance of hierarchical micro- and nanostructured, hydrophilic, superhydrophobic, and superoleophobic surfaces in microelectronic systems was examined by Gogolides et al. [20]. They demonstrated that incorporating surface wetting characteristics in sensors, microfluidics, and labs on chip offers new functional devices and systems. Fabrication of hydrophilic poly(dimethylsiloxane) with periodic wrinkling surface and its applications were presented by Lee et al. [21]. They indicated that the proposed wrinkling surface was hydrophilic and the measured contact angle was about 62 degrees. Moreover, it was observed from the simple cell culture test that the fabricated wrinkling surface was more effective for cell spreading and adhesion than the native poly(dimethylsiloxane) substrate.

5.2.1 Mechanics of Droplet Attachment on Hydrophobic Surfaces

Droplet sliding and rolling at the surface are associated with the net adhesion force at the surface. Although inclination of the hydrophilic surface increases the lateral component of the gravitational force for the droplet slipping and/or rolling at the surface, droplet contact angle hysteresis is critically important for the dynamic movement of the droplet at the surface. However, unlike hydrophobic surfaces, the contact angle hysteresis and the wetted area of the droplet remain high for hydrophilic surfaces. In addition, the fluid acceleration due to the flow current generated in the droplet during the inclination of the surface contributes to both deformation of the droplet shape and adhesion force between the droplet and the inclined surface. The rolling and/or sliding of the droplet is not observed during the experiments for the droplet volumes considered $(5 \, \mu L \leq \forall \leq 25 \, \mu L)$ in the present study. The advancing and receding angles of droplet with different sizes and solid-surface inclination angles are given in Table 5.1.

TABLE 5.1 Advancing (θ_A), Receding (θ_R) and Hysteresis ($\theta_R-\theta_A$) of Water Droplets on Inclined Hydrophilic Surfaces for Various Droplet Sizes and Inclination Angles [1]

	$V_d = 5\ \mu L$				$V_d = 10\ \mu L$		
δ (Deg)	θ_a (Deg)	θ_R (Deg)	$(\theta_a-\theta_R)$ (Deg)	δ (Deg)	θ_a (Deg)	θ_R (Deg)	$(\theta_a-\theta_R)$ (Deg)
10	80.79	80.15	0.64	10	88.48	78.72	9.76
20	87.34	78.61	8.73	20	89.67	76.98	12.69
30	88.14	77.6	10.54	30	90.71	74.48	16.23
40	90	76.35	13.65	40	91.79	72.5	19.29
50	91.31	75.43	15.88	50	92.9	69.69	23.21
60	92.62	72.65	19.97	60	94.37	68.7	25.67
70	92.78	70.38	22.4	70	95.67	66.8	28.87
80	93.47	69.16	24.31	80	96.82	64.45	32.37
90	94.6	68.55	26.05	90	97.62	62.25	35.37

	$V_d = 15\ \mu L$				$V_d = 20\ \mu L$		
δ (Deg)	θ_a (Deg)	θ_R (Deg)	$(\theta_a-\theta_R)$ (Deg)	δ (Deg)	θ_a (Deg)	θ_R (Deg)	$(\theta_a-\theta_R)$ (Deg)
10	82.21	78.88	3.33	10	88.85	82.93	5.92
20	84.35	75.7	8.65	20	90.2	79.91	10.29
30	87.18	73.39	13.79	30	92	77.3	14.7
40	89.19	70.2	18.99	40	93.5	74.6	18.9
50	91.88	67.83	24.05	50	95.45	71.6	23.85
60	93.62	65.18	28.44	60	97	68.4	28.6
70	95.89	61.59	34.3	70	98.7	65.6	33.1
80	98.62	59.83	38.79	80	100.2	63	37.2
90	100.62	56.63	43.99	90	101.73	59.97	41.76

	$V_d = 25\ \mu L$		
δ (Deg)	θ_a (Deg)	θ_R (Deg)	$(\theta_a-\theta_R)$ (Deg)
10	88.85	77.42	11.43
20	90.81	73.7	17.11
30	93	70.3	22.7
40	94.62	66.71	27.91
50	96.6	63	33.6
60	98.5	60	38.5
70	100.5	56	44.5
80	102.4	52.5	49.9
90	104.27	48.92	55.35

The droplet contact angle hysteresis ($\theta_R-\theta_A$) increases with inclination angle and droplet size, which in turn alters the lateral adhesion force and the droplet bulging. The lateral adhesion force can be formulated heuristically by approximating the three-phase contact line with a single ellipse [22]. In addition, the experimentally obtained polynomial function can be incorporated [23] for the dependence of the contact angle on the position along the three-phase contact line to formulate the lateral adhesion force, which can be written as:

$$F_L = \frac{24}{\pi^3}\gamma_{LS}d(cos\theta_R - cos\theta_A) \tag{5.1}$$

where γ_{LS} is the surface tension of the liquid on the solid surface, d is the droplet diameter prior to deformation (same area of the ellipse), θ_R is the receding angle, and θ_A is the advancing angle. Since the solid surface is considered to be smooth, the roughness parameter is not incorporated in the adhesion force equation (Eq. 5.1), similar to the Young–Dupre equation [24]. Under the static condition of the droplet attachment at the surface (Fig. 5.1), the net lateral force becomes the same as the lateral force components resulting from the sum of the gravitational force, shear force at the wetted surface, and the adhesion force due to surface tension. In the case of inclination of the hydrophilic surface, the droplet undergoes the elastic deformation while the net force remains constant. In this case, the line of action of the net force changes inside the droplet volume. As the adhesion force remains higher than the gravitational and shear forces generated on the droplet due to the inclination of the surface, the

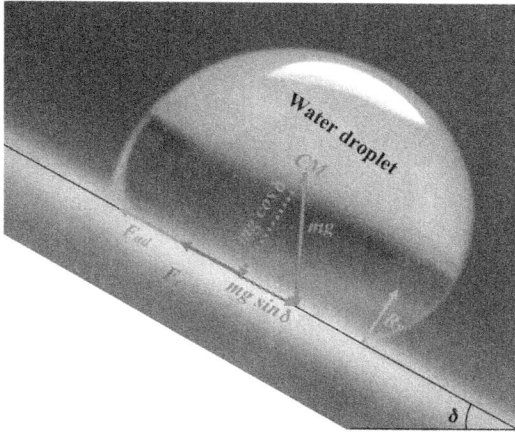

FIGURE 5.1 High-speed camera image of a droplet on inclined hydrophilic surface and force diagram [1].

droplet attaches to the surface. The force balance along the lateral direction at the inclined surface gives rise to:

$$F_{ad} + F_\tau + mg\,sin\delta \cong \frac{24}{\pi^3}\gamma_{LV}d(cos\theta_R - cos\theta_A) \qquad (5.2)$$

where F_{ad} is the adhesion force, F_τ ($F_\tau = A_w\mu(\partial V/\partial s)$, A_w is the wetted area, μ is the fluid viscosity, $\partial V/\partial s$ is the rate of fluid strain, and s is direction normal to the inclined surface) is the shear force acting at the wetted surface due to the flow developed in the droplet liquid during the inclination, m is the droplet mass, g is the gravity, and δ is the inclination angle of the surface (Fig. 5.1). Here, the force generated due to the local and convective acceleration of the fluid inside the droplet during the inclination is considered to be negligibly small as compared to the gravitational acceleration of the droplet.

Therefore, the force acting on the droplet because of gravity and fluid acceleration is considered to be the same as the droplet weight. The rearrangement of Eq. (5.2) yields:

$$F_{ad} \cong \frac{24}{\pi^3}\gamma_{LV}d(cos\theta_R - cos\theta_A) - F_\tau - mg\,sin\delta \qquad (5.3)$$

Eq. (5.3) is used to obtain the adhesion force acting on the droplet at the inclined hydrophilic surface. Since the droplet mass remains the same during the inclination and only changes with the droplet volume, a dimensionless force ratio is introduced to account for the droplet volume change. The dimensionless force ratio represents the ratio of the gravitational force over the surface tension force for a given droplet volume, i.e., $MN = 2\gamma R^2/3\sigma$, where MN is the Merve number. Fig. 5.2 shows the adhesion force obtained from Eq. (5.3) with the MN for various inclination angles. It should be noted that the advancing and receding angles of the droplet for a given inclination angle are taken from the experimental data, which are given in Table 5.1. Since the contact angle hysteresis increases with both droplet size and inclination angle of the surface, the adhesion force increases for large droplets and high inclination angles. Although droplet weight lowers the adhesion force (Eq. 5.3), the overall increase in the adhesion force is associated with increased: (1) circumference of the wetted area and (2) droplet hysteresis with large droplet sizes. In addition, increasing the inclination angle of the surface enhances the value of $mg\,sin\delta$ in Eq. (5.3); hence, increasing the droplet hysteresis and wetted circumference overcomes the increase of the lateral component of the droplet weight.

The shear force, due to flow field generated inside the droplet during the droplet inclination, can be determined from the velocity distribution inside the droplet. In this case, the flow field can be obtained from the numerical simulations incorporating the appropriate boundary conditions [1]. This is shown in Figs. 5.3 and 5.4 for various inclination angles and two droplet sizes. The shear force generated at the wetted surface due to flow field inside

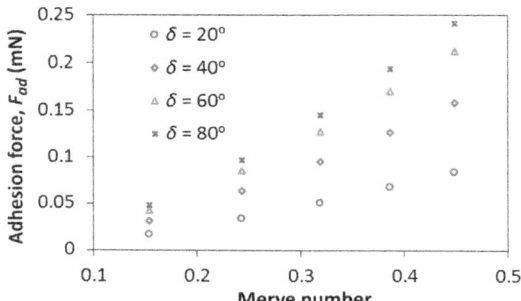

FIGURE 5.2 Adhesion force with Merve number ($MN = 2\gamma R^2/3\sigma$) for various inclination angles of the hydrophobic surface [1].

the droplet changes with the droplet size and inclination angles; in which case, it increases with increasing droplet size and the inclination angle. The maximum shear force generated is in the order of 8×10^{-7} N, which is significantly less than the adhesion force (0.25×10^{-3} N). Consequently, the influence of shear force on the droplet adhesion is negligible, which is true for all the droplet sizes considered in the analysis. The flow field developed inside the droplet forms a single circulation cell for all inclination angles of the surface and droplet sizes (Figs. 5.3 and 5.4). However, the maximum flow velocity slightly varies inside the droplet; in which case, for a 10 μL droplet and inclination angle $\delta = 20$ degrees, the maximum velocity is in the order of 2.7×10^{-3} m/s while for a 10 μL droplet and inclination angle $\delta = 80$ degrees, it is in the order of 4.52×10^{-3} m/s. Similarly, the maximum velocity for a 20 μL droplet is in the order of 2.7×10^{-3} m/s at inclination angle $\delta = 20$ degrees and it is 4.65×10^{-3} m/s at inclination angle $\delta = 80$ degrees. Consequently, change of the maximum velocity is notable due to the droplet size, but the inclination angle has a significant effect on the maximum velocity inside the droplet.

Since the droplet geometry changes with the inclination angle, the droplet contact length and the puddle thickness is formulated analytically, as is given below, in terms of the droplet contact angle and the inclination angle. On the other hand, consider a water droplet on a hydrophilic surface as shown in Fig. 5.5. The following geometric relations can be written for the droplet puddle thickness (h):

$$sin\theta = \frac{\ell}{2R} \text{ and } cos\theta = \frac{R-h}{R} \text{ and } R = \frac{\ell}{2sin\theta} \qquad (5.4)$$

where θ is the droplet contact angle on the hydrophilic surface, R is the imaginary radius, and l is the droplet contact length and equals to the contact area diameter. The droplet puddle thickness (h) yields:

$$h = R(1 - cos\theta) \qquad (5.5)$$

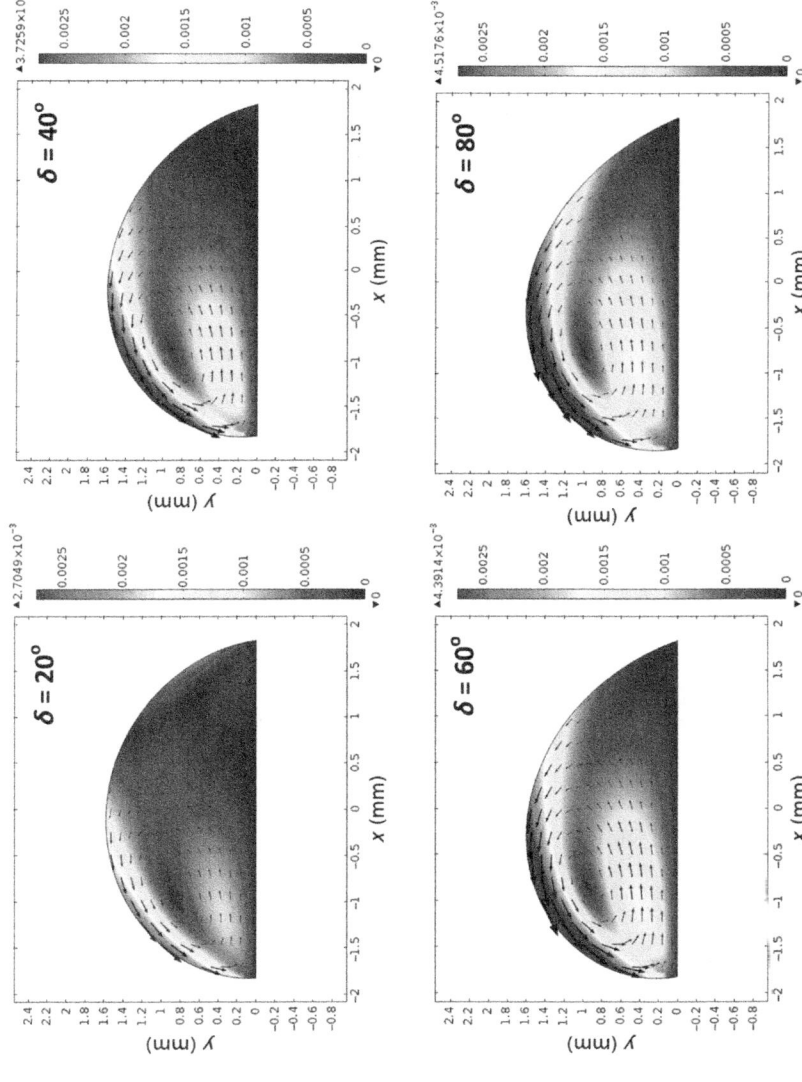

FIGURE 5.3 Velocity contours inside 10 μL droplet for different inclination angles [1].

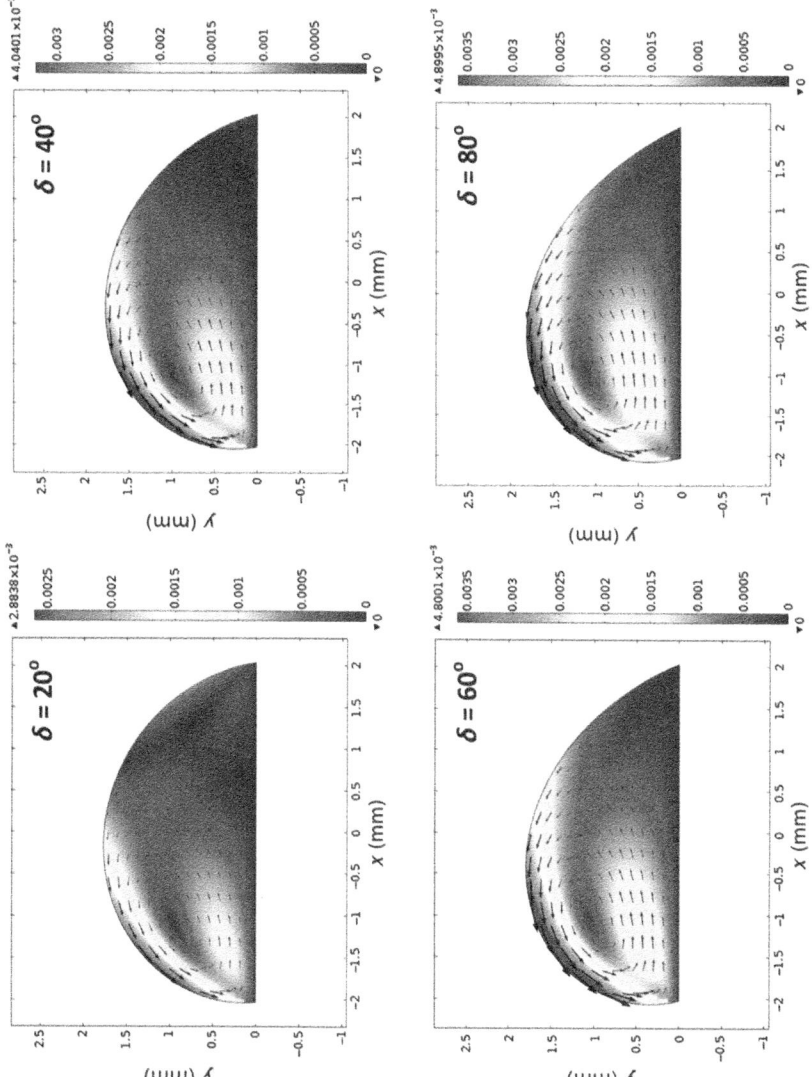

FIGURE 5.4 Velocity contours inside 20 μL droplet for different inclination angles [1].

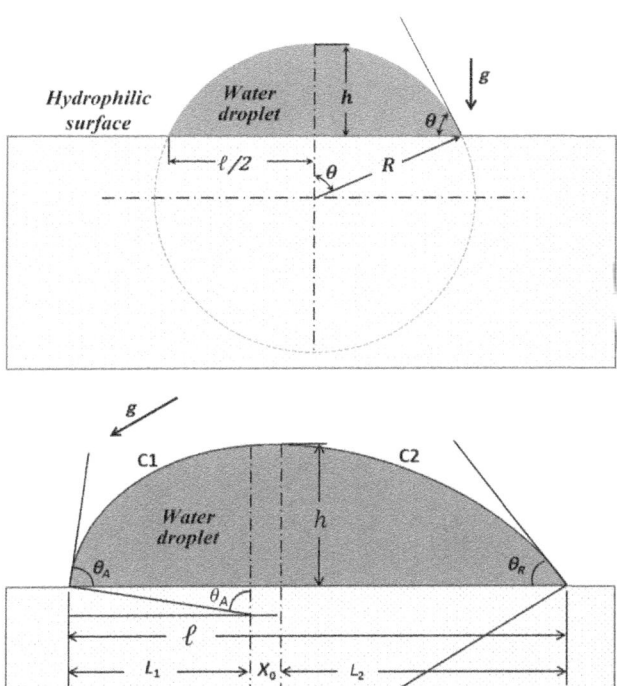

FIGURE 5.5 Droplet geometry used in the analytical formulation for the puddle height (h) and wetted length (l) of the droplet angles [1].

Eq. (5.5) is used to compute the droplet puddle height for various droplet sizes.

In the case of the inclined surfaces, the droplet undergoes an elastic deformation (bulging). The droplet advancing (θ_A) and receding (θ_R) angles are changed during the inclination of the surface as shown in Fig. 5.5.

In line with the previous study, the following relations can be written:

$$\frac{L_1}{sin\theta_1} - \frac{L_1}{sin\theta_1}cos\theta_1 = \frac{L_2}{sin\theta_2} - \frac{L_2}{sin\theta_2}cos\theta_2 \qquad (5.6)$$

Simplifying, L_1 can be expressed as:

$$L_1 = L_2 L_f \qquad (5.7)$$

where

$$L_f = \frac{sin\theta_1(1 - cos\theta_2)}{sin\theta_2(1 - cos\theta_1)} \qquad (5.8)$$

Since $L_1 + L_f = \ell$, the variables L_1 and L_2 can be expressed as:

$$L_1 = \frac{\ell L_f}{1 + L_f} \text{ and } L_1 = \frac{\ell}{1 + L_f} \tag{5.9}$$

Eq. (5.9) is used to determine the contact length of the droplet on the inclined surface.

The puddle thickness depends on the droplet contact angle on the surface and the contact length of the droplet. Consequently, it changes with the droplet radius and the inclination angle of the surface. On the other hand, the droplet size is critical for bulging and reforming of the droplet shape under the surface inclination; in which case, the capillarity length ($\kappa^{-1} = \sqrt{\sigma/\rho g}$, where κ^{-1} is the capillarity length, σ is the surface tension, ρ is the density, and g is the center of gravity) associated with the droplet becomes critical. In this case, the droplet diameter less than the capillarity length remain rather spherical droplet on the hydrophilic surface with small bulging on the inclined surface. The droplets with larger diameter than the capillarity length undergo bulging and form a large puddle. Fig. 5.6 shows the normalized puddle thickness (h/κ^{-1}), where h is the puddle thickness, κ^{-1} is the capillarity length ($\kappa^{-1} = \sqrt{\sigma/\rho g}$) of the droplet with Merve number ($MN = \gamma_w a^2/4\sigma$, where γ_w is the specific weight of water, a is the droplet diameter and σ is the surface tension of water). Normalized puddle thickness increases with increasing MN. This indicates that increasing droplet size enhances the puddle thickness of the droplet on the hydrophilic surface. Moreover, the values of the puddle thickness remain less than unity while showing that the droplet height is less than the capillarity length. This is related to the hydrophilic characteristics of the surface; in which case, the droplet contact angle remains less than 90 degrees for all cases. In addition, the change of puddle thickness with inclination angle of the surface remains small. Although bulging takes place with increasing inclination angle

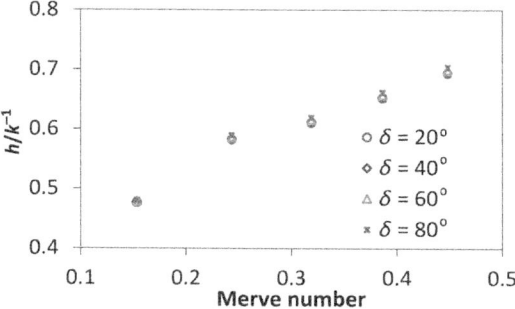

FIGURE 5.6 Normalized puddle thickness ((h/κ^{-1}), where h is the puddle thickness, κ^{-1} is the capillarity length, and $\kappa^{-1} = \sqrt{\sigma/\rho g}$) with Merve number ($= 2\gamma R^2/3\sigma$) for various inclination angles of hydrophilic surface [1].

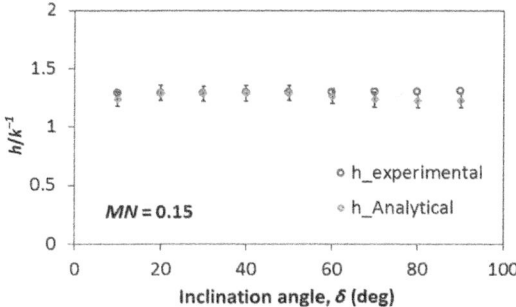

FIGURE 5.7 Normalized puddle thickness (Eq. 5.5) obtained from calculations and experiment for Merve number = 0.15 (5 μL droplet volume) [1].

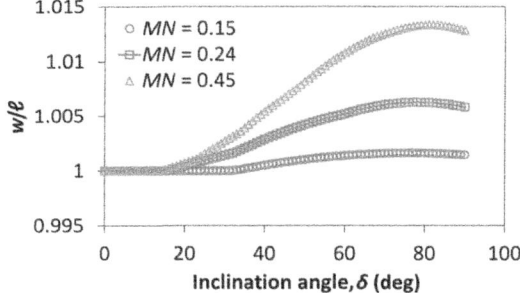

FIGURE 5.8 Normalized lateral extension of droplet (w/l, where w is the lateral length of the droplet and l is the wetted length) with inclination angle of hydrophilic surface for various Merve numbers [1].

(Fig. 5.6), the change of the puddle thickness remains almost the same for all inclination angles.

Fig. 5.7 shows the puddle thicknesses, determined from the analytical expression presented above and obtained from the measurements, with inclination angle of the surface for $MN = 0.15$, which corresponds to a 5 μL droplet volume. The difference between both findings is considerably small for all the inclination angles of the surface considered. In the case of lateral deformation of the droplet, which is parallel to the inclined surface, the change of the droplet length in the lateral direction is considerably large. This can be observed from Fig. 5.8, in which normalized lateral droplet length (w/l, where w is the lateral length of the droplet and l is the droplet wetted contact length) is shown with different inclination angle of the surface for three values of MN. Inclination of the hydrophobic surface results in bulging of the droplet laterally, which is more pronounced for large values of MN (corresponding to large droplets). The lateral extension of the droplet becomes significant as the inclination angle of the surface increases.

However, a further increase in the inclination angle of the surface results in a small change of the lateral extension of the droplet, that is, the lateral extension of the droplet almost ceases after 60 degrees of inclination angle for $MN \le 0.24$ and it is 80 degrees for $MN \ge 0.45$. The lateral extension of the droplet on the hydrophobic surface modifies the flow field and the maximum velocity inside the droplet (Figs. 5.3 and 5.4). However, the contact length (wetted length) of the droplet remains same for individual droplet regardless of the droplet size. Consequently, inclination of the hydrophobic surface does not result in sliding of the droplet for all sizes considered in the analysis. This can be associated with the strong adhesion force generated between the droplet and the inclined hydrophilic surface.

5.2.2 Water-Droplet Attachment on Hydrophobic Surfaces: Force Balance and Heating

The water-droplet attachment and dynamics on the inclined hydrophobic surface is important in terms of heat transfer during condensation or evaporation [25,26]. The influence of the gravitational force on the flow field and heat transfer inside the droplet remains critical for condensation and boiling applications. This is due to the fact that the gravitational force is important for reshaping the sessile droplet on the inclined hydrophobic surfaces. This in turn alters the heat transfer rates and modifies the Marangoni and buoyant forces inside the droplet while altering the Nusselt number for different inclination angles. In the present study, water-droplet attachment (pinning) on an inclined hydrophobic surface is considered and heat transfer from the inclined hydrophobic surface to the pinned droplet is examined. The droplet shape and the heat transfer rates are evaluated for various inclination angles and the droplet volumes. The following analysis is presented for droplet volume varied from 10 to 25 μL and inclination angle changed from 0 to 180 degrees in line with the previous study [2]. In addition, the experiment study on monitoring of the droplet shape on the inclined hydrophobic surface is also included in accordance with the previous study [2].

The governing equations of flow and heat transfer were provided in the previous study [2]; hence only the boundary conditions for the flow and heat transfer simulations are provided here in in line with the previous study [2]. Fig. 5.9 shows the boundary conditions used in the simulations. The constant pressure boundary is assumed at the droplet outside; in which case, external pressure of the droplet is set at atmospheric pressure. In addition, stagnant air is considered at the droplet outer surface, which yields zero velocity of the air. The natural convection ($h = 10$ W/m^{2}K) and the radiation boundary condition is considered at the interface of the droplet surface with air ambient temperature of 300K. Since the hydrophobic surface is maintained at constant temperature, a constant temperature heat transfer boundary condition is adopted at the interface between the droplet bottom and the

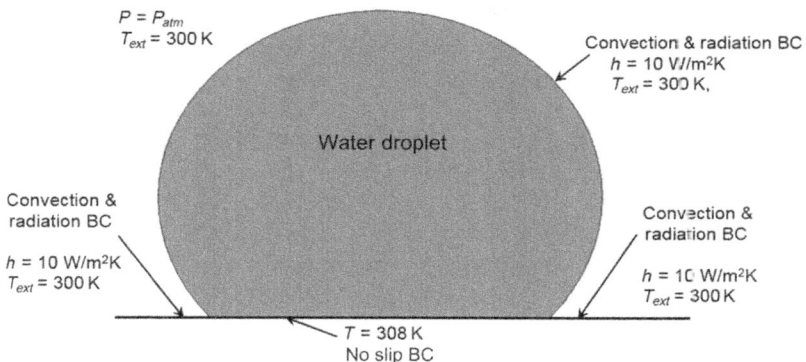

FIGURE 5.9   Boundary conditions incorporated into the simulations on the hydrophobic surface [2].

hydrophobic surface. In addition, a no-slip boundary is adopted at droplet and surface interface.

Since the time taken for the experiment and the simulation of experimental conditions is short, that is, in the order of 30 seconds, the evaporation from the droplet surface is neglected in the simulations. This was verified during the experiment. Droplet images were taken after 30 and 100 seconds and compared with the images of the droplet at the onset of the experiment. In both images the droplet has identical diameter and height. Therefore, the omission of evaporation from the droplet surface is justified. COMSOL Multiphysics software [27] was incorporated to simulate the flow field inside the droplet. Simulations were carried out incorporating 10, 15, 20, and 25 μL water-droplet volumes and the inclination angle of the hydrophobic surface within the range of 0 degree $\leq \delta \leq$ 180 degrees. A laminar isothermal two-phase flow model was used during the simulations. This model solves numerically continuity and momentum equations simultaneously to obtain the flow field in the solution domain. Three-dimensional simulation of the flow field is very expensive because of the excessive mesh requirements for the accurate solutions. The comparison of the flow field obtained from 2D and 3D simulations were provided in the previous study [4,28]; in which, it was reported that 2D and 3D simulations result in similar velocity predictions inside the droplet. Consequently, 2D simulation of the flow field is adopted in the analysis. In the numerical approach, finer meshes are located in the region where the fluxes are high. Mesh independence tests are conducted for each droplet contact angle considered in the simulations. The implicit scheme with a backward difference approximation is used and unconditionally stable solutions are ensured [29]. The selection of time step is critical to ensure the accuracy of the scheme; here, it is in the order of 10^{-4} seconds. The residuals of flow parameters are set as $\left| \psi^k - \psi^{k-1} \right| \leq 10^{-8}$.

To validate the predictions the experiment is carried out incorporating the hydrophobic surface and a high-speed camera. The experimental details and measurement of flow velocities inside the droplet are as follows in line with the previous study [2]:

The solution-crystallized polycarbonate surface had hydrophobic characteristics and was used in the experiments. To texture the surface via surface crystallization, a bare polycarbonate wafer with 3 mm thickness was cleaned and immersed in acetone for 4 minutes in accordance with the procedure stated in an earlier study [30]. The hierarchical texture was obtained after solution crystallization of the polycarbonate surface. The crystallized surface composed of micro/nanospherules and fibrils. A goniometer (Kyowa model—DM 501) was used to conduct sessile drop tests on solution-crystallized polycarbonate to measure the water-droplet contact angle. In this case, the droplet static contact angle was measured and found to vary from 134 to 136 degrees. High-precision drop shape analysis (HPDSA) was adopted during the measurements in accordance with the previous study [31]. In order to examine the droplet pinning on the hydrophobic surface, a test rig was designed and built. The experimental arrangements enabled to test the dynamic behavior of the droplets on the inclined hydrophobic surface. A Dantec Dynamics (SpeedSense 9040) high-speed camera was used to record the droplet shape on the hydrophobic surface at different inclination angles.

To validate the flow field inside the droplet, desalinated water was mixed with 3% (by volume) of the hollow glasses with the nominal size of 5 μm. The droplet of this mixture was used to from a droplet located on the 25 degrees inclined solution-crystallized surface, which had a constant temperature (308K). The droplet volume was 25 μL and initially the laser plane-illumination beam was introduced, which passed through the droplet cross-section. Particle image velocimetry (PIV) was used to record the particles motion across the droplet cross-section. The focus setting of PIV lens was adjusted to focus at the illuminated plane of the droplet cross-section while simultaneously realizing the PIV measurements. It should be noted that the width of the illuminated plane was slightly larger than the width of the focused plane at the droplet cross-section due to the lens aberration correction of the PIV system. Nevertheless, the difference between the widths of both planes is negligibly small. To simulate the flow field developed inside the droplet, simulations were carried out while resembling the water-hollow glass mixture. In this case, the governing momentum equation was solved using the discrete-phase model after incorporating the hollow glass particles in the solution domain. The energy equation was solved by assuming a slurry-single fluid, which resembled water and a hollow glass mixture. Because the hollow glass particle concentration was low (3%) in the carrier fluid (water), the effective thermal properties were incorporated in the energy equation. The formulation of the governing equations and numerical discretization is not given herein, but the details can be found in a previous study [32]. In the simulations, the initial and the boundary conditions for energy and momentum equations were set mimicking the experimental conditions. The use of hollow glass particles facilitated to monitor and trace the particle velocities inside the droplet during the constant temperature heating from the hydrophobic surface. Experiments were repeated 12 times and based on the distribution of the experimental data the confidence level of 95% was obtained. The mean of the data distribution was within ± 1.75 of the standard deviation of the distribution of a single measurement from that distribution. The experimental uncertainty analysis revealed that the uncertainty less than 4% was resulted for the velocity measurements. The flow velocities that were measured from PIV and predicted from the multiphysics code, and the PIV images of the hollow glass particles inside the droplet are provided in Table 5.2. The particle velocities predicted from the multiphysics code and the experiments are in good agreement. However, the small discrepancies between both results are related to computational errors such as round-off errors, and the experimental errors based on measurement uncertainties.

On the other hand, the solvent crystalized surface possesses hierarchical texture with micro/nanosize spherules and fibrils and the wetting of the surface is governed by the Cassie and Baxter state. In this case, the resulting surface texture gives rise to the air pockets, which are trapped in the texture while resulting in the Cassie–Baxter state on the crystallized polycarbonate surface. The apparent contact angle including surface roughness can provide more realistic formulation of the liquid-droplet contact angle [33]. A liquid droplet has liquid–solid and liquid–vapor interfaces and the apparent contact angle equation incorporates the contributions of these interfaces. Hence, the equation for the apparent contact angle becomes [33] $cos\theta_c = f_1 cos\theta_1 + f_2 cos\theta_2$, where θ_c is the apparent contact angle, f_1 is the surface fraction of liquid–solid interface, f_2 is the surface fraction of liquid–vapor interface, θ_1 is the contact angle for liquid–solid interface, and θ_1 is the contact angle for liquid–vapor interface. For the air–liquid interface, f_1 can be represented as f, which is the solid fraction, and air fraction (f_2) becomes $(1-f)$. The parameter f ranges from 0 to 1; in which case, $f = 0$ is the case where the liquid droplet is not in contact with the surface and $f = 1$ is the case where the surface is completely wetted. However, the contact mode changes from Cassie–Baxter state to Wenzel state [34] when the surface texture becomes sparse or when the droplets impact the surface with high velocity [35]. The solid fraction of the wetted surface area (f) can be associated with the surface roughness ratio (r). Therefore, the solid fraction of the surface can be formulated from the projected area of spherules over the total projected area of the textured surface. This corresponds to the fraction of the area covered by the spherules at the crystallized surface. The fraction of the area wetted by the liquid can be written

TABLE 5.2 Predicted Velocity Field and PIV Images for Hollow Glass Particles in the Droplet

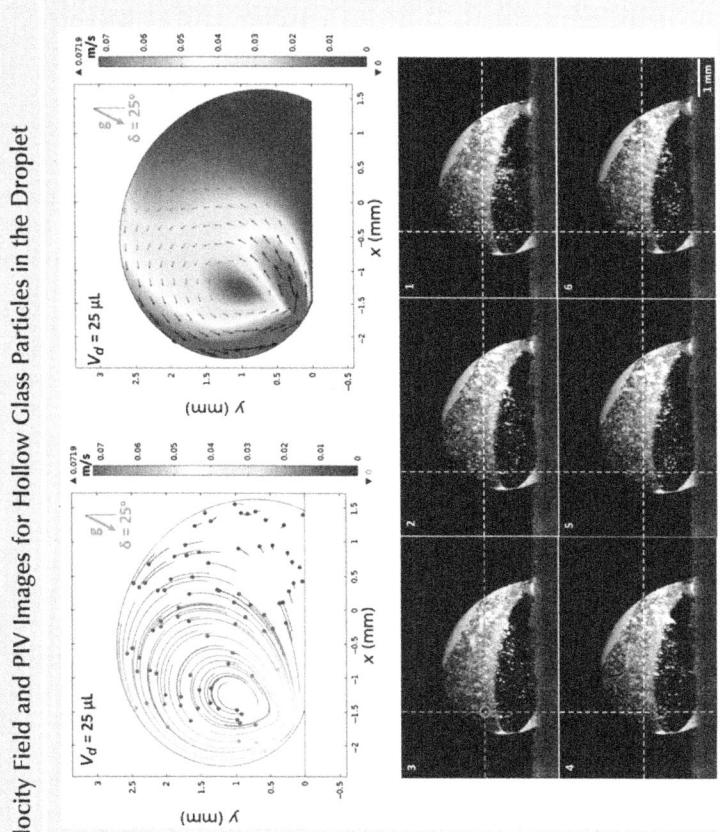

Particle #	X (mm)	Y (mm)	Simulations V (m/s)	Experiment V (m/s)
1	− 1.736133456	0.514478445	0.013209282	0.0133
2	− 1.723148108	0.761199713	0.010100600	0.01
3	− 1.943898797	0.917023659	0.013503169	0.014
4	− 1.801060081	1.254642367	0.009094937	0.01
5	− 1.320602894	1.825996995	0.007347681	0.0072
6	− 0.93104291	1.890923619	0.006587902	0.0066
7	− 0.632380247	1.683158278	0.006124617	0.00614
8	− 0.424614906	1.280612946	0.007150121	0.00714
9	− 0.37267375	0.930009127	0.008558929	0.0082
10	− 0.424614906	0.709258437	0.009489877	0.0095
11	− 0.775218964	0.579405189	0.013201891	0.0123
12	− 1.086867094	0.410595775	0.017002025	0.0165

Flow velocity predicted from simulations and obtained from PIV data. The inclination angle is $\delta = 25$ degrees and droplet volume is 25 μL [2].

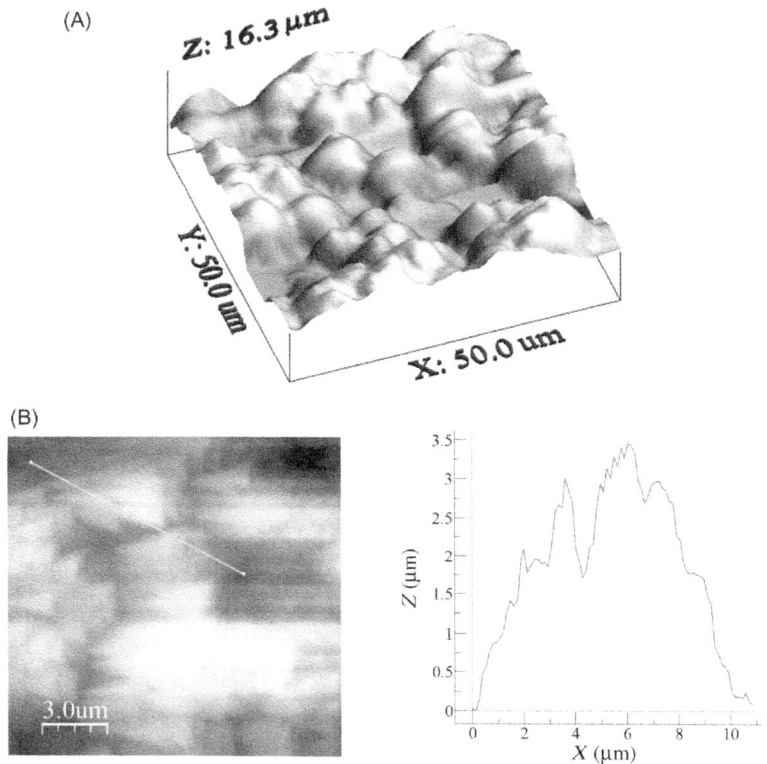

FIGURE 5.10 AFM images and line scan of crystallized polycarbonate surface: (A) 3D image of surface, and (B) line scan on textured surface [2].

as $=$ ((*Projected Total Textured Area*) − (*Area of Liquid_Air Interface*))/ (*Projected Total Textured Area*). The data related to the areas of the textured and the air−liquid interfaces are determined from the AFM images. Fig. 5.10A shows an AFM image of the 3D crystallized surface while Fig. 5.10B shows the AFM line scan of the crystallized surface. The presence of spherules and fibrils is evident on the crystallized surface. The spherules form close-packed texture gaps that are filled by the air captured below the droplet meniscus on the surface. Some small oscillations on the peaks of spherules represent the submicron-size fibrils. This situation can be observed from the surface line scan (Fig. 5.10B). The average surface roughness is in the order of 3.4 μm. The solid fraction for the solvent-induced crystallized polycarbonate surface is determined to be within the range of $0.4 \le f \le 0.6$.

Several tests were carried out for the water-droplet contact angle measurements on the crystallized surface to ensure the correct measurement of the droplet contact angle. Table 5.3 gives the droplet receding and advancing

TABLE 5.3 Receding (θ_R) and Advancing (θ_A) Contact Angle Measurements for Crystallized Polycarbonate Surface for Various Inclination Angles (δ)

	$V_d = 10\ \mu L$		$V_d = 15\ \mu L$	
δ (Deg)	θ_R (Deg)	θ_A (Deg)	θ_R (Deg)	θ_A (Deg)
0	134.09	134.76	136.41	135.95
30	121.61	146.73	119.29	156.65
60	110.78	155.03	103.75	166.92
90	105.22	152.32	95.25	162.8
120	107.68	146.07	95.33	150.46
150	112.38	132.4	105.64	135.48
180	120.76	122.05	117.11	121.31
	$V_d = 20\ \mu L$		$V_d = 25\ \mu L$	
0	136.74	138.53	135	135.83
30	109.58	156.01	109.62	159.44
60	92.68	166.76	86.7	169.6
90	82.38	159.6	77.22	166.72
120	86.19	145.14	80.81	149.17
150	93.41	131.8	94.71	131.55
180	109.9	110.34	111.66	108.79

The measurement errors lies between ±2 degrees [2].

angles and contact angle hysteresis for various angles of inclination. The contact angle hysteresis is determined from $\theta_{hysteresis} = \theta_{Advancing} - \theta_{Receding}$, where $\theta_{Advancing}$ is the advancing angle and $\theta_{Receding}$ is the receding angle. The attainment of large values of the contact angle hysteresis indicates the presence of large adhesion force, which acts along the three-phase contact line at the interface of the droplet on the crystallized surface.

In general, the buoyant convection can cause the rolling off of the droplets from the hydrophobic surface, due to instability, which is true for droplet volumes $\forall < 0.11\ \mu L$ [36]. Thus, the droplet volumes considered in the analysis ($10\ \mu L \leq \forall \leq 25\ \mu L$) are larger than those reported in the previous study [36]. In addition, the droplet rolling is also not observed during the experiments. On the other hand, the force required to detach a droplet from the surface is the same as the force needed to overcome the work of pinning [37]. The adhesion composes of the lateral force (in the plane of the hydrophobic surface) and the normal force (normal to the hydrophobic surface). The normal force becomes important when the hydrophobic surface is extremely inclined ($\delta \geq 90$ degrees). The pressure of air captured within the surface texture, below the droplet bottom meniscus, remains slightly higher than the atmospheric pressure because of the curvy nature of the droplet meniscus. It should be noted that droplet meniscus forms a curvy arc on the top of the texture. Once the droplet is inclined on the surface, the curvature of the meniscus arc changes while altering the pressure of the trapped air in the texture. Depending on the angle of inclination of the hydrophobic surface, the air volume in the texture can either be increased or decreased slightly. Once the inclination angle increases more than 90 degrees, the air volume in the texture increases while reducing the trapped air pressure and generating Magdeburg-like forces. These forces are related to $\sim(1-f)\Delta P A_{total}$, where f is the solid fraction, ΔP is the trapped air pressure change within a single texture gap (cavity), and A_{total} is the total area of textured surface. The formulation of Magdeburg-like forces and trapped air pressure change in the texture gap is given briefly as follows: air is trapped in between the droplet meniscus when the water droplet is located on the textured hydrophobic surface. If some texture gaps (isolated gaps) are packed (not connected with air in the other texture gaps), pressure in the isolated gaps remains different than those of texture connected gaps. This, in turn, results in sealing of air trapped from atmospheric air because of droplet meniscus. As the droplet inclines, the meniscus arc changes slightly over the texture height giving rise change of the pressure in the trapped air while causing pressure force to be acting on the droplet meniscus. In case of expansion of the trapped air, due to slight volume change during the change of the geometric position of the droplet meniscus arc, a suction pressure is generated, which in turn results in Magdeburg-like forces acting on the droplet meniscus. This contributes to the adhesion of the droplet on the surface. Since microsize spherules are formed on the crystallized polycarbonate surface, the spherules are presented as round textures in Fig. 5.13A and B. The droplet meniscus height prior to bending can be formulated after incorporating the horizontal force balance. Consider Fig. 5.11A, the vertical force balance yields:

$$\rho g \pi a^2 h_d + \rho g\, \Delta \forall_d = F_\gamma sin\theta \qquad (5.10)$$

where mg is the specific weight, h_d is the droplet height, F_γ ($F_\gamma = 2\pi a \gamma$) is the surface tension force, and Δ_d is the volume of inflection. Consider that $\Delta \forall_d$ is a half volume of an ellipsoid, then Eq. (5.10) becomes:

$$\rho g \pi a^2 h_d + \frac{1}{2}\rho g\, \frac{4\pi}{3} a^2 \chi = 2\pi a \gamma sin\theta \qquad (5.11)$$

The air volume inside the closed packed gap can be approximated by a half ellipsoid. The meniscus height across a single textured packed is shown schematically in Fig. 5.11A and B.

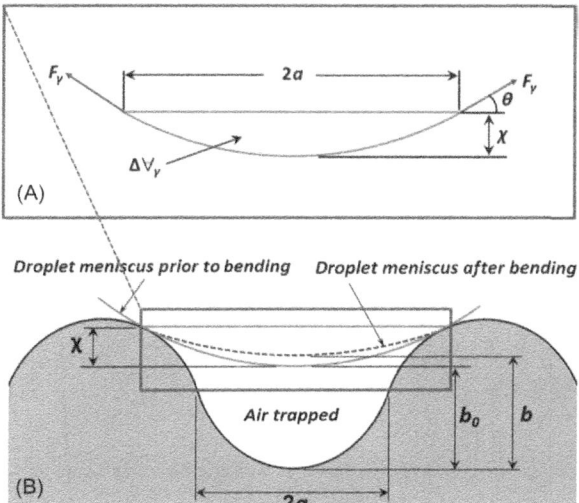

FIGURE 5.11 A schematic view of a close-packed texture gap: (A) droplet meniscus across texture gap prior to inclination of hydrophobic surface, and (B) droplet meniscus across texture gap prior to and after inclination of hydrophobic surface [2].

Since $sin\theta \approx sin\theta = \sqrt{\chi^2/(a^2 + \chi^2)}$, and divide by $\rho g \pi a^2$:

$$\frac{2}{3} \chi - \frac{2\gamma \chi}{\rho g a} \sqrt{\frac{1}{(\chi^2 + a^2)}} + h_d = 0 \qquad (5.12)$$

The solution of Eq. (5.12) yields the functional relation between χ and the droplet height (h_d) in terms of fluid properties and lateral distance of the droplet meniscus across two consecutive texture pillars, where air is trapped. The value of χ is in the order of 2.5 nm for a 2 mm radius droplet and the spacing of closely spaced two spherules of 8 µm apart in the lateral direction (similar to Fig. 5.11A).

Consider the droplet meniscus prior to and after inclination (Fig. 5.11B). The air-trapped volume in a single gap can be formulated after approximating it to half of an ellipsoid. Considering air is an ideal gas, the pressure drop in the air trapped during the geometric change of droplet meniscus, due to hydrophobic surface inclination, can be formulated as follows. Consider the equation of state for air:

$$P = \rho R T \qquad (5.13)$$

where P is the pressure, ρ is the density, R is the gas constant for air, and T is the temperature within the closed packed gap.

$$P = RT \frac{m}{\forall} \qquad (5.14)$$

where $\forall$ is the air volume and m is the air mass within the closed packed gap. In the differential form:

$$dP = -RTm\frac{d\forall}{\forall^2} \text{ or } \Delta P = -RTm \int_{\forall_1}^{\forall_2} \frac{d\forall}{\forall^2} \tag{5.15}$$

Now, consider the half of an ellipsoid resembling air-trapped volume. The differential form of this volume yields:

$$d\forall = \frac{4}{3}a^2 db \tag{5.16}$$

Combining Eqs. (5.15) and (5.16) leads to:

$$\Delta P = -RTm \int \frac{\frac{4}{6}\pi a^2 db}{\left(\frac{4}{6}\pi a^2 b\right)^2} \text{ or } \Delta P = \frac{6RTm}{\pi a^2}\left[\frac{1}{b} - \frac{1}{b_0}\right] \tag{5.17}$$

The pressure force generated in a single packed texture gap is:

$$F = \pi a^2 \Delta P \tag{5.18}$$

The number of closed packed gaps on the crystallized polycarbonate surface is to be incorporated to find the total pressure force acting on the droplet meniscus. Therefore, the approximate total pressure force is:

$$F_T = n\pi a^2 \Delta P \tag{5.19}$$

where n is the number of packed texture gaps on the crystallized polycarbonate surface. Combining Eqs. (5.17) and (5.19) results in the total pressure force, which becomes:

$$F_T = 6nRTm\left[\frac{1}{b} - \frac{1}{b_0}\right] \tag{5.20}$$

The area ratio of closed packed gaps sites over the total area of the crystallized polycarbonate surface is assessed using the texture height landscape image (Fig. 2.11B). On average, the area ratio is estimated to be in the order of 17%.

The pressure force (Magdeburg force) variation with the change of the height of the droplet meniscus within the air trapped in the closed packed is shown in Fig. 5.12. The pressure force increases significantly with small change of the droplet meniscus height in the close-packed texture gap. It should be noted that the area ratio estimated from the texture height landscape for the close-packed texture gaps is incorporated in Fig. 5.12. In addition, the average closed packed gap height is estimated as 3.4 μm from AFM data. The Magdeburg-like forces contribute to the adhesion of the droplet on the inclined textured surface. However, at some texture locations pressure may remain as the atmospheric pressure because of the full connection of the texture gaps to the free atmosphere. In this case, the

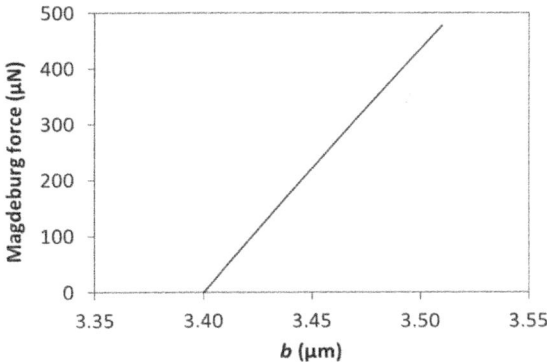

FIGURE 5.12 Magdeburg force with the droplet meniscus height due to pressure change inside the trapped air [2].

forces generated within the texture gap, due to pressure drop during the volume increase of the trapped air within the texture, become zero. On the other hand, the lateral adhesion force is formulated heuristically by approximating the three-phase contact line with a single ellipse and using the experimentally obtained polynomial function for the dependence of the contact angle on the position along the three-phase contact line [22]. Therefore, the resulting equation for pinning force becomes [22,23] $F_{\gamma L} = (24/\pi^3)\gamma_{LV}D(cos\theta_R - cos\theta_A)$, where γ_{LV} is the surface tension of the liquid on the solid surface, D is the droplet diameter prior to deformation (same area of the ellipse), θ_R is the receding angle, and θ_A is the advancing angle.

The polycarbonate surface is crystallized and possesses a hierarchical texture; therefore, the roughness parameter can be introduced in the adhesion force equation pertinent to the Young–Dupre equation [24]. Consequently, the pinning force equation can be written including the roughness parameter, that is, $F_{\gamma L} = (24/\pi^3)\gamma_{LV}Df(cos\theta_R - cos\theta_A)$, where f is the solid–surface fraction (solid–liquid contact fraction) as obtained from the AFM data, which varies within the range of 0.4–0.6. The force diagram resembling the droplet attachment on the surface is shown in Fig. 5.13. The droplet pinning results in zero net lateral force in the plane of the hydrophobic surface. The net lateral force is comprised of the gravitational force component in the plane of the hydrophobic surface, shear force at the wetted surface, the adhesion force, and the Magdeburg-like forces. The force balance in the lateral direction of the inclined hydrophobic surface yields $mgsin\delta - \{(24/\pi)\gamma_{LV}Df_1(cos\theta_R - cos\theta_A) + F_{mL} + F_\tau\} = 0$, where F_τ is the shear force generated by the flow field inside the droplet ($F_\tau = A_w\mu\partial V/\partial s$, A_w is the wetted area, μ is the fluid viscosity, and $\partial V/\partial s$ is the rate of fluid strain), which is the same order as the shear force acting on the wetted surface due to flow circulation in the droplet; F_{mL} is the lateral component of

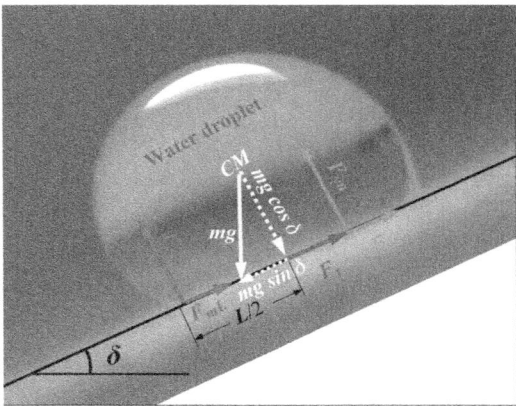

FIGURE 5.13 Optical image of the droplet and schematic view of the forces acting on the inclined hydrophobic surface [2].

the Magdeburg-like force; m is the droplet mass; g is the gravity; and δ is the inclination angle of the surface (Fig. 5.13).

It should be noted that the force generated due to fluid acceleration inside the droplet because of bulging (during inclination of the surface) and heat transfer is assumed to be negligibly small as compared to the gravitational acceleration; therefore, the force acting on the droplet due to gravity and fluid acceleration is considered to be the same as the component of the droplet weight in the lateral direction. For the pinning droplet, the normal component of the net force, which is normal to the surface, also remains zero during the inclination of the surface. The total normal force acting on the hydrophobic surface is associated with the normal component of the forces due to tension force, gravity, and Magdeburg-like affects. Therefore, the force balance normal to the surface yields $F_{\gamma n} + F_{mn} - mg\cos\delta = 0$, where $F_{\gamma n}$ is the normal component of the surface tension force and F_{mn} is the normal component of the Magdeburg-like force generated due to the trapped air volume expansion within the texture because of the minute change of the droplet meniscus curvature during inclination. Since the surface tension force has two components, namely lateral and normal forces, the normal component can be written as $F_{\gamma n} = \sqrt{F_\gamma^2 - F_{\gamma L}^2}$, where $F_{\gamma n}$ is the normal component of the surface tension force, F_γ is the overall surface tension force, and $F_{\gamma L}$ is the lateral component of the surface tension force. After incorporating the corrected length of the three-phase contact line and introducing the solid-surface fraction, the surface tension force becomes $F_{\gamma L} = (24/\pi)\gamma_{LV}Df$.

Fig. 5.14 shows the lateral component of the adhesion force with the inclination angle of the surface for different droplet sizes. It should be noted that the advancing and receding angles of the droplet for a given inclination

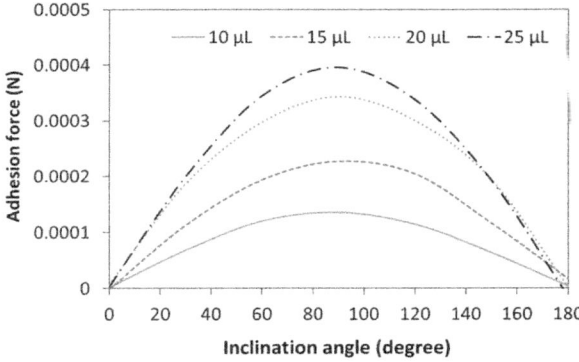

FIGURE 5.14 Lateral component of adhesion force with inclination angle for various droplet volumes [2].

TABLE 5.4 Water-Droplet Contact Angle and Hysteresis Prior to Rolling on the Hydrophobic Surface [2]

Droplet volume (µL)	0.52	4.2	14	33.5
Contact angle (deg)	160	158	156	156
Contact angle hysteresis (deg)	1	2	2.9	2.9

angle are taken from the experimental data (Table 5.4). The contact angle hysteresis changes slightly with droplet size because of the increased bulging of the droplet with increasing inclination of the hydrophobic surface (Table 5.4).

The adhesion force first increases with increasing inclination angle, which is more pronounced for the large volume droplets. This behavior is associated with the increased three-phase contact line and changing the receding and advancing angles of the droplet with increasing droplet volume (Table 5.4) despite the fact that the mass of the droplet increases, which tends to reduce the adhesion force. The lateral adhesion force reaches its maximum at the inclination angle of 90 degrees. In this case, because of the excessive droplet bulging in the direction of gravity, the Magdeburg-like forces become important for droplet adhesion on the hydrophobic surface. When comparing the lateral adhesion force and the droplet weight when the inclination angle is 90 degrees, the lateral component of the adhesion force remains slightly less than the droplet weight. However, the shear force (8×10^{-12} N) generated due to the rate of fluid strain in the region of the droplet bottom is significantly smaller than that of the lateral adhesion force (4×10^{-5} N). Thus, to sustain the droplet

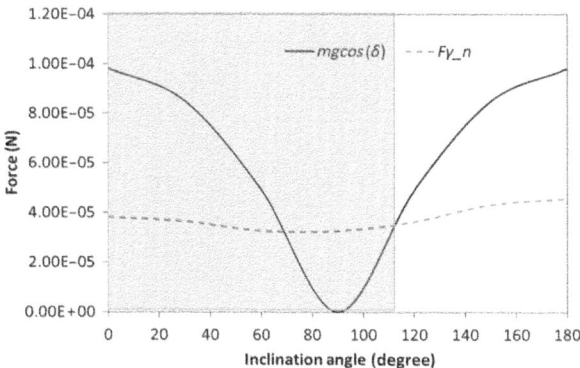

FIGURE 5.15 Normal component of adhesion force ($F_{\gamma n}$) and gravitational force ($mgcos\ (\delta)$) normal to hydrophobic surface with inclination angle. The droplet volume is 10 μL [2].

pinning at 90 degrees, the force generated becomes higher that the lateral adhesion force. The contribution of Magdeburg-like force in the lateral direction is considerable, which is in the order of 10^{-4} N. The variation of the normal component of the surface tension and droplet weight in the normal direction is shown in Fig. 5.15 with the inclination angle of the hydrophobic surface for a 10 μL droplet volume. The normal component of the surface tension force overcomes the normal component of the gravitational force ($mgcos\delta$) in between the inclination angle of 90 degrees $\leq \delta \leq 180$ degrees; however, as the inclination angle increases further, the normal component of the gravitational force remains higher than the normal component of the surface tension force. Consequently, the Magdeburg-like forces cause droplet pinning on the hydrophobic surface for the inclination angle beyond 90 degrees.

Two deformation parameters are introduced to assess the geometric changes of the droplet due to bulging under the various inclinations of the hydrophobic surface. These include the vertical and lateral deformation parameters. The vertical deformation parameter corresponds to the ratio of droplet height (h) over the wetted length (l) of the droplet on the hydrophobic surface while the lateral deformation parameter is the ratio of the maximum width (w_{th}) of the droplet over the wetted length (l) of the droplet on the hydrophobic surface. It should be noted that the wetted length of the droplet on the hydrophobic surface changes with the droplet volume. Fig. 5.16 shows the vertical and lateral deformation parameters of the droplet with the inclination angle for various droplet volumes. Both deformation parameters increase with the inclination angle. In addition, increasing droplet volume enhances both deformation parameters. This behavior is attributed to the excessive bulging of the large volume droplets with increasing inclination angle. The vertical deformation parameter remains larger than that of the lateral deformation parameter, which is more pronounced for large volume

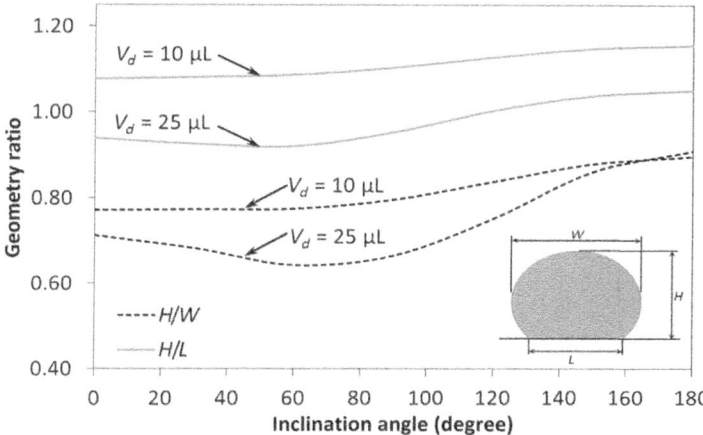

FIGURE 5.16 Vertical and lateral deformation parameters of the droplet with the inclination angle for various droplet volumes [2].

parameters. Consequently, inclination of the surface results in greater reduction of the droplet height than the droplet width.

On the other hand, in the analysis, the internal fluidity of the droplet with heat transfer from the inclined hydrophobic surface is considered. In this case, the situation with a constant-temperature heat source from the droplet bottom is considered to resemble the experimental conditions. Heat transfer modifies the flow field inside the droplet. The characteristic timescale associated with the thermal diffusion has vital importance for the flow stability inside the droplet fluid. The Rayleigh number (Ra) mainly influences the flow stability in the buoyant convection [36]. The inclusion of the Marangoni effect in the simulations further influences the droplet instability. The Marangoni and the buoyant currents can result in a droplet rolling off from the hydrophobic surface during the heating period depending on the characteristic time for the stable flow. This situation may occur for the small droplet diameters [36]. However, for a stable flow conditions, the ratio of square of the characteristic droplet diameter (D) over the thermal diffusivity of the droplet fluid (α_T) should remain greater than unity ($D^2/\alpha_T > 1$) [36]. The characteristic time is estimated in the order of 30 seconds for the present case.

Fig. 5.17 shows the velocity and temperature variations along the vertical line inside the droplets for various heating durations. The droplet contact angle is 135 degrees and the hydrophobic surface is kept horizontal in Fig. 5.17. Temperature (Fig. 5.17A) and velocity (Fig. 5.17B) varies significantly with time in the early heating period. As the time progresses variation of temperature along the vertical line collapses into a single curve. This behavior is also true for the velocity variation. The time for identical temperature profiles along the vertical line corresponds to 30 seconds, which satisfies the condition

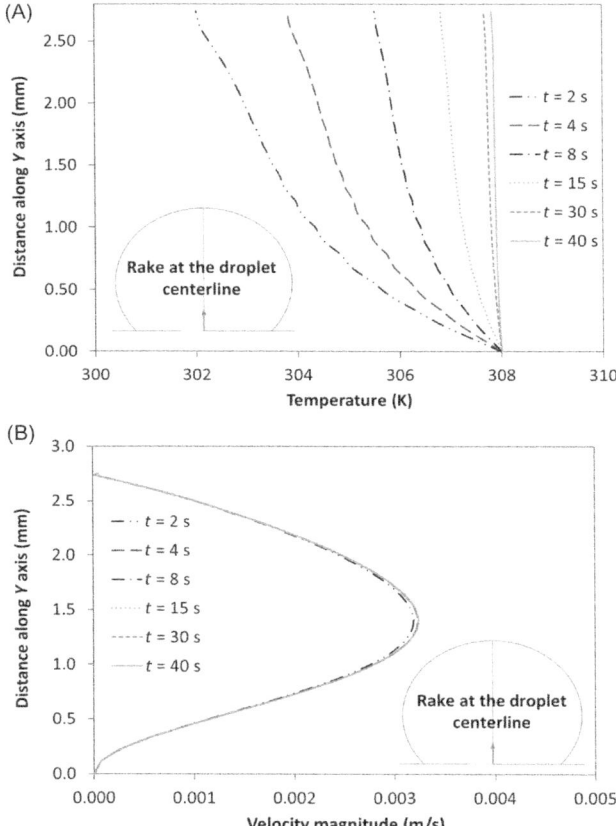

FIGURE 5.17 Temperature and velocity variation along the vertical distance in the droplet: (A) temperature variation for various times, and (B) velocity variation for various times. Droplet volume is 25 μL on the horizontal hydrophobic surface [2].

of $D^2/\alpha_T > 1$. Therefore, the time selected for presenting the flow and temperature fields satisfy the condition for the stable flow inside the droplet.

Fig. 5.18 shows the flow field inside the droplets on a horizontal hydrophobic surface together with the optical images of the droplets, prior to inclination, for various droplet volumes. Two counter-rotating circulation cells are formed inside the droplet. The formation of counter-rotating cells is associated with the Marangoni and buoyancy currents developed inside the droplet during the heating period. The velocity ratio of Marangoni flow over the natural convection is the same order of Marangoni over the Rayleigh

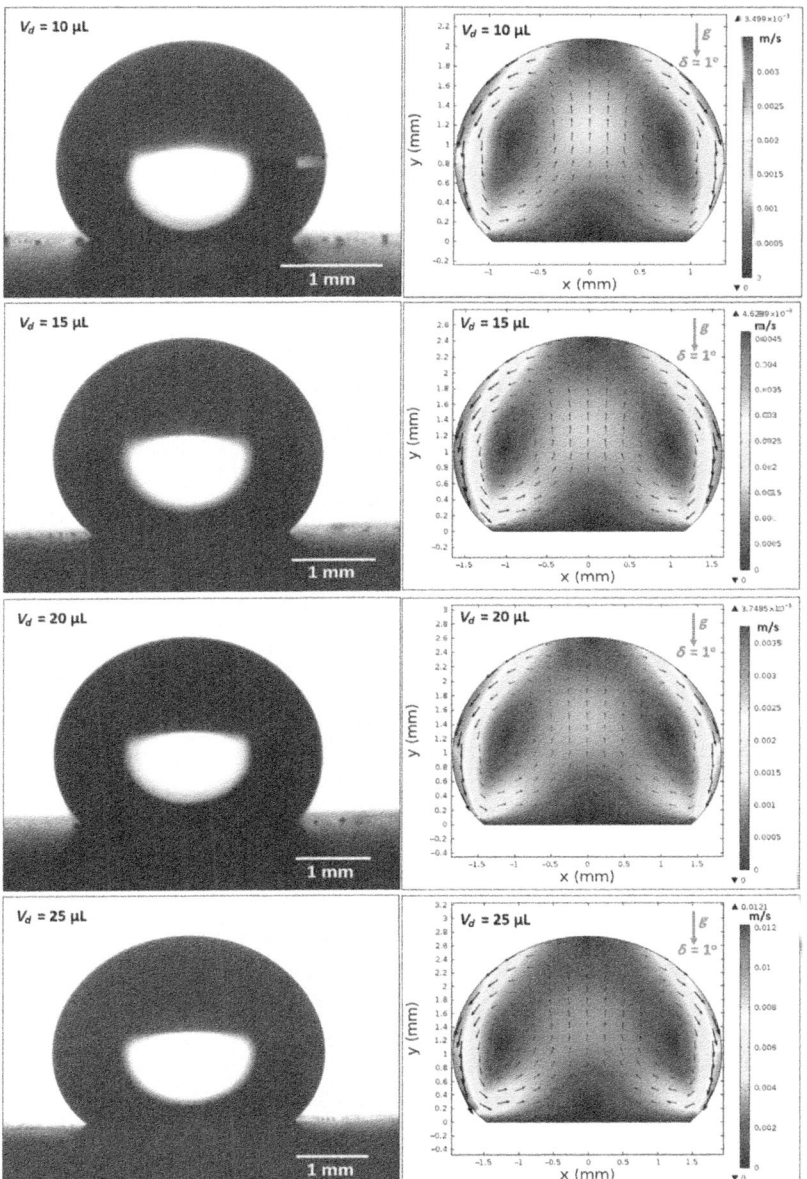

FIGURE 5.18 Optical images of droplets (on the left) and velocity contours (on the right) inside droplet for different volumes at 0 degree inclination angle [2].

number $((\partial\sigma/\partial T)\Delta T/\alpha_t\rho ga^2)$ [36]; in which case, the velocity ratio remains greater than unity for a droplet characteristic diameter $\geq 2 \times 10^{-3}$ m. Consequently, in the present case, the Marangoni current becomes larger than the buoyancy current while causing formation of the counter-rotating circulation cells, which is consistent with the previous findings [4,7,28]. The influence of Marangoni current on the flow field is more pronounced in the upper region of the droplet where the circulation cell centers are located. Increasing the droplet size slightly alters the circulation cell center and velocity field inside the droplet; in which case, the maximum velocity reduces with increasing droplet size.

Fig. 5.19 shows the temperature contours inside the droplet for the horizontal hydrophobic surface and various droplet sizes. Since the heating takes place from the droplet bottom, heat transfer results in development of a high-temperature region in the close region of the droplet bottom. Hence, the convection current carries the heated fluid from the droplet bottom toward the fluid interior. In addition, flow circulation contributes to heat transfer toward the fluid interior and the attainment of the high-temperature region in the outer region of the circulation cell. This is more pronounced for the large droplet volumes. However, within the circulation cells, the heated fluid is not carried toward the circulation cell centers. Moreover, the heating is enhanced by both diffusion and convection. The large extension of the high-temperature region takes place in the central region of the droplet bottom. This occurs because of: (1) heat transfer by heat diffusion in the lateral and normal directions, and (2) convection heat transfer, that is, heat carried by the convection current developed within the circulation cell exterior.

Since the temperature gradient remains low in the lateral direction at the central location of the droplet bottom, heat diffusion takes place along the normal direction because of a relatively lower temperature gradient along the normal direction as compared to that of the lateral direction. The heated fluid carried by the convection current slightly lowers the fluid temperature in the region of droplet edges. In addition, convection of cooling of the surface contributes to the attainment of low temperatures in the edge regions.

Figs. 5.20 and 5.21 show velocity and temperature contours when the hydrophobic surface is inclined at 90 degrees for various droplet volumes. Inclination of the hydrophobic surface alters the droplet shape significantly, which in turn modifies the flow field. The pressure distribution in the droplet changes because of the change of the gravity vector. In this case, a single circulation cell is developed inside the droplet, which extends almost covering the droplet interior. The circulation cell center and the velocity magnitude in the flow field changes with the droplet size. However, the maximum velocity inside the droplet attains lower values for 90 degrees inclined surface that that corresponding to horizontally located droplet. This is more pronounced for the large volume droplet. Temperature field (Fig. 5.21) also differs from that shown in Fig. 5.19. In this case, temperature increases along the outer

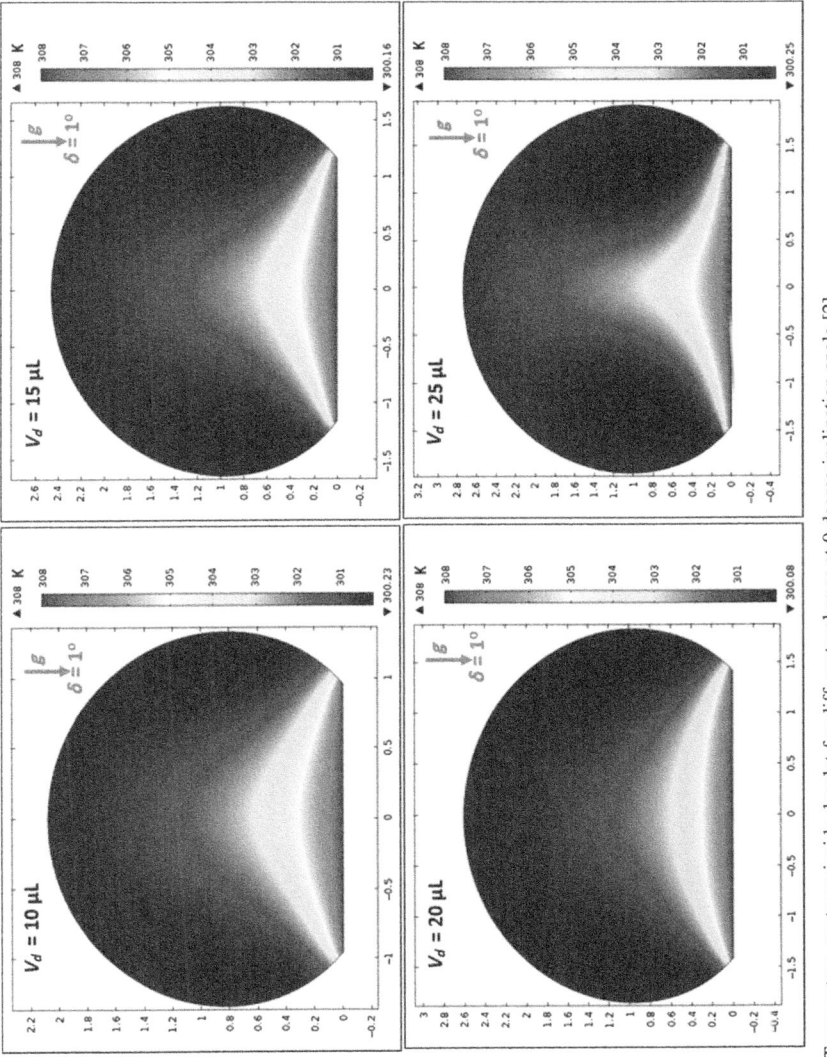

FIGURE 5.19 Temperature contours inside droplet for different volumes at 0 degree inclination angle [2].

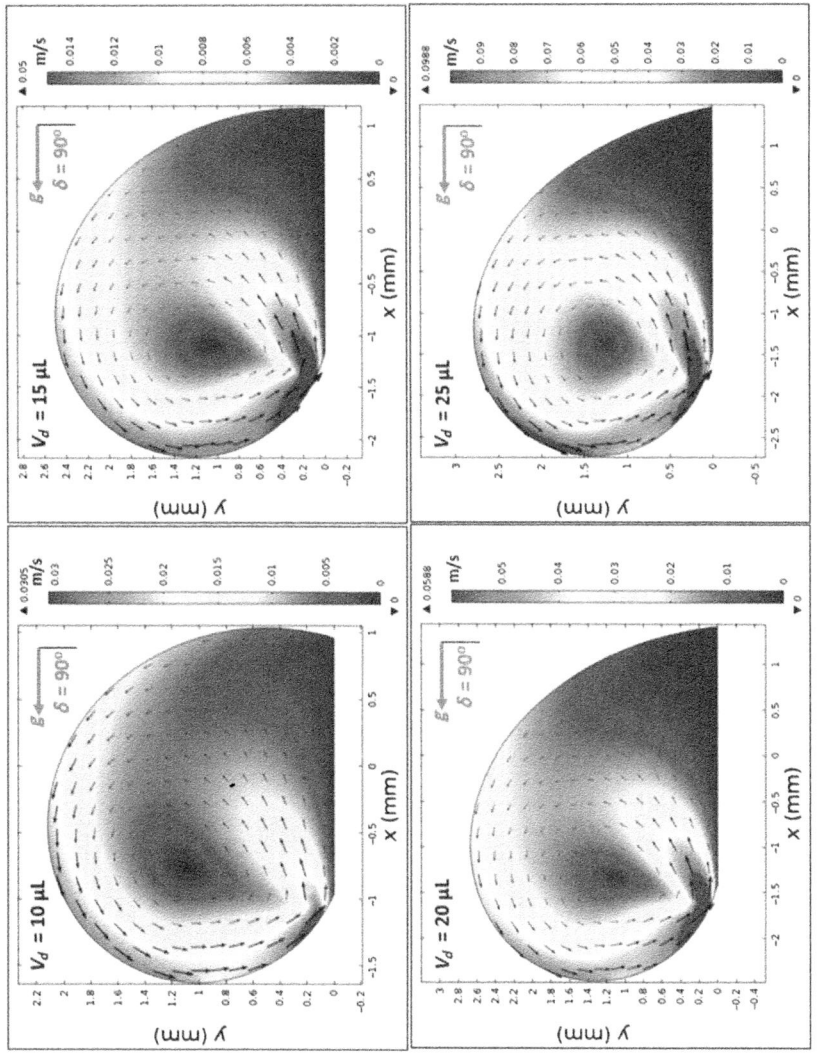

FIGURE 5.20 Velocity contours inside droplet for different volumes at 90 degrees inclination angle [2].

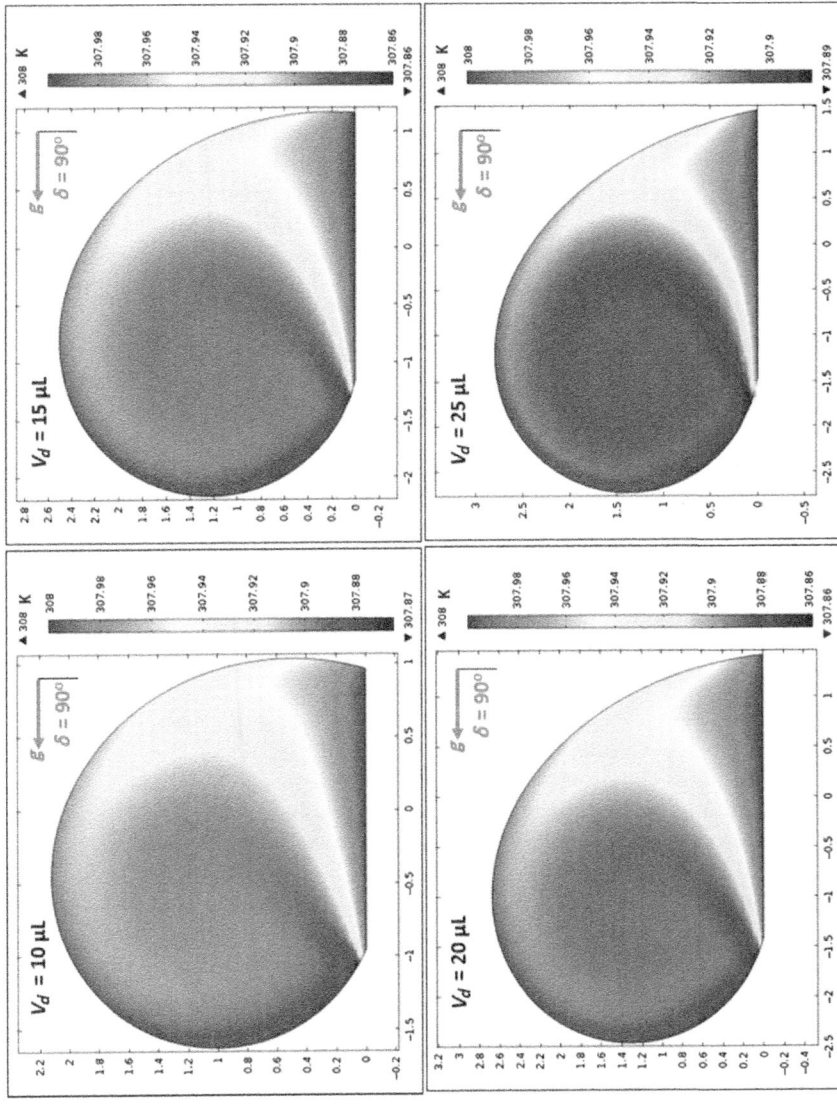

FIGURE 5.21 Temperature contours inside droplet for different volumes at 90 degrees inclination angle [2].

region of the circulation cell. Consequently, thermal diffusion and temperature increase in the inner region of the circulation cell is not significant. However, thermal diffusion and the convection current are responsible for the extension of temperature field in the close region of the outer boundary of the circulation cell. When the droplet is inclined 180 degrees from its horizontal position, the droplet shape changes significantly, which in turn modifies the pressure distribution inside the droplet. Consequently, the flow and temperature fields differ significantly from that of the horizontal droplet.

The velocity and temperature contours are shown in Figs. 5.22 and 5.23 for the 180 degrees-rotated droplet. The droplet shapes are also shown in Fig. 5.22 for comparison. Two counter-rotating circulation cells are formed inside the droplet and the location of the cell center moves toward the droplet top edge unlike that shown in Fig. 5.19. Consequently, change of the cell structure in the droplet modifies the heat transfer by convection and diffusion and temperature distribution inside the droplet (Fig. 5.22). Since the magnitude of the flow velocity remains low, flow acceleration due to convection ($V\partial V/\partial s$), local ($\partial V/\partial t$), and radial ($-V^2/r$, r being the droplet radius) effects remains low. The scale analysis for the acceleration depicts that the maximum radial accelerations is in the order of 10^{-5} m²/s, the convective acceleration is close to 0.005 m²/s, and the local acceleration is about in the order of 10^{-4} m²/s. In this case, convection acceleration takes over the flow acceleration inside droplet during inclination of the hydrophobic surface. The shear stress and shear force due to the rate of fluid strain in the bottom region of the droplet are in the order of 3.3×10^{-6} N/m² and 8×10^{-12} N, respectively, considerably smaller than that of the adhesion force (Fig. 5.14). Therefore, the influence of shear force developed in the droplet bottom region on the droplet pinning is negligibly small.

Fig. 5.24 shows the Nusselt number variation with the inclination angle of the droplet for various droplet sizes. It should be noted that the droplet pins during the inclination of the hydrophobic surface. The Nusselt number changes with the inclination angle of the hydrophobic surface; in which case, it reaches almost its peak value for the inclination angle of 65 degrees and remains high for the inclination angle within a range of 45 degrees $\leq \delta \leq 90$ degrees. This behavior is attributed to the flow field developed inside the droplet, which influences the convection heat transfer from the droplet bottom toward the droplet interior. It should be noted that a single and large circulation cell is developed inside the droplet for the range of inclination angles in which the Nusselt number remains high. Consequently, heat carried by the convection current in the outer region of the circulation cell is mainly responsible for the heat transfer enhancement from the hydrophobic surface. In addition, heat diffusion in the region of the circulation cell outer boundary also contributes to this enhancement. As the inclination angle increases

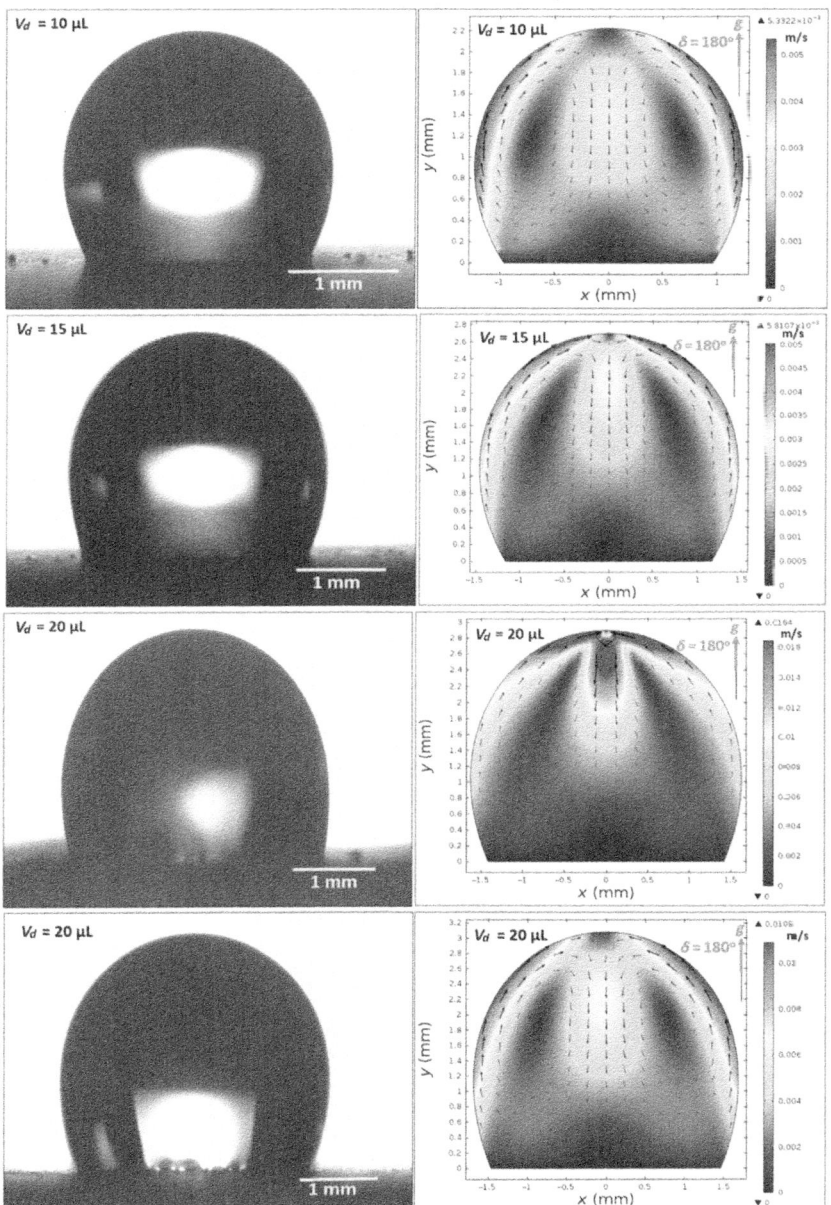

FIGURE 5.22 Optical images of droplets (on the left) and velocity contours (on the right) for different droplet volume at 180 degrees inclination angle [2].

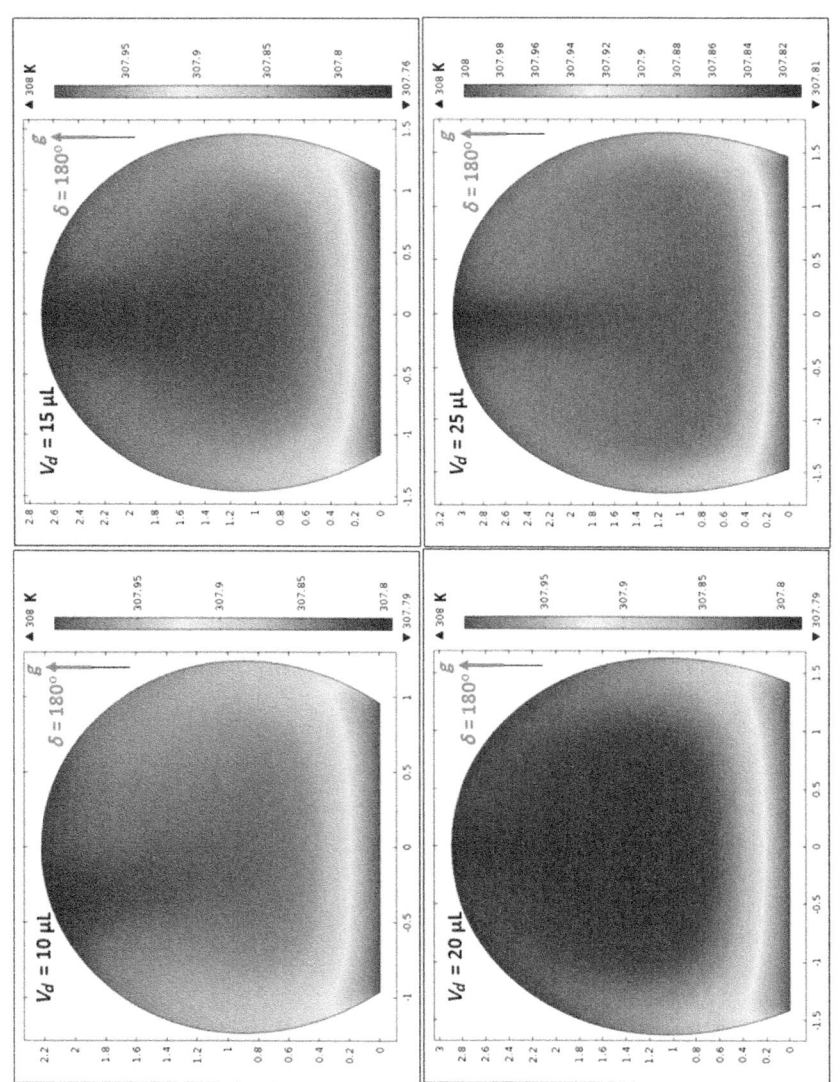

FIGURE 5.23 Temperature contours inside droplets for different volumes at 180 degrees inclination angle [2].

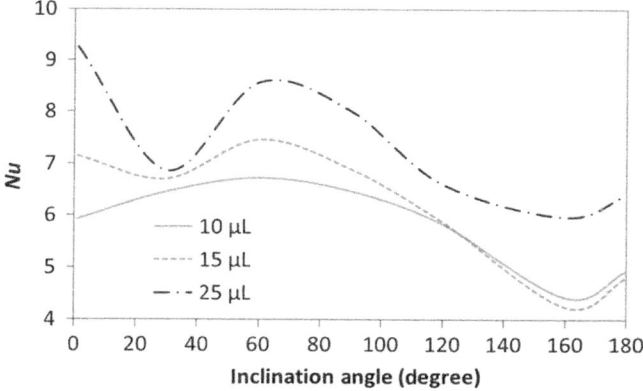

FIGURE 5.24 Nusselt number with inclination angle of hydrophobic surface for various droplet volumes [2].

further, the Nusselt number reduces. The formation of two counter-rotating circulation cells in the region away from the heated hydrophobic surface (Fig. 5.19) gives rise to reduced heat transfer rates from the surface, which is particularly true for the inclination angle of 180 degrees. Increasing droplet volume improves the heat transfer and enhances the Nusselt number. This behavior is related to the fluid bulk temperature inside the droplet, which attains low values for the large volume droplets. Consequently, temperature difference between the hydrophobic surface and the droplet fluid becomes larger for large droplets than that of small droplets while improving the Nusselt number.

5.3   DROPLET SLIDING AND ROLLING ON SURFACES

The dynamic motion of water droplets remains critical for removing dust particles on hydrophobic surfaces. The contact line dynamics of droplets govern droplet behavior—either rolling off or sliding on the surfaces. Droplet rolling off is mainly associated with the contact angle hysteresis and inclination angle of the surface. In the case of hydrophobic surfaces with high contact angle hysteresis, the adhesion force remains high and droplet attaches at the surface even at high angle of inclinations [38]. The droplet attachment is results due to the force balance along the surface inclination. The sum of adhesion force, which is related to $\pi r \sigma (cos\theta_R - cos\theta_A)$, where r is the contact line radius, θ_R is the receding angle of droplet, θ_A is the advancing angle of droplet, and σ is the surface tension of droplet liquid and shear force, which is $\pi r^2 \mu (dV/dn) -$, where μ is the fluid viscosity and dV/dn is the rate of fluid strain normal to the contact surface, becomes larger than the gravitational force, which is associated with $mgsin\alpha$, where m is the droplet mass, g is the center of gravity, and α is the inclination angle of the surface. It should be noted when the surface is inclined the flow field is generated in the droplet and shear stress is formed at the interface of the contact area, which generates the frictional force at the interface of the droplet and the solid surface. On the other hand, when gravitational force overcomes the adhesion and frictional force, droplet movement at the inclined surface is influenced by the droplet bulging/puddling due to gravitational effect [39]. In this case, the dynamic motion of the droplet is governed by the rotational and sliding modes of the droplet movement. However, the droplet size is critical for the bulging and forming the droplet shape. In this case, the capillarity length ($\kappa^{-1} = \sqrt{\sigma/\rho g}$, where κ^{-1} is the capillarity length, σ is the surface tension, ρ is the density, and g is the center of gravity) associated with the droplet becomes important. The droplet diameter less than the capillarity length remain as the spherical droplet during its motion on the inclined surface. The droplets with larger diameter than the capillarity length undergo bulging and form a puddle. The puddle size can be determined from the force balance between the capillarity and gravitational forces. This yields the droplet puddle thickness (h) in the order of $\sqrt{2(1-cos\theta)(\sigma/\rho g)}$, where θ is the droplet contact angle of the droplet [40]. Under the low contact angle hysteresis, the droplet with larger diameter than the capillarity length can roll and slide because of interfacial friction on the inclined hydrophobic surface. The center of mass of the droplet changes during rolling on the hydrophobic surface because of the adhesion force at the droplet-solid-surface interface, that is, droplet advancing and receding angles change during the droplet rolling. This gives rise to wobbling of the droplet, that is, transiently changing of the droplet height during the rolling. Consequently, wobbling affects the internal fluidity of the droplet via volume evolving during the rolling, which in turn change the rotational characteristics of the droplet. Therefore, the droplet motion on the inclined hydrophobic surface with low contact angle hysteresis remains critical for self-cleaning of surfaces.

In the present section, the case study for the droplet dynamics on the inclined surface is presented for various droplet sizes and inclination angles in line with the previous study [3]. The data obtained from the experimental study for droplet rolling on the inclined surface via high-speed camera are provided. The force balance of the droplet along the contact line is considered and the droplet rotational velocity is formulated. The internal fluidity of the droplet was simulated during the rolling and predictions of the rolling and translation velocities, which were then compared to those obtained from the experiments. The air drag and shear force along the contact line are predicted and incorporated in the analysis. The flow field inside the droplet and the maximum and the minimum droplet puddle height were predicted numerically and compared with their counterparts obtained from the experiment. The findings are presented in line with the previous study [3].

5.3.1   Experimental

A polycarbonate wafer and acetone, as a solution for crystallization, were used to prepare the hydrophobic surface. Since a solution-crystallized surface possesses hydrophobic characteristics, it was produced and used in the experiments. To texture the surface via surface crystallization, a bare polycarbonate wafer was cleaned and immersed into acetone for 4 minutes in accordance with the procedure stated in the early study [30]. The surface of the polycarbonate wafer was crystallized and the hieratical surface texture was obtained, which was comprised of micro/nanospherulites and poles. In order to improve the surface hydrophobicity and reduce the contact angle hysteresis on the solvent-induced crystallized polycarbonate surface, synthesized silica particles were deposited at the surface. To synthesize silica particles, the method adopted was similar to that reported in the previous study [41]. The process is shortly described herein. Tetraethyl orthosilicate (TEOS), isobutyltrimethoxysilane (OTES), ethanol, and ammonium hydroxide were used in the synthesizing process. In this case, 14.4 mL of ethanol, 1 mL of ultrapure water, and 25 mL of ammonium hydroxide were mixed and stirred for 12 minutes. Later, 1 mL TEOS diluted in 4 mL of ethanol was added in the mixture. Following 30 minutes after this process, 0.5 mL of TEOS diluted in 4 mL ethanol was added. After 5 minutes, a modifier silane molecule was added in a molar ratio of 3:4 with respect to the second edition of TEOS. The final mixture was stirred for 20 hours at room temperature, and later centrifuged and washed with ethanol to complete the removal of reactants. The solvent casting was applied to deposit the solution onto PDMS copied and replicated surfaces. Upon vacuum drying until all solvent was evaporated, characterization of resulting surfaces was carried out.

Jeol 6460 SEM and AFM in contact mode were used to examine the surface topology and surface texture of the crystallized polycarbonate wafer. The tip was made of silicon-nitride probes ($r = 20-60$ nm) with a manufacturer-specified force constant, k, of 0.12 N/m. The micrographs of solution-crystallized polycarbonate surface obtained from scanning electron microscope are shown in Fig. 5.25A and B. The surface composed of spherules-like (Fig. 5.25A) and whisker-like structures (Fig. 5.25B), which resulted in hydrophobic characteristics. The average surface roughness of the crystallized polycarbonate surface was in the order of 2.8 μm. Fig. 5.25C and D shows the deposited functionalized silica particles on the solvent crystallized polycarbonate surface. Fig. 5.26A and B show the AFM images and line scan of the solvent crystalized and functionalized silica particle-deposited surfaces used in the current experiments of droplet rolling. It is evident that the surface composes of hierarchical texture with micro/nanopoles. Since the solvent-induced crystalized surface has textures comprise of micro/nanopoles, water-droplet does not wet the surface completely. The AFM image of the texture pole heights (Fig. 5.26A and B) are also analyzed in details.

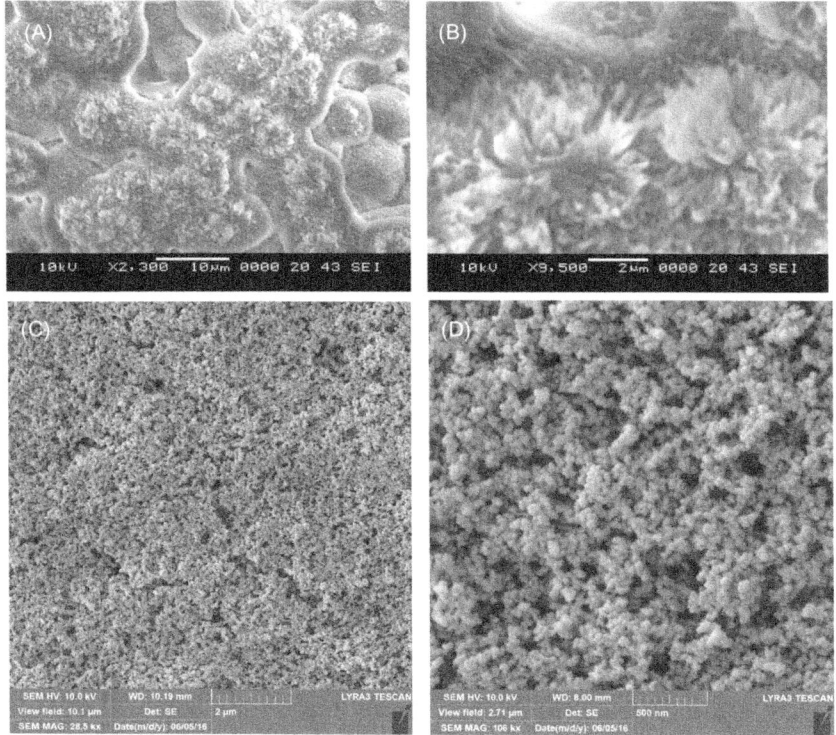

FIGURE 5.25 SEM micrographs of crystallized polycarbonate surface and deposited functionalized silica particles: (A) spherules on crystallized surface, (B) whisker-like structures on spherules, (C) deposited synthesized silica particles, and (D) agglomerated synthesized silica particles [3].

The solid fraction for the solvent-induced crystallized polycarbonate surface is determined to be within the range of $0.4 \le f \le 0.6$. Since the functionalized silica particles created the lotus effect on the solvent crystallized polycarbonate surface, several tests were carried out for the water-droplet contact angle measurements on the workpiece surface. A goniometer Kyowa (model—DM 501) was used to conduct sessile drop tests for the measurement of droplet contact angle. In this case the droplet static contact angle was measured, which was varied within 145–150 degrees.

A test rig was designed and built to test the dynamic behavior of the droplets on the inclined nonwetting surface. In this case, a high-speed camera (Dantec) was used to record the rolling and sliding motion of the droplet of various sizes located on different inclination angles of the hydrophobic surface. Fig. 5.27 shows images of the droplet of $V_d = 14$ μL on a 45 degrees inclined hydrophobic surface together with the images obtained from the simulations incorporating the identical boundary conditions of the experiments.

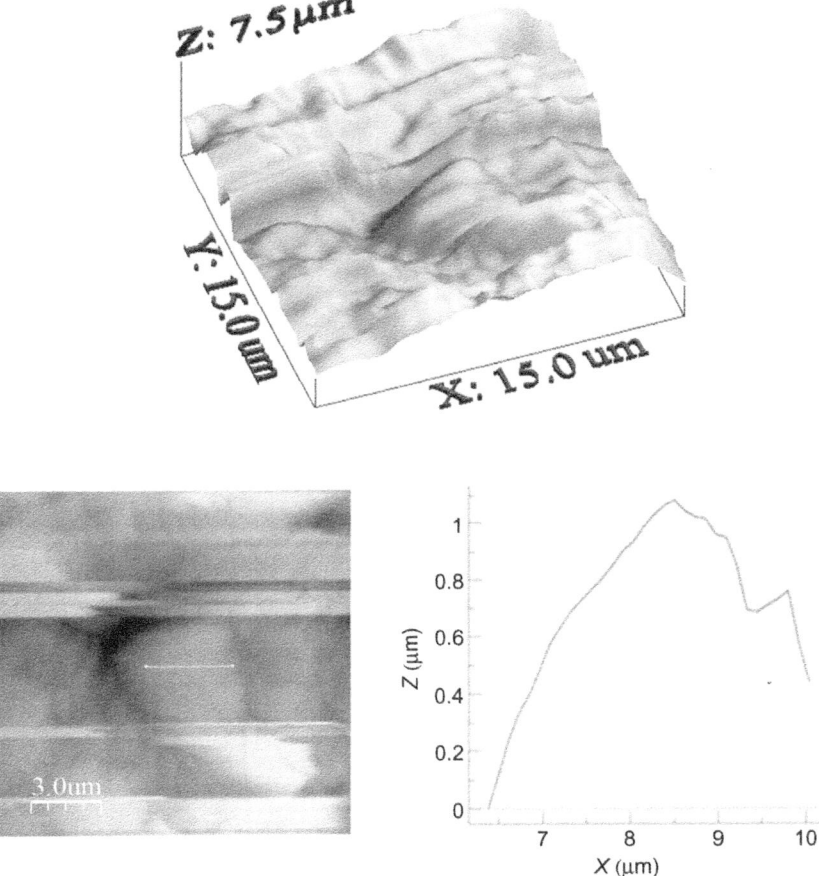

FIGURE 5.26 AFM image of crystalized surface: (A) 3D image of the crystallized surface, and (B) line scan over spherules at the surface [3].

In order to validate the model study, the particle image velocimetry (Dantec PIV) was used to monitor the particle velocities inside the droplet. In this case, a sessile droplet of $V_d = 40\,\mu L$ volume was used to simulate flow field numerically and monitor the flow velocities using PIV. In this case, the hollow glass particles were used in the droplet water to monitor and trace the particle velocities inside the droplet during the heating process. The governing equation of momentum was solved after incorporating the hollow glass particles in the solution domain via using the discrete-phase model. The energy equation was solved after assuming slurry single fluid resembling water and the hollow glass particles mixture. Since the hollow glass particle concentration is low (3%) in the carrier fluid (water),

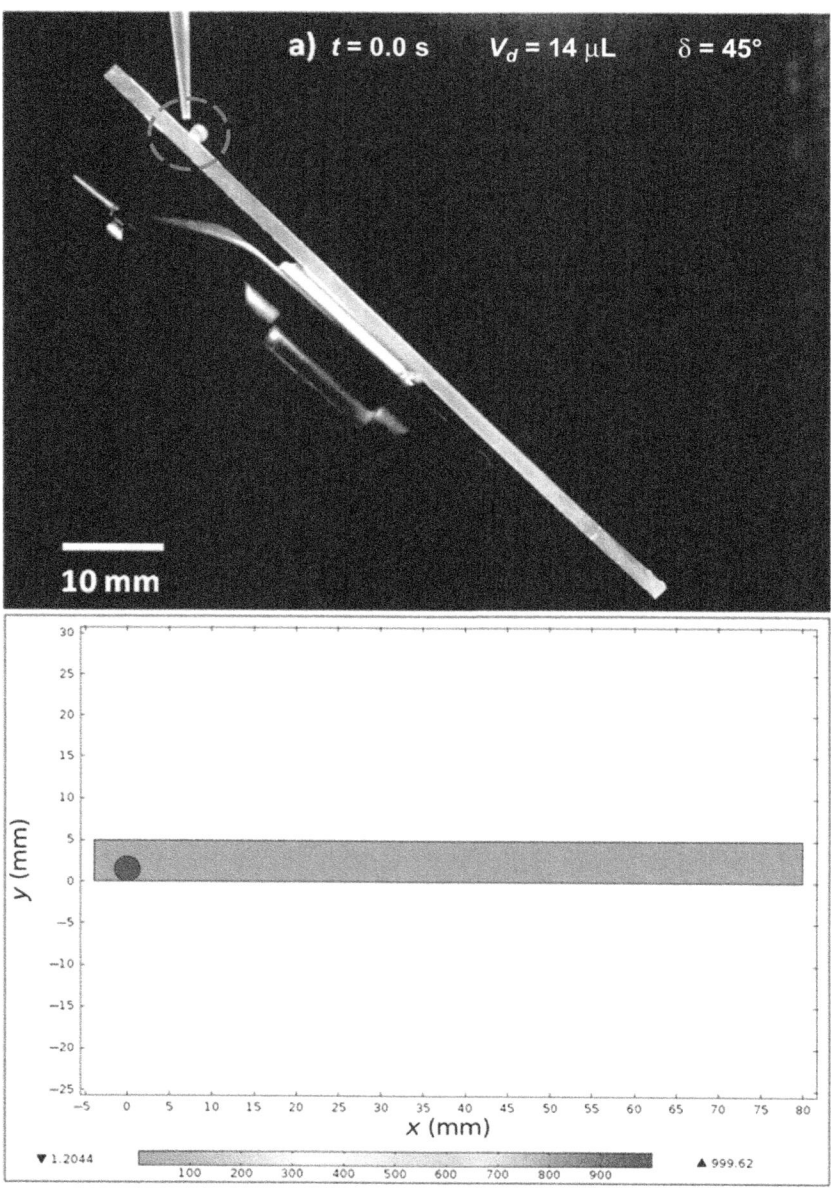

FIGURE 5.27 A typical high-speed camera image of droplet on the inclined surface and the corresponding simulation condition. The gravity vector is rotated with $\delta = 45$ degrees in the simulations. The circle indicates the droplet location in the experiments [3].

effective thermal properties are incorporated in the energy equation. The incompressible flow field is considered to formulate the governing equations of flow and heat transfer. The field equations are solved numerically in line with the experimental conditions. The coupled flow and thermal fields are considered simultaneously in the simulations. The formulation of the governing equations and numerical discretization is not given herein, but the details can be found in the previous study [42]. The flow velocities measured from PIV and predicted from the multiphysics code are in good agreement [3]. However, the small discrepancies between both results are related to computational errors, such as round-off errors, and the experimental error based on the measurement repeatability, which is in the order of 6%.

5.3.2 Numerical Modeling of Droplets on Hydrophobic Surfaces

The numerical simulation conditions are briefly provided herein; however, the details can be found in [3]. The internal fluidity of a droplet is simulated in line with the experimental conditions of the inclined hydrophobic surface. Multiphysics software [27] is incorporated to simulate the flow field inside the droplet. A laminar isothermal two-phase flow, moving mesh interface is used during the simulations. In this case, arbitrary Lagrangian–Eulerian (ALE) formulation is incorporated in the computations [43]. The solution domain and meshes are shown schematically in Fig. 5.28. Since temperature is kept constant during the experiments, an isothermal flow is assumed in the simulations.

The governing equations of flow are as follows.

In the air and water domains, the incompressible Navier–Stokes equations are solved:

$$\rho\left(\frac{\partial V}{\partial t} + (V - V_M) \cdot \nabla V\right) = -\nabla p$$

$$+ \nabla\left[\mu\left(\nabla V + (\nabla V)^T\right) - \frac{2}{3}\mu(\nabla \cdot V)\right] + F$$

$$(5.21)$$

where V is the fluid velocity, V_M is the mesh velocity, p is the pressure, μ is the dynamic viscosity, and F is the body force per unit volume. In line with the experiments, it is assumed that isothermal condition yields in the solution domain.

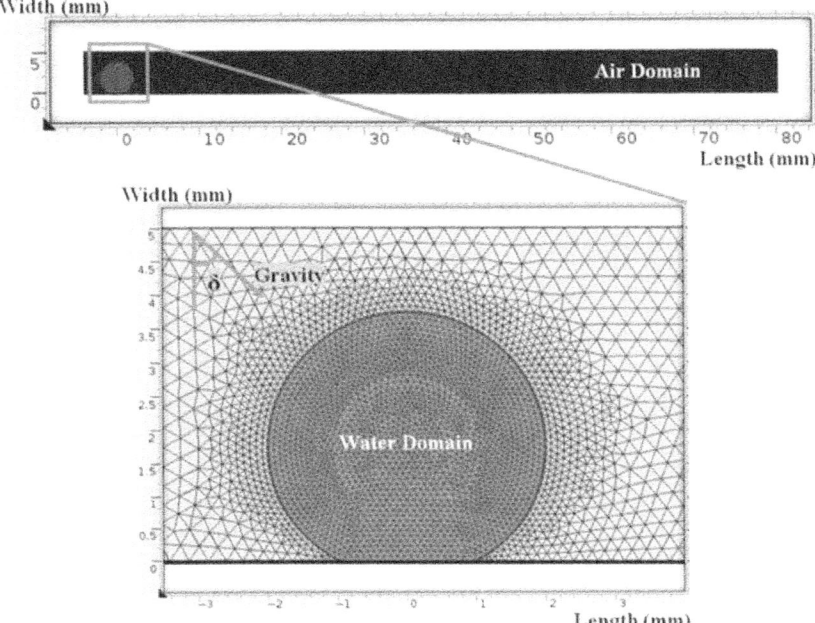

FIGURE 5.28 A typical mesh used in the numerical computations [3].

The continuity equation yields:

$$\nabla \cdot V = 0 \tag{5.22}$$

5.3.2.1 Initial Conditions

Initially water at stagnant condition is considered inside the droplet; in which case, the flow velocity is set to zero and the pressure is set to the Laplace pressure.

5.3.2.2 Boundary Conditions

The constant pressure boundary is assumed at the droplet outside; in which case, the external pressure of the droplet is set at atmospheric pressure. In addition, stagnant air is considered at the droplet outer surface, which yields zero velocity of the air initially and air drag force is generated during the droplet movement on the surface. The volume force is applied on the droplet domain in the direction of the inclination angle. To track the evolution of the water−air interface, a fluid−fluid interface is applied to the free boundary of the water droplet:

$$n \cdot (\tau_w - \tau_a) = \sigma(\nabla \cdot n)n - \nabla\sigma \tag{5.23}$$

and

$$\tau_{w,a} = -p_{w,a}I + \mu_{w,a}\left(\nabla V_{w,a} + \left(\nabla V_{w,a}\right)^T\right) \tag{5.24}$$

where n is the normal vector, σ is the surface tension, and τ is the total stress tensor. Subscripts w and a are for water and air, respectively.

This boundary condition can be decomposed into a normal component:

$$n \cdot \tau_w \cdot n - n \cdot \tau_a \cdot n = \sigma(\nabla \cdot n) \tag{5.25}$$

and a tangential component:

$$n \cdot \tau_w \cdot t - n \cdot \tau_a \cdot t = \nabla \sigma \cdot t \tag{5.26}$$

The term on the right-hand side of Eq. (5.25) is the force per unit area due to local curvature of the interface while the term on the right-hand side is a tangential stress associated with the gradients in the surface tension coefficient. A mesh velocity equal to the fluid velocity is imposed on the interface:

$$V_M = V \tag{5.27}$$

Free deformation is considered in the fluid flow domain to account for the movement of the water–air interface. Navier Slip boundary condition is considered at the bottom boundary in which it enforces the slip condition ($V.n_{wall} = 0$) and takes into account a frictional force ($F_{fr} = -(\mu/\beta)V$), where β is the slip length ($\beta = \xi/2$) and ξ is the mesh element size. Open boundary conditions are specified at the left and right edges while a symmetry boundary condition is considered at the top boundary of the computational domain. Since the time taken for the experiment and the simulation of experimental conditions is short, the evaporation from the droplet surface is neglected in the simulations. This situation is verified during the experiment and the sessile droplet images were taken after 30 and 100 seconds periods and compared with the images from the start of the experiment. It is noted that both images have identical diameter and height of the droplet. In the numerical approach, finer meshes are located in the region near the water–air interface. Mesh independence tests were conducted for each droplet contact angle considered in the simulations. After the mesh independence tests, as an example, the mesh size comprising of 17,114 cells was selected to realize the simulations (for 33.5 µL droplet). The governing equations of flow are discretized using the backward Euler finite difference method. The selection of time step is critical to ensure the accuracy of the scheme; in which case, it is in the order of 10^{-4} seconds. The residuals of flow parameters are set as $\left|\psi^k - \psi^{k-1}\right| \leq 10^{-8}$.

5.3.3 Findings of Experiments and Simulations

The findings from the experimental and simulation studies are presented here in in line with the previous study [3]. Rolling of the water droplet on the hydrophobic surface can be formulated through the force balance along the contact line in the rolling direction. Since various droplet sizes are considered in the present study ($0.52\ \mu L \leq \forall \leq 33.5\ \mu L$), the contact angle hysteresis due to each droplet size is measured. In general, the contact angle hysteresis varies within $1-3$ degrees for a pinned droplet on the hydrophobic surface. Consequently, the functionalized silica deposition results in considerably low contact angle hysteresis by creating the lotus effect. On the other hand, the force required to detach and initiate the rolling of a droplet from the surface is the same as the force needed to overcome the work of adhesion. On the other hand, in the case of inclination of the surface, the droplet undergoes elastic deformation and the total net force changes according to the droplet acceleration. The droplet puddling and wobbling during the rolling on the hydrophobic surface modify the line of action of the net force on the droplet. In addition, they cause variation of dynamic droplet hysteresis ($\theta_R - \theta_A$, where θ_R is the receding angle and θ_A is advancing of the droplet) during the rolling, which in turn alters the droplet adhesion force.

In accordance Fig. 5.29, taking the moment about the center of mass of the droplet yields:

$$mgsin\delta - F_{ad} - F_\tau - D_a = \frac{2}{5}mR\omega^2 \qquad (5.28)$$

where m is the droplet mass; δ is the inclination angle of the hydrophobic surface; F_{ad}, F_τ, and D_a are the adhesion, shear, and air drag forces, respectively; R is the droplet radius; and ω is the angle of rotation. In relation to the previous study [22], the lateral adhesion force is formulated heuristically by approximating the three-phase contact line with a single ellipse and using the experimentally obtained polynomial function for the dependence of the contact angle on the position along the three-phase contact line. The resulting equation is [23]: $F_{ad} = (24/\pi)\gamma_{LV}D(cos\theta_R - cos\theta_A)$, where γ_{LV} is the surface tension of the liquid on the solid surface, D is the droplet diameter prior to deformation (same area of the ellipse), θ_R is the receding angle, and θ_A is the advancing angle. Since the solid surface is textured, the roughness parameter can be incorporated into the lateral adhesion force equation and in line with the Young–Dupre equation [24]. In this case, the adhesion force yields $F_{ad} = (24/\pi)\gamma_{LV}Df(cos\theta_R - cos\theta_A)$, where f is the solid–surface fraction (solid–liquid contact fraction).

The shear force is generated during the droplet rolling because of the rate of fluid strain formed along the contact line between the water droplet and the hydrophobic surface. The shear stress can be written as:

$$F_\tau = A_w(\mu\frac{dV}{dy}) \qquad (5.29)$$

where A_w is the contact area ($A_w = \pi l^2/4$ and l is the contact length between the droplet and the surface), μ is the droplet fluid viscosity, V is flow

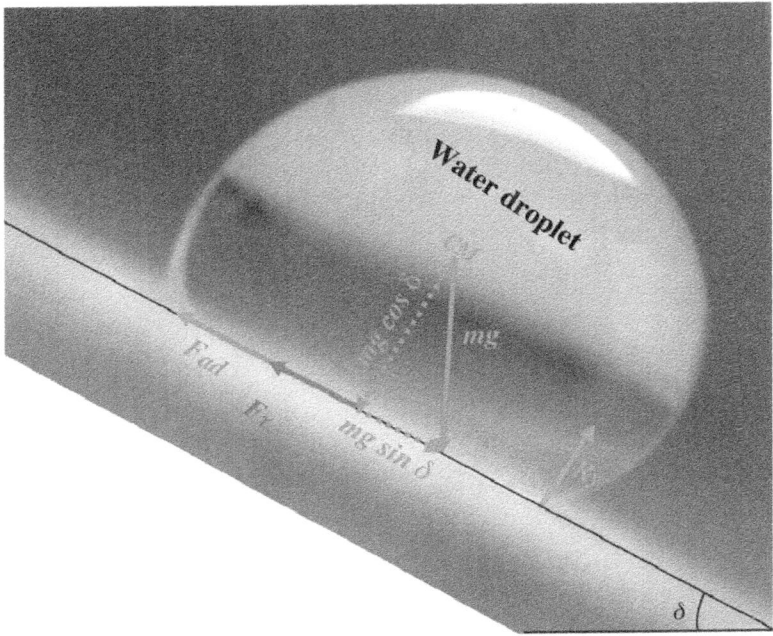

FIGURE 5.29 High-speed camera image of droplet on inclined surface and force diagram [3].

velocity, and y is distance normal to the contact surface. It should be noted that the rate of fluid strain (dV/dy) is obtained from the simulation data.

The drag force due to air resistance during the droplet rolling is related to the pressure drag and the frictional drag. However, the simplified form of the drag force for the spherical body due to air resistance can be related to $D \sim 1/2 C_d \rho_a A_c U_T$, where C_d is the drag coefficient, which is 0.5 for sphere; ρ_a is the air density; A_c is droplet cross-sectional area; and U_T is the droplet transverse speed along the inclined surface. The transverse speed of the droplet is obtained from the simulations. The rearrangement of Eq. (5.28) yields:

$$\omega = \sqrt{\frac{5}{2mR}\left(mg sin\delta - \frac{24}{\pi}\sigma f(cos\theta_R - cos\theta_A) - \mu A_w \frac{\partial u}{\partial y} - \frac{1}{2}C_d\rho_a A_c U_T^2\right)}$$

(5.30)

The rotational speed of the droplet obtained from Eq. (5.14) and measured from high-speed camera records as well as predicted from the simulations were compared. The findings show that the rotational speed obtained from Eq. (5.30) and measured from high-speed camera data are in good agreement. Since the various droplet sizes are considered in the analysis, the droplet size can be presented in the form of a dimensionless number, which corresponds to the ratio of the droplet weight over the surface tension force prior to initiation of the droplet rolling, that is, $MN = 2\gamma R^2/3\sigma$, where MN is the Merve number.

Fig. 5.30 shows the rotational speed of the droplet, obtained from the experiments and predicted from the simulations, with Merve number. Increasing MN reduces the rotational speed, which indicates that the increasing droplet size lowers the rotational speed of the droplet. This behavior is attributed to the increased adhesion and drag forces with increasing droplet radius, in line with Eq. (5.30), despite the fact that the body force under gravity increases with increasing droplet radius. Therefore, the magnitude of force opposing the droplet rolling becomes larger than the gravitational force driving the droplet sliding and rolling on the inclined hydrophobic surface. This situation can also be seen from Fig. 5.31, in which the rotational Bond number $(\rho\omega^2 R^3/8\sigma)$ with MN is shown. Since the rotational Bond number is proportional to R^3 and ω^2, increasing MN gives rise to an increase in the rotational Bond number. The droplet shape changes under the influence of

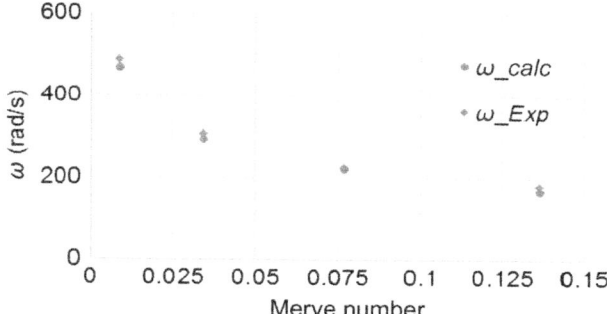

FIGURE 5.30 Droplet rotational speed of droplet obtained from experiment and calculations with MN [3].

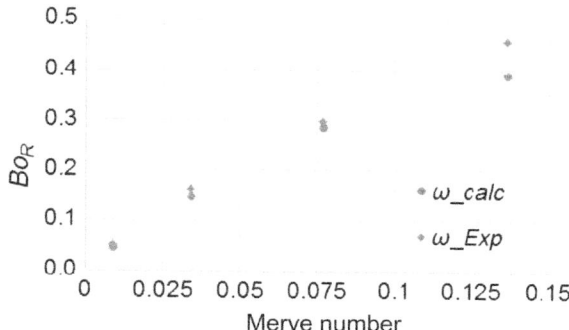

FIGURE 5.31 Droplet rotational Bond number obtained from experiment and calculations with MN [3].

gravity and the surface tension. The exact shape of the sessile droplet depends on the balance between the gravitational force, which is in favor of increasing contact area, and capillarity force, which opposes the droplet bulging and reduces the contact area between the droplet and the underlying surface. The gravitational force lowers the center of mass of the droplet by an amount δ and the difference of energy from a perfect sphere, which is tangent to the plane of the solid surface, can be approximated by $\sigma\delta^2 \cong \rho g R^3 \delta$ [44], where R is the droplet radius and σ is the surface tension of the droplet liquid. The contact length (l) between the droplet and the solid surface due to droplet bulging is related to $l = \sqrt{R\delta}$. Minimization of energy difference and using the contact length results in $\rho g R^3 \delta \sim \sigma l^4 / R^2$. This further leads to the contact length in the form $l \cong R^2 \sqrt{\sigma/\rho g}$, which is similar to that reported in the previous study [44].

The term $\sqrt{\sigma/\rho g}$ represents the capillary length. Moreover, after the mathematical arrangements, the amount of shift of the center of mass of the droplet (δ) can be reduced to $\sim R^3 \sqrt{\sigma/\rho g}$. Consequently, for the droplet of radius R larger than the capillarity length, the gravitational force flattens the droplet into a paddle. However, the droplet of radius R, which is smaller than the capillary length ($R < \sqrt{\sigma/\rho g}$), remains quasispherical. For the large droplets, the puddle thickness (h) can be expressed as $h \cong \sqrt{2(1 - cos\theta)(\sigma/\rho g)}$, where θ is the droplet contact angle [40]. On the other hand, for the rolling droplet, the puddle thickness varies because of the elastic response of the droplet, which modifies the droplet internal fluidity; in which case, the contact line dynamics modify the receding and advancing angles of the droplet while altering the adhesion force ($F_{ad} = (24/\pi)\gamma_{LV}D(cos\theta_R - cos\theta_A)$) between the droplet and the surface. This changes the force balance at the droplet—solid interface and gives rise to the droplet wobbling on the solid surface during the rolling. Fig. 5.32 shows the temporal variation of droplet height obtained from the experiment and simulations for various volumes of droplets while Fig. 5.33 shows droplet height variation with time for various angle of inclination during the rolling of the droplets on the inclined surface. The droplet height changes with time, which is more pronounced for the large volume droplets. The change is almost 24% of the total height of the droplet during the early rolling period of the droplet.

As the rolling progresses, the radial acceleration (Figs. 5.34 and 5.35) and, consequently, the rotational speed of the droplet, increases. The dynamic contact angle hysteresis ($\theta_R - \theta_A$, where is the receding and advancing angles of the droplet during rolling) reduces while reducing the adhesion force (F_{ad}) on the inclined surface. This lowers the droplet wobbling on the inclined surface. Small volume droplets with a diameter in the range of capillarity length behave like a quasisolid sphere; in which case, the maximum and minimum height of the droplet remains very close during the rolling. This situation can also be seen from Fig. 5.36, in which the dimensionless

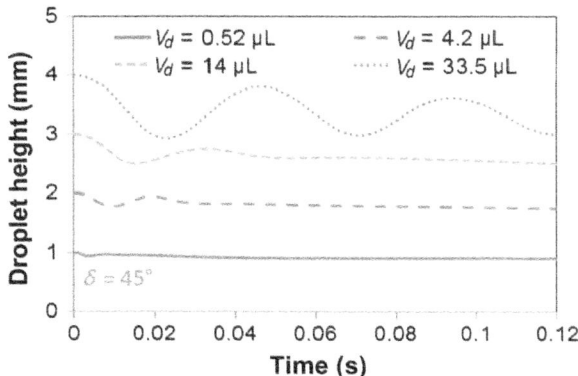

FIGURE 5.32 Droplet height (puddle) with time during rolling for various droplet volumes (V_d) [3].

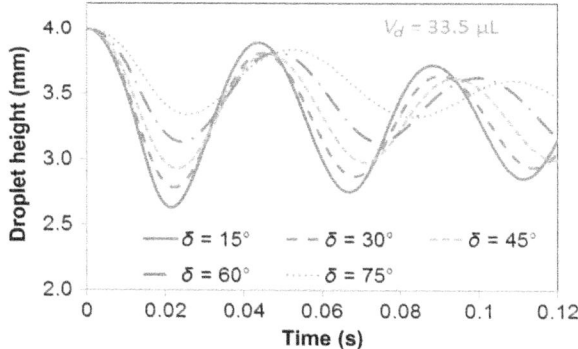

FIGURE 5.33 Droplet height (puddle) with time during rolling for various inclination angles and $V_d = 33.5 \, \mu L$ [3].

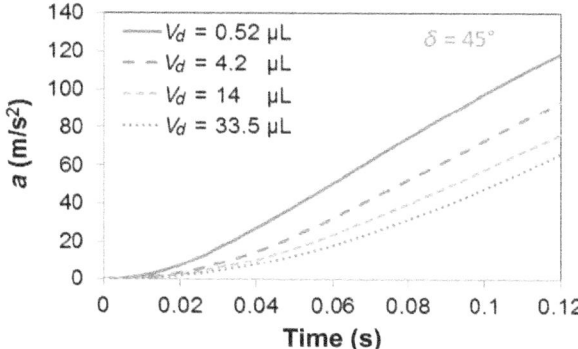

FIGURE 5.34 Droplet rotational acceleration with time for various droplet volumes (V_d) and inclination angle $\delta = 45$ degrees [3].

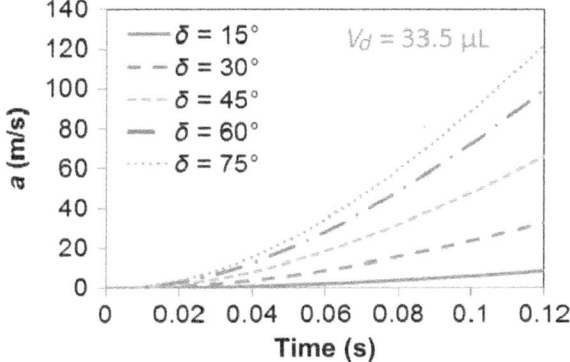

FIGURE 5.35 Droplet rotational acceleration with time for various inclination angle of surface and droplet volume $V_d = 33.5$ μL [3].

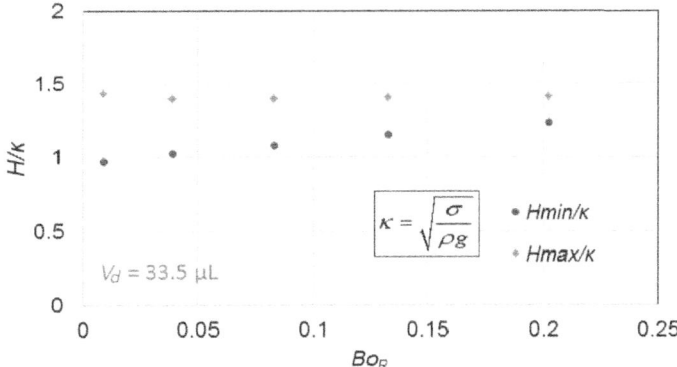

FIGURE 5.36 Normalized maximum and minimum droplet heights with rotational Bond number for droplet volume $V_d = 33.5$ μL [3].

maximum and minimum heights of the droplet with rotational Bond number ($\rho\omega^2 R^3/8\sigma$) are given.

The maximum and minimum droplet heights are nondimensionalized with the capillarity length ($\sqrt{\sigma/\rho g}$). The difference between the maximum and the minimum heights remains small for small values of the Bond number ($Bo_R < 0.02$). The rotational Bond number is proportional to the cubic power of the radius (R^3); therefore, the droplet wobbling ceases for the small droplets. However, the rotational Bond number increases for the large droplets, which in turn gives rise to the enhancement of difference in the maximum and the minimum droplet heights during the rolling. The relation between the dimensionless droplet height and the Bond number is rather parabolic for small values of the rotational Bond number ($Bo_R < 0.02$); however, the relationship becomes almost linear as the Bond number increases ($Bo > 0.02$). For example, for water droplet of volume $V_d = 0.52\,\mu\text{L}$ ($Bo_R = 0.005$), the difference between the droplet maximum and minimum heights is in the order 1%; however, it is almost 40% for the droplet volume of $V_d = 33.5\,\mu\text{L}$ ($Bo_R = 0.075$). Therefore, for small diameter droplets, the effect of the droplet rotation and the droplet radius on the droplet geometric symmetry becomes nonlinear, despite the fact that the deviation from the droplet geometric symmetry is small. On the other hand, the difference between the droplet maximum and minimum heights remain large as the inclination angle of the surface changes. This situation can be seen from Fig. 5.37, in which the maximum and the minimum droplet height with inclination angle of the surface are shown.

In the early rolling period, droplet rotational speed remains low irrespective of the inclination angle. This gives rise to large wobbling of the droplet. As the droplet rotational speed increases with progressing time, the difference between the maximum and the minimum height of the droplet reduces. Hence, increasing the inclination angle reduces the droplet wobbling and the difference between the maximum and minimum heights of the droplet. In this case, the droplet rotational speed increases while lowering the droplet height difference. This situation can also be seen from Fig. 5.37. Since the rotational Bond number is related to ω^2, the difference between the maximum and the minimum droplet height remains low for large values of the rotational Bond number. On the other hand, the ratio of translational speed (U_T) over the rotational speed (ωR) influences the droplet wobbling during rolling [45]; however, in the current study, the rotational and translational speeds have similar orders, that is, $\omega R/U_T \sim 0.9$. In addition, the ratio of rotational speed over the translational speed is critical for the relative magnitude of the dynamic pressure generated between the droplet center and the droplet ambient pressure [45]. In this case, for the

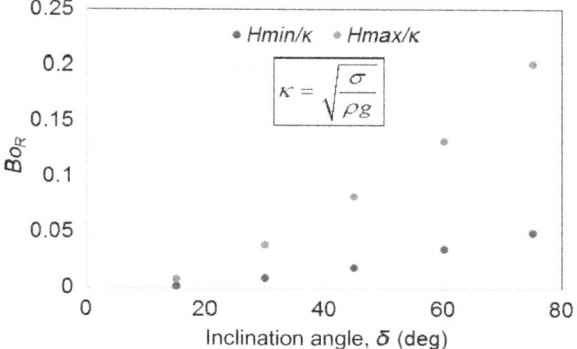

FIGURE 5.37 Normalized maximum and minimum droplet heights with inclination angle of surface [3].

condition $\varphi = \Delta\rho\omega^2 R^2/\rho_a U^2 \gg 1$ (where $\Delta\rho$ is the density difference between the droplet liquid and the droplet ambient gas, ρ_a is the droplet ambient gas density), the dynamic pressure associated with the translation speed does not have a significant effect on the droplet wobbling consistent with the previous finding [46], that is, φ is in the order of 900; therefore, the effect of the dynamic pressure is not significant on the droplet wobbling.

Fig. 5.38 shows the velocity contours for the droplet of volume $V_d = 14 \ \mu L$ for 40 mm located on the inclined surface while Fig. 5.39 shows locations of droplet of various sizes on the inclined surface with the same time period ($t = 0.12$ seconds). It should be noted that the inclination angle of the hydrophobic surface is selected as 45 degrees to demonstrate the flow field. In general, a single circulation cell is developed inside the droplet during rolling. The center of the circulation cell almost coincides with the droplet mass center. The presence of the regular flow patterns indicates that the flow inside the droplet is laminar in nature. The flow Reynolds number based on the average velocity and the contact length (l) is in the range of 32–400, which changes at different droplet locations on the hydrophobic surface. The fluid velocity in the contact region of the rolling droplet changes on the inclined hydrophobic surface. This is particularly true in the region of the first and the last contact points of the droplet on the hydrophobic surface, which can be seen from the velocity contours in Fig. 5.38. For example, for the droplet volume ($V_d = 14 \ \mu L$ and $R = 1.5$ mm) and the droplet location of 40 mm on the inclined surface (rotational speed, $\omega = 225$ rad/s), the maximum fluid velocity is in the order of 0.31 m/s while the fluid velocities in the region of the first and the last contact points on the surface are in the order of 0.2 and 0.1 m/s. Therefore, the viscous dissipation takes place in this region.

In addition, despite the small diameter of the droplet, the flow field inside the droplet does not follow the velocity field of the solid body rotation; consequently, the elasto-viscous behavior of the fluid during the droplet rolling results in the viscous dissipation inside the droplet. The local capillary number ($Ca = \mu V/\sigma$) varies in the fluid within the range of 0.002 and 0.0027. In this case, increasing the rotational Bond number ($\rho\omega^2 R^3/8\sigma$) increases the flow velocity and, consequently, the viscous dissipation inside the droplet, according to $\sim \mu \int_{Vd}(\nabla \mathbf{u})^2 dV$, where $\mathbf{u}$ is the velocity vector in the droplet fluid and V is the volume. The progression of rolling droplet shape is evident at different locations on the hydrophobic surface; in which case, the round shape evolves ellipsoid shape in the droplet, which is consistent with that reported in the previous study [47]. The rate of fluid strain in the close region of the first and second contact points of the droplet on the surface is high, which results in the shear stress along the contact line. However, due to the low rate of fluid strain and its localized effect, the shear stress and shear force become small; in which case, the shear stress is in the order of

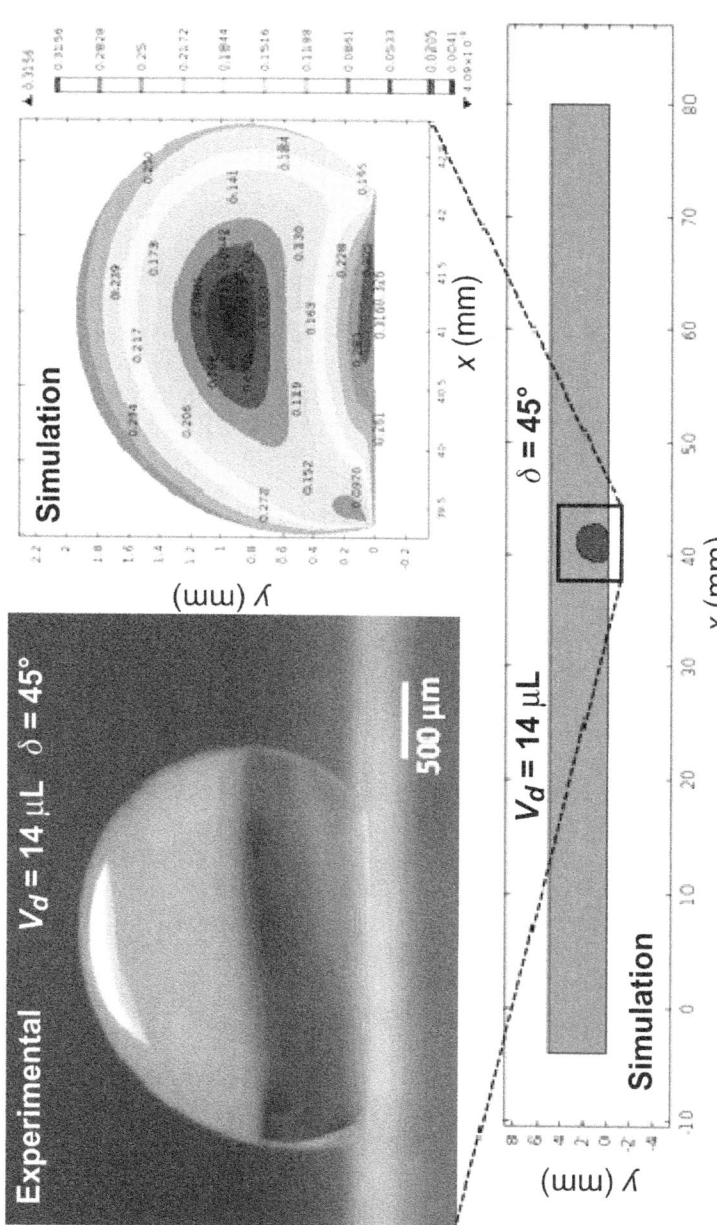

FIGURE 5.38 Velocity field inside droplet at location 40 mm on the inclined surface [3].

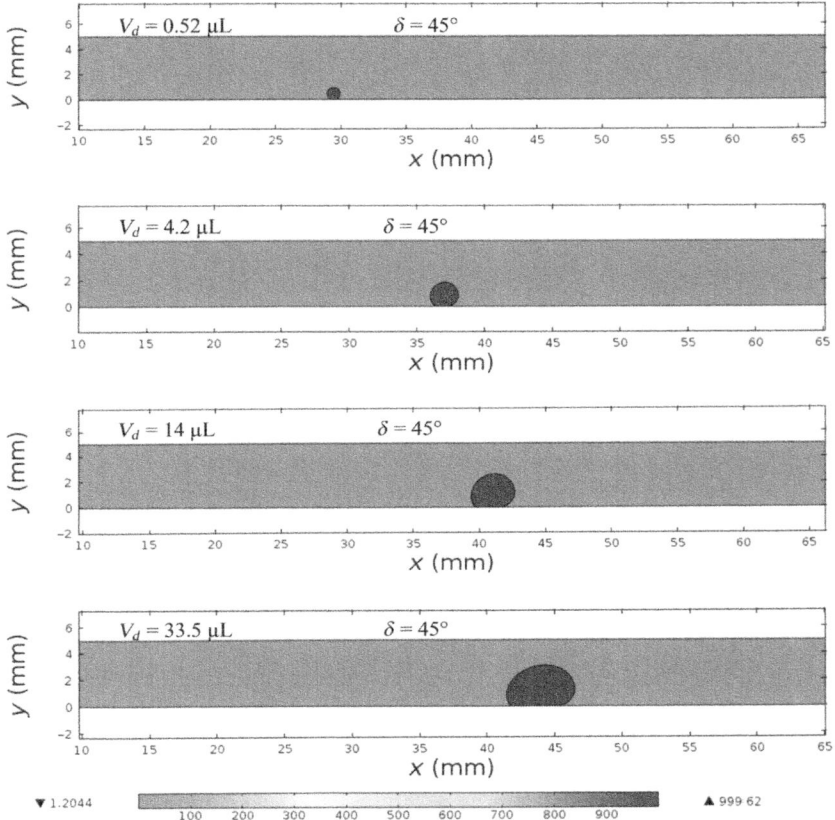

FIGURE 5.39 Locations of different size droplets after 0.12 s when the inclination angle $\delta = 45$ degrees. The red color (light gray in print version) represents the air zone while the blue color (dark gray in print version) represents the water zone [3].

0.03 N/m^2 and the shear force acting on the contact surface is in the order of 5×10^{-8} N.

Although the distance traveled by the small droplet ($V_d = 0.52$ μL) appears to be shorter than the other droplet sizes, the rotational speed of the droplet remains higher than those of the large size droplets. This finding reveals that small size droplet attains higher rotational acceleration than those of the large droplet sizes (Figs. 5.34 and 5.35). This behavior is attributed to the reduced wobbling of the small size droplet, which gives rise to a smaller contact length ($l \cong 0.0003$ m) than the large droplets ($V_d = 33.5$ μL and $l \cong 0.002$ m), that is, the adhesion and the shear forces remain smaller for the small droplets than those of the large droplets. Consequently, high rolling and small sliding velocities result in less dissipative energy of the droplet via resulting small adhesion work and frictional work due to low fluid shear rate.

5.4 DROPLET DYNAMICS ON SURFACES WITH REVERSIBLE-EXCHANGE WETTING STATES

Reversible exchange of the wetting state of surfaces finds applications in advanced multifunctional systems [48] and is important for extraction, separation, surface chemistry, life science, and organic solvents [49]. One of the methods for achieving reversible exchange of wetting state is applying a thin coating of phase-change material on a hydrophobic surface [50]. The surface remains hydrophobic in the solid phase of the phase-change material while it becomes hydrophilic in liquid phase of the phase-change material [50]. The wettability in terms of spreading rate (S) of the phase-change material in a liquid phase plays a crucial role to secure the formation of continuous film on the hydrophobic surface. Since the surface remains hydrophilic in the liquid phase of phase-change material, the interfacial and surface free energy of the liquid phase and the solid surface should satisfy the condition for the spreading rate $S > 0$. Therefore, the selection of phase material becomes important to give rise to a positive spreading rate on the hydrophobic surface in the liquid phase. One of the candidates of the phase-change material fulfilling the requirements of positive spreading rate is n-octadecane, which has low solidus and liquidus temperatures and positive spreading rate for most hydrophobic surfaces in the liquid phase. Consequently, the water-droplet mobility on the hydrophobic surface with reversible exchange of wetting state incorporating the phase-change material on the surface becomes fruitful. The reversible exchange of wetting state was studied for silicon nanowires [50]; however, the process can be extended to include hydrophobic surfaces from different techniques such as solution crystallization, surface ablation, electrospinning, and others. Hence, in the present section, the dynamic behavior of a water droplet on the reversible exchange of the wetting state of the inclined hydrophobic surface is analyzed in the line of the previous study [6]. Initially, solution crystallization of polycarbonate surface is carried out using acetone. The resulting surface morphologies are examined using the analytical tools. In order to increase the water-droplet contact angle and reduce the contact angle hysteresis, the functionalized silica particles are deposited on the crystallized polycarbonate surface and the resulting surface. To achieve reversible exchange of wetting state, a thin coating of n-octadecane, as a phase-change material, is introduced on the functionalized silica particle-deposited surface. The surface remains hydrophobic when n-octadecane is in solid phase while the wetting state changes to hydrophilic when n-octadecane liquefies on the functionalized silica particle-deposited surface. In this case, the surface characteristics for reversible exchange of wetting state are examined using the analytical tools and the high-speed camera to record the data for droplet rolling and sliding behavior on the inclined hydrophobic surface with reversible-exchange wetting state.

5.4.1   Experimental

The experimental study is presented in light of the previous study [6]. A polycarbonate wafer (3 × 20 mm × 20 mm, thickness, width, and length) and acetone were used for solution crystallization. The polycarbonate wafers were immersed in 60% concentrated acetone in acetone−water mixture for 4 minutes in line with the previous study [51]. This process resulted in a hydrophobic surface with water-droplet contact angle of $\theta_w = 130$ degrees and contact angle hysteresis of 36 degrees. In order to improve the contact angle and reduce the contact angle hysteresis, the functionalized silica particles were deposited onto crystallized polycarbonate surface. The silica nanosize particles were produced through synthesizing process. Tetraethyl orthosilicate (TEOS), 3-aminopropyltrimethoxysilane (AMPTS), isobutyltrimethoxysilane (OTES), ethanol, and ammonium hydroxide were used in the synthesizing process in line with the previous study [41]. In this case, 14.4 mL of ethanol, 1 mL of ultrapure water, and 25 mL of ammonium hydroxide were mixed and stirred for 12 minutes. Later, 1 mL TEOS diluted in 4 mL ethanol was added in the mixture. Following 25 minutes after this process, 0.5 mL of TEOS diluted in 4 mL ethanol was added. After 5 minutes, a modifier silane molecule was added in a molar ratio of 3:4 with respect to the second edition of TEOS. The final mixture was stirred for 18 hours at room temperature, and later centrifuged and washed with ethanol for complete removal of reactants. The solvent casting applied deposition of the functionalized silica particles onto the solution-crystallized polycarbonate surface. Upon vacuum drying until all solvent was evaporated, characterization of resulting surfaces was carried out. This arrangement gave rise to the water contact angle in the order of 160 degrees with contact angle hysteresis of 2 degrees. The surface morphologies and texture characteristics of the solution-crystallized and functionalized silica particle-deposited surfaces were characterized with a focused ion beam field emission dual-beam scanning electron microscope.

The coating of n-octadecane, as a phase-change material, was introduced onto the functionalized silica particle-deposited crystallized polycarbonate surface. The dip-coating technique was used for the n-octadecane coating of the surface while keeping n-octadecane in the liquid phase at constant temperature (306 degrees) during the coating process. The liquid film thickness of n-octadecane on the surface was measured using the ellipsometer (Model: M-2000 Manufacturer: J.A. Woolam Co., USA). The liquid n-octadecane film thickness measurement relied on the change in polarization state as defined by the quantities: (1) amplitude ratio (Ψ), and (2) phase difference (Δ), that is, $tan\Psi e^{i\Delta} = R_p/R_s$, where R_p and R_s are the Fresnel reflection coefficients for the p- and s-polarized light, respectively. A spectroscopic ellipsometer provided the measured data for Ψ and Δ value for each wavelength of incident optical radiation and generated the spectrum accordingly.

The measured liquid n-octadecane film thickness was in the order of 1.5 μm ± 20 nm.

A goniometer (Kyowa, model—DM 501) was used to conduct sessile drop tests for the measurement of the droplet contact angle. Desalinated water was used in the sessile drop experiments, and during the measurements, the droplet volume was controlled by an automatic dispensing system with a volume step resolution of 0.1 μL. Still images were captured, and contact angle measurements were performed 1 second after deposition of the water droplet on the surface in line with a previous study [52]. The dynamic behavior of the water droplets was recorded using a high-speed camera (Model: SpeedSense 9040, Manufacturer: Dantec Dynamic, Denmark). The experiments were repeated 10 times to observe the repeatability of the high-speed recorded data, and it was observed that the error related to translational velocity of the droplet measurement was in the order of 3% while it was in the order of 4% for the instantaneous contact angle and contact angle hysteresis measurements.

5.4.2 Surface Morphology and Reversible Wetting State on Surfaces

Fig. 5.40 shows SEM micrographs of solution-crystallized polycarbonate prior and after the deposition of functionalized silica particles. The solution crystallization resulted in hierarchically distributed micro/nanosize spherules and fibrils on the entire surface (Fig. 5.40A).

The fibrils partially cover the top of the spherule surface. The average roughness of the surface is in the order of 3.4 μm. The crystallized surface demonstrates the hydrophobic behavior and gives rise to the sessile droplet

FIGURE 5.40 SEM micrographs of solution-crystallized polycarbonate surface prior to and after functionalized silica particles deposition: (A) Crystallized polycarbonate surface composing of globules and fibrils while functionalized silica particle-deposited surface resulting in spongy-like structures [6].

contact angle in the order of 130 degrees and contact angle hysteresis of 36 degrees, which are similar to those reported in the previous study [51]. However, high contact angle hysteresis enhances the droplet pinning on the surface while reducing the droplet mobility. In order to lower the contact angle hysteresis and generate the lotus effect on the crystallized polycarbonate surface, functionalized silica particles are deposited on the surface. The functionalized silica particles have a 30 nm average size and are closely spaced on the surface (Fig. 5.40B). In addition, a few pore-like structures are formed on the surface (Fig. 5.40B), which are associated with agglomeration of the functionalized silica particles. It should be noted that TEOS is used during the synthesizing of silica particles and the surface roughness of the particles can be modified slightly by the functionalized shells and the size of the cells increases slightly in the region of porous-like structures (spongy texture) [41]. This behavior is associated with the condensing of monomer units during functionalizing of the silica particles. In this case, the units can grow at a faster rate than the nucleation rate, which in turn gives rise to the agglomeration of functionalized silica particles. After the functionalized silica particle deposition, the average surface roughness remains in the order of 3.4 μm, which is the same as that of the crystallized polycarbonate surface. The sessile droplet contact angle is measured on the various locations of the functionalized silica particle-deposited surface and the findings revealed that the contact angle remains almost the same on the surface, which is in the order of 160 degrees, and the contact angle hysteresis is 2 degrees. Consequently, deposition of functionalized silica particles improves the water-droplet contact angle significantly and lowers the contact angle hysteresis on the surface. In order to develop reversible exchange of wetting state on the functionalized deposited silica particles, the thin film of n-octadecane is coated. Since the liquidus and solidus temperatures of n-octadecane are low ($T_{liquidus} = 303.15K$ and $T_{solidus} = 301.15K$), the coating remains solid at temperatures less or equal to the solidus temperature and forms a uniform thickness of thin liquid film for temperatures greater or equal to the liquidus temperature. The formation of the continuous liquid film of n-octadecane on the functionalized silica particle-deposited surface depends on the spreading rate of the liquid phase on the surface. The condition for the liquid phase of n-octadecane encapsulating the texture of functionalized silica particles should satisfy $\theta_{L-s(a)} < cos^{-1}(1/r)$ and the spreading rate $S_{L-s(a)} \equiv -\gamma_{L-s}(r - 1/r)$ or $S_{L-s(a)} \geq 0$ [26,27]. Here, $\theta_{L-s(a)}$ is the contact angle of the droplet on the liquid phase of n-octadecane film ($\theta_{L-s(a)} = 85degrees$), which is deposited on the functionalized silica particles; r ($=0.82$) is the ratio of the texture area to the projected area of the solid; γ_{L-s} is the interfacial energy between the liquid phase of the change material and the functionalized silica-deposited surface; and $S_{L-s(a)}$ is the spreading rate of liquid n-octadecane film on the functionalized silica particle-deposited surface in air. The first condition is satisfied, since $1/r = 1.22$.

The spreading rate of the liquid phase of n-octadecane on the textured surface is in the order of $S_{ls(a)} = 2.95$ mN/m, which is greater than zero. It should be noted that the surface energy of the liquid phase of n-octadecane is 21.6 mN/m [53]. Therefore, the liquid phase of the n-octadecane film encapsulates the texture of the functionalized silica particle-deposited surface. However, the solidified n-octadecane film is decomposed into the solid flakes at the surface of the functionalized silica particles (Fig. 5.41A). This, in turn, allows the functionalized silica particles to emerge to the free surface of the coating, which can be observed from (Fig. 5.41A). The solid phase n-octadecane flakes are irregularly distributed on the surface (Fig. 5.41B). Consequently, the n-octadecane coating partly covers the functionalized silica particle-deposited surface; in which case, the area ratio of the emerging functionalized silica particles over the area of n-octadecane coating on the surface is in the order of 32%. The resulting surface with the presence of the solid phase of n-octadecane flakes alters the wetting state of the surface from hydrophilic to hydrophobic. In this case, the water-droplet contact angle becomes in the order of 140 degrees and the contact angle hysteresis is 8 degrees. Upon melting of the solid flakes of n-octadecane on the functionalized silica-deposited surface, via increasing surface temperature above the liquidus temperature of n-octadecane, a constant thickness of the liquid film of n-octadecane is formed on the surface and the wetting state changes from hydrophobic to hydrophilic. The film thickness of n-octadecane in the liquid phase is measured using the ellipsometer

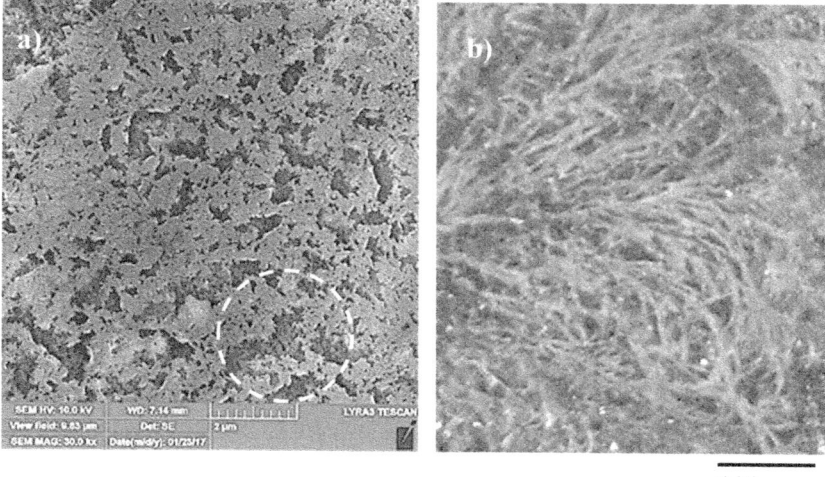

100 μm

FIGURE 5.41 SEM micrograph and optical image of resolidified n-octadecane layer on functionalized silica particle-deposited surface: (A) SEM micrograph of n-octadecane composing of solid flakes and emerging silica particles (dotted circle), (B) optical image of solid n-octadecane demonstrating an irregular pattern of solid flakes [6].

and the liquid film thickness is in the order of $1.5 \, \mu m \pm 20 \, nm$. Moreover, repeating melting and resolidification of n-octadecane film results in the reversible exchange of the wetting state on the functionalized silica particle-deposited surface.

5.4.3 Water-Droplet Dynamics on Solid-Phase n-Octadecane-Coated Hydrophobic Surfaces

The droplet dynamics, including rolling and sliding, are related to the droplet contact angle, the contact angle hysteresis, and the inclination angle of the hydrophobic surface. The analysis and findings are presented in line with the previous study [6]. The large contact angle and low contact angle hysteresis are favorable for droplet rolling and sliding on the inclined surface. The large contact angle hysteresis gives rise to droplet pinning on the inclined surface [38]. In this case, the adhesion force developed laterally along the three-phase contact line slows the droplet motion or pins the droplet on the surface. In addition, the internally generated shear force over the wetted surface at the droplet—solid interface, due to flow field developed inside the droplet during dynamic motion, contributes to the droplet pinning on the inclined surface. On the other hand, when the gravitational force acting on the droplet becomes larger than those of the adhesion and frictional forces generated on the inclined surface, the droplet can move via rolling and/or sliding. The bulging/puddling of the droplet under gravitational force influences the droplet motion on the inclined surface [39]. In this case, the size of the droplet remains important for the droplet bulging/puddling on the inclined hydrophobic surface. However, small diameter droplets remain spherical during rolling/sliding of the droplet on the inclined surface. The criterion for the droplet size is related to capillarity length ($\kappa^{-1} = \sqrt{\sigma/\rho g}$, where κ^{-1} is the capillarity length, σ is the surface tension, ρ is the density, and g is the acceleration due to gravity); in which case, the droplet diameter less than the capillarity length becomes spherical as it rolls/slides on the inclined hydrophobic surface. However, the droplets with diameter larger than the capillarity length can also roll and slide. In this case, the droplet center of mass changes inside the droplet during rolling/sliding on the hydrophobic surface, which gives rise to the change of droplet height while causing the droplet wobbling. The droplet wobbling alters the rotational dynamics of the droplet and affects the droplet internal fluidity. The gravitational force lowers the droplet center of mass by a distance λ during droplet wobbling. The gravitational and surface energies are of the same order at equilibrium, which can be expressed as $\rho g R^3 \delta \sim \gamma_f l^4/R^2$ [54], where ρ is the droplet fluid density; R is the droplet radius; γ_f is the surface tension of the droplet liquid; and l is the length of the contact zone of the droplet on the surface, which is related to $l = \sqrt{R\lambda}$ [44]. The term $\gamma_f l^4/R^2$ is associated with the increase of surface energy, if the droplet deformed due to wobbling is compared to a liquid sphere located on the horizontal solid surface. Introducing the length of contact zone ($l = \sqrt{R\lambda}$) into surface energy ($\gamma_f l^4/R^2$) and equating with the gravitational energy ($\rho g R^3 \delta$) results in the droplet center of mass (λ), which becomes $\lambda \sim R^3(\rho g/\gamma_f)$ or $\lambda \sim R^3 \kappa^2$, where $\kappa^{-1} = \sqrt{\gamma_f/\rho g}$ represents the capillary length. The length of contact zone becomes $l \cong R^2 \sqrt{\rho g/\gamma_f}$ or $l \cong R^2 \kappa$ which is similar to the estimation reported in the previous study [54]. Moreover, a droplet with a diameter larger than the capillarity length undergoes bulging while forming a puddle. This occurs because of the force balance between the capillarity and gravitational forces. The puddle height of a droplet (h) was formulated previously [40] in the form of droplet contact angle, fluid density, center of gravity, and surface tension of the droplet fluid, which takes the form $\sqrt{2(1 - cos\theta)(\sigma/\rho g)}$, where θ is the droplet contact angle. On the other hand, the droplet on an inclined hydrophobic surface suffers from the elastic deformation due to wobbling, which in turn modifies the line of action of the net force inside the droplet. Droplet wobbling also alters the dynamic hysteresis of the droplet ($\theta_R - \theta_A$), where θ_R is the receding angle and θ_A is the advancing angle of the droplet during rolling, while influencing the droplet adhesion on the hydrophobic surface. The force balance for a steadily rolling droplet around the center of mass can be written as [6]: $mgsin\delta - F_{ad} - F_\tau - D_a = (2/5)mR\omega^2$, where m is the droplet mass; δ is the inclination angle of the hydrophobic surface; F_{ad} and F_τ are the adhesion and shear forces between the surface and droplet during rolling, respectively; D_a is the air drag force; R is the droplet radius; and ω is the angle of rotation.

A flow field is generated inside the droplet during droplet movement, which in turn results in a shear strain and shear stress acting on the contact area between the water droplet and the hydrophobic surface. Therefore, the resulting shear force can be written as $F_\tau = A_w(\mu(dV_f/dy))$, where A_w is the contact area ($A_w = \pi r^2$, where r is the radius of the contact area); μ is the viscosity of the droplet fluid; V_f is the flow velocity inside the droplet; and y is the distance normal to the contact surface. The droplet movement on the hydrophobic surface results in a drag force due to air resistance, which is associated with the form drag (pressure drag) and frictional drag. For a spherical body at low air velocity, the drag force, because of air resistance, is the function of the flow Reynolds number. The drag force can be simplified in the form of $D \cong 1/2 C_d \rho_a A_c U_T^2$, where C_d is the drag coefficient [55] and U_T is the air velocity opposing the droplet during droplet motion on the hydrophobic surface. The air flow velocity can be the same order as the translation velocity of the droplet on the inclined hydrophobic surface, that is, $U_T \cong V$, where V is the translational velocity of the droplet. The rotational speed of the droplet becomes:

$$\omega = \sqrt{((5/2mR)(mgsin\delta - (24/\pi^3)f\gamma D(cos\theta_R - cos\theta_A) - \mu_r A_w(\partial V_f/\partial y))/1 + (5/4m)C_d \rho_a A_c R)}.$$

In order to obtain the rotational speed, the contact angle hysteresis on the inclined surface needs to be known. In this case, high-speed camera images are used to determine the contact angle hysteresis of the droplet when moving on the inclined hydrophobic surface.

Fig. 5.42 shows the high-speed camera images of the droplet at various locations on the hydrophobic surface in the presence of solid phase n-octadecane coating on the surface of the functionalized silica particles. In addition, the contact angle hysteresis is shown in Fig. 5.43 along the inclined surface for various droplet volumes. It should be noted that the inclination angle of the hydrophobic surface is 45 degrees. On the other hand, energy dissipation takes place during droplet movement on the inclined surface. The energy losses are related to the fluid friction across the wetted surface area, body deformation during wobbling, and resistance due to adhesion force. The shear stress generated due to the rate of fluid strain developed inside the droplet at the contact area of the hydrophobic surface is small [56]; however, it contributes to the energy dissipation during the droplet motion on the inclined hydrophobic surface. The energy conservation can yield the formulation of droplet velocity on the surface. The change of the droplet potential energy remains the same as the sum of the energy dissipation and kinetic energy change during the droplet transition on the hydrophobic surface. Therefore, it satisfies the equation $\Delta E_{Tot} = \Delta E_{loss} + \Delta E_{kinetic}$, where ΔE_{tot} ($\Delta E_{tot} = mg\Delta h$, where m is the mass of the droplet; g is the gravitational acceleration; and Δh is the elevation between the droplet location and the reference level) represents the potential energy change of the droplet along the inclined hydrophobic surface. ΔE_{loss} is the dissipation energy, which can be written in the form of $\Delta E_{loss} = \Delta E_{deformation} + \Delta E_{adhesion} + \Delta E_{shear} + \Delta E_{air\text{-}drag}$, where $\Delta E_{deformation}$ is the elastic deformation of the droplet during wobbling, $\Delta E_{adhesion}$ is the work done against droplet adhesion due to dynamic contact angle hysteresis, and $\Delta E_{air\text{-}drag}$ is the work done against air drag during the movement of the droplet.

The energy dissipation during deformation of the droplet due to wobbling can be described as $\Delta E_{deformation} = \forall_p \Delta P = \forall_p \gamma_L ((D_{h_1} - D_{h_2})/(D_h D_{h_2}))$, where $\forall p$ is the droplet volume; ΔP is the pressure change inside the droplet during wobbling; γ_L is the surface tension of the droplet fluid; D_{h_1} is the instant hydraulic diameter of the droplet at a location on the inclined hydrophobic surface; and D_{h_1} and D_{h_2} are the changes in the hydraulic diameter of the water droplet along the distance ΔL on the inclined hydrophobic surface due to wobbling. The energy dissipation due to the adhesion force can be described as $\Delta E_{adhesion} = (24/\pi^3)\gamma_L Df(cos\theta_R - cos\theta_A)\Delta L$, where θ_A is the dynamic advancing angle and θ_R is the dynamic receding angle, which change along the distance ΔL during rolling/sliding of the droplet. The energy dissipation due to fluid friction, because of the rate of fluid strain, can be described as $\Delta E_{shear} = A_w(\mu_t(dV_f/dy))\Delta L$, where A_w is the contact area ($A_w = \pi r^2$, where r is the contact area radius); μ_t is the droplet fluid

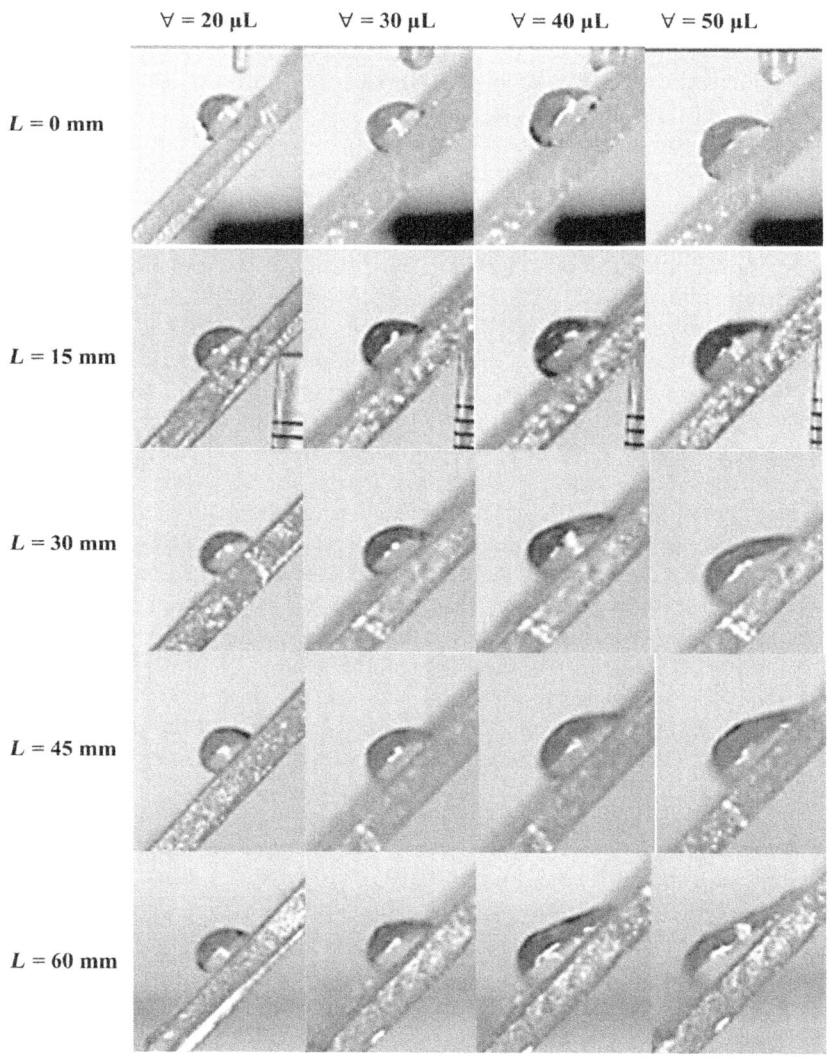

FIGURE 5.42   Images of droplets on solid phase of *n*-octadecane-coated functionalized silica-deposited surface for different droplet volumes and at different locations on the inclined surface [6].

viscosity; V_f is the flow velocity of the droplet; and y is the distance normal to the contact surface. The energy dissipated due to the air drag can be presented as $\Delta E_{air-drag} = (\Delta L/2)C_d \rho_a A_c U_T^2$, where the tangential velocity (U_T) is determined from the angular rotation of the droplet after introducing the hydraulic radius ($D_H/2$ is the instant hydraulic diameter). The energy balance

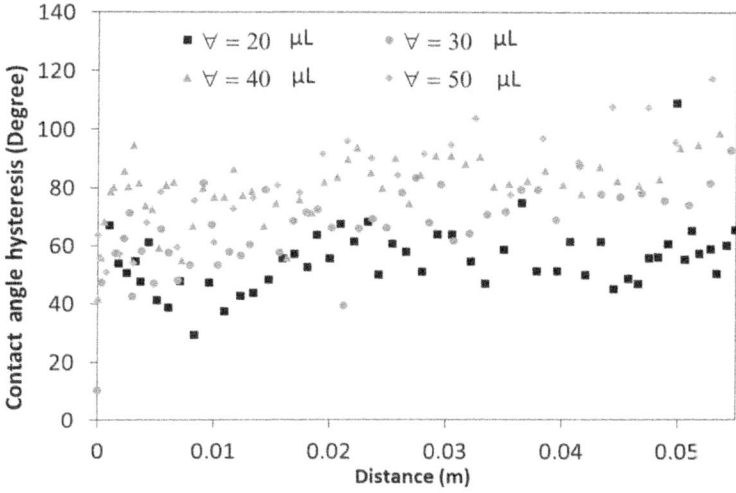

FIGURE 5.43   Water-droplet contact angle hysteresis ($\theta_R - \theta_A$, where R represents the receding and A corresponds to advancing angles) along the inclined solid phase of n-octadecane-coated functionalized silica-deposited surface for various droplet volumes [6].

of the droplet ($\Delta E_{Tot} = \Delta E_{loss} + \Delta E_{kinetic}$) allows formulation of the droplet velocity on the inclined hydrophobic surface along the distance Δl:

$$
V = \sqrt{
\begin{array}{c}
2g[\Delta L\sin\alpha - \dfrac{1}{mg}\dfrac{24}{\pi^3}\gamma_L Df \Delta L(\cos\theta_R - \cos\theta_A) \\[2mm]
- \dfrac{4\gamma_L}{\rho g \Delta L}\left(\dfrac{D_{h_1} - D_{h_2}}{D_{h_1} D_{h_2}}\right) - \dfrac{1}{mg}A_w\left(\mu_t\dfrac{dV_f}{dy}\right)\Delta L - \dfrac{\Delta L}{mg}C_d\rho_a A_c U_T^2]
\end{array}
}
$$

(5.31)

In order to assess the droplet translational and rotational velocities on the n-octadecane-coated inclined hydrophobic surface, the findings of Eq. (5.31) are compared with high-speed camera data.

Fig. 5.44 shows the translational velocity of the droplet obtained from Eq. (5.31) and the experiment incorporating the high-speed camera data on the inclined hydrophobic surface for various droplet sizes. The high-speed camera recordings were repeated ten times and the images of the droplet at various locations were compared in terms of droplet advancing, receding angles, droplet wetting length on the surface, droplet width, and droplet height. The experimental errors related to the measurement are: (1) in terms of variations of droplet height, droplet wetting length on the surface, and droplet width is in the order of 3%; and (2) in terms variations of advancing and receding angles is in the order of 4%. Since the work done against

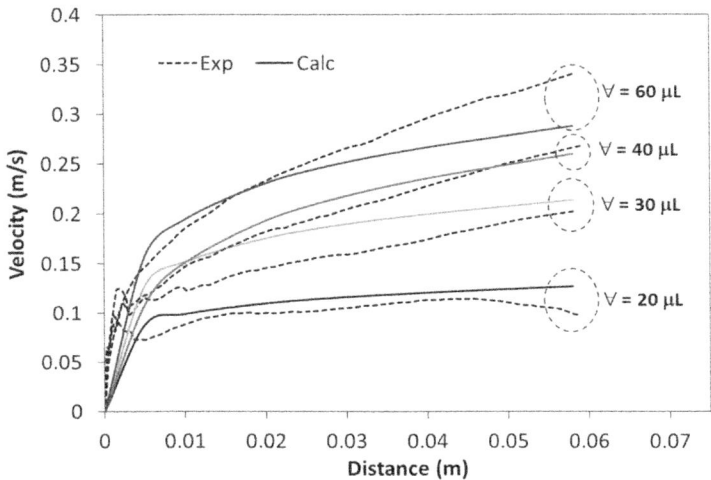

FIGURE 5.44 Water-droplet translational velocity along the inclined solid phase of n-octade-cane-coated functionalized silica-deposited surface for various droplet volumes [6].

the adhesion force is associated with the advancing and receding angles, the error related to the measurement of advancing and receding angles of the droplet at each location of the inclined hydrophobic surface is adapted in the calculation of translational velocity of the droplet.

Scale analysis was carried out to assess the influence of each term in Eq. (5.31) on the droplet velocity. In this case, the work done against adhesion is considered as the reference for the scale analysis of the work of resistance. In this case, the ratio of potential energy change over the adhesion work is in the order of 1.662, the ratio of deformation work during rolling and sliding of the droplet over the adhesion work is in the order of 0.0072, the ratio of shear work due to the rate of fluid strain inside the droplet over the adhesion work is in the order of 0.0002, and the work done against the air drag is in the order of 0.0016. Hence, the work done against the droplet adhesion is the largest in terms of the energy dissipation during the rolling and sliding of the droplet. The droplet velocity increases rapidly and reaches almost the terminal velocity after the small distance from the starting point on the inclined hydrophobic surface. The difference in the experimental and predictions from Eq. (5.31) is small; however, this difference is mainly associated with the instant advancing and receding angles of the droplet measured on the inclined hydrophobic surface via using the high-speed camera data. The surface of the solidified n-octadecane coating comprises of solid flakes and locally exposed functionalized silica particles (Fig. 5.41A). The area covered by flakes or by exposed functionalized silica particles, emerging besides the solid flakes, varies arbitrarily over the total surface area of the resolidified n-octadecane coating (Fig. 5.41B). Since the flakes alone are the solidified *n*-octadecane on the surface, they demonstrate the hydrophilic behavior. This situation is tested via measuring the droplet contact angle on the solidified thick layer of *n*-octadecane film (700 μm). In this case, the water-droplet contact angle remained in the order of 75 degrees and the contact angle hysteresis is in the order of 24 degrees on the surface of the solid phase of the *n*-octadecane-thick layer. Consequently, the area of the coverage of the solid *n*-octadecane flakes plays a critical role on the water-droplet contact angle and contact angle hysteresis during the droplet movement. Hence, the droplet movement takes place within the hydrophilic and hydrophobic regions on the inclined surface. The presence of varying wetting state on the inclined surface influences the droplet dynamic, which is consistent with the previous study. It should be noted that the dynamic contact angle hysteresis is measured by incorporating the method reported in the previous study [57]. The receding angle increases while advancing angle reduces in the region where the solid *n*-octadecane is highly populated on the surface; in contrast, advancing and receding angles becomes almost similar along the region where the area of the emerged functionalized silica particles dominates the area on the inclined surface. Increasing droplet volume gives rise to increasing differences between the receding and advancing angles of the droplet on the inclined surface (Fig. 5.41).

Fig. 5.43 shows the rotational speed of the droplet along the inclined surface for various droplet volumes. The scale analysis of the terms in Eq. (5.31) with reference to the adhesion force demonstrates that the adhesion force along the lateral direction, which is along the inclined surface, dominates over the viscous dissipation and air drag terms. However, the magnitude of the adhesion force varies along the lateral distance on the inclined surface. This situation can be seen from Fig. 5.44 in which the adhesion force along the lateral direction is shown for various droplet volumes. The variation in the adhesion force is associated with the droplet contact angle hysteresis (receding and advancing angles), which changes along the inclined surface (Fig. 5.43). Consequently, the adhesion force increases for large differences between the values of advancing and receding angles and it reduces for small differences of the receding and advancing angles. The rotational speed varies along the inclined surface.

The values of the rotational speed demonstrate that the droplet rolls with different rotational speeds on the inclined surface; however, zero value of the rotational speed corresponds to droplet sliding rather than rolling on the surface. Consequently, the droplet does not roll for high values of the adhesion force but slides along the inclined hydrophobic surface. Because of the nonuniform distribution of the solid *n*-octadecane flakes on the surface, rolling and sliding of the droplet occur randomly along the inclined surface. In addition, during the rolling, the variation of the contact angle hysteresis influences the droplet wobbling; in which case, reduction in the contact angle hysteresis lessens the droplet wobbling on the inclined surface while resulting in small variation of the droplet height. This is more pronounced for the small droplet volumes, which can be seen from Fig. 5.42. In the early droplet rolling period, the droplet rotational speed remains low on the hydrophobic surface (Fig. 5.45) and the low rotational speed gives rise to the large wobbling of the droplet [3]. The droplet height varies at different locations on the inclined surface, which is more pronounced for large volume droplet (Fig. 5.42). However, the difference between the maximum and minimum droplet heights remains low for large rotational Bond numbers ($\rho\omega^2 R^3/8\gamma_f$, where ρ is the water density, R is the droplet radius, ω is the angle of rotation, and γ_f is the surface tension). In the present study, the rotational and translational speeds are of close orders when in the region where the droplet rolls on the inclined surface, that is, $\omega R/V \sim 0.95$. The ratio of the rotational speed over the translational speed is critical for the relative magnitude of the dynamic pressure, which is generated between the droplet center and the ambient pressure around the droplet [45]. In this case, the dynamic pressure associated with the rotational speed does not have a significant effect on the droplet wobbling for the condition $\varphi = \Delta\rho\omega^2 R^2/\rho_a V^2 \gg 1$ (where $\Delta\rho$ is the density difference between the droplet liquid and droplet ambient gas and ρ_a is the droplet ambient gas density) [45]. In the current study, φ is estimated to be in the order of 850; therefore, the dynamic pressure does not significantly affect droplet wobbling during droplet rolling. The translational velocity of the droplet along the inclined surface composes of rotational and

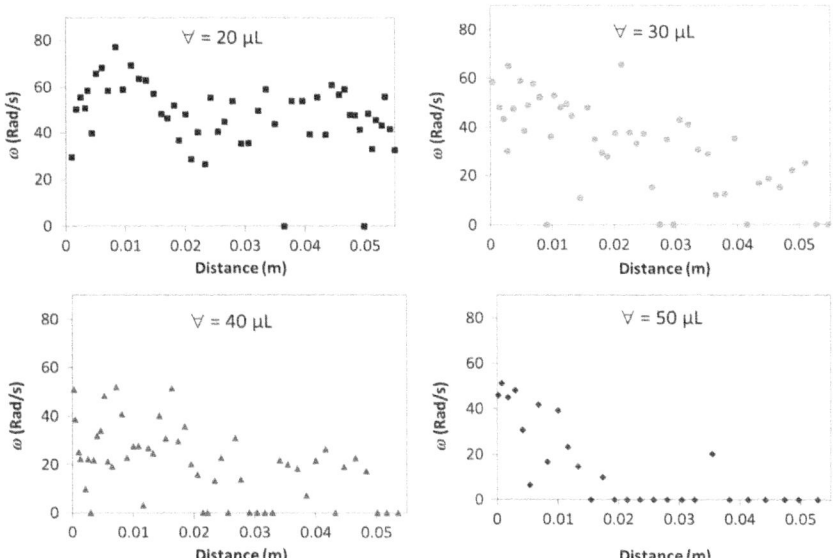

FIGURE 5.45 Water-droplet rotational velocity along the inclined solid phase of *n*-octadecane-coated functionalized silica-deposited surface for various droplet volumes [6].

sliding velocities; in which case, the translational velocity can be written as $V = V_{rot} + V_{slide}$. The rotational velocity can be simplified as $V_{rot} = R_H \omega$, where R_H is the instantaneous hydraulic radius of the droplet on the inclined surface and ω is the rotational speed.

Fig. 5.47 shows the sliding velocity of the droplet along the inclined surface for various droplet volumes. The sliding velocity of the droplet attains slightly larger values for large volume droplet than that of the small volume

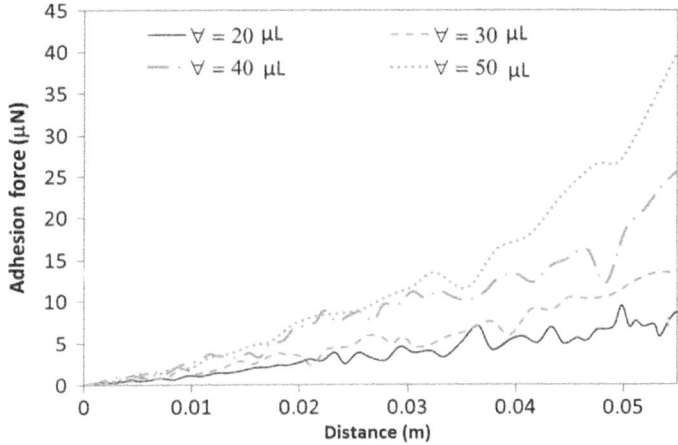

FIGURE 5.46 Water-droplet adhesion force along the inclined solid phase of *n*-octadecane-coated functionalized silica-deposited surface for various droplet volumes [6].

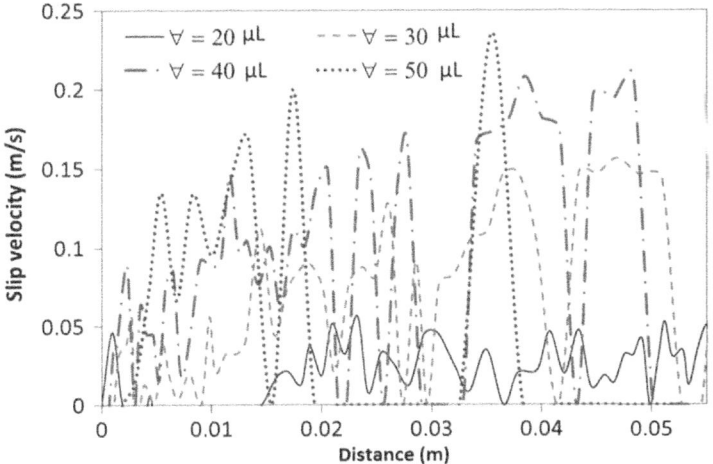

FIGURE 5.47 Water-droplet slip velocity along the inclined solid phase of *n*-octadecane-coated functionalized silica-deposited surface for various droplet volumes [6].

droplet. The sliding results in a small liquid tail behind the large volume droplet on the surface (Fig. 5.42), particularly in the region where the droplet sliding dominates, that is, continues sliding on the surface gives rise to strong adhesion of the droplet on the surface, due to extended length of the three-phase contact line of the droplet, while forming a droplet tail on the surface in the direction opposing to the direction of motion (Fig. 5.46).

5.5 DROPLET HEAT TRANSFER

Heat transfer in a sessile droplet on a solid surface initiates internal flow under the combination of the Marangoni and the buoyancy forces, which is critical for the surface heating and cooling applications. Depending on the fluid properties such as surface tension, density, and droplet contact angle, the flow current inside the droplet changes while affecting the heat transfer characteristics in the droplet. On the other hand, for droplets with high contact angles, the characteristic time scale associated with the thermal diffusion is critical for flow stability in the droplet. In this case, the ratio of a^2/α, where a is the characteristic droplet diameter and α is the thermal diffusivity of the fluid inside the droplet, becomes important and it must remain greater than unity to secure the flow stability [36] The rolling of droplet may occur due to the thermo-capillarity effect; however, the contact line morphology influences droplet pinning and droplet mobility on the surface [58]. The degree of roll-off angle, which is associated with the angle of inclination at which the droplet initiates rolling at the surface, is important when the surface is inclined [59]. In addition, the length of the contact line over which the pinning occurs is related to the droplet radius and fraction of protected area of the surface as occupied by the droplet. Consequently, the force balance tangential to the surface needs to be satisfied for the sessile droplet pinning at the surface. Evaporation of the droplet takes place when the heat transfer toward the droplet fluid remains high for long duration of heating interaction [59]; in which case, the droplet size decays gradually and the rate of evaporation increases with progressing time. In the case of short heating durations, the sessile droplet does not suffer extensive evaporation and the flow field inside the droplet is governed by the combination of the Marangoni- and the buoyancy-driven flows. Since the rate of heat transfer and sessile droplet contact angle influence the droplet internal flow while generating a complicated flow field, investigation of heat transfer and resulting flow field characteristics inside the sessile droplet becomes essential for practical applications. However, in this section low-temperature heating and short heating situations are considered; in which case, the evaporation from the droplet surface is ignored in the analysis. The cases for the droplet heat introduced in this section are presented in light of the previous studies [4,5,7,8]. A large number of research studies have been carried out to examine heat transfer characteristics associated with sessile droplets. Internal flow in a water droplet on a vibrating hydrophobic surface was investigated by Kim and Lim [60]. They demonstrated that the shape of a water droplet on a vibrating hydrophobic surface depends on the resonance mode and, additionally, the flow inside the droplet depends on the vibration frequency. Internal flow patterns of an evaporating multicomponent droplet on a flat surface were studied by He and Qiu [61]. They indicated that in the transition stage, the buoyancy effect was comparable to the Marangoni effect as the liquid in the droplet continuously evaporated; however, the buoyancy effect dominated as the heating progressed. Investigation of internal fluidity in a water droplet during sliding on hydrophobic surfaces was carried out by Sakai et al. [62]. They indicated that on a normal hydrophobic surface, both slipping and rolling controlled the velocity of the droplet during sliding. In addition, the advancing velocity became high when the slip velocity was large while the contact area was small. Microdroplet movement on a hydrophobic microchannel wall was examined by Hao and Cheng [63]. They introduced the Lattice Boltzmann method to simulate the deformation of the droplet for predicting the dynamic contact angle of the moving microdroplet. Thermocapillary convection inside a stationary sessile water droplet on a horizontal surface was investigated by Pradhan and Panigrahi [64]. The findings revealed that the convection pattern inside the droplets changed with the droplet size, and, for small droplets, the fluid flowed from the hot side to the cold side along the contact line leading to formation of two recirculation bubbles. Moreover, the buoyancy effect became predominant for large droplets, and the fluid flowed from the hot side to the cold side along the apex of the droplet. Flow and heat transfer in a liquid droplet sliding underneath a hydrophobic surface was studied by Sikarwar et al. [65]. They showed that the function for the figure of merit, represented by the ratio of average Nusselt number to the friction coefficient, increased with the contact angle. Investigation of superhydrophobicity on impinging droplet heat transfer was carried out by Rosengarten and Tschaut [66]. They indicated that significant reductions occurred in both the instantaneous heat transfer rates and the overall cooling effect of the droplets, which impinged on the superhydrophobic surfaces. They demonstrated that, in the range of droplet velocities considered, there was little dependency of the heat transfer or fluid flow with impact velocity due to the dominance of inertial forces. The effect of slip on a circulation inside a droplet was examined by Thalakkottor and Mohseni [67]. They suggested that active manipulation of velocity slip, for example, through actuation of hydrophobicity, could be employed to control droplet circulation and consequently its mixing rate. The analysis for droplet-hot wall interactions was carried out by Chatzikyriakou et al. [68]. They demonstrated that as a droplet approached a hot surface, a vapor layer was formed due to evaporation from the droplet, which acted like a cushion and could prevent contact between the liquid and the hot surface. Thereby, the droplet could rebound from the vapor film rather than hitting and wetting the surface.

5.5.1 Heat Transfer and Internal Fluidity of Droplet on Hydrophilic and Hydrophobic Surfaces

Investigation of the effects of heating and cooling rates on the internal fluidity of the droplet for various contact angles is of interest for self-cleaning applications of surfaces. In this section, the heat transfer characteristic and the flow field developed inside the sessile droplet are presented, in the line of the previous work [5], incorporating two cases: (1) heat transfer from hydrophobic surface to the sessile droplet and (2) heat transfer from sessile droplet to the surface. Since the time required for the flow stability in a droplet is estimated to be slightly less than 100 seconds, the analyses were continued for a heating period of over 100 seconds. In addition, in the analysis, evaporation from the droplet surface was avoided in the simulations because of the short heating period.

Since water is used to form droplets, in the simulations, incompressible flow field situation was incorporated to formulate the governing equations of flow and heat transfer. The governing equations were solved numerically in line with the experimental conditions. The time required for the flow stability in the droplet is in the order of 100 seconds; thus the simulations were carried out slightly more than this timeframe. The evaporation of the droplet was also avoided in the analysis due to the short timeframe considered. However, the experiments were carried out to measure the mass loss from the droplet during 200 seconds. It was observed that the mass loss from the droplet is negligibly small during this period. Consequently, neglecting the evaporation in the analysis is justifiable by the experimental findings. The coupled flow and thermal fields were considered simultaneously in the simulations. The continuity equation for transient flow is:

$$\frac{\partial \rho}{\partial t} + \nabla.(\rho V) = 0 \qquad (5.32)$$

where ρ is the water density and V is the liquid velocity.

For natural convection, the density variation is mainly caused by the thermal expansion of the fluid and can be expressed from the Boussinesq approximation as:

$$\rho = \rho_o[1 - \beta(T - T_o)] \qquad (5.33)$$

where β is the thermal expansion of the water. The momentum equation can be written as:

$$\rho\left(\frac{\partial V}{\partial t} + V \cdot \nabla V\right) = -\rho_o\beta(T - T_O)\,\vec{g} - \nabla(p - p_o)$$
$$+ \nabla\left[\mu\left(\nabla V + (\nabla V)^T\right) - \frac{2}{3}\mu(\nabla \cdot V)\right] \qquad (5.34)$$

where p is the pressure, μ is the dynamic viscosity of the liquid, g is the gravity, and p_o is the hydrostatic pressure corresponding to density ρ_o and temperature T_o.

The flow field should satisfy the energy balance according to:

$$\rho\, C_p \frac{\partial T}{\partial t} + \rho\, C_p\, V \cdot \nabla T = \nabla \cdot (k\nabla T) \qquad (5.35)$$

where C_p is the specific heat capacity and k is the thermal conductivity.

The relative contribution between natural convection and Marangoni convection can be written in terms of the Bond number (B_o), which is proportional to the ratio of buoyancy force and the surface tension force, that is: $B_o = \beta g \rho a^2 / |\, d\sigma/dT\,|$, where ($\sigma$) is the surface tension and a is the characteristic diameter (length scale), which is considered as $a = \forall / \pi R^2$, where $\forall$ is the volume of the droplet and R is the wetting radius. It should be noted that when $B_o < 1$, the internal flow is dominated mainly by the Marangoni convection.

Natural convection is driven by the buoyancy force, which overcomes viscous resistance in the flow. To characterize the strength of natural convection, the Grasshoff number (Gr) can be introduced, which is given by $Gr = \beta g \Delta T a^3 / \nu^2$, where ($\Delta T$) is the temperature difference between the droplet and its ambient and (ν) is the kinematic viscosity of water. It should be noted that the effect of natural convection can be ignored when $Gr < 2400$. The Marangoni number (Ma) can be used to describe the intensity of the Marangoni convection, which is $Ma = |\, d\sigma/dT\,| \Delta T a / \mu \alpha$, where α is the thermal diffusivity.

Efforts are made to simulate the flow field inside the droplet in 3D configuration; however, it requires expensive computational efforts because of the extremely large meshes and significantly long runtime. In order to visualize the flow field, one case is simulated in line with the experimental droplet size (Fig. 5.48A) and findings are shown in Fig. 5.48B and C. It is evident that the 2D configuration results in similar flow behavior to that obtained from the 3D simulation.

The initial and boundary conditions for the governing equations of flow and heat transfer are as follows.

5.5.1.1 Initial Conditions

Two heating situations are considered in the analysis: (1) water temperature is assumed to be at 308K initially and the substrate surface is at 300K, and the droplet is formed at the substrate surface (case 1), and (2) water temperature is at 300K and the substrate surface is kept at 308K, and the droplet is formed (case 2). Consequently, two cases have different initial temperature settings.

Case 1:

Initially, it is assumed that the droplet has a uniform temperature at 308K and the substrate surface is at 300K. The stagnant water is assumed in the droplet, that is, the initial velocity is set to zero in the droplet.

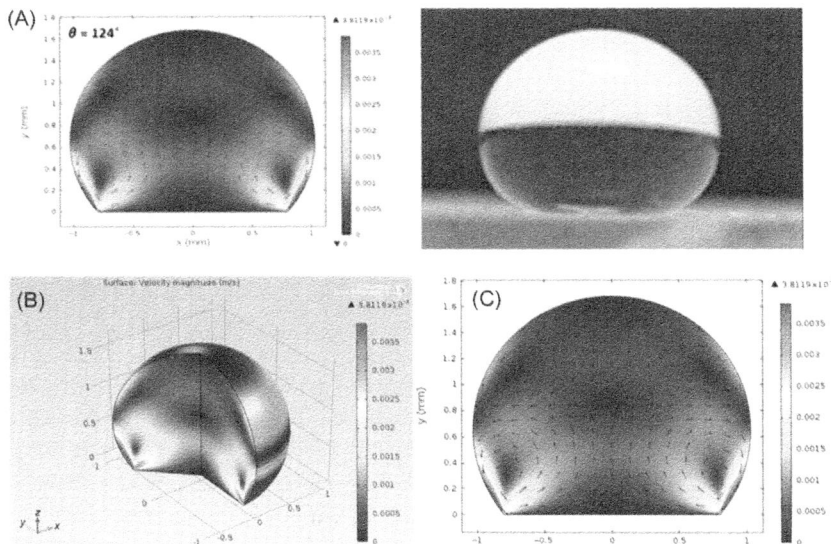

FIGURE 5.48 Images of the droplet simulated: (A) optical image of actual droplet and simulated image of the same droplet for $\theta = 124$ degrees, (B) velocity field obtained from 3D simulation of the same droplet, (C) flow field and velocity vector in the droplet obtained from 2D simulation of the same droplet [5].

Case 2:

Initially, it is assumed that the droplet has a uniform temperature at 300K and the substrate surface is at 308K. The stagnant water is assumed in the droplet, that is, the initial velocity is set to zero in the droplet.

5.5.1.2 Boundary Conditions

It is considered that the substrate surface temperature remains the same as its initial temperature; in which case, the substrate surface acts like a constant temperature (300K) heat sink for the first case, while it is a heat source at constant temperature (308K) for case 2. The natural convection boundary condition is adopted at the droplet surface with surrounding air temperature of 298K. No-slip flow boundary is incorporated at the droplet bottom.

Temperature-dependent properties of water were incorporated in the simulations. COMSOL multiphysics [27] code incorporates the finite element method, which provides a means of spatial and temporal discretization of the governing equations of flow and heat transfer. The differential operators in the governing equations were discretized using the backward Euler finite difference method. The implicit scheme with a backward difference approximation for the time used in the study was unconditionally stable [27]; however, the accuracy of the scheme is governed by the size of the time step, which is considered to be 10^{-4} seconds and the residuals of flow parameters are set as 10^{-8} during the simulations.

5.5.1.3 Findings

The findings from the thermal analysis of a sessile droplet on hydrophilic and hydrophobic surfaces are presented in the line of the previous study [5]. Heat transfer conditions are altered in such a way that heat transfer from droplet to the solid surface and vice versa is incorporated through setting the boundary conditions. This arrangement provides the different flow field development inside the droplet. In addition, the effect of droplet contact angle on the heat transfer characteristics is analyzed for two heating cases.

Since the characteristic timescale related to the thermal diffusion is crucial for the flow stability in the droplet, the ratio of the square of the characteristic droplet diameter (a) over the thermal diffusivity of the fluid (α) should remain greater than unity ($a^2/\alpha > 1$) for the stable flow situation [36]. In this case, the characteristic time is estimated as in the order of 100 seconds. To demonstrate the importance of the characteristic time in the flow field, velocity and temperature variations along the vertical rake are plotted in the droplet for various durations. Figs. 5.49A and 5.47B show temperature and velocity distributions along the vertical rake in the droplet for the droplet contact angle 124 degrees, respectively. The temperature and velocity profiles vary significantly in the early heating period, and as the heating period progresses, temperature and velocity profiles settle in such a way that at two late heating periods they almost collapse into a single curve. This occurs for the heating period in the order of 100 seconds, which satisfies the condition $a^2/\alpha > 1$. Consequently, the heating time of 100 seconds is considered to demonstrate flow and temperature fields in the droplet. The contact length corresponding to the droplet pinning is related to the droplet radius and projected area of the contacted surface by the droplet [59]. Since the stationary droplet is incorporated in the simulations resembling the experimental conditions for the actual droplet, care is taken to satisfy the requirements of droplet pinning at the surface. Therefore, the tangential force balance is considered to be secured to satisfy the droplet pinning at the hydrophobic surface, which is also implemented in the simulations.

Fig. 5.50 shows the velocity contours inside the droplet for various droplet contact angles and the heating period of 100 seconds, while Fig. 5.51 shows the temperature contours inside the droplet for the same conditions of Fig. 5.50. In the simulations, two heating situations are considered: (1) heat transfer takes place from the droplet to the substrate surface, that is, initially, the fluid temperature is set at 308K and the surface temperature remains 300K; and (2) heat transfer occurs from surface to liquid, since surface temperature is kept at 308K, while the initial droplet temperature is 300K. It should be noted that the droplet volume remains constant for all the contact

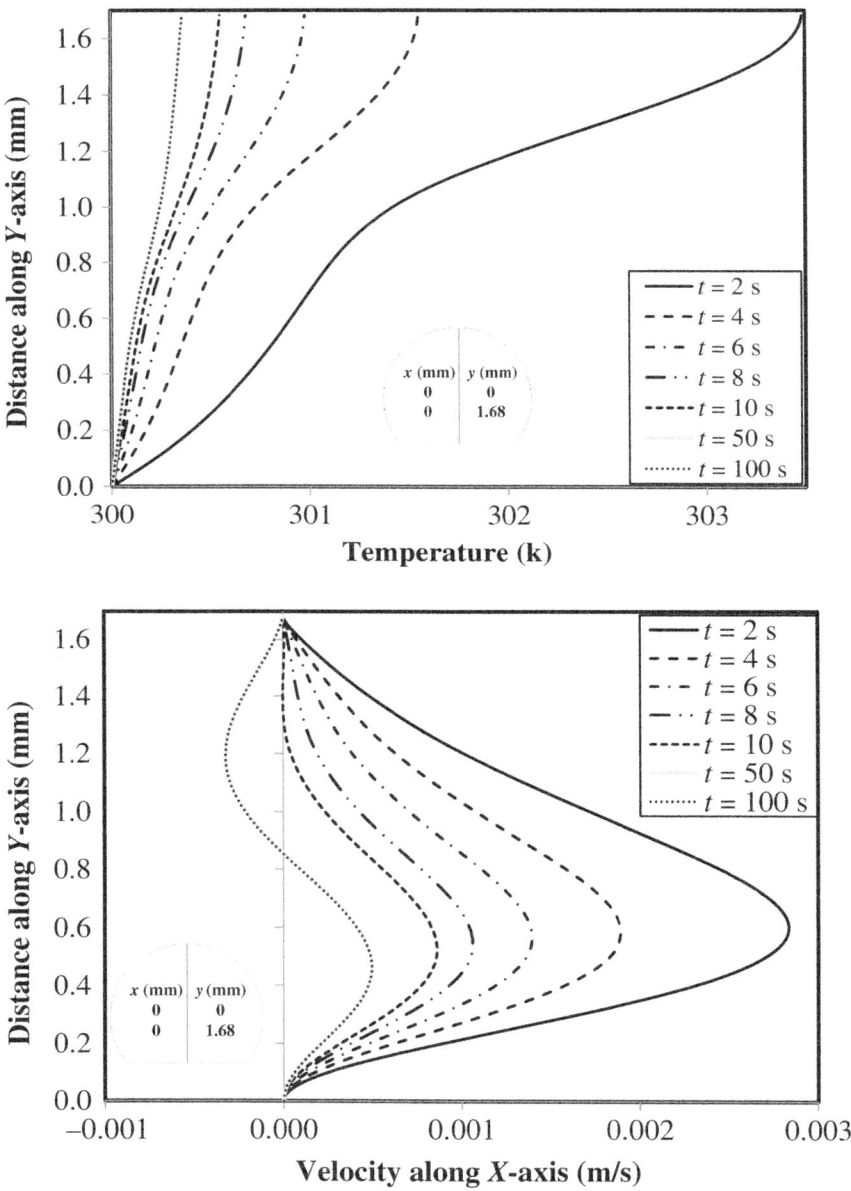

FIGURE 5.49 Temperature and velocity variation along the horizontal rake located at the symmetry line in the droplet for various heating durations: (A) Temperature distribution, and (B) velocity distribution [5].

Heat transfer from droplet to surface
(Case 1)

Heat transfer from surface to droplet
(Case 2)

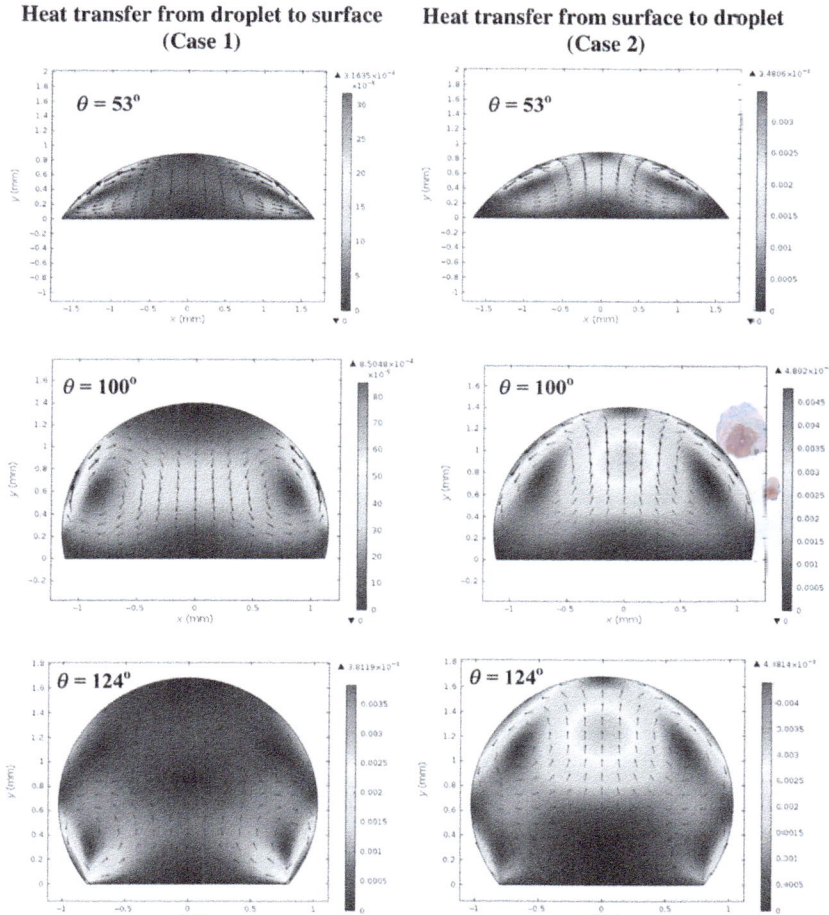

FIGURE 5.50 Velocity contours inside the droplet for various droplet contact angles and two heating situations (case 1 and case 2) [5].

angles in the simulations for both cases of hydrophilic and hydrophobic surfaces. The Marangoni and the buoyancy force generated in the droplet gives rise to two circulation cells in the droplet for contact angles less than 90 degrees, which resembles the case for droplet at the hydrophilic surface. In the case of heat transfer from droplet to surface (case 1), the counter clock wise circulation is developed in the right lobe of the droplet and it becomes the clockwise in the left lobe of the droplet for the hydrophilic surface. In this case, the flow direction is always from droplet top toward the droplet

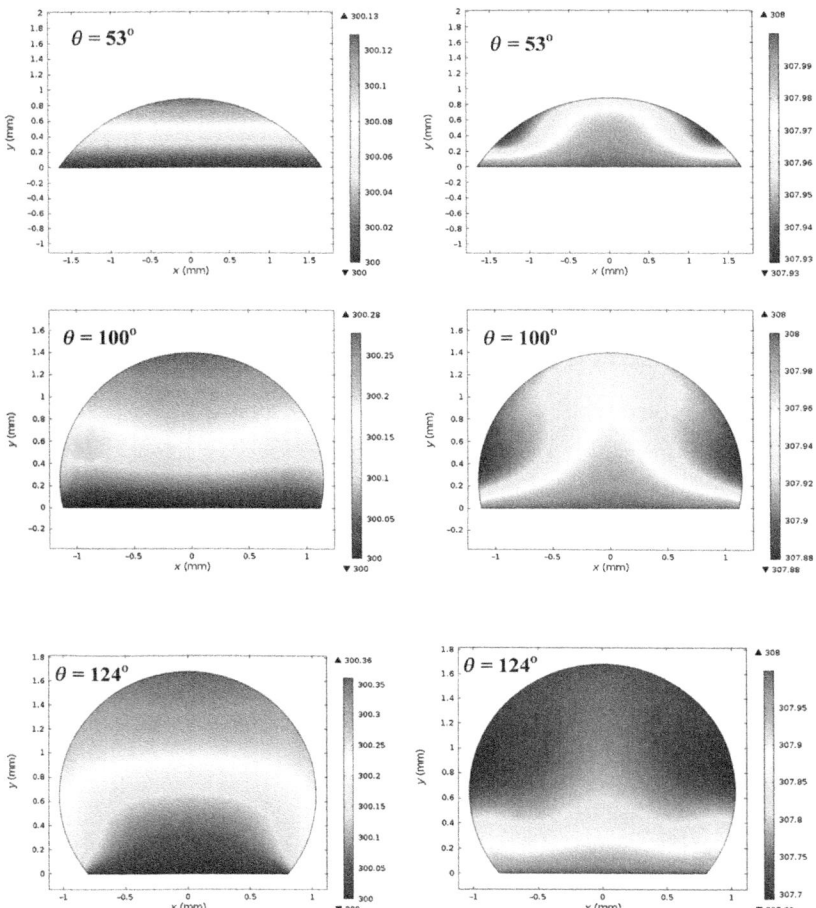

FIGURE 5.51 Temperature contours inside the droplet for various droplet contact angles and two heating situations (case 1 and case 2) [5].

bottom. This is associated with the Marangoni current over the buoyancy current; in which case, Marangoni current dominates the buoyancy current generated due to the density variation in the droplet. It should be noted that the ratio of the Marangoni force over the buoyancy force is the same order of the ratio of the Marangoni number ($Ma = (\partial \sigma / \partial T)(a\Delta T / \upsilon \alpha)$, where a represents the characteristics diameter of the droplet, σ is the surface tension, T is the temperature, υ is the kinematic viscosity) over the Rayleigh number ($Ra = \alpha_t g a^3 \Delta T / \upsilon \alpha$, where α_t is the thermal expansion coefficient and g is the center of gravity) [36]. In this case, the velocity ratio due to Marangoni force over the buoyancy force is $(\partial \sigma / \partial T)\Delta T / \alpha_t \rho g a^2$. Since the droplet size simulated has a small characteristic diameter ($a \approx 4 \times 10^{-3}$ m), the denominator becomes much smaller than the numerator in the velocity ratio $((\partial\sigma/\partial T)\Delta T/\alpha_t\rho g a^2)$ for the droplet size considered in the simulations; therefore, Marangoni current dominates over the convection current developed in the bottom region of the droplet where the temperature change is significantly small, that is, solid surface at low temperature (300K) acts like a heat sink altering the flow field in the droplet.

In the case of heat transfer from the solid-surface to the droplet (case 2), two circulation cells are formed in the droplet for the hydrophilic surface ($\theta \leq 90$ degrees, where θ is the droplet contact angle), similar to the case 1. However, the rotational direction of the circulation cells are opposite as compared to those occurring for the case 1. The sign of the surface tension gradient ($d\sigma/dT$) at the droplet surface changes because of the direction of the heat transfer. Although temperature remains high at the bottom of the droplet (Fig. 5.51), temperature variation is small in this region while causing thickening of the thermal boundary layer so that flow acceleration under the buoyancy force remains low in the thick thermal boundary layer. This gives rise to considerably small velocities in this region (Fig. 5.50). In the case of hydrophobic surface, where the droplet contact angle is large such as $\theta = 110$ degrees, the flow field is completely modified as compared to those corresponding to the hydrophilic surface; in which case, four circulation cells are formed in the droplet. The neighboring circulation cells are counter rotating, and the circulation cells formed in the upper region of the droplet are larger than those in the bottom region. This is associated with the flow interaction because of the Marangoni and the buoyancy forces. In this case, increasing droplet contact angle results in increased droplet height and the temperature gradient in the droplet. Consequently, the surface tension gradient ($d\sigma/dT$) increases at the free surface of the droplet and enhances the Marangoni force in the flow field. This, in turn, causes the attainment of higher velocities in the flow field as compared to that corresponding to the hydrophilic surface. This situation can be observed after comparing the velocity scale in the figures. In addition, increasing the temperature gradient enhances the heat diffusion in the droplet causing the buoyancy force enhancement in the flow. Consequently, enhanced Marangoni and buoyancy forces give rise to complicated flow field in the droplet, that is, forming four circulation cells in the droplet. In the case of heat transfer from droplet to the substrate surface (case 1), the rotation of the circulation cells are opposite to those corresponding to the heat transfer from surface to the droplet (case 2). Moreover, the direction of heat transfer alters the maximum value of the velocity in the droplet; in which case, the heat transfer from the droplet to the substrate surface gives rise to relatively lower value of the maximum velocity as compared to that corresponding to the heat transfer from the substrate surface to the droplet.

Fig. 5.52A and B shows the Nusselt number, based on the averaged heat transfer coefficient, and the Bond number with the contact angle of the

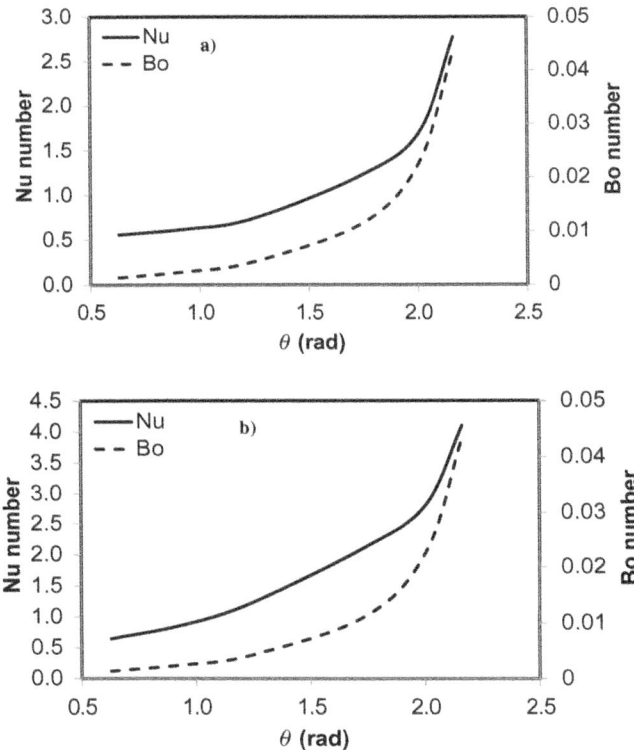

FIGURE 5.52 Variation of the Nusselt and Bond numbers with contact angle of the droplet [5].

droplets for two heating situations (case 1) and (case 2), respectively. Increasing contact angle enhances both the Nusselt number and the Bond number, which is more pronounced for the hydrophobic surfaces where the contact angle $\theta \geq 1.91$ rad or 110 degrees. The increased Nusselt and Bond numbers are associated with the heat transfer enhancement and increased surface tension gradient at the droplet free surface with increasing contact angle for the high contact angle droplet. Heat transfer enhancement results in extension of high-temperature region toward the droplet interior, which can be observed from Fig. 5.51. In addition, increasing droplet height at large contact angles contributes to increased temperature gradient in the droplet while enhancing the heat diffusion inside the droplet. The bond number remains almost the same for the cases of heat transfer from the droplet toward the surface (case 1) and from surface toward the droplet (case 2). This is attributed to the Force balance between the Marangoni and buoyancy forces inside the droplet for both cases of heating configuration. In the case of heat transfer from droplet and the surface (case 1), the maximum value of the Nusselt number is close to 2.7; however, the maximum value of the

Nusselt number attains about 4.2 for the case corresponding to the heat transfer from surface to the droplet (case 2). This indicates that the Nusselt number enhancement occurs when the heat transfer takes place from surface to the liquid droplets. This is attributed to one or all of the following: (1) the high-temperature gradient is developed in the vicinity of the droplet bottom, which is close to the substrate surface, and (2) the buoyancy-driven flow current enhances in the close region of the substrate surface inside droplet. The enhancement of the Marangoni flow and buoyancy flow currents can also be seen from Fig. 5.50 for the case of heat transfer from the substrate surface toward the droplet (case 1), in which the maximum velocity remains higher than that of the case corresponding to the heat transfer from the droplet toward the substrate surface (case 2). Moreover, the gradual increase of the Bond and the Nusselt number for the small contact angles indicates the thermal diffusion toward the solid-surface (for case 1) or from the solid surface toward the droplet (case 2) is small because of low convection and Marangoni currents, as observed from the flow velocity in the droplet. The maximum magnitude of velocity is in the order of 4.8 mm/s. In the case of large contact angles, a combination of the Marangoni- and buoyancy-driven flow enhances the flow current and velocity in the droplet, which causes the heat transfer and Nusselt number enhancement. Since a similar trend is observed for the Bond number and the Nusselt number in terms of the droplet contact angle, a new relation is introduced to correlate the Nusselt number in terms of the ratio of the Bond number over the contact angle. The new ratio reflects the force ratio due to the Marangoni and buoyancy forces, and the ratio of interfacial energies ($\theta = cos^{-1}(\gamma_{sg} - \gamma_{sl})/\gamma_{lg}$, where γ_{sg} is the solid−vapor interfacial energy, γ_{sl} is the solid−liquid interfacial energy, and γ_{lg} is the liquid−vapor interfacial energy [69]). Consequently, the new ratio is called *Ayse* number.

Fig. 5.53 shows the Nusselt number variation with *Ayse* number for two cases of heat transfer situations (cases 1 and 2). The Nusselt number increases with increasing *Ayse* number and the Nusselt number increase depends on the heat transfer cases (case 1 and 2). Since the Nusselt number attains higher values for the heat transfer from the substrate surface to the droplet (case 2) than that corresponding to case 1, the rise of the Nusselt number with the *Ayse* number is sharper for case 1 as compared to case 2, which is more pronounced for the *Ayse* number in the range of $Ayse \leq 0.065$ where the surface is hydrophilic. This indicates that the effect of the combination of the Bond number and the droplet contact angle on the Nusselt number is significant. In the case where the *Ayse* number ≥ 0.65, where the surface remains hydrophobic, the slope the variation of the Nusselt number with *Ayse* number remains almost constant for both cases of heating situations (case 1 and case 2). The analytical expression is developed between the Nusselt number and the *Ayse* number for two cases of the heat transfer situation. The relation is in the linear form, which is $Nusselt = a \times Ayse + b$,

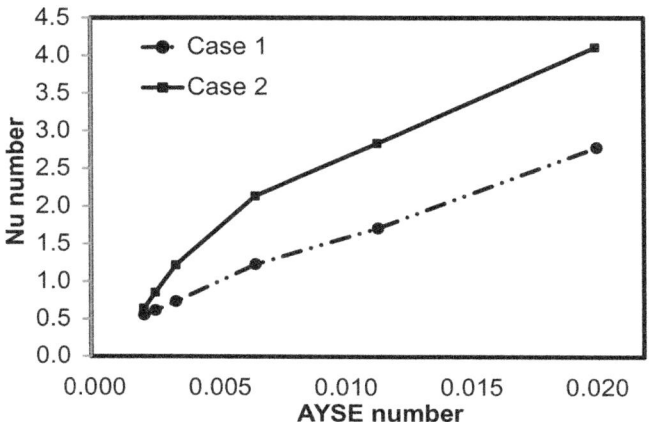

FIGURE 5.53 Variation of the Nusselt number with *Ayse* number [5].

TABLE 5.5 The Constants of Equation Derived Between the Nusselt Number and the *Ayse* Number [5]

		a	b
Heat transfer from droplet to substrate surface (case 1)	Hydrophilic	143.5	0.25
	Hydrophobic	111.5	0.45
Heat transfer from substrate surface to droplet (case 2)	Hydrophilic	457	−0.32
	Hydrophilic	146	1.18

where a and b are the constants. These are given in Table 5.5 according to the hydrophobicity or hydrophilicity of the substrate surface. Consequently, the new number developed provides useful information about the heat transfer rates from the droplet toward the surface or vice versa for hydrophobic and hydrophilic surfaces.

5.5.2 Local Heating of Droplet by a Radiation Source

The convection current generated, due to the combination of Marangoni and buoyancy flows, is significantly influenced by both the rate of heat transfer from/toward the droplet and the contact angle of the droplet. The internal fluidity of the droplet influences the droplet mobility on hydrophobic surfaces because of the local, convective, and radial accelerations of the flow under the Marangoni and buoyant forces. The nonuniform and localized nonmechanical contact heating of droplet surfaces can further alter the flow filed inside the droplet while influencing the droplet mobility at the surface. One of the methods to create a nonmechanical contact heating is the radiative heating of the droplet surface by a localized point heat source. In this case, the variation of density and the surface tension of the droplet fluid with temperature causes localized flow acceleration in the droplet, which in turn alters the momentum and force balance across the droplet. Consequently, the internal fluidity of the droplet on the hydrophobic surface subjected to a localized radiative heating is introduced in this section and the findings are discussed in line with the previous study [4].

5.5.2.1 Heating and Flow Analysis

Since water is used to form droplets, an incompressible flow field is incorporated to formulate the governing equations of flow and heat transfer. The governing equations are solved numerically in line with the experimental conditions. Since the heating duration is in the order of 10 seconds, the evaporation of the droplet is avoided in the analysis. The size change of the water droplet due to the mass loss via evaporation is measured experimentally during 50 seconds of heating. The evaporation of the droplet is calculated and the amount of water evaporated from the droplet is estimated as 0.000434 grams after 50 seconds of heating for the droplet volume of 100 μL ($D = 5.7$ mm) under ambient conditions of 298K temperature and 85% of air relative humidity. The estimations reveal that the mass of water evaporated from the droplet corresponds to 0.43% of the total mass of a 100 μL water droplet; therefore, neglecting the evaporation in the analysis is justifiable by the experimental findings. The coupled flow and thermal fields are considered simultaneously in the simulations. The governing equations of flow are given in the previous section; therefore, they are not presented herein.

In the numerical analysis, COMSOL Multiphysics finite element code [27] is used to solve governing equations of flow and heat transfer for a stationary droplet incorporating the experimental conditions.

The efforts are made to simulate the flow field inside the droplet in 3D configuration; however, simulation requires expensive computational efforts because of the involvement of the extremely large number of meshes and significantly long computational duration to secure the converge results. Therefore, 3D simulation is avoided in the present study.

The initial and boundary conditions for the governing equations of flow and heat transfer are as follows.

5.5.2.1.1 Initial Conditions

Initially, the water-droplet fluid and hydrophobic plate temperatures are assumed to be the same at 300K and the substrate surface is at 300K.

5.5.2.1.2 Boundary Conditions

The boundary conditions are shown in Fig. 5.54. Radiative intensity distribution is considered at the left surface of the water droplet.

Fig. 5.55B is used to determine the radiation intensity distribution along the droplet side surface. In this case, the solid angle (ω) is determined from:

$$dw = sin\theta\ d\theta\ d\varphi \tag{5.36}$$

or

$$\int_h dw = \int_0^{2\pi} \int_0^{\pi/2} sin\theta\ d\theta\ d\varphi = 2\pi \int_0^{\pi/2} sin\theta\ d\theta\ d\varphi = 2\pi\ sr \tag{5.37}$$

where subscript h refers to the integration over the hemisphere. The unit of the solid angle is the steradian, which analogous to radians for plane angles. The radiation intensity at which emission from dA_1 passes through dA_n (Fig. 5.55) can be expressed in terms of spectral intensity I_λ of the emitted radiation, where λ is the wavelength.

The spectral intensity in W/m^2 can be written as [70]:

$$I_{\lambda,e}(\lambda, \theta, \varphi) = \frac{dq}{dA_1 cos\theta \cdot dw \cdot d\lambda} \tag{5.38}$$

Eq. (5.38) is used to compute the radiation intensity distribution along the surface of the droplet. In this case, gray body analysis is adopted to estimate the emitted radiative power from the heater tip surface toward the droplet side surface. The natural convection boundary condition ($h = 10$ W/m^{2}K) is considered around the surface of the droplet surface surrounded with ambient air temperature at 300K. The conjugate heat transfer is incorporated across the interface between the droplet liquid and the hydrophobic surface. In this case, the continuity of heat flux and temperature are assumed across the interface. Moreover, the slip length of the hydrophobic

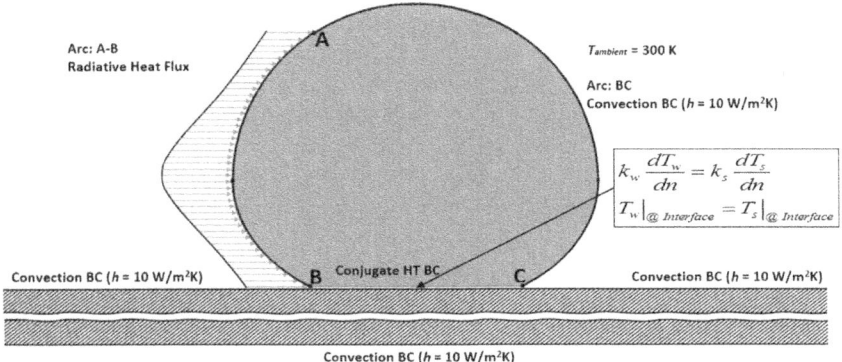

FIGURE 5.54 Boundary conditions incorporated in the numerical simulations [4].

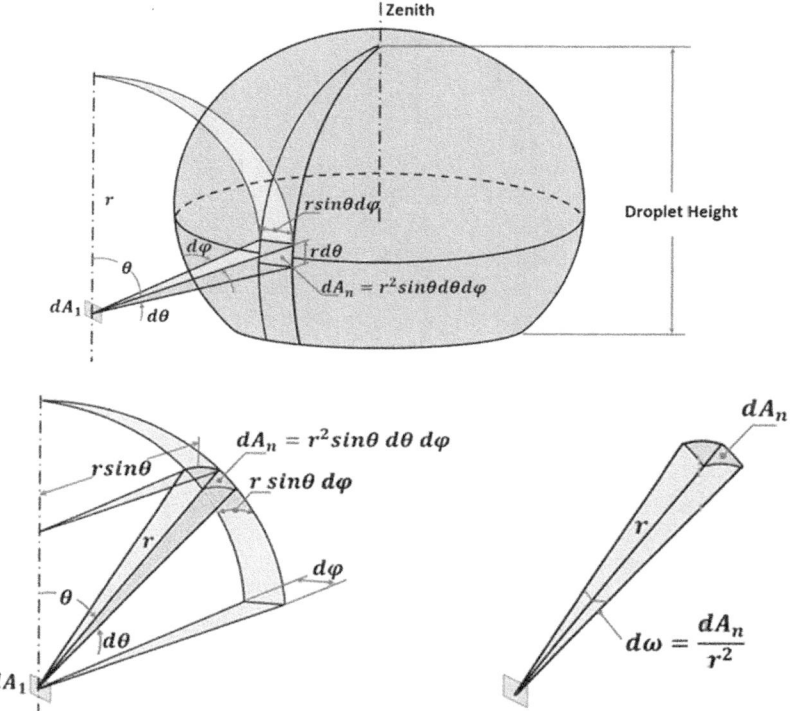

FIGURE 5.55 Geometric configurations and thermal radiation heating along the droplet surface [4].

surface is selected according to the contact angle of the droplet. This flow boundary condition is incorporated at the interface between the droplet bottom and the hydrophobic surface. The free surface of the hydrophobic plate is assumed to be exposed to the air ambient and natural convection boundary is assumed along the plate surface with the ambient air temperature at 300K.

Temperature-dependent properties of water are incorporated in the simulations. COMSOL multiphysics [27] incorporate the finite element method, which provides means of spatial and temporal discretization of the governing equations of flow and heat transfer. The differential operators in the governing equations are discretized using the backward Euler finite difference method. The implicit scheme with a backward difference approximation for the time used in the study was unconditionally stable [71]; however, the accuracy of the scheme is governed by the size of the time step, which is considered to be 10^{-4} seconds and the residuals of flow parameters are set as 10^{-8} during the simulations. The findings of the grid independent reveal that mesh sizes of 4823 and 8518 result in identical temperature and velocity profiles. Consequently, the mesh size of 4823 is selected to simulate the flow field and temperature distribution.

5.5.2.2 Experimental and Validation of Flow-Field Predictions

Polycarbonate (PC) wafers of 2.5 mm thickness were used as the workpieces. After ultrasonic cleaning of the PC wafers, they were immersed into liquid acetone for 2, 4, 6, 8, and 10 minute durations. In order to select the immersion durations for crystallization of PC wafers, tests were repeated several times. The immersion durations resulting in crystal structures toward forming a surface texture for hydrophobic behavior were selected; in which case, the immersion duration of 4 minutes was selected. The characterization of the solvent-induced workpiece surfaces was carried out using SEM and AFM microscopes. Jeol 6460 electron microscopy was used for SEM examinations and the Agilent AFM/SPM microscope in contact mode was used to analyze the surface texture. The sample surfaces were gold coated prior to SEM examinations to increase the conductivity and reduce the electron charging at the surface. The AFM tip was made of silicon-nitride probes ($r = 20-60$ nm) with a manufacturer-specified force constant, k, of 0.12 N/m. The SEM micrographs are shown in Fig. 5.56A and B for the textured surface. Fig. 5.56C and D shows 3D images and line scanning of the textured surface obtained from AFM. The average surface roughness of crystallized polycarbonate surface was in the order of 2.8 μm. A goniometer Kyowa (model— DM 501) was used to conduct sessile drop tests for the measurement of droplet contact angle. Deionized water was used in sessile drop experiments and droplet volume was controlled with an automatic dispensing system. The images of droplets were taken after 1 second of deposition of water droplet on the textured surface. The sessile drop method was used to measure the droplet static contact angle on the textured surfaces, which was varied from 110 to 130 degrees with the contact angle hysteresis of 46 degrees. Since the contact angle hysteresis is high (in the order of 46 degrees), the synthesized silica particles were deposited on the textured surface to reduce the droplet contact hysteresis. The procedure adopted for synthesizing silica particles was similar to that reported in the previous study [41]. The process is briskly described herein. Tetraethyl orthosilicate (TEOS), isobuthytrimethoxysilane (OTES), ethanol, and ammonium hydroxide were used in the synthesizing process. In this case, 14.4 mL of ethanol, 1 mL of deionized water of 18.2 MΩ resistivity and 25 mL of ammonium hydroxide were mixed and stirred for 12 minutes. Later, 1 mL TEOS diluted in 4 mL of ethanol was added in the mixture. Following 30 minutes after this process, 0.5 mL of TEOS diluted in 4 mL ethanol was added. After 5 minutes, a modifier silane molecule was added in a molar ratio of 3:4 with respect to the second edition of TEOS. The final mixture was stirred for 20 hours at room temperature, and later centrifuged and washed with ethanol to complete the removal of

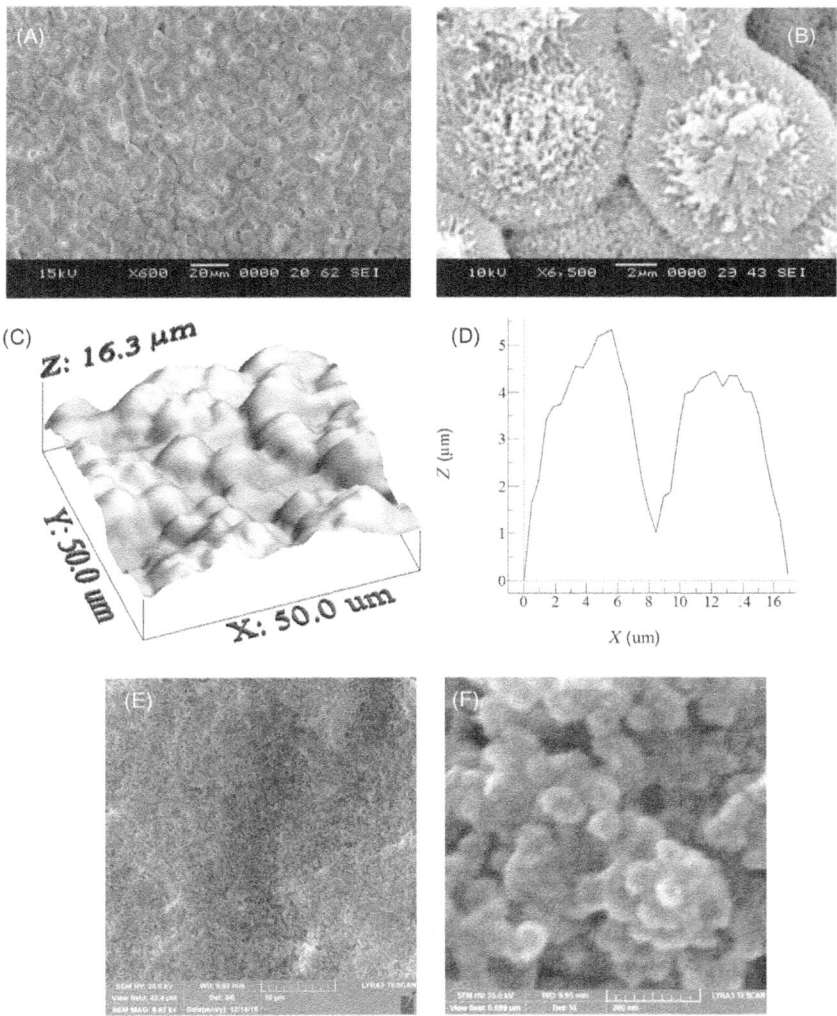

FIGURE 5.56 SEM micrograph and AFM microimages of solution-crystallized surface: (A) SEM micrograph of crystallized surface, (B) globes and whisker-like structures on crystallized surface, (C) 3D AFM image of crystallized surface, (D) line scan of surface on AFM image, (E) SEM micrograph of deposited functionalized silica particles, and (F) SEM image of close view of functionalized silica particles [4].

reactants. The solvent casting was applied to deposit the solution onto textured surfaces. Upon vacuum drying until all solvent was evaporated, characterization of resulting surfaces was carried out. The water droplet contact angle of the order of 140−150 degrees resulted and the contact angle hysteresis reduced to the order of 2 degrees. Fig. 5.56E and F shows the SEM micrographs of the textured surface after synthesized silica particle-deposited. The resulting surface was used in the heating experiments.

A test rig was developed to carry out the radiative heating of the water droplet on the hydrophobic surface and high-speed camera was used to monitor the water droplet rolling off on the surface. Fig. 5.57 shows the schematic view of the droplet heating. The voltage and current characteristics of the heater bar were adjusted to a constant temperature of 355K at the heater tip surface. The heater bar was made from a circular iron with outer diameter of 2 mm and a conical tip and located 0.4 mm away from the droplet side surface (Fig. 5.57). The heater tip bar diameter was in the order of 0.85 mm. The outer surface of the heater bar was insulated and the tip of the heater diameter as polished prior to the experiments. The beam tracing method was adopted to obtain the radiation intensity distribution around the droplet surface, which is shown in Fig. 5.55. It should be noted that the gray body analysis was used to estimate the emitted radiative power intensity from the heater tip surface toward the droplet side surface. In the experiments, initially, temperature and pressure of the droplet and its environment were kept at 300K and 101.32 kPa.

The Dantec PIV system was used to assess the flow field and measure the velocity distribution within the 50 μL water droplet. The hollow glass particles of 3%, by volume, were mixed with water prior to the droplet formation for the particle velocity measurements.

The droplet initiates rolling on the functionalized silica particle-deposited hydrophobic surface when the droplet is subjected to a thermal loading via radiative heating from one side. High droplet rolling speed on the hydrophobic surface makes it difficult to monitor the flow field inside the droplet

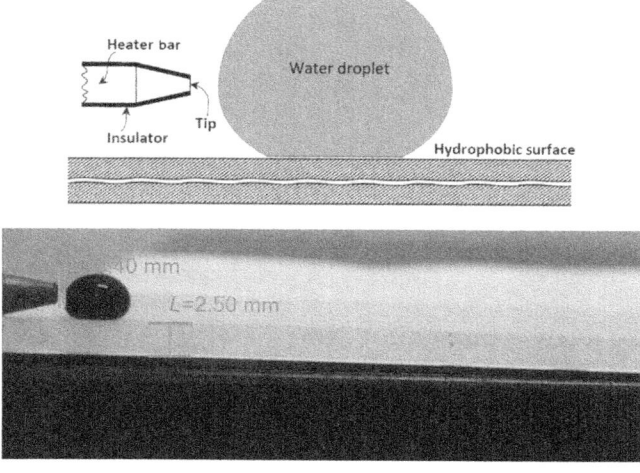

FIGURE 5.57   Schematic view and image of the droplet and side heating arrangements [4].

with accuracy by using the PIV. However, the solution-crystallized polycarbonate surface without functionalized silica particle deposition has high contact angle hysteresis and locating the water droplet on the solution-crystallized polycarbonate surface (SCPS) pins the droplet on the surface. This is because of the increased adhesion between the water droplet and crystallized surface. Consequently, minimizing the lotus effect gives rise to high contact angle hysteresis at the surface while the droplet contact angle remains high. This arrangement suppresses the droplet rolling at the hydrophobic surface. Therefore, the internal fluidity of a stationary (sessile) droplet with high contact angle can be experimented and monitored via PIV when subjected to the thermal radiation heating from one side of the droplet. The hollow glass particles were mixed in the droplet water to monitor and trace the particle velocities inside the droplet during the heating process. The governing equation of momentum is solved after incorporating the hollow glass particles in the solution domain via using the discrete-phase model. The energy equation is solved after assuming a slurry single fluid resembling water and the hollow glass particle mixture. Since the hollow glass particle concentration is low (3%) in the carrier fluid (water), the effective thermal properties are incorporated in the energy equation. The incompressible flow field is considered to formulate the governing equations of flow and heat transfer. The field equations were solved numerically in line with the experimental conditions. The coupled flow and thermal fields were considered simultaneously in the simulations. The formulation of the governing equations and numerical discretization is not given herein, but the details can be found in the previous study [32].

Fig. 5.58 shows typical hollow glass images inside the droplet and particle tracking at different heating durations. The flow velocities measured from PIV and predicted from the multiphysics code are given in Table 5.6.

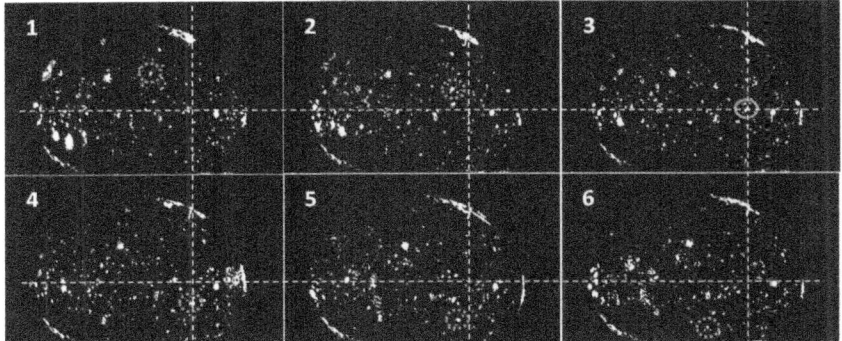

FIGURE 5.58 PIV images for particle tracing during side surface heating of a sessile droplet located on SCPS. The time interval between each frame is 0.2 seconds. The particle velocity determined from the PIV images is in the order of 0.0158 m/s and its counterpart predicted from the simulation is 0.0160 m/s [4].

TABLE 5.6 Velocity Predictions and Measured From Particle Image Velocimetry Data [4]

Simulations

Experiment — Particle Initial Location

Experiment — Particle Final Location

$V_p = 0.00115$ fm/s

Particle #	X (mm)	Y (mm)	Experiment U (m/s)	Simulations U (m/s)
1	−0.43687	0.50633	0.00139	0.00112
2	1.32257	1.29628	0.00082	0.00081
3	0.55656	0.60208	0.00138	0.00150
4	−1.27469	0.72177	0.00142	0.00135
5	−1.76542	2.15804	0.00120	0.00137
6	−0.38899	2.30167	0.00083	0.00089
7	−0.02992	1.60747	0.00010	0.00009
8	−0.44884	3.05571	0.00151	0.00155
9	0.08977	3.37888	0.00183	0.00176
10	0.83184	2.40939	0.00115	0.00163
11	−1.64573	0.75768	0.00127	0.00122
12	1.43029	0.60208	0.00132	0.00126
13	1.83723	1.69125	0.00120	0.00129
14	0.60443	3.09162	0.00158	0.00160
15	−0.95153	1.81094	0.00081	0.00076
16	0.01795	0.84146	0.00100	0.00106

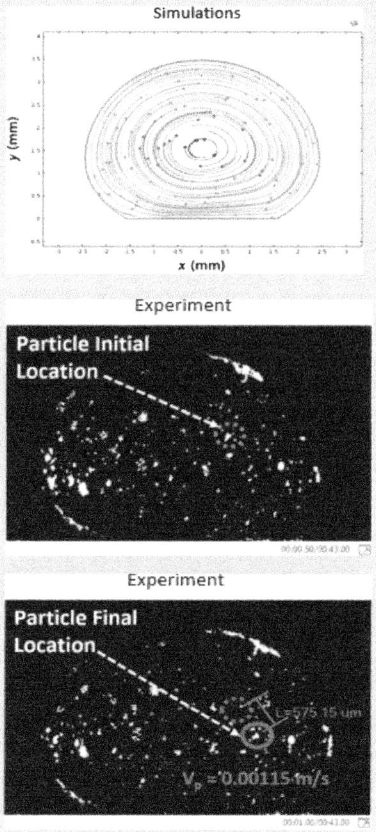

The average error estimated from the velocity data is in the order of 8.9%. The occurrence of the large error, particularly for particle# 10 ($\sim$41%), is the outlier of the error distribution and can be neglected. It is evident that the particle velocities predicted from the multiphysics code and the experiment are in good agreement. However, the small discrepancies between both results are related to the computational errors, such as round-off errors and experimental error based on the measurement repeatability, which is in the order of 6%.

5.5.2.3 Findings

Heat transfer in a sessile droplet on a hydrophobic surface gives rise to internal flow generation inside the droplet [28]. In general, the Marangoni and buoyancy currents govern the internal fluidity of the droplet. However, the surface tension gradient ($d\sigma/dT$) and the density variation in the droplet fluid under the thermal load determine the flow field inside the droplet. In some situations, flow acceleration overcomes the adhesion work required for the droplet attachment on the hydrophobic surface, which in turn gives rise to initiation of the droplet rolling off on the hydrophobic surface under the thermalcapillary effects. Moreover, fluid inside the sessile droplet with large contact angle can suffer from flow instability because of the thermal diffusion [71]. The characteristic time for the flow instability of the droplet is associated with the ratio of droplet diameter square over the thermal diffusivity of the fluid (a^2/α, where a is the characteristic droplet diameter and α is the thermal diffusivity of the fluid inside the droplet) [36]. In this case, for stable flow situation inside the droplet, a^2/α remains greater than unity inside the droplet [58]. Although various droplet diameters are considered in the simulations, the minimum droplet characteristic diameter incorporated in the simulations is in the order of 3 mm, which gives rise to the characteristic time of the order of 63 seconds. To exhibit the importance of the characteristic time in the flow field, velocity and temperature variations along the vertical rake inside the droplet are plotted in Figs. 5.59 and 5.60 for various heating durations. The contact angle of the droplet in Figs. 5.59 and 5.60 is 150 degrees. In the early heating period, temperature and velocity vary considerably along the vertical rake. However, the flow and temperature fields start settling along the rake while resulting in similar velocity and temperature profiles along the rake for the heating duration approaches to 10 seconds. The contact line morphology of the interface, between the droplet and the hydrophobic surface, influences droplet pinning and droplet mobility on the hydrophobic surface [58]. On the other hand, the length of the contact line in which the pinning occurs is associated with the droplet radius and fraction of projected area of the surface as occupied by the droplet [58]. The tangential force balance along the contact line at the surface defines the state of the sessile droplet pinning or rolling at the surface. Moreover, evaporation

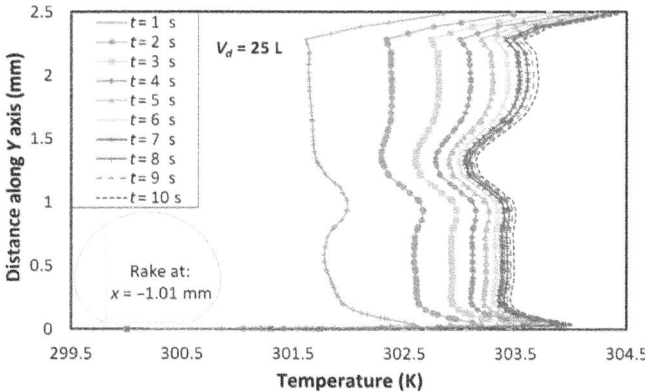

FIGURE 5.59   Temperature distribution along the vertical rake for different heating periods [4].

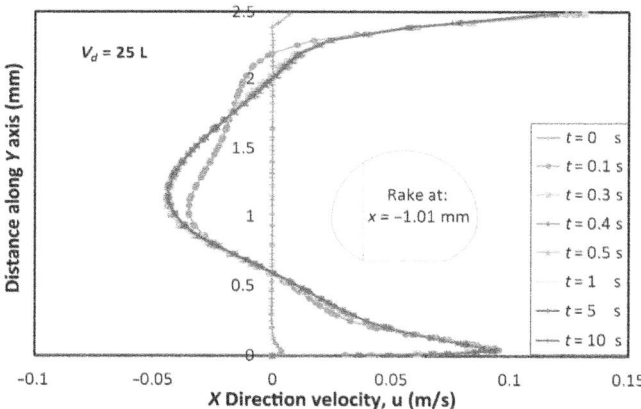

FIGURE 5.60   Velocity distribution along the vertical rake for different heating periods [4].

of the droplet occurs when the heating of the droplet continuous at a high rate for long duration of heating interaction [59]. In this case, the droplet size decays gradually and the rate of evaporation increases with progressing time. However, in the case of heating at low rates for short durations, the droplet does not undergo extensive evaporation; in which case, the flow field inside the droplet is governed by the combination of the Marangoni and the buoyancy-driven flows. Consequently, a care is taken to avoid droplet evaporation during the heating period, that is, in the experiments, the droplet size is monitored via high-speed camera during the heating period. Therefore, the tracing of the droplet images up to 60 seconds of heating reveals that evaporation only alters the droplet size within 2%. In this case, the droplet evaporation is neglected in the simulations in line with the experimental observations.

Figs. 5.60 and 5.61 show goniometer images of the droplet and temperature field inside the corresponding droplet for various droplet volumes. The
temperature contours corresponds to the 10 seconds of the heating duration.
The actual geometry and the contact angle of the droplets are incorporated in
the simulations. The radiative heating from the left hand side of the droplet
surface is considered in the simulations in line with the experimental heating
conditions (Fig. 5.57). Fluid temperature remains high in the region where
the radiative intensity incidents at the droplet surface. The bulging of the droplet occurs, under the force balance between the pressure and the surface tension
forces, for the large volume droplets. This, in turn, changes both the droplet
surface profile and the droplet surface area. Consequently, temperature gradient
(dT/dn, where n is the normal direction to the surface) and surface tension gradient ($d\sigma/dT$, where σ is the surface tension of the droplet fluid) along the
heated surface changes with the droplet size. In addition, the droplet expansion
in the direction of the solid-surface takes place because of the droplet bulging.
Therefore, the heat transfer in the large volume droplet differs than that of the
small volume droplets. This alters temperature field in the droplet fluid with
changing droplet size. Moreover, surface tension gradient ($d\sigma/dT$) plays a critical role for the intensity of the Marangoni current; in which case, increasing

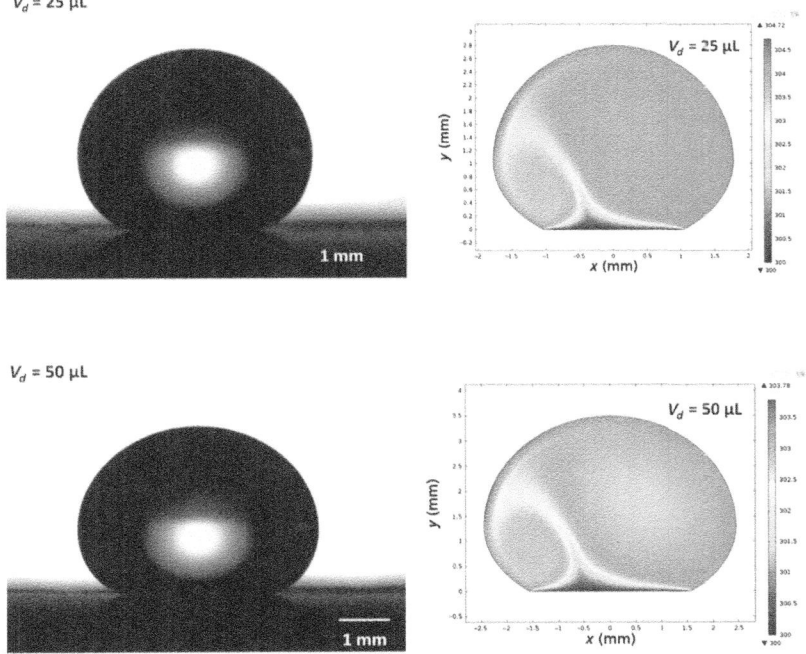

FIGURE 5.61   Droplet image and temperature (K) contours inside the droplet for droplet
volumes (25 and 50 μL) after 10 s of heating [4].

surface tension gradient enhances the Marangoni current intensity inside the droplet [28]. Since the droplet is heated from the side surface, the buoyancy force due to density variation takes place in the region of the droplet side. Moreover, droplet temperature in the surface vicinity exposed to thermal irradiation remains high and becomes almost uniform along the arc as seen by the thermal irradiation. This, in turn, lowers the surface tension gradient ($d\sigma/dT$) along this arc and suppresses the Marangoni current in this region. Consequently, the buoyancy current dominates over the Marangoni current in this region and the combination of the buoyancy and the Marangoni currents modifies the size of the circulation cell, which becomes small. Since the convection current changes locally inside the droplet because of the variation of the surface tension gradient and density in the heated region, nonuniform-like heating takes place in the droplet fluid. Therefore, heat transfer by convection current and thermal diffusion inside the droplet results in locally distributed high-temperature regions in the droplet fluid. This situation changes with the droplet volume, that is, altering the droplet volume changes the maximum fluid temperature and the size of the locally heated regions inside the droplet.

Fig. 5.62 shows velocity contours inside the droplet fluid for various droplet diameters. The velocity fields are produced after 10 seconds of

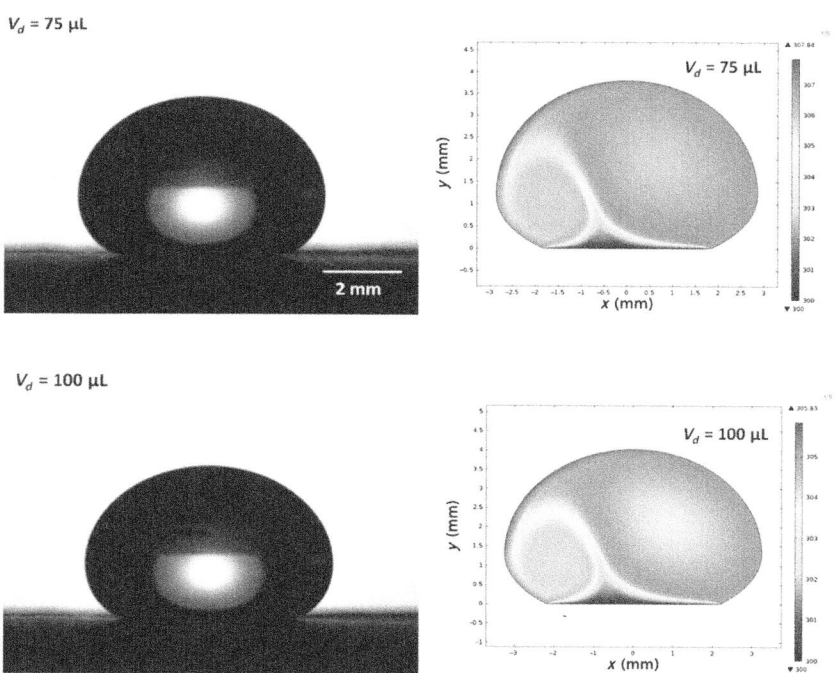

FIGURE 5.62 Velocity (m/s) contours inside the droplet for various droplet sizes after 10 s of heating [4].

heating similar to those shown in Figs. 5.61 and 5.63. Two counter-rotating circulation cells are formed inside the droplet and the size of the circulation cell remains smaller in the region of the heated droplet surface. This is true for all droplet volumes incorporated in the simulations. The center of rotation of the circulation cells also remains almost the same for all droplet volumes despite the fact that the bulging of the droplet increases with droplet size, that is, the expansion of the droplet along the radial direction does not influence significantly the location of the circulation cells. The formation of the counter-rotating circulation cells is associated with the mixing of the Marangoni and the buoyancy currents. Since density variation is high in the region close to the side surface of the droplet due to radiative heating, the buoyancy current dominates in this region. On the other hand, the surface tension gradient is high along the droplet surface in the region of droplet top, due to temperature difference along the droplet surface, Marangoni convection dominates over the buoyancy current in this region. Consequently, the shear layer developed between these currents gives rise to the formation of the counter-rotating circulation cells inside the droplet. On the other hand, the ratio of the Marangoni force, due to the Marangoni current, over the buoyancy force, due to the buoyancy current, is the same order of the ratio of the Marangoni number $(Ma = (\partial \sigma / \partial T)(a \Delta T / \mu \alpha))$, where a represents the characteristic diameter of the droplet, σ is the surface tension, T is the

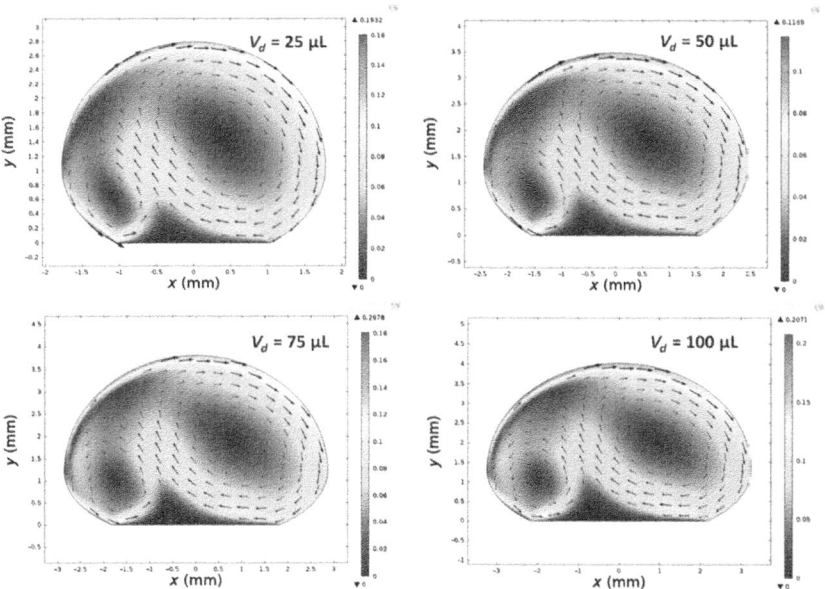

FIGURE 5.63 Droplet image and temperature (K) contours inside the droplet for droplet volumes (75 and 100 μL) after 10 s of heating [4].

temperature, and μ is the dynamic viscosity) over the Rayleigh number ($Ra = \alpha_t ga^3 \Delta T / v\alpha$, where α_t is the thermal expansion coefficient and g is the center of gravity) [36]. Therefore, the velocity ratio due to the Marangoni current over the buoyancy current becomes $(\partial \sigma / \partial T) \Delta T / \alpha_t \rho ga^2$. Since the droplet size incorporated in the simulations has a small characteristic diameter ($a \approx 4 \times 10^{-3} - 6 \times 10^{-3}$ m), the denominator becomes much smaller than the numerator in the velocity ratio ($(\partial \sigma / \partial T) \Delta T / \alpha_t \rho ga^2$). Hence, the Marangoni current dominates over the convection current developed in the top region of the droplet where the temperature change is smaller than that of the region irradiated by the thermal heat source. Relatively higher flow velocity occurs in the top region of the droplet than that corresponding to the side region of the droplet where the radiative heating taking place. This situation is true for all the droplet sizes incorporated in the simulations. The maximum velocity is in the order of 0.2678 m/s, which occurs for 75 μL droplet. The fluid acceleration due to the local effect ($\partial U / \partial t$), the convection effect ($U \partial U / \partial s$), and radial effect because of the fluid rotation ($-U^2/r$, r being the droplet radius) is considered inside the droplet. The scale analysis of the acceleration is carried out and the findings revealed that maximum values of the local, convection, and radial accelerations for 50 μL volume of droplet are in the order of 7.5×10^{-3} m²/s, 0.7 m²/s, and 1.65 m²/s, respectively. The radial acceleration of the fluid dominates over the local and convection acceleration inside the droplet. The radial acceleration of the fluid is associated with the fluid circulation inside the droplet, which becomes large in the outer region of the droplet, particularly in the region close to the top of the droplet.

The sum of the acceleration vector over the entire droplet volume results in the average fluid acceleration in the order of 0.84 m²/s for 100 μL volume of the droplet, 0.95 m²/s for 75 μL volume of the droplet, 1.12 m²/s for 50 μL volume of the droplet and 2.46 m²/s for 25 μL volume of the droplet. The acceleration increases with reducing droplet volume, which is more pronounced for 25 μL volume of the droplet; in which case, it is almost twice that of the 50 μL volume of the droplet. On the other hand, the flow acceleration can cause rolling off of the droplet on the hydrophobic surface because of flow instability [44]. Moreover, it was demonstrated that the force required to roll off the droplet on the surface is the same as the force needed to overcome the work of adhesion [37]. In line with the previous study [22], the lateral adhesion force is formulated heuristically by approximating the three-phase contact line with a single ellipse and using the experimentally obtained polynomial function for the dependence of the contact angle on the position along the three-phase contact line. The resulting equation is [23]: $F_L = (24/\pi)\gamma_{LV}a(cos\theta_R - cos\theta_A)$, where γ_{LV} is the surface tension of the liquid on the solid surface, a is the droplet diameter prior to deformation (same area of the ellipse), θ_R is the receding angle, and θ_A is the advancing angle. The droplet advancing and receding angles on the hydrophobic surface can be obtained from the dynamic contact angle measurements. The droplet contact angle and contact angle hysteresis, which are incorporated in the analysis, are given in Table 5.7.

Since the solid surface is textured, the roughness parameter can be incorporated into the adhesion force in line with the Young—Dupre Equation [24]. In this case, it yields $F_L = (24/\pi)\gamma_{LV}Df(cos\vartheta_R - cos\theta_A)$, where f is the solid-surface fraction (solid/liquid contact fraction). The solid fraction of the surface f can be defined through the projected area of poles over the total project area of the textured surface. This corresponds to the fraction of the area covered by the poles at the textured surface. This can also be written in terms of the fraction of the area wetted by the liquid, that is, f = ((Projected Total Textured Area) − (Area of Liquid_Air Interface))/ (Projected Total Textured Area). Several tests are carried out for the water-droplet contact angle measurements on the solvent-induced textured surface and the AFM image of the texture pole heights (Fig. 5.56D) are also analyzed in details. The solid fraction for the solvent-induced crystallized polycarbonate surface is determined to be within the range of $0.4 \leq f \leq 0.6$. The lateral force is associated with the adhesion force of the droplet at the surface. Under the droplet attachment (pinning) conditions at the surface, the lateral force becomes the same as the net force resulting from the sum of the shear force at the wetted surface, due to rate of fluid strain at the contact area of the droplet between the water and the hydrophobic surface, and the adhesion force. In this case, the droplet rolling off can initiate, when the net force is generated among the fluid inertial force, due to fluid acceleration inside the droplet, the shear force at the liquid—solid interfacial wall of the droplet, and the adhesion force. Therefore, the condition for the net force causing the droplet roll off from the hydrophobic surface can be written as $F_a - (F_{ad} + F_\tau) > 0$, where F_a is the fluid inertial force, due to flow acceleration in the droplet fluid along the line parallel to the surface, F_{ad} is the adhesion force, and F_τ, ($F_\tau = A_w\mu(\partial V/\partial n)$, A_w is the wetted area, μ is the fluid viscosity, $\partial V/\partial n$ is the rate of fluid strain normal to the interfacial surface) is the shear force acting at the wetted surface. It should be noted that

TABLE 5.7 Contact Angle of Water Droplets and Contact Angle Hysteresis (CAH) With Droplet Sizes [4]

Droplet Volume (μL)	θ (Deg)	CAH (Deg)
25	152.43	1.14
50	149.14	2.06
75	150.61	1.02
100	145.87	2.36

the rate of fluid strain (shear rate) in the vicinity of the droplet liquid–solid surface interface is low because of the side heating rather than heating from the droplet bottom [28]. In addition, the droplet fluid viscosity and the interfacial area between the droplet and the solid surface are considerably small. The shear force remains low for all the droplet volumes considered in the present study, that is, it is in the order of 10^{-9} N. Therefore, the droplet mobility toward rolling off from the surface is initiated by the difference between the fluid inertial force and the adhesion force. The adhesion force is in the order of 1.72×10^{-5} N for the droplet volume of 50 μL, and it is 1.73×10^{-6} N for the droplet volume of 25 μL. However, the fluid inertial force, due to flow acceleration inside the droplet, is in the order of 3.75×10^{-5} N for the droplet volume of 50 μL, and it is 8.13×10^{-6} N for the droplet volume of 25 μL. Therefore, the force difference is in the order of 2.03×10^{-5} N for the droplet volume of 50 μL while it is in the order of 6.58×10^{-5} N for the droplet volume of 25 μL. Therefore, the droplet rolls off from the hydrophobic surface for both volumes of 50 and 25 μL. This situation is also observed from the experiment (Fig. 5.64). Consequently, the side heating of the water droplet on the hydrophobic surface results in droplet rolling, that is, this situation is predicted from the simulations and observed from the experiments.

The Bond number ($B_o = \beta g \rho a^2 / | \, d\sigma/dT \, |$) depends on the properties of the droplet fluid, droplet size, and surface tension gradient. The surface tension gradient changes with time due to heating and it changes with time for each volume. Fig. 5.65 shows variation of the normalized Bond number with heating time for various droplet volumes. The Bond number reduces with increasing heating duration in nonlinear form because of the change of the surface tension gradient with temperature. The rate of reduction of the Bond number differs with the droplet volume. Moreover, the change of the droplet diameter and droplet fluid properties contributes to the change of the Bond number with the droplet size. Since the Bond number ($B_o = \beta g \rho a^2 / | \, d\sigma/dT \, |$) varies with the droplet volume and heating time (Fig. 5.65), the Merve number [72] is used to account for the ratio of the droplet weight to the droplet fluid surface tension force. The correlation among the Nusselt, the Bond, and the Merve numbers is introduced. Although the inertia force in the Weber number is associated with the square of the length scale of the droplet geometry, the Weber number is related to the dynamic motion of the droplet because of the inertia force association. In addition, the net force, due to the difference between the gravitational and buoyancy forces, over the surface tension force is incorporated into the Bond number. However, for a sessile droplet, the assessment of heat transfer can be made based on the droplet mass, which is directly associated with the weight force alone. Therefore, the Merve number, which is different than the Bond and the Weber numbers, is used to examine the Nusselt number variation with the droplet size. The Merve number takes the form

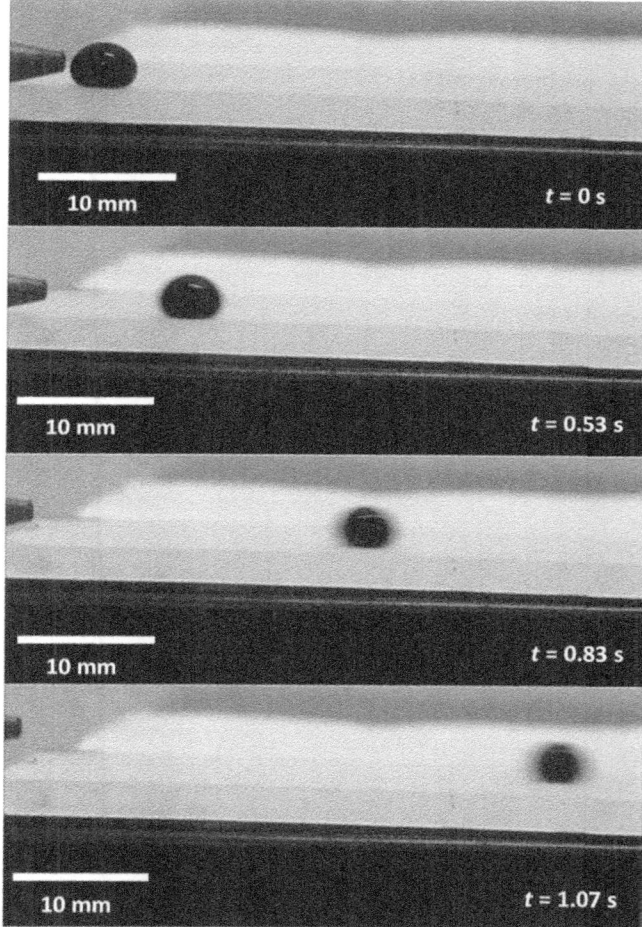

FIGURE 5.64 Droplet rolling off on hydrophobic surface at different time frames [4].

$(MN = \gamma_w a^2/4\sigma)$, where γ_w is the specific weight of water, a is the droplet characteristic length, and σ is the surface tension.

Fig. 5.66 shows the variation of the Nusselt and the Bond numbers with the Merve number. The bond and the Nusselt numbers increase with increasing Merve number. The Bond number remains less than unity while showing that the Marangoni current dominates over the buoyancy current in the fluid inside the droplet. In this case, the thermal radiation heating of the droplet side gives rise to the flow field resulting in high surface tension gradient $(d\sigma/dT)$. This creates the Marangoni current and forms the bases for the driving mechanism of the flow field in the side the droplet; consequently, the Marangoni current is the main contributor to the formation of the circulation

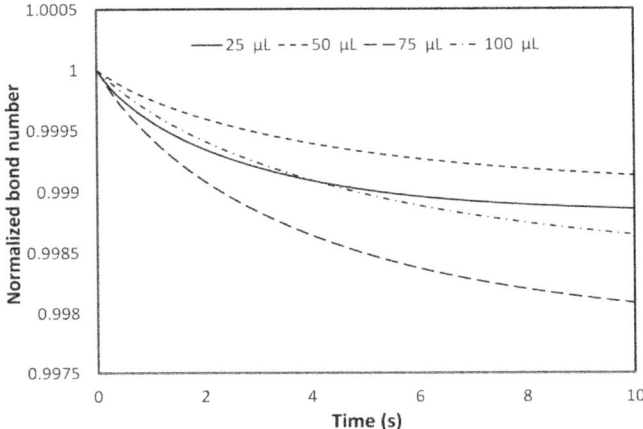

FIGURE 5.65 Temporal variation of the normalized Bond number (Bo/Bo_{max}) for various droplet volumes [4].

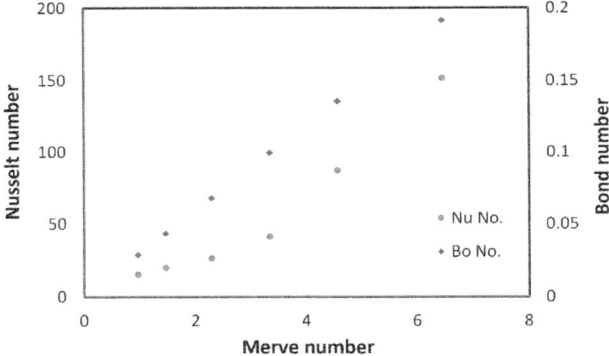

FIGURE 5.66 Variation of the Nusselt and Bond numbers with the Merve number ($\gamma_w a^2/4\sigma$) [4].

cells inside the droplet. The Nusselt number variation with the Merve number results in two distinct regions. The first region extends to $MN \leq 2.5$ in which the Nusselt number remains at low values. In second region, $MN \geq$ 2.5, the Nusselt number increases linearly with increasing the Merve number. This indicates that the droplet size has considerable effect on the heat transfer rates from the radiative source to the water droplet, that is, the ratio of gravitational force over the surface tension force becomes important in terms of the heat transfer rates.

Moreover, the Bond number remains small for small values of the Merve number ($MN \leq 2.5$). In this case, the contribution of buoyant force to the convection current in the vicinity of the solid-water contact region becomes small. Consequently, this suppresses the convection heat transfer from the contact region toward the solid bulk of the hydrophobic surface while causing the attainment of low-temperature gradients in the interfacial region between the droplet fluid and hydrophobic surface. Therefore, the heat transfer to the solid bulk reduces while resulting in the small buoyant current in this region. In general, density variation is high in the region of the droplet side where the temperature remains high (Figs. 5.61 and 5.62). This gives rise to mixing of buoyancy and Marangoni currents in this region while enhancing the convective heat transfer rates in this region, which is more pronounced for the large volume droplets, that is, the heat transfer rates in the surface region of the heated section of the droplet are enhanced with the increasing droplet volume while enhancing the Nusselt number.

5.5.3   Droplet Heat Transfer on Hydrophobic Micropost Arrays

A one-step, cost-effective process is preferable to create hydrophobic characteristics at the surface. One of the well-recognized techniques used to generate hierarchical textures is lithographic micropatterning at the surface. The lithographically patterned micropost arrays provide hydrophobic characteristics of the surface and can be easily copied and replicated to cover large areas [73,74]. In addition, replicating micropost arrays by optically transparent materials such as PDMS enables the use of a textured surface in solar energy-harvesting devices such as photovoltaic panels. On the other hand, temperature difference occurs between water droplets, such as a rain droplet, and textured surfaces in outdoor environments. This is particularly true in the Middle East, where the average temperature difference between rain droplet and outdoor surface is in the order of 8°C in typical spring months of Saudi Arabia. This, in turn, results in heat transfer from surface to water droplet while generating flow within the droplet and modifying droplet mobility on the textured surface. Heat transfer and droplet internal flow characteristics on surfaces composing of micropost arrays are important for self-cleaning application of surfaces. The influence of thermocapillary and buoyant forces on the internal fluidity of the droplet changes with the surface texture features despite the surfaces possesses hydrophobic characteristics with the same or similar contact angles. The flow dynamics inside the water droplet located on the hydrophobic surface with irregular texture patterns are investigated previously after simplifying the influence of air traps, in between the pattern pitches and the droplet, on the heat transfer rates. The simplification can be justified because of the difficulties in representing the topology of the surface texture in terms of mathematical functions. However, the geometric configuration of the surface can be easily formulated for regular texture patterns such as micropost arrays. In this case, heat transfer analysis incorporating the texture geometry becomes feasible. In addition, optically transparent surfaces with micropost arrays have the potential to be used as a protective layer of optically active device surfaces, such as photovoltaic panels. On the other hand, in natural environmental conditions, the temperature difference between the rain droplets and the solid surface initiates heat transfer from the solid surface toward the water droplets located on the solid surface. Moreover, the heat transfer characteristics of a water droplet on surfaces with micropost arrays are important in terms of droplet mobility and self-cleaning applications. Consequently, in the present section, heat transfer and internal flow in water droplet on the hydrophobic surface of micropost arrays is presented in light of the previous study [7].

5.5.3.1 Flow and Heat Transfer Analysis

Desalinated water droplets were incorporated and the governing equations for the incompressible flow adopted inside the droplet in the analysis. The governing equations were solved numerically in line with the experimental conditions. The time required for the flow stability inside the droplet was estimated in the order of 30 seconds; therefore, the transient simulations were carried out up to 40 seconds. Since the simulation period is short, it is assumed that the evaporation of the droplet can be avoided. However, to verify this assumption, the experiments were carried out to measure the liquid mass loss from the droplet during 200 seconds under the similar conditions of the experiment. In this case, the optical images of the water droplet are taken every 20 seconds to monitor the volume change of the droplet during the experiments; in which case, the change of the droplet size was judged from the geometric feature of the droplet at various times. The droplet images reveal that the droplet volume does not change notably over a period of 200 seconds. In addition, evaporation of the droplet is calculated and the amount of water evaporated from the droplet is estimated as 0.0002244 grams after 30 seconds for a droplet volume of 80 μL ($D = 5.35$ mm) under the ambient conditions of 300K temperature and 85% of the air relative humidity. The estimation reveals that the mass of water evaporated from the droplet corresponds to 0.28% of the total mass of an 80 μL water droplet. Consequently, the assumption of neglecting the evaporation from the droplet during the measurements is justifiable.

The governing flow and thermal fields are coupled simultaneously in the simulations. The continuity equation for the incompressible flow is:

$$\nabla . V = 0 \tag{5.39}$$

where V is the liquid velocity.

The momentum equation can be written as:

$$\rho\left(\frac{\partial V}{\partial t} + V \cdot \nabla V\right) = -\rho_o \beta (T - T_O)\,\vec{g} - \nabla(p - p_o) + \nabla\left[\mu\left(\nabla V + (\nabla V)^T\right)\right] \tag{5.40}$$

where p is the pressure, μ is the dynamic viscosity of the liquid, g is the gravity, and p_o is the hydrostatic pressure corresponding to density ρ_o and temperature T_o. It should be noted that the density variation is mainly caused by the thermal expansion of the fluid and can be expressed from Boussinesq approximation as:

$$\rho = \rho_o[1 - \beta(T - T_o)] \tag{5.41}$$

where β is the thermal expansion of the water.

The flow field should satisfy the energy balance according to:

$$\rho\, C_p \frac{\partial T}{\partial t} + \rho\, C_p\, V \cdot \nabla T = \nabla \cdot (k \nabla T) \tag{5.42}$$

where C_p is the specific heat capacity and k is the thermal conductivity.

5.5.3.1.1 Initial Conditions

Initially, it is assumed that the droplet has a uniform temperature at 300K and the textured micropost arrays are at 308K. It should be noted that the air in the pitch (spacing) between the micropost arrays is assumed to at 300K initially and air is considered to be an ideal gas and equation of state is considered for the thermodynamic relations for air. The consideration of uniform temperature for the micropost array surface at 308K is because of a local winter temperature in the Kingdom of Saudi Arabia and the droplet temperature is assumed to be 300K initially, which is also a typical winter rain-droplet temperature. The stagnant water is assumed in the droplet, that is, the initial velocity is set to zero in the droplet.

5.5.3.1.2 Boundary Conditions

Fig. 5.67 shows the SEM image of the micropost and optical image of the water droplet on the hydrophobic micropost-textured surface, while Fig. 5.68 shows the relevant boundary conditions incorporated in the simulations. The natural convection boundary condition with the heat transfer coefficient $h = 15$ W/m^{2}K is adopted along the free surface of the water droplet with the surrounding air ambient temperature of 300K. In addition, slip boundary is considered for flow velocity at the free surface of the droplet ($V.n_{wall} = 0$). The conjugate heat transfer is assumed between micropost pitch of the textured surface and the water-droplet bottom, which is shown in Fig. 5.68. In this case, no-slip flow boundary is incorporated at the droplet bottom interfacing with the micropost-textured surface. To account for the Cassie & Baxter state due to nonwetting characteristics of the micropost arrays, the slip boundary is incorporated between droplet bottom and the pitches of the micropost array (Fig. 5.68) where air is captured after droplet formation on the micropost array surface. In this case, the Navier slip boundary condition ($V.n_{wall} = 0$) is considered at the bottom boundary of the water droplet across the micropost pitch, where air gap is present. Moreover, continuity of heat flux boundary condition is considered at the interface between air in the micropost pitch and water-droplet bottom surface. It should be noted that during the experiments, the relative air humidity is checked and it is read as in the order of 85%.

Since the droplet is partially bulged across the micropost pitch, an arc is introduced to account for the bulging, which bases on the vertical force balance across the micropost pitch. This situation is incorporated in the droplet geometry and the point of inflection is set accordingly. For the formulation of the inflection, consider micropost pitch as shown in Figs. 5.69 and 5.70. The force balance along the vertical axis due to gravity and surface tension of a droplet on the small distance micropost yields:

$$mg \frac{\pi}{4} b^2 h_d + mg \, \Delta V_d = F_\gamma sin\theta \qquad (5.43)$$

where mg is the specific weight, h_d is the droplet height, F_γ ($F_\gamma = \pi d\gamma$) is the surface tension force, and ΔV_d is the volume of inflection. Considering ΔV_d is the half volume of an ellipsoid, Eq. (5.43) becomes:

$$mg \frac{\pi}{4} b^2 h_d + mg \, \frac{4\pi}{2 \times 3} \left(\frac{b}{2}\right)^2 \chi = \pi b \gamma sin\theta \text{ or } \frac{mg}{4} h_d + mg \, \frac{4}{64} \frac{1}{} \chi = \frac{\gamma}{b} sin\theta$$

$$(5.44)$$

And the height of the droop yields:

$$\chi = \frac{6\gamma}{mgb} sin\theta - \frac{3}{2} h_d \qquad (5.45)$$

Since, $sin\theta \approx \chi/(b/2)$, then:

$$\chi = \frac{6\gamma}{mgd} \frac{2\chi}{b} - \frac{3}{2} h_d \text{ and } \chi = \frac{3h_d}{2\left(\frac{12\gamma}{mgb^2} - 1\right)} \qquad (5.46)$$

Inserting values for $h = 0.002$ m and $d = b = 10 \, \mu m$, results $\chi = 0.034 \, \mu m$.

The local deformation of small droplets is governed by surface effects rather than vertical force balance [14]. The maximum droop of the droplet (δ) occurs in the center of the four neighboring microposts. The spherical droplet geometry can be considered similar to that shown in Fig. 5.71, if the distance between the micropost is large [14]. In this case, the maximum droop of the droplet (δ) is at the midpoint of the diagonal line (ℓ), as shown in Fig. 5.69, and it can be written as $\ell = \sqrt{2}(a + b)$.

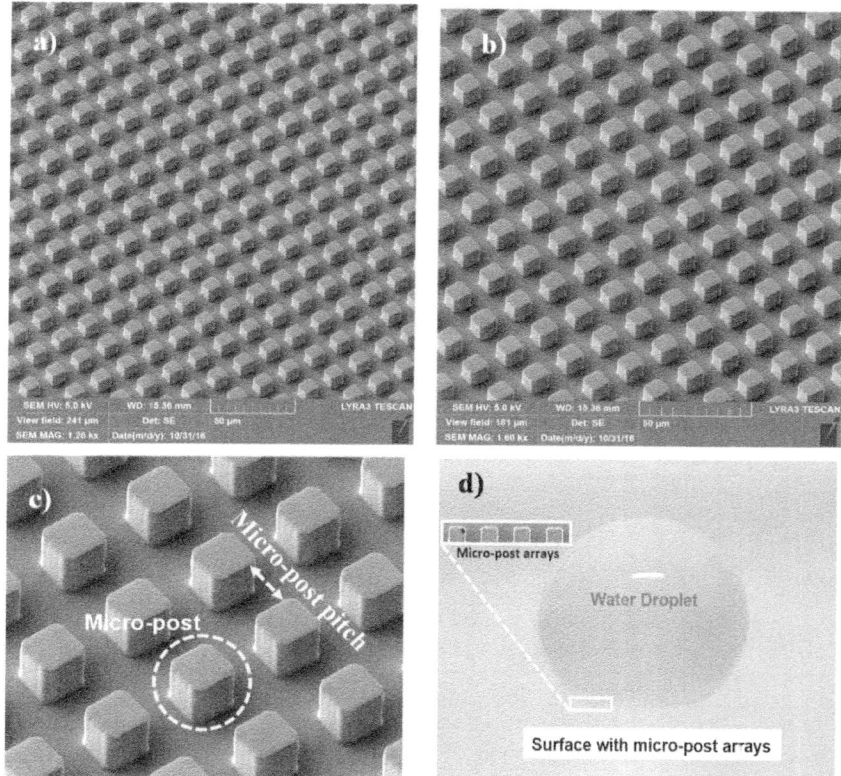

FIGURE 5.67 SEM micrographs of micropost arrays on surfaces and optical image of water droplet: (A) micropost array on silicon wafer, (B) replicated micropost arrays on PDMS, (C) close view of micropost arrays on surface, and (D) water droplet on micropost array ($a = 10\,\mu m$, $b = 10\,\mu m$, $h = 10\,\mu m$) [7].

The geometric relation between droop (δ) and size of the micropost, according to Fig. 5.71, yields:

$$R^2 = \left(\frac{1}{\sqrt{2}}(a+b)\right)^2 + (R-\delta)^2 \text{ or } (R-\delta)^2 = R^2 - \frac{1}{2}(a+b)^2 \qquad (5.47)$$

Therefore, δ yields,

$$\delta = R - \sqrt{R^2 - \frac{1}{2}(a+b)^2} \qquad (5.48)$$

On the other hand, the temperature-dependent properties of water are incorporated in the simulations given in Table 5.8. COMSOL multiphysics code [27] incorporates the finite element method, which provides a means of spatial and temporal discretization of the governing equations of flow and

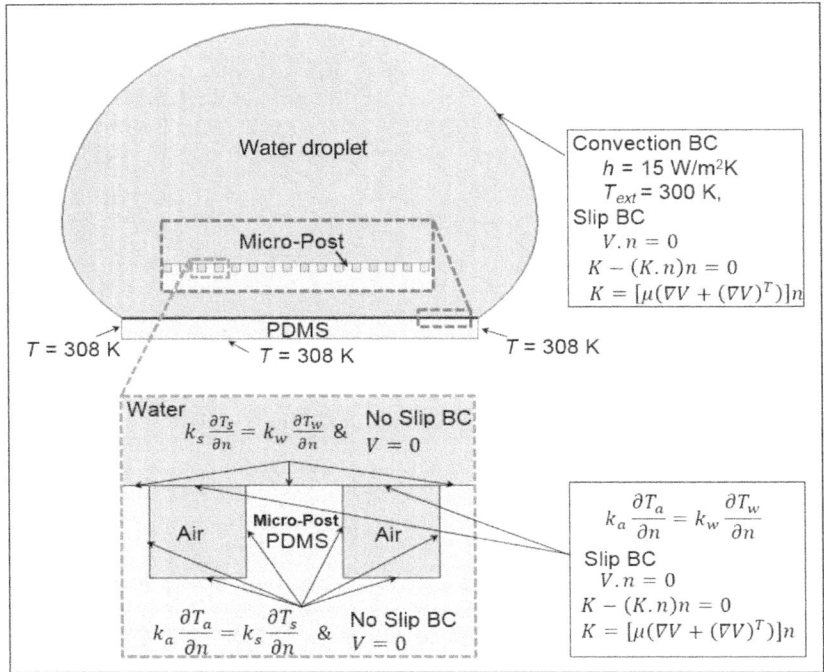

FIGURE 5.68 A schematic view of water droplet and micropost arrays with boundary conditions [7].

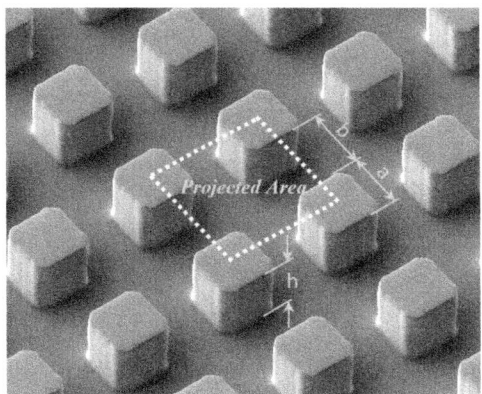

FIGURE 5.69 Micropost arrays and geometric configurations [7].

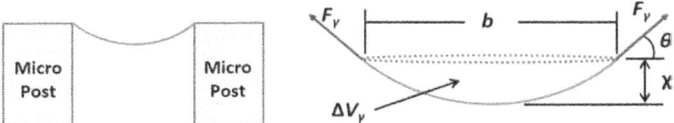

FIGURE 5.70 A schematic view of micropost pitch and bulging droplet, with radius R, in between the microposts with small bitch length [7].

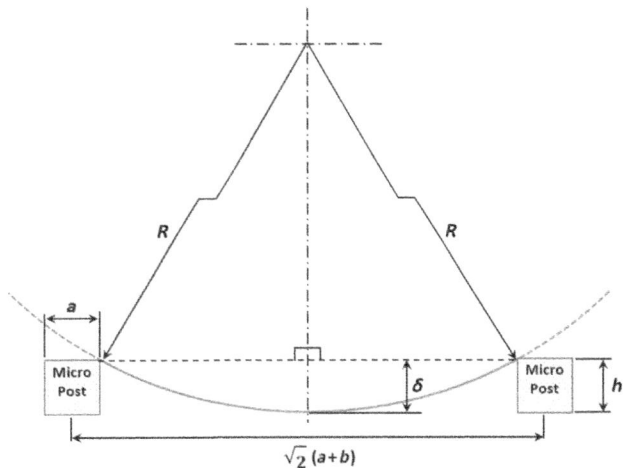

FIGURE 5.71 schematic view of micropost pitch and bulging droplet, with radius R, in between the microposts with large bitch length [7].

heat transfer. The differential operators in the governing equations are discretized using the backward Euler finite difference method. The implicit scheme with a backward difference approximation for the time used in the study was unconditionally stable; however, the accuracy of the scheme is governed by the size of the time step, which is considered to be 10^{-4} seconds and the residuals of flow parameters are set as 10^{-8} during the simulations. Moreover, in the simulation, the cross-sections of the droplets are constructed, which are identical to the optical images captured from the experiments. The findings of the grid-independent tests reveal that mesh sizes of 8936 and 12,640 result in identical temperature profiles. Consequently, the mesh size of 8936 is selected to simulate the flow field and temperature distribution. Fig. 5.72 shows the meshes used in the solution domain.

The 3D simulation of the flow and temperature fields are attempted initially; however, the computational efforts required for the simulation become extremely expensive in terms of memory size and runtime; in

TABLE 5.8 Temperature-Dependent Properties of Water

Property	Equation	Units
Density	$\rho = -3.0115 \times 10^{-6}T^3 + 9.6272 \times 10^{-6}T^2$ $-0.11052T + 1022.4$	kg/m^3
Kinematic viscosity	$\mu = 3.8208 \times 10^{-2}(T-252.33)^{-1}$	Pa · s
Specific heat capacity	$C_p = 1.7850 \times 10^{-7}T^3 - 1.9149 \times 10^{-4}T^2x$ $+ 6.7953 \times 10^{-2}T - 3.7559$	kJ/kgK
Thermal conductivity	$k = 4.2365 \times 10^{-9}T^3 - 1.144 \times 10^{-5}T^2x$ $+ 7.1959 \times 10^{-3}T - 0.63262$	W/mK

which case, the mesh size used becomes significantly larger (800,000) as compared to that of the mesh used for the 3D simulations (8936). Nevertheless, one case of 3D simulation of the flow field succeeded for a water droplet of 60 μL volume. Fig. 5.73A and B shows 3D and 2D simulation results for a droplet of 60 μL volume. It can be observed that the 2D configuration results in similar flow field to that of the 3D simulation results and the difference in the maximum velocity is in the order of 0.06% for both simulations. Consequently, 2D simulations are incorporated in the current study.

One of the goals of the simulation study is to incorporate the thermalcapillarity effect and observe the flow field inside the water droplet with various diameters on micropost-textured surfaces in relation to self-cleaning applications. In the Dammam area of Saudi Arabia, the temperature difference between the rain droplets and the surfaces in open environments is in the range of 8°C and the droplet diameter varies within 2−5 mm on the surface. The droplets corresponding to this diameter range have the volumes in the range of 20−80 μL and these volumes are considered in the analysis.

5.5.3.2 Experimental

Silicon wafers with 1 mm thickness is used to texture micropost arrays at the surface. The lithographic technique was used to generate micropost arrays with square cross-section. The typical micrographs of the micropost arrays obtained from SEM are shown in Fig. 5.67A. In order to secure the optical transmittance, the micropost-textured surface was replicated by PDMS. Fig. 5.67B shows the SEM micrographs of the replicated surface. The lithographic textured silicon wafer surface was identical to that replicated by PDMS as seen in Fig. 5.67A and B. To examine the surface topology micropost-replicated surface, Jeol 6460 SEM was used.

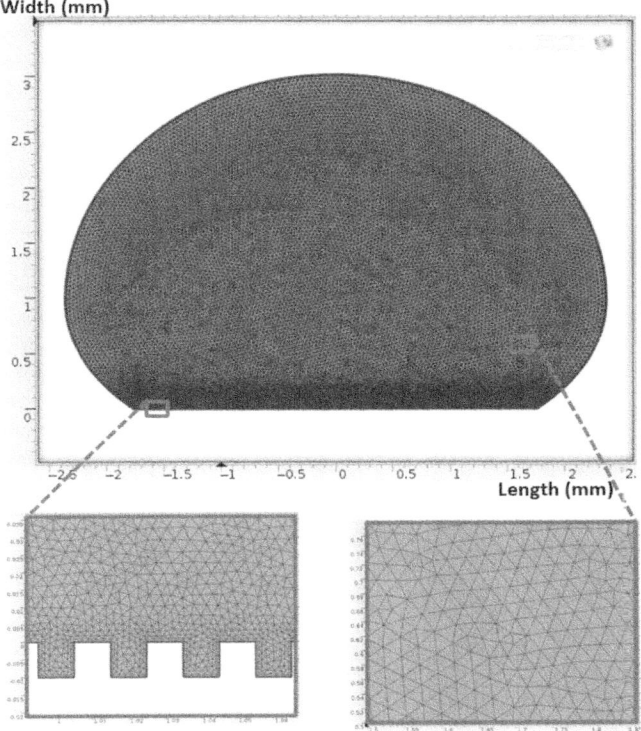

FIGURE 5.72   Mesh used in the simulations [7].

The water-droplet apparent contact angles and hysteresis were measured incorporating the sessile drop method using a Kyowa (model—DM 501) contact angle goniometer. Fig. 5.74 shows the contact angles and optical images of the droplet on the plane PDMS and PDMS-replicated surfaces, which were later incorporated in the simulations.

Desalinated water was kept at 300K in a container for 2 hours prior to the experiments. The PDMS-replicated micropost array surface was maintained at 308K for 1 hour using a controlled heating unit prior to the experiments. The surrounding air temperature and relative air humidity were recorded during the experiments and the variation of the environmental temperature and humidity were found to be negligible, which were in the order of $298 \pm 1K$ and $85\% \pm 0.5\%$, respectively. The Dantec Dynamics PIV System (Dantec Dynamics SpeedSense 9040) was used to monitor the velocity of the hollow glass particles inside the water droplet. The sample rate was 100 frames per second and the resolution was 960×720 pixels. The curvature effect of the droplet on the images recorded was corrected optically. In this case, the optical displacement is

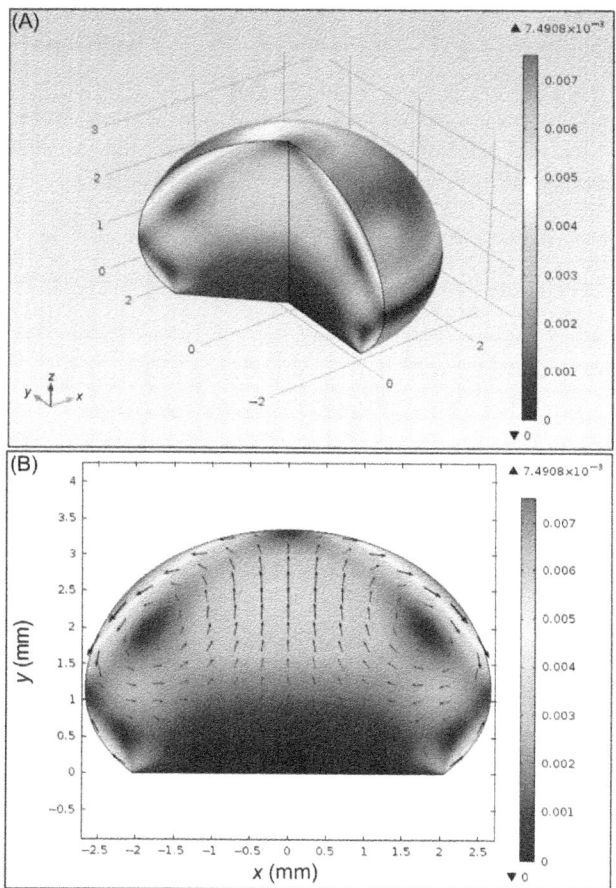

FIGURE 5.73 Temperature field obtained from 2D and 3D simulations for 60 μL droplet volume after 40 seconds of heating: (A) 3D simulation result, and (B) 2D simulation result [7].

incorporated for the correction of the particle velocity measurement inside the droplet (Fig. 5.75) in accordance with the previous study [36]. Liquid in the droplet acts like a lens and modifies the original location of the particles in the droplet. The light rays along the optical axis becomes stigmatic to the first order. The image of a point $P(0, r)$ in the midsection is shifted at point $P'(p', r')$ as shown in Fig. 5.72 in accordance with the Snell−Descartes law [75]. In this case, the location of $P'(p', r')$ can be expressed as [36]:

$$p' = g(r) = -\frac{n_{water}}{n_{air}} r sin\left[arcsin r - arcsin\left(\frac{n_{water}}{n_{air}} r\right)\right] \quad (5.49)$$

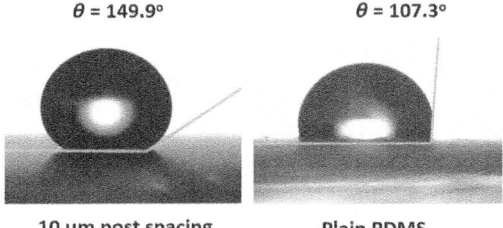

$\theta = 149.9°$ $\theta = 107.3°$

10 μm post spacing   **Plain PDMS**

FIGURE 5.74 Water-droplet contact angle of replicated PDMS surface for plain and micropost arrays with 10 μm postpitch (spacing) [7].

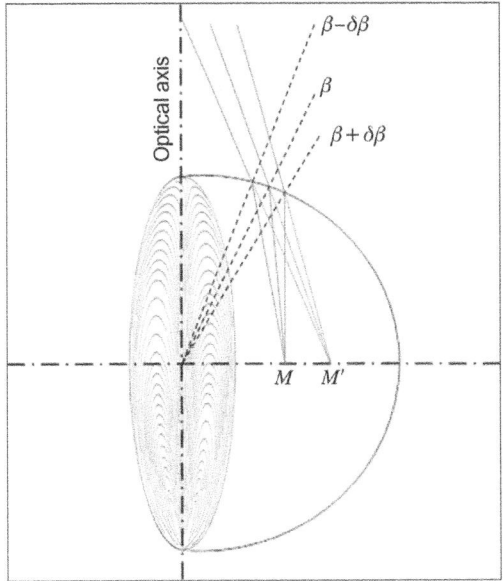

FIGURE 5.75 A schematic view of ray-tracing diagram for a spherical liquid lens [7].

and

$$r' = f(r) = \frac{n_{water}}{n_{air}} r \cos \left[arcsinr - arcsin \left(\frac{n_{water}}{n_{air}} r \right) \right] \qquad (5.50)$$

where r is the distance from the particle to the optical axis, r' is the distance from the particle to the optical axis, p' is the distance from the image to the midsection plane of the droplet, n_{water} and n_{air} are the refraction index of the water and air, respectively. A correction of almost 7% of the radius for $r = 0.85 \times R_d$ is estimated, where R_d is the droplet radius, after introducing the data recorded from the optical images and the microscope object lens setting.

The uncertainty of the displacement, in terms of pixels, was estimated as 0.03, which was in agreement with the value reported in the previous study [76].

Particle velocimetry is incorporated to assess the flow field inside the droplet for the validation of the numerical predictions. Since PIV was used to monitor the particle velocities inside the droplet, it is considered that the fluid inside the droplet composes of the mixture of water and hollow glass particles with a size of 10 μm. The governing equation of momentum is solved after incorporating the particles in the solution domain resembling the hollow glass particles in the droplet. The energy equation is solved after assuming slurry single fluid resembling water and the hollow glass particles mixture. Since the hollow glass particle concentration is low (1%) in the carrier fluid (water), effective thermal properties are incorporated into the energy equation. The incompressible flow field is considered to formulate the governing equations of flow and heat transfer. The field equations are solved numerically in line with the experimental conditions. The coupled flow and thermal fields are considered simultaneously in the simulations. The conservation equations for the mixture of water and hollow glass particles are not given herein, but they are referred to the previous study [72]. Table 5.9 gives the properties for hollow glass particles used in the simulations. To validate the predictions, the Dantec PIV system was used to measure the velocity distribution within the 40 μL water droplet. The hollow glass particles of 1%, by volume, were mixed with water prior to the droplet formation for the particle velocity measurements. Fig. 5.76 shows the typical hollow glass images inside the droplet. The flow velocities measured from PIV and predicted from the multiphysics code are given in Table 5.10. It is evident that the particle velocities predicted from the multiphysics code and the experiment are in good agreement. However, the small discrepancies between both results are related to the computational

TABLE 5.9 Properties of Hollow Glass Spheres at 300K [7]

	Hollow Glass Spheres
Mean particle size (μm)	10
Size distribution (μm)	10
Particle shape	Spherical
Density (kg/m³)	1400
Melting point (°C)	740
Thermal conductivity (W/mK)	1.14
Specific heat capacity (kJ/kgK)	0.83

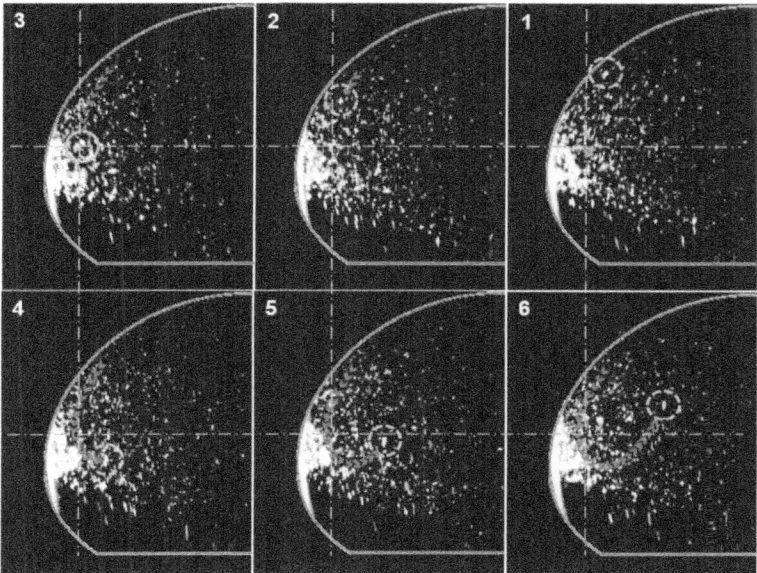

FIGURE 5.76 PIV images for hollow glass particles in 40 µL droplet volume. Each frame is represented at 300 ms during the heating period [7].

errors, such as round-off errors and experimental error based on the measurement repeatability, which is in the order of 5%.

5.5.3.3 Findings

The spreading coefficient of water on replicated PDMS surface depends on the interfacial tension between PDMS, water, and air according to $S_{sw(a)} = \gamma_{wa} - \gamma_{sa} - \gamma_{sw}$, where γ_{wa} is the surface tension force of water in air, γ_{sa} is the surface free energy of PDMS in air, and γ_{sw} is the interfacial tension between the water and PDMS. Since the surface free energy of PDMS is in the order of 21.3 mJ/m^2 [77], the surface tension of water is 0.0712 mJ/m^2 and interfacial tension between water and PDMS is 40 mJ/m^2 [78], and the spreading coefficient of water on PDMS is $S_{sw(a)} = -61.229$ mJ/m^2; therefore, water does not spread on the PDMS solid surface ($S_{sw(a)} < 0$) [58]. The replicated PDMS surface has regular micropost arrays (Fig. 5.67B) that form a texture at the surface. The wetting characteristics and droplet mobility at the surface depend on the contact line at the water-droplet perimeter. In this case, three possible states can be possible for water droplets on the textured surface depending on the total interfacial energy, which remains at the lowest level [58]. According to these states, the textured surface is: (1) partially wetted by water and partially remained dry, or (2) partially impregnated, or (3) impregnated with encapsulation [58].

TABLE 5.10 Velocity of Particles Predicted and Obtained From PIV Data at Different Locations Inside the Droplet [7]

Particle #	X (mm)	Y (mm)	Experiment V (m/s)	Simulations V (m/s)
1	−1.0321	2.7295	0.0041	0.0045
2	−1.1249	2.6947	0.0050	0.0048
3	−1.2408	2.6019	0.0048	0.0047
4	−1.4148	2.4511	0.0044	0.0046
5	−1.6583	2.3120	0.0055	0.0055
6	−1.8554	2.1149	0.0055	0.0055
7	−2.0178	1.8481	0.0045	0.0046
8	−2.1337	1.6394	0.0040	0.0039
9	−2.1569	1.3727	0.0025	0.0023
10	−2.0410	1.1292	0.0021	0.0020
11	−1.7279	0.9784	0.0026	0.0025
12	−1.4264	1.1060	0.0020	0.0023
13	−1.1596	1.2915	0.0025	0.0022
14	−0.9741	1.4770	0.0026	0.0025

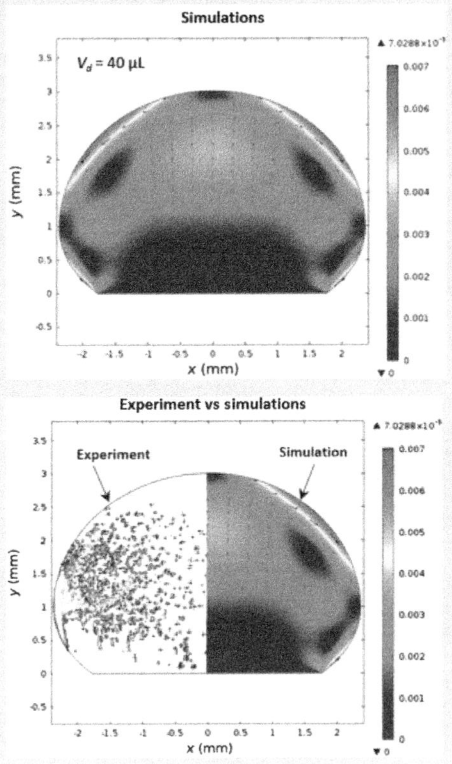

Simulations

$V_d = 40\ \mu L$

Experiment vs simulations

Water wets locally on the textured PDMS surface without spreading over the surface due to negative value of the spreading coefficient. However, the state of the wetting characteristics of water depends on the energy equation, which relates the interfacial energy with the texture parameters [58]. In this case, the relation between the surface energy and micropost geometry with partial wetted by water and partially remained dry can be written as [58]:

$$S_{sw} < -\gamma_{wa} \frac{(r-1)}{(r-\phi_s)} \tag{5.51}$$

where ϕ_s is the fraction of the projected area of the surface, which is occupied by the solid and r is the ratio of the total surface area to the projected area of the solid. Consider the micropost arrays shown in Fig. 5.70. The fraction of the projected area of the surface (ϕ_s) yields:

$$\phi_s = \frac{A_{soild}}{A_{Projected}} = \frac{a^2}{(a+b)^2} \tag{5.52}$$

and the ratio of the total surface area to the projected area of the solid becomes:

$$r = \frac{(a+b)^2 + 4ah}{(a+b)^2} = 1 + \frac{4ah}{(a+b)^2} \tag{5.53}$$

For example, for the micropost arrays with $a = b = h = 10\,\mu m$ (a is the postwidth, b is postpitch, and h is postheight, Fig. 5.70), $\phi_s = 0.25$ and $r = 2$. Inserting into Eq. (5.51) yields $-\gamma_{wa}(r-1)/(r-\phi_s) = 0.6897\,\text{mJ/m}^2$, which is greater than $S_{sw(a)} = -61.229\,\text{mJ/m}^2$. Therefore, micropost arrays result in partial wetting of water, that is, the textured surface was partially wetted by water and partially remained dry. In order to determine the water contact angle on the micropost arrays, the following equation is considered [79]:

$$cos\theta_r = r\phi_s \cdot cos\theta_0 - 1 + \phi_s \tag{5.54}$$

where θ_o is the water-droplet contact angle on a plain PDMS surface, which is in the order of 108 degrees (Fig. 5.74). Introducing the fraction of the projected area of the surface that is occupied by the solid:

$$\phi_A = 1 - \phi_s = 1 - \frac{a^2}{(a+b)^2} \tag{5.55}$$

The water-droplet contact angle on the micropost array surface yields:

$$cos\theta_r = rcos\theta_0 - \phi_A + (rcos\theta_0 + 1) \tag{5.56}$$

This yields the water-droplet contact in the order of 154.7 degrees for the micropost array with $a = b = h = 10\,\mu m$, which is very close to water-droplet contact angle measured 149.9 ± 5 degrees. Therefore, micropost arrays result in hydrophobic characteristics on the replicated PDMS surface with a micropost size of $a = b = h = 10\,\mu$m. However, increasing the micropost array pitch gives rise to wetting of the micropost arrays surface because of the occurrence of Wenzel state [14]. If the droop is larger than the micropost height (h), the droplet contacts at the surface of the micropost pitch and Cassie and Baxter state change to Wenzel state [14]. The condition for the Wenzel state yields $R - \sqrt{R^2 - (1/2)(a+b)^2} > h$ where R is the droplet radius [14]. In this case, increasing micropost array pitch results in Wenzel state.

On the other hand, heat transfer from a solid surface toward the water droplet can give rise to the flow instability inside the droplet, and the time scale to analyze the flow field becomes critical for the stable flow. The characteristic time for the stable flow is related to the thermal diffusion [36], and it takes the from a^2/α, where a is the characteristic droplet diameter (within the range of $a \approx 2 \times 10^{-3}$ m) and α is the thermal diffusivity of droplet fluid ($\alpha = 1.41 \times 10^{-7}$ m²/s). The characteristic time estimated for the flow stability inside water droplet is in the order of 30 seconds. Fig. 5.77A shows velocity variation along the central rake of the droplet for various heating durations. Velocity profiles vary along the rake with time and it becomes almost identical after 30 seconds of heating. Similarly, temperature profiles demonstrate the same behavior of the velocity profiles along the rake. This can be seen from Fig. 5.77B, in which the temperature variation along the central rake of the droplet is shown for various heating periods. Consequently, flow inside the droplet becomes almost stable, reaching a quasi-steady state after 30 seconds of heating. Moreover, the Marangoni and buoyancy currents shape the flow field inside the droplet. The ratio of Marangoni current over the buoyant current is in the same order of Marangoni number over the Rayleigh number. It should be noted that the Marangoni number can be written as $Ma = (\partial\gamma_w/\partial T)(a\Delta T/\mu\alpha)$, where a represents the characteristic diameter of the droplet, γ_w is the surface tension, T is the temperature, and μ is the dynamic viscosity. The Rayleigh number is $Ra = \alpha_t g a^3 \Delta T/\upsilon\alpha$, where α_t is the thermal expansion coefficient, υ is the kinematic viscosity and g is the center of gravity [36]. The velocity ratio due to Marangoni force over the buoyancy force can be written as $(\partial\gamma_w/\partial T)\Delta T/\alpha_t \rho g a^2$ [36]. Hence, occurrence of large values of surface tension gradient $(\partial\gamma_w/\partial T)$ in a small droplet diameter (where a *is small*) increases the Marangoni current-induced flow velocity in the droplet fluid. Alternatively, buoyancy-driven flow current dominates over Marangoni current when the surface tension gradient becomes small in a large droplet (where a is large). Since the strength of buoyancy current is assessed by the Grashoff number ($Gr = \beta g \Delta T L^3/\upsilon^2$) and buoyancy flow dominates the flow field for $Gr > 2400$ [80]. The intensity of the Marangoni convection is determined by the Marangoni number ($Ma = (\partial\gamma_w/\partial T)(a\Delta T/\mu\alpha)$); in which case, the Marangoni current dominates the flow field for $Ma > 100$ [80]. On the

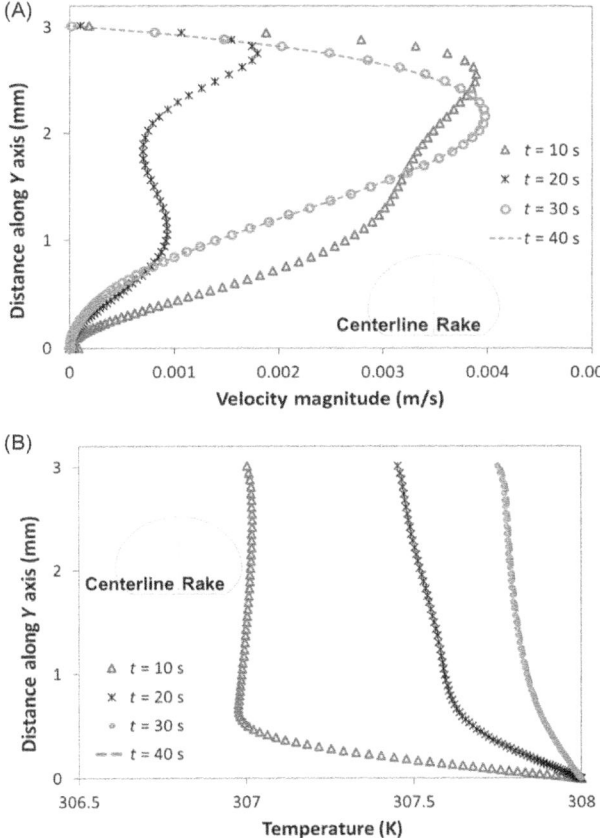

FIGURE 5.77 Velocity (A) and temperature (B) variations along the central rake for various heating times [7].

other hand, the relative contribution between natural convection and Marangoni convection can be expressed in terms of the Bond number ($B_o = \beta g \rho L^3 / (d\gamma_w / dT)$), which is proportional to the ratio of buoyancy force and the surface tension force, that is: $B_o < 1$ results in Marangoni domination in the flow field of the droplet fluid.

Fig. 5.78 shows velocity contours inside the droplet for various droplet volumes after 40 seconds of heating ($t \geq a^2 / \alpha \cong 40$ seconds). The micropost size considered in the simulations is $a = b = h = 10\ \mu m$ (Fig. 5.67B). The sessile droplet images obtained from the experiments are also included in Fig. 5.78 for the comparison of the droplet shape and size. It should be noted that droplet water temperature was kept at 300K and PDMS surface with micropost arrays was set to 308K, while mimicking the local winter temperature in the Kingdom of Saudi Arabia. In addition, the droplet initial

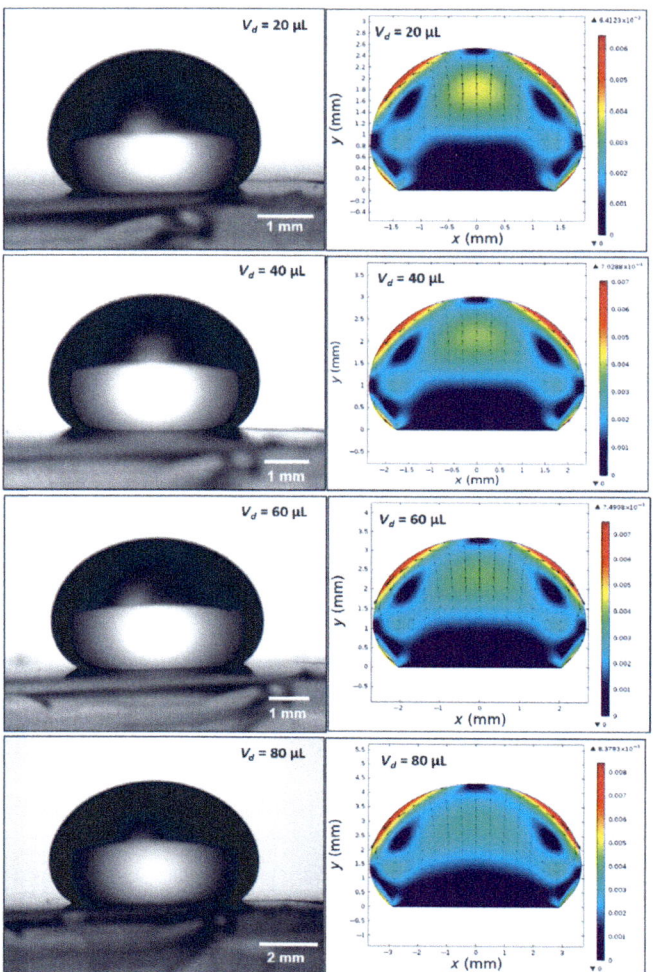

FIGURE 5.78 Water-droplet image and velocity contours inside the droplet for different droplet volumes after 40 s of heating [7].

temperature is considered to be 300K, which mimics a typical winter rain-droplet temperature. Two counter-rotating circulation cells are formed in the upper part of the droplet. The convection current generated at the droplet bottom, due to heat transfer from solid surface toward the droplet, shifts the circulation cells in the upper section of the droplet. The Marangoni current generated in the surface vicinity of the droplet influences the location of the center of the circulation cells. This is more pronounced for the large volume droplet (80 μL). The maximum flow velocity occurs in the region close to the droplet upper surface where the Marangoni current dominates the flow field. Increasing the droplet size increases the value of the maximum velocity inside the droplet and the value of the maximum velocity is in the order of 8.4×10^{-3} m/s, which occurs for 80 µL droplet volume. The attainment of high velocity for the large droplet volume is associated with a large value of the surface tension gradient $(\partial \gamma_w / \partial T)$. In this case, the surface tension gradient remains high because of the large droplet height (paddle height) corresponding to the droplet volume, that is, the local temperature gradient increases while enhancing the surface tension gradient. The maximum velocity increase is almost 23% between the droplet sizes of 20 and 80 µL.

Fig. 5.79 shows isothermal lines inside the droplet for various droplet volumes after 40 seconds of heating. Temperature decays gradually from the droplet bottom to droplet top; however, it remains high in the region close to the droplet bottom because of the thermal diffusion. Temperature remains low in the region where the circulation cells are formed. In addition, the high-temperature region in the droplet extends toward the droplet top, which is attributed to the thermal diffusion and buoyancy current that carries heated fluid from the droplet bottom toward the droplet top along the droplet central rake.

In this case, a shear layer developed between the buoyancy current and the outer edge of the circulation cells prevents mixing of hot fluid with

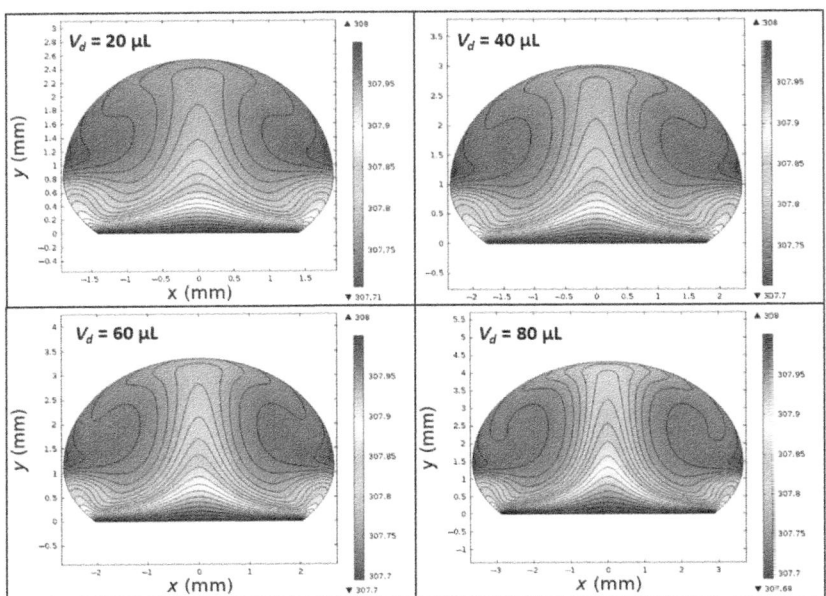

FIGURE 5.79 Temperature contours inside the droplet for different droplet volumes after 40 s of heating [7].

low-temperature fluid in the circulation cells. Consequently, fluid temperature remains low in the region covered by the circulation cells. In the region where temperature remains high, flow velocity remains low (Fig. 5.75), which in turn results in low Reynolds and Grashoff numbers, which are in the order of 0.85 and 420, respectively. Therefore, thermal diffusion is primarily responsible for the attainment of high-temperature regions inside the droplet.

Fig. 5.80 shows the temperature distribution along the horizontal rake, which is 10 μm above the droplet bottom, for 40 seconds of heating of 40 μL volume droplet. In order to assess the effect of micropost texture structure on temperature distribution in the close region of the droplet bottom, the temperature distribution along the same rake for plane surface without micropost arrays is also shown for the identical droplet volume and contact angle. Temperature remains slightly higher for the plane surface as compared to that

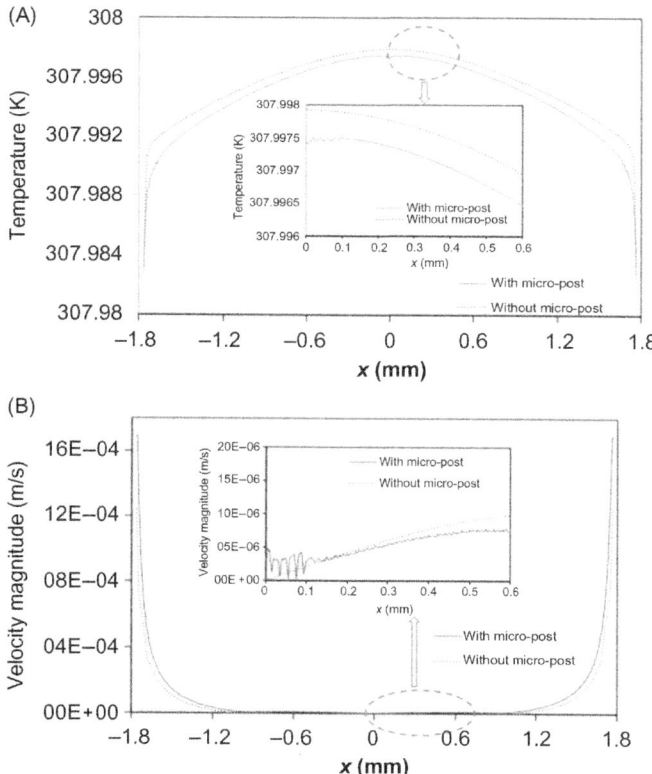

FIGURE 5.80 (A) Temperature variation, (B) velocity magnitude, along the horizontal rake inside the water droplet of 40 μL after 40 s of heating. The horizontal rake is located 10 μm above the micropost array surface inside the droplet. The small figures show the close view of temperature and velocity variations along the horizontal rake [7].

of the micropost arrays. However, the close examination of temperature distribution across a small length along the horizontal rake reveals the presence of temperature oscillation (small figure in Fig. 5.80A). This is attributed to the contact line of water droplet with the micropost arrays at the droplet bottom. In this case, the temperature remains high along the contact line between the droplet and micropost arrays while it reduces across the micropost gaps where the air is captured. However, the amplitude of temperature oscillation (difference in the maximum and minimum temperature) remains high in the close region of the central rake. This behavior is attributed to the high heat transfer rates from micropost arrays toward the droplet fluid in this region, that is, the temperature remains high in the central region of the droplet bottom, which can also be seen from Fig. 5.79. Moreover, temperature oscillation results in oscillation in velocity along the horizontal rake, which is shown in the small figure in Fig. 5.80B. Consequently, the temperature difference between the micropost gap and the micropost surface gives rise to the development of the convection current in the close region of the droplet bottom. The convection current generated carries the heated fluid from the droplet bottom toward the droplet interior while improving the heat transfer rates via convection current and lowering temperature in the bottom region of the droplet.

Temperature contours inside the micropost pitch is shown in Fig. 5.81 for four heating periods, which helps to examine the influence of micropost

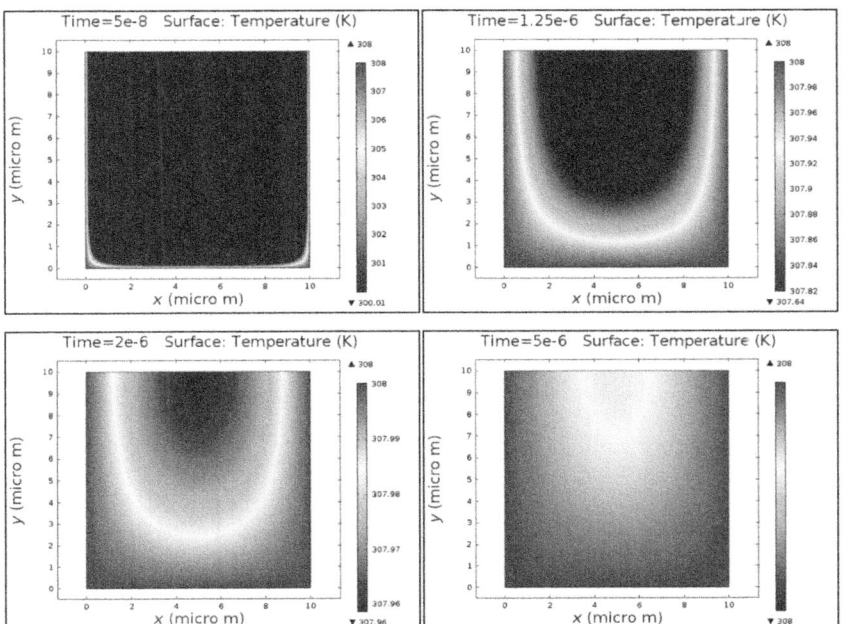

FIGURE 5.81 Temperature contours inside the micropost pitch for various heating periods [7].

arrays pitch on the heating process. Temperature rises rapidly in the micropost pitch. Increasing of temperature from the bottom of pitch reveals that the heating is governed by the combination of conduction and convection in this region. Air temperature within the pitch increases to reach close to 308K, which is the micropost wall temperature, within 5 microseconds. Temperature and velocity contours are shown in Fig. 5.82 inside the micropost pitch after 5 microseconds of heating. The heated air causes the development of a circulation cell within the micropost pitch while it remains almost stagnant at the pitch bottom.

Since the Bond number represents the ratio of buoyancy force over the Marangoni force, the Bond number changes as the buoyancy and Marangoni forces vary during the heating period. The variation of the Bond number with temperature is shown along the heating period in Fig. 5.83. The Bond number varies along the heating period because of temperature variation in the droplet. This is mainly because of the buoyancy and Marangoni forces, which depend on temperature. On the other hand, the assessment of heat transfer for a sessile droplet can be made incorporating the droplet mass, which is related to the droplet weight and the weight remains constant for a given droplet size. Therefore, the Bond and the Nusselt numbers may not be correlated correctly to represent the influence of the droplet volume on the heat transfer rates. In addition, the Weber number is the ratio of droplet inertia force over the surface tension force and it resembles the force ratio when droplet is in motion, such as rolling or linearly moving. Since the droplet

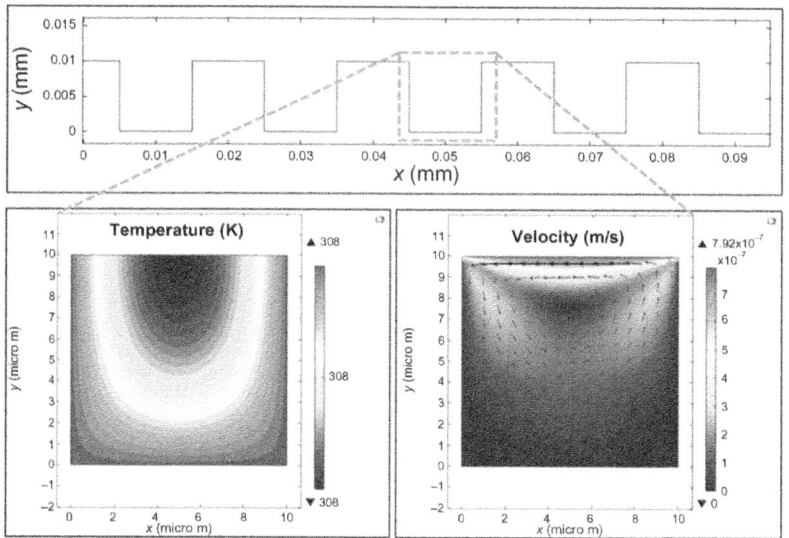

FIGURE 5.82 Temperature and velocity contours inside the micropost pitch after 5 μs of heating [7].

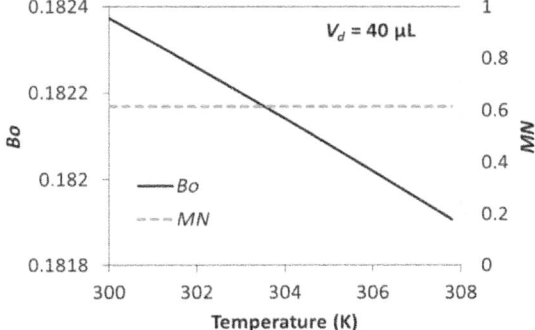

FIGURE 5.83 Bond and Merve number variation with temperature for 40 μL droplet volume [7].

remains stationary on the surface, which composes of micropost arrays, a dimensionless ratio (number) uniquely representing the droplet size is introduced. The new number represents the ratio of gravitational force over the surface tension force and it is called the Merve number. The new Merve number takes the form: $\rho_w g a^2/4\gamma_w$, where ρ_w is the density of water, g is the acceleration due to gravity, a is the droplet characteristic diameter, and γ_w is the surface tension of water.

Fig. 5.84A shows the variation of the Nusselt and the Bond numbers with the Merve number. Each point represented by the Merve number in Fig. 5.84A corresponds to different water-droplet volume. The bond number increases with increasing Merve number and the values of the Bond number remain less than unity for all values of MN. This indicates that the Marangoni current dominates over the buoyancy current in the droplet interior. The Nusselt and Bond number increases with increasing Merve number. This behavior is attributed to increased contact line between the droplet and micropost arrays for large size droplets. Since the Merve number is proportional to the square of droplet diameter ($\rho_w g a^2/4\gamma_w$, where a is the droplet characteristic diameter), increasing droplet size increases the Merve number. This situation is also seen from Fig. 5.84B in which the dimensionless contact line (l/k^{-1}), where l is the contact line length, k^{-1} ($\kappa^{-1} = \sqrt{\gamma/\rho g}$, where γ is the surface tension, ρ is the density of the fluid, and g is the acceleration due to gravity) is the capillary length with the Merve number is shown. Consequently, increasing droplet volume gives rise to droplet bulging on the surface of the micropost arrays while increasing the contact line between the droplet and the micropost arrays. In addition, the Nusselt number variation with the Merve number is also shown in Fig. 5.81A for the plane PDMS surface, that is, surface without micropost arrays. The Nusselt number has lower values for the plane surface compared to that of the surface composing of the micropost arrays. This behavior is associated with the velocity oscillation in the close region of the droplet bottom because of temperature oscillation in this region. Consequently, the convection

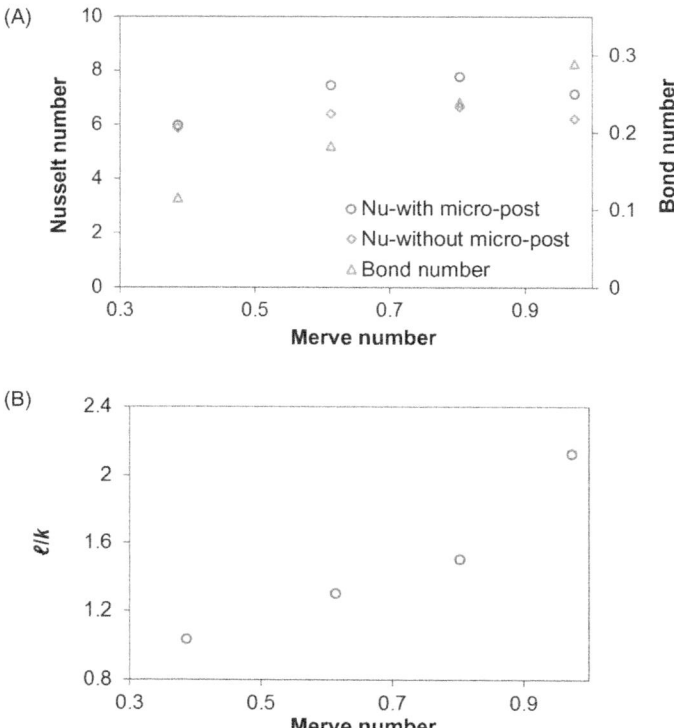

FIGURE 5.84 (A) Nusselt and Bond numbers with Merve number for cases with and without micropost arrays created at the surface, and (B) droplet contact (wetting) length at the surface of micropost arrays [7].

current generated in the droplet bottom transports heat form the interface of the droplet and the micropost arrays toward the droplet interior while enhancing the heat transfer rates at the interface.

5.5.4 Heating of a Ferro-Liquid Droplet Heat on Water Surface

The oil-based ferro-liquid droplets on surfaces remained interest because of various applications in sensor and medical technologies [81−84]. Automatic magnetic manipulation of droplets on the solid-surface using a superhydropho-bic electromagnet needle was studied by Yang et al. [81]. They indicated that the electromagnetic needle-based droplet manipulation was facile, cost-effective, and programmable, and would significantly enhance the functional-ity of magnetic droplet platforms, and thus broaden their applicability to numerous fields including the life science. Investigation of oil droplet separa-tion and manipulation via using asymmetric nanoorifice-induced DC dielectro-phoretic method was carried out by Zhao and Li [82]. They discussed the positive and negative dielectrophoretic behaviors of the droplets varying with the electrical conductivity of the suspending solution and demonstrated the separation of different types of droplets of similar sizes with different contents. The fluid dynamics and mass transfer in a single droplet in liquid/liquid systems was investigated by Wegener et al. [83]. They presented a chart to initiate a selection process starting from simpler cases to cases with increasing complexity of the extraction system, especially addressing the behavior of the systems, which was dominated by Marangoni instabilities. A study on the formation of the micromagnetofluidics of ferro-liquid droplet in a T-junction was carried out by Zhang et al. [84]. They observed three typical flow regimes including the slug, slug-dripping transition, and dripping flow. The external magnetic field promoted the transition from the slug-dripping transition flow to the slug flow, prolonged the generation cycle of ferro-liquid droplets and hindered the ferro-liquid droplet formation. As the magnetic flux density increased both the increasing rate of the length for the thread tip and the expanding rate of the dispersed neck decreased, and the peak value of the minimum width of the dispersed neck increased. The ferro-liquid droplet size increased with increasing the magnetic flux density while decreased with increasing the two-phase flow rate ratio and the capillary number. The dynamics of magnetic modulation of ferro-liquid droplets for digital microfluidic applications was studied by Sen et al. [85]. They observed that, for dipoles placed upstream of the junction, the droplet formation was suppressed at some higher dipole strengths, and this value was found to increase with increasing capillary number. Droplet time period was also found to increase with increasing dipole strength, along with the droplet size, that is, an increase in droplet volume. The magnetic fluid droplet deformation in electrostatic field was examined by Cimbala et al. [86]. They indicated that the suspended particles started to aggregate during the tests. In addition, it was observed that the further deformation development depended on aggregates shape and location.

The ferro-liquid base droplet heating is presented in light of the previous work (Abdullah Ferro Liquid). The heat transfer from stagnant water to oil-based ferro-liquid droplet is simulated in line with the experiments and the condition for the droplet floating is assessed via the vertical force balance. The interfacial tension between the ferro-liquid droplet and water is determined and slip height is formulated. The transformer oil-based ferro-liquid, which is developed previously [87], is incorporated in the heating simulations and the force balance analysis is carried out for various volumes of ferro-liquid droplet floating in water. The relevant analysis and findings are presented in line with the previous study [8].

5.5.4.1 Heating and Flow Analysis

Flow and temperature fields in transformer oil-based sessile ferro-liquid droplet partially immersed into water are simulated in line with the experimental conditions. Although the droplet is partially immersed into water, the dynamic force interactions among the air, ferro-liquid, and water is considered to be weak. This is because of the fact that the ferro-liquid droplet remains stationary in the solution domain while it is partially exposed to air and partially immersed into water in line with the experiment. Therefore, the governing equations of flow and heat transfer are adopted to formulate velocity and temperature fields in water and in the sessile ferro-liquid droplet partially immersed in water. Since the air ambient does not have a dynamic interaction with a sessile droplet, except convection heat transfer, three-phase simulation incorporating the three point model is avoided via introducing the convection heat transfer boundary condition along the droplet free surface, which is exposed to air ambient. Consequently, the convection heat transfer boundary condition is adopted along the free surface of the ferro-liquid droplet and water, which are exposed to the ambient air. The mathematical formulation of the governing equations of heat transfer and fluid flow can be found in the previous study [8]; however, the momentum equation is given herein.

The density variation results in a convection current, which is mainly caused by the thermal expansion of the fluid. This can be expressed from the Boussinesq approximation:

$$\rho = \rho_o[1 - \beta(T - T_o)] \tag{5.57}$$

where β is the thermal expansion of the droplet fluid. The momentum equation can be written as:

$$\rho\left(\frac{\partial V}{\partial t} + V \cdot \nabla V\right) = -\rho_o\beta(T - T_O)\,\vec{g} - \nabla(p - p_o)$$
$$+ \nabla\left[\mu(\nabla V + (\nabla V)^T) - \frac{2}{3}\mu(\nabla \cdot V)\right] \tag{5.58}$$

where p is the pressure, μ is the dynamic viscosity of the liquid, g is the gravitational acceleration and p_o is the hydrostatic pressure corresponding to density ρ_o and temperature T_o.

In order to solve the conservation equations, initial and boundary conditions are introduced in line with the experimental conditions.

5.5.4.1.1 Initial Condition

Initially water and the droplet at stagnant conditions are considered; therefore, the flow velocity is set to zero, the pressure is fixed to the Laplace pressure, and uniform temperature is incorporated inside water and the droplet, which is the same as the ambient temperature (300K). Initially, a thermal equilibrium between the droplet and water is assumed. The water container surface is considered to be at 308K in line with the experimental condition, which resembles a constant temperature heat source.

5.5.4.1.2   Boundary Condition

Fig. 5.85 shows the boundary conditions and schematic view of partially immersed ferro-liquid droplet in water incorporated in the simulations in line with the experimental conditions. The droplet geometric features are extracted from the droplet images obtained from the high-precision camera (Dantec Dynamics (SpeedSense 9040)) and reconstructed for the numerical simulations. The constant pressure boundary is assumed in the droplet and water surface ambient (droplet and water outside surfaces), which is as the atmospheric pressure. In addition, stagnant air is considered at the droplet and water outer surfaces, which yields zero velocity of the air. The combination of natural convection and radiation boundary condition is incorporated at the interface between the droplet free surface and its surrounding air ambient, which has a uniform temperature of 300K. The natural convection and radiation boundary condition is considered for the free surface of water, which is not occupied by the droplet, that is, the natural convection and radiation boundary condition is adapted at water free surface exposing to the surrounding ambient air. Since the water container walls are maintained at constant temperature (308K), a constant temperature boundary condition is adopted at the interface between water and container walls. The conjugate boundary condition is adapted along the interface between droplet bottom and water; in which case, continuity of heat flux and temperature is incorporated. In addition, slip boundary is adopted at the interface of the droplet and water. In this case, the slip height is determined from equating shear stresses, due to the droplet bottom and water, at the interface.

The time taken for simulating the experimental conditions is in the order of 500 seconds. Since the container wall temperatures are low (308K), which is not much different than the ambient temperature (300K), and experiment is carried out in the laboratory environment with relative humidity 85%, the

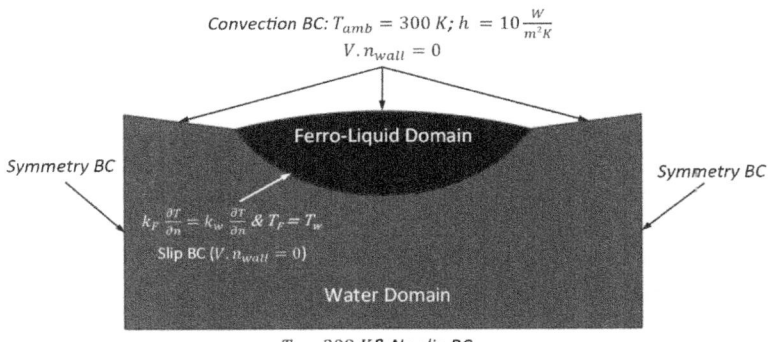

FIGURE 5.85   A schematic view of transformer oil-based ferro-liquid droplet and water and the boundary conditions incorporated in the simulations [8].

evaporation from the droplet and water surfaces is assumed to be negligible in the simulations. This assumption is verified during the experiments; in which case, the droplet and surrounding water images were taken after 40 and 500 seconds periods and these images were compared with the image corresponding to the droplet at the onset of the experiment. It was noticed that droplet and surrounding water in both images had identical geometric sizes such as identical diameters for the droplet and identical heights for water. Hence, the neglecting evaporation from the droplet and surrounding water surface is justifiable during the heating period (500 seconds).

COMSOL Multiphysics software [27] was used to simulate the flow field and temperature distribution inside the droplet. Simulations were carried out incorporating four sizes of the droplet including 15, 30, 45, and 60 μL. The contact angle of the droplet was measured within the range of $\theta = 10-14$ degrees and it remained almost the same for all the droplet volumes incorporated. The variation of the droplet geometries was recorded using the high-speed camera (Dantec Dynamics (SpeedSense 9040)). The data for the droplet geometries corresponding to the different volumes were analyzed to monitor the geometric changes during the heating period. It was observed that the droplet geometry and water level remained the same during the heating period, which was true for all droplet volumes considered. Consequently, the droplet geometries obtained from the high-speed camera were reconstructed to resemble the actual droplet geometry. These geometries were incorporated in the simulations in line with the experimental conditions. The laminar two-phase flow model (water and transformer oil-based ferro-liquid) coupled with heat transfer in fluid was used during the simulations. The 3D simulation of the flow field and temperature distribution was very expensive because of the excessive mesh requirements for the accurate solutions. However, one case was simulated incorporating the 3D droplet for comparison. Fig. 5.86 shows velocity and temperature contours inside water and droplet obtained from 3D and 2D simulations. It is evident that the velocity and temperature predictions of 3D simulation (Fig. 5.86A) are similar to those obtained from 2D simulations (Fig. 5.86B). Consequently, 2D simulation of the flow field was incorporated in the analysis. In the numerical approach, finer grid points were placed in the region where the fluxes are high. The grid independence tests were carried out to secure the grid-independent solutions.

After the accomplishing the grid independence tests, the mesh size comprising of 20,738 cells was selected and adapted in the simulations. The governing equations of flow were discretized using the backward Euler finite difference method. The selection of time step was critical to ensure the accuracy of the scheme; in which case, it was set in the order of 10^{-4} seconds. The residuals of flow parameters were set as $\left|\psi^k - \psi^{k-1}\right| \leq 10^{-8}$. Tables 5.8 and 5.11 give the data used in the simulations for water and transformer oil-based ferro-liquid, respectively [27,76].

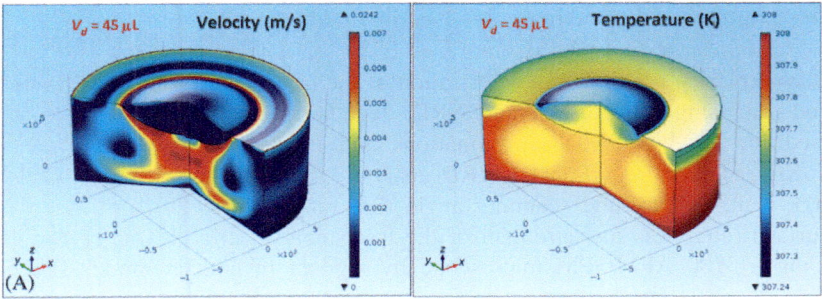

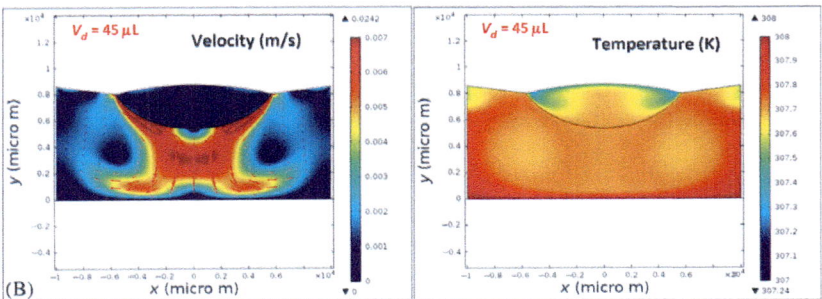

FIGURE 5.86 Comparison of velocity and temperature contours for 45 μL droplet after 500 s heating: (A) 3D prediction, and (B) 2D prediction [8].

5.5.4.2 Validation Study for Velocity Field

The transformer oil-based ferro-liquid was used to form a droplet on the free surface of water in a container. The transformer oil was chemically stabile at temperatures up to 410K and the ferro-liquid was prepared in line with the previous study [87]. In this case, iron oxide nanoparticles stabilized with oleic acid (0.5% volume) was mixed with the transformer oil and later the mixture was placed on ultrasonic shaker for 20 minutes. The droplets of 15, 30, 45, and 60 μL were formed separately on water surface in a container. Due to the density differences between the transformer oil (875 kg/m^3 at 300K) and water (996.5 kg/m^3 at 300K), the droplet formed was partially immersed into water, that is, it floated on water surface. The droplet contact angle was measured using the optical photographic technique, which was varied within 12−14 degrees depending on the droplet volume. A test rig was developed to set a constant temperature (308K) heating at the water container walls. The high-speed camera (Dantec Dynamics (SpeedSense 9040)) was used to monitor and record the droplet geometric features and dynamics during the heating period over 500 seconds. During the high-speed recording, the sample rate was kept at 100 frames per second, which provided image resolution of 960 × 720 pixels.

TABLE 5.11 Temperature-Dependent Properties of Transformer Oil [27,76]

Property	Equation	Units
Density	$\rho = -6.4053 \times 10^{-5} T^2 - 0.58176T + 1055.0461$	kg/m³
Kinematic viscosity	Temperature range (K)	Pa · s
	243–273	$\mu = 7.5704 \times 10^{-7} T^4 - 8.40477 \times 10^{-4} T^3 + 0.3499 T^2 - 64.7409T + 4492.2023$	
	273–373	$\mu = -1.94631023 \times 10^{-11} T^5 + 3.32419673 \times 10^{-8} T^4 - 2.2727 \times 10^{-5} T^3 + 0.0077 T^2 - 1.3323T + 91.4525$	
Specific heat capacity	Temperature range (K)	kJ/kgK
	223–293	$C_p = -2.36742424 \times 10^{-5} T^4 + 0.02567 T^3 - 10.3058 T^2 + 1816.7621T - 117056.38$	
	293–373	$C_p = 3.125 \times 10^{-4} T^3 - 0.3354 T^2 + 123.0442T - 13408.1491$	
Thermal conductivity	$k = -8.0497 \times 10^{-5} T + 0.1343$	W/mK

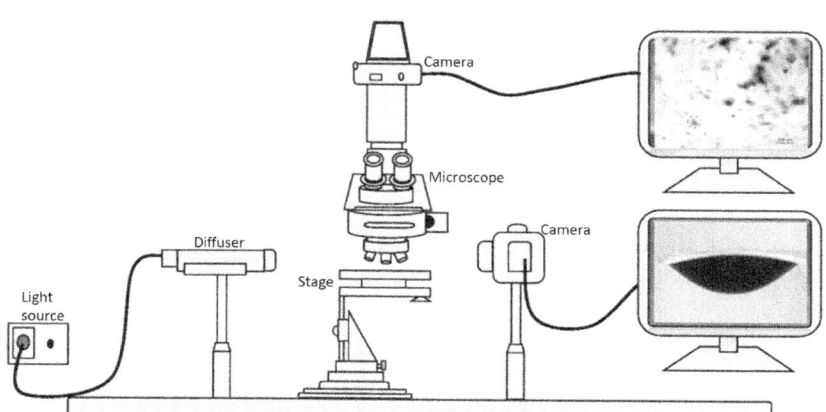

FIGURE 5.87 Schematic view of microscopic imaging system for droplet [8].

To validate the flow field inside the droplet, motion of the particles inside the droplet are recorded using microscopic system which consisting of a microscope and the recording camera system. Fig. 5.87 shows the schematic view of the microscopic system. The optical correction was introduced to avoid the curvature effect of the droplet during the optical data analysis, in line with the previous study [36]. The carbon nanotubes were mixed (1% volume) with the transformer oil-based ferro-liquid. The mixture was placed on a shaker for 20 minutes to ensure almost uniform dispersion of carbon nanotubes inside the fluid prior to formation of the droplet on water surface. The motion of the carbon nanotubes was recorded on the microscope. However, the laser beam was illuminated on the droplet to locate the cross-sectional plane where the motion of carbon-nanotube clusters was recorded.

In addition, the simulations were carried out to predict the flow field developed inside the droplet while resembling the transformer oil-based ferro-liquid and carbon-nanotube mixture. In this case, the discrete-phase model was used to solve the governing equations of flow while the carbon-nanotube clusters were present in the solution domain. Since the carbon-nanotube concentration was low in the droplet fluid (1% by volume), a slurry-single fluid model was assumed in the solution of governing equations. Consequently, the effective thermal properties were used in the solution of the equations. The details of the formulation and numerical solution are found in the previous study [32]. In the simulations, the initial and the boundary conditions for momentum and energy equations were set while mimicking the experimental conditions. The trajectory of the carbon-nanotube clusters provided tracing of the flow velocities inside the droplet. Experiments were repeated several times and experimental data distribution resulting in the confidence level of 95% was secured. In this case, the mean distribution of the data was within $\pm$ 1.75 of the standard deviation of the distribution of a single measurement. The experimental uncertainty involved with the experimental measurements was in the order of 5%. The flow velocities predicted and measured from the microscopic system are given in Table 5.12.

The findings show that the velocity predictions agreed well with the data obtained from the measurements; nevertheless, the discrepancies between the findings are small and can be attributed to the computational errors, such as round-off errors and experimental uncertainties.

5.5.4.3 Findings

The findings of ferro-liquid droplet are presented in line with the previous study [8]. The slip length at the interface between the droplet fluid and water is important for the dynamic behavior of the droplet on the water surface [58]. The velocity of the droplet fluid at the water surface is the same order of the slip velocity and it is related to the slip length. Since water surface possesses the smooth surface characteristics, the slip length (b) can be approximated by $b \sim (\mu_w/\mu_p)h_t$, where μ_w is the dynamic viscosity of water, μ_p is the dynamic viscosity of the droplet liquid, and h_t is the water

TABLE 5.12 Flow Velocity Predicted From Simulations and Obtained From Experiments at Various Locations Inside 45 μL Droplet [8]

Particle #	X (mm)	Y (mm)	Simulation V (m/s)	Experiment V (m/s)	Simulation (velocity contours (m/s))
1	1.399	8.517	0.000776	0.000780	
2	2.575	8.489	0.001386	0.001385	
3	3.526	8.294	0.001498	0.00150	
4	4.505	8.042	0.001075	0.001070	
5	3.806	6.726	0.000929	0.000930	
6	2.686	6.419	0.000668	0.000665	
7	1.791	6.447	0.000566	0.000550	Experiment (CNT particle tracing)
8	0.784	6.586	0.000637	0.000630	
9	0.336	7.118	0.000796	0.000800	
10	0.028	8.154	0.000546	0.000550	
11	−0.196	8.517	0.000247	0.000240	

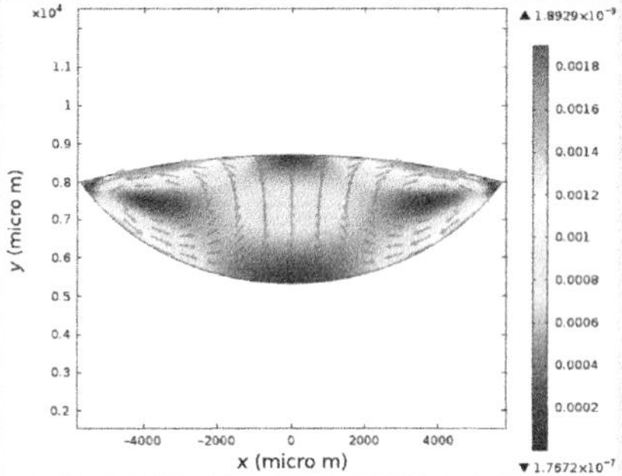

Location #1

t = 0 s

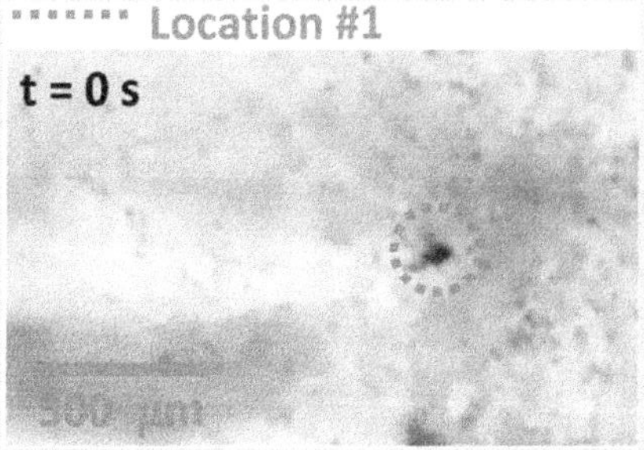

12	−1.595	8.377	0.000785	0.000785 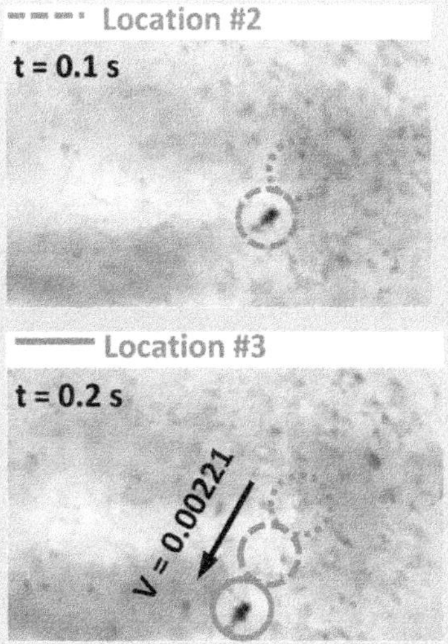
13	−3.106	8.321	0.001370	0.001375
14	−4.729	7.986	0.000875	0.000870
15	−4.002	6.754	0.001032	0.001035
16	−2.994	6.335	0.000791	0.000790
17	−2.295	5.999	0.000641	0.000645
18	−0.756	5.747	0.000289	0.000290
19	−0.224	5.999	0.000376	0.000375
20	−3.750	7.454	0.000052	0.000050

Location #2
t = 0.1 s

Location #3
t = 0.2 s
V = 0.00221

film height (thickness) [58]. The dynamic viscosity of water is 0.7644×10^{-3} Pa·s at 305K and dynamic viscosity of the droplet fluid at the same temperature is in the order of 50×10^{-3} Pa·s [87]. The water film thickness is in the order of 5 mm; hence, the slip length becomes 0.07664 mm. The slip velocity (u_s) at the water interface below the droplet bottom is related to the slip length in the form of $u_s = (b/\mu_w)\tau_{wi}$, where τ_{wi} is the shear stress at the interface [58]. The shear stress at the droplet interface is in the order of $\sim \mu_p((V - u_f)/l_m)$, where V is the velocity of the liquid in the droplet bottom, u_f is the velocity of water at the interface, μ_p is the dynamic viscosity of the droplet liquid, and l_m is the distance from the interface to the droplet center of mass [58]. The geometric relations between the base radius of the droplet (r_o) and height of center of mass (l_m) can be written as $r_o/l_m = (4/3)sin\theta(2 + cos\theta)/(1 + cos\theta)^2$, where θ is the droplet contact angle [58]. The shear stress at the water interface is in the order of $\sim \mu_w((u_f - u_s)/h_t)$. Since the shear stresses at water and droplet fluid interface are the same, equating the shear stresses at the interface results in:

$$\frac{V}{u_f} \sim 1 + \frac{\mu_p}{\mu_w}\frac{l_m}{h_t}(1 - \frac{b}{h_t}) \tag{5.59}$$

Since b/h_t is in the order of 0.015, Eq. (5.59) predicts the same velocity ratio as that in the previous study [58]. Nevertheless, the ratio of $(\mu_p/\mu_w)(l_m/h_t) < 1$ and it gives rise to $u_f > V$, which agrees with the predictions. Moreover, from the high-speed camera data, it can be seen that the droplet remains stationary during the heating period. The force balance enables formulating the pinning force for the droplet on the surface. In this case, the surface tension force of the droplet on the water surface can be formulated through the analogy of the three-phase contact line around the droplet bottom. The surface tension force becomes:

$$F_\gamma = 2\pi r_o \gamma_{w-p} \tag{5.60}$$

where γ_{w-p} is the interfacial surface energy between the droplet liquid and water and r_o is the radius of the droplet liquid cap at the meniscus (rim) of the droplet bottom. Consider that the initial shape of the droplet on water surface resembles the spherical cap with height h_o and diameter R_o, then the radius of the three-phase contact line can be written as $r_o = \sqrt{(2R_o - h)h}$, where h is the spherical cap height, R_o is the initial radius of the spherical cap and it is related to initial height of the cap (h_o) and spherical droplet radius, $r_d = (4/9)h_o((R_o^3/h_o^3) + 1)$, here r_d is the droplet radius with spherical geometric feature prior to dispensing on to the water surface. The details of formulation of the spherical cap are given the appendix. The interfacial surface tension between the droplet and water can be obtained from Young's equation [88]; in which case, the interfacial surface tension becomes $\gamma_{w-p} = \gamma_{w(a)} - \gamma_{p(a)}cos\alpha$. Inserting the values of the surface tension of the

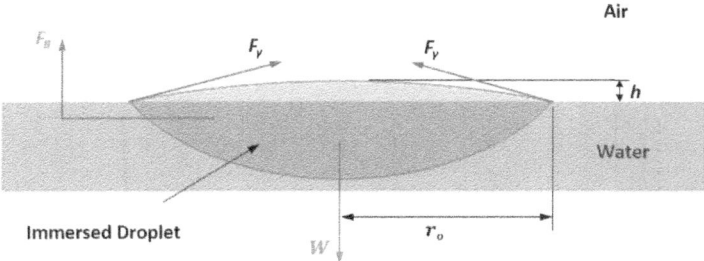

FIGURE 5.88 A schematic view of transformer oil-based ferro-liquid droplet formed on the water and the force diagram [8].

droplet liquid in air ($\gamma_{p(a)}$), which is 30 mN/m [89], and surface tension of water in air ($\gamma_{w(a)}$), which is 72 mN/m, and the water-droplet contact angle ($\alpha = 12$ degrees at the droplet rim), the interfacial energy (γ_{w-p}) yields 42.66 mN/m. The force balance along the vertical direction remains critical to access the immersing (sinking) state of the droplet on the water surface. The vertical force balance in relation to partial immersing of the droplet at equilibrium in the water yields:

$$F_\gamma sin(\theta_c + \alpha) + F_B - W = 0 \tag{5.61}$$

where θ_c is the filling angle, which defines the position of the contact ring reference to the vertical axis (Fig. 5.88); α is the contact angle of the water-droplet cap at the droplet bottom rim; F_B is the buoyancy force; and W is the weight of the droplet. It should be noted that the vertical acceleration of the droplet ceases once the droplet comes into equilibrium on the water surface. The details of the formulation of surface tension force are given herein.

After assuming the semispherical geometry of the transformer oil-based ferro-liquid droplet on water surface, the vertical force balance gives rise to the droplet immersion in water. In line with Fig. 5.88, in which the droplet geometry—resembling a semispherical cap—and force diagram is shown schematically, the geometric features of the droplet can be formulated. The droplet radius on the meniscus of the three-phase contact line is:

$$r = \sqrt{h(2r_o - h)} \tag{5.62}$$

where h is the droplet height and r_o is the radius of the spherical cap on the liquid n-octadecane surface.

However, the spherical droplet volume after spreading on water surface becomes a semispherical cap. The initial volume of the droplet occupying a semispherical cap yields:

$$\forall = \frac{\pi}{3}h^2(3r_o - h_o) \tag{5.63}$$

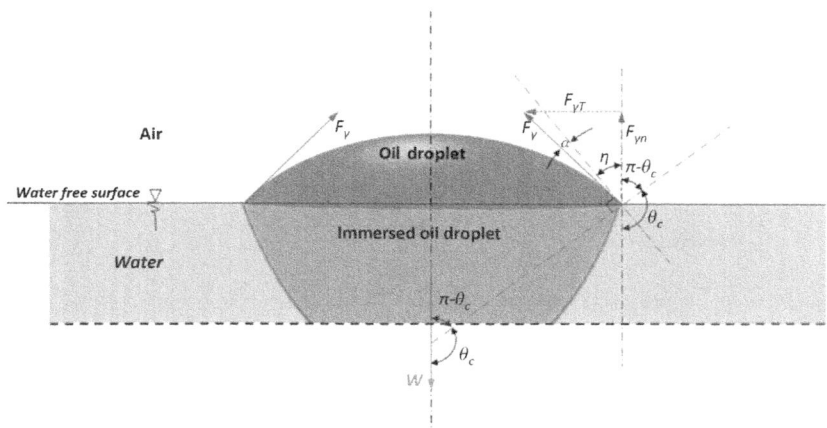

FIGURE 5.89 A schematic view of transformer oil-based ferro-liquid droplet on water when located horizontally and surface tension force diagram [8].

where $r_o = \sqrt{h_o(2R_o - h_o)}$ and h_o is the maximum droplet height, which corresponds to the spherical cap onset of its formation on the water surface.

However, droplet immerses into water depending on the force balance. In this case, the droplet volume remaining above water becomes $\forall = (\pi/3)h^2(3r - h)$.

The volume of displaced water is:

$$\forall_{disp} = \forall_{in} - \forall_{new} \text{ or } \forall_{disp} = \frac{\pi}{3}\left[h_o^2(3r_o - h_o) - h^2(3r_o - h)\right] \qquad (5.64)$$

The length of the three-phase contact line is: $\ell = 2\pi r$.

Consider Fig. 5.89. The vertical force balance for immersing water droplet occupying a semispherical cap yields:

$$\sum F_y = F_B + F_\gamma - W = 0 \qquad (5.65)$$

The buoyancy force is:

$$F_B = \rho_{pcm}g\forall_{disp} = \frac{\pi}{3}\rho_{pcm}g\left[h_o^2(3r_o - h_o) - h^2(3r_o - h)\right] \qquad (5.66)$$

The surface tension force is:

$$F_\gamma = 2\pi\gamma_{w-pcm}\ell = 2\pi\gamma_{w-pcm}\sqrt{h(2r_o - h)} \qquad (5.67)$$

The vertical component of the surface tension force yields:

$$F_{v-\gamma} = F_\gamma sin(\theta_c + \alpha) \qquad (5.68)$$

where θ_c is the filling angle, which defines the position of the contact ring reference to the vertical axis (Fig. 5.89), and α is the contact angle of the water-droplet cap at the ridge rim.

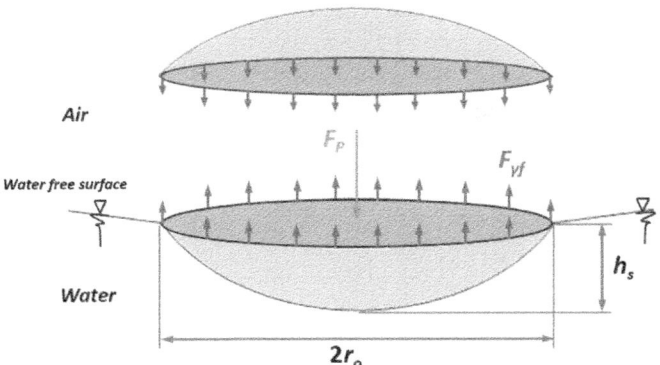

FIGURE 5.90 A schematic view of droplet cross-section and forces acting on the droplet interior [8].

The droplet weight is:

$$W = \rho_w g \nabla_{in} = \frac{\pi}{3}\rho_w g h_o^2(3r_o - h_o) \tag{5.69}$$

Inserting the forces into the vertical force balance equation yields:

$$\frac{\pi}{3}\rho_{pcm}g\left[h_o^2(3r_o - h_o) - h^2(3r_o - h)\right]$$
$$+ 2\pi\gamma_{w-pcm}\sqrt{h(2r_o - h)}\sin(\theta_c + \alpha) - \frac{\pi}{3}\rho_w g h_o^2(3r_o - h_o) = 0 \tag{5.70}$$

Since the term $(\pi/3)h_o^2(3r_o - h_o)$ is constant for a fixed droplet volume, that is, $C = (\pi/3)h_o^2(3r_o - h_o)$, the force balance equation reduces to:

$$\left\{\rho_w g(C - 3\pi h_o^2(r_o - h_o^2)] + 2\pi\sqrt{(2r_o - h_o)h_o}\gamma_{w-p}\sin(\theta_c + \alpha) - \rho_p g C\right\} = 0 \tag{5.71}$$

After considering the water-droplet volume is spherical prior to spreading on the water surface and equating the volume of the spherical droplet with that of the semispherical cap, we have:

$$\frac{4\pi}{3}r_d^3 = \frac{\pi}{3}h_o^2(3r_o - h_o) \tag{5.72}$$

where r_d represents the spherical droplet radius. Rearrangement results in:

$$r_o = \frac{h_o}{3}\left(\frac{4r_d^3}{h_o^3} + 1\right) \tag{5.73}$$

Here, Eq. (5.73) can be used to replace r_o in Eq. (5.71); in which case, the force balance equation can be formulated in terms of the droplet radius when the droplet is spherical prior to spreading and forming a spherical cap.

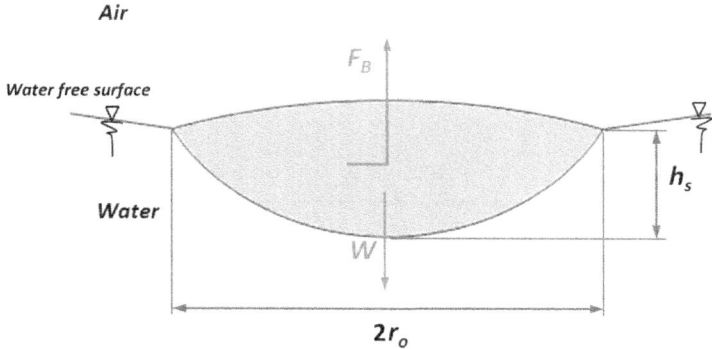

FIGURE 5.91   A schematic view of vertical forces action on the droplet immersed into water. Here, F_B and W are the buoyancy and weight forces, respectively [8].

Consider Fig. 5.90. The surface tension force is related to the pressure force inside the noncircular droplet, which is immersed into water, that is:

The vertical force balance with in the droplet yields:

$$F_P = F_{\gamma_f} \tag{5.74}$$

where F_p is the pressure force and F_{γ_f} is the surface tension force or $\Delta PA = S \cdot \gamma_f$, where ΔP is the pressure difference between the droplet and its surrounding, A is the horizontal cross-sectional area of the droplet, and S is the circumference of the droplet at three-phase contact line. Using the analogy of the vertical force balance according to Fig. 5.91:

$$\Delta P = \frac{6\gamma_f h_s^2}{4SR_o^3 + h_s^3} \tag{5.75}$$

It should be noted that the immersion takes place when the droplet weight (W) overcomes the sum of buoyancy and the vertical component of the surface tension force. After assuming that the geometric feature of the droplet cap can be simplified as the sphere cap, then, the force balance yields:

$$\left\{ \rho_w g \left(C - 3\pi h_o^2 (r_o - h_o^2) \right) + 2\pi \sqrt{(2r_o - h_o)h_o} \gamma_{w-p} \sin(\theta_c + \alpha) - \rho_p g C \right\} = 0 \tag{5.76}$$

where $C = 4\pi h_o [h_o^2 ((R_o^3/h_o^3) + 1) - 1]$. Here, ρ_w is the density of water, ρ_p is the density of the droplet fluid, h_o is the spherical cap height, r_o is the diameter of the droplet spherical cap on water surface and g is the gravitational acceleration. It should be noted that for a fixed volume of droplet, r_o, and h_o remain constant for known volume of droplet. Formulation of the droplet force balance is given in the appendix. For 30 μL droplet volume, the droplet height inside water after immersion is in the order of 2 mm as obtained from

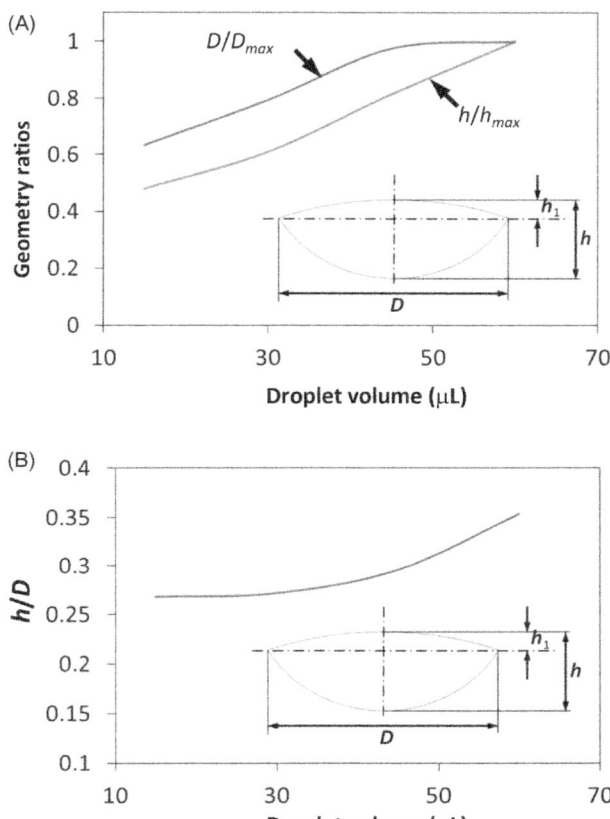

FIGURE 5.92 Geometric ratios of droplet in water for various droplet volumes: (A) D/D_{max} and h/h_{max} (D_{max} and h_{max} corresponds to a droplet diameter of 60 μL), and (B) h/D ratio [8].

the high-speed camera, while it is 1940 μm as predicted from the vertical force balance (Eq. 5.76). Hence, the prediction of the droplet height agrees well with the experimental data.

Fig. 5.92 shows the droplet geometric ratios with the droplet volume. The normalized droplet diameter (D/D_{max}, where D and D_{max} are the diameter and the maximum diameter of the droplet along the three-phase contact line on the water surface, respectively, and D_{max} corresponds to the droplet diameter when the droplet volume is 60 μL) increases sharply up to the droplet volume 45 μL and the normalized droplet diameter does not increase noticeably beyond this volume (Fig. 5.92A). In this case, the normalized droplet height (h/h_{max}, where h and h_{max} are the droplet height and maximum droplet height corresponding to droplet volume 60 μL, respectively) increases steadily with increasing droplet volume (Fig. 5.92A). In this case, increasing

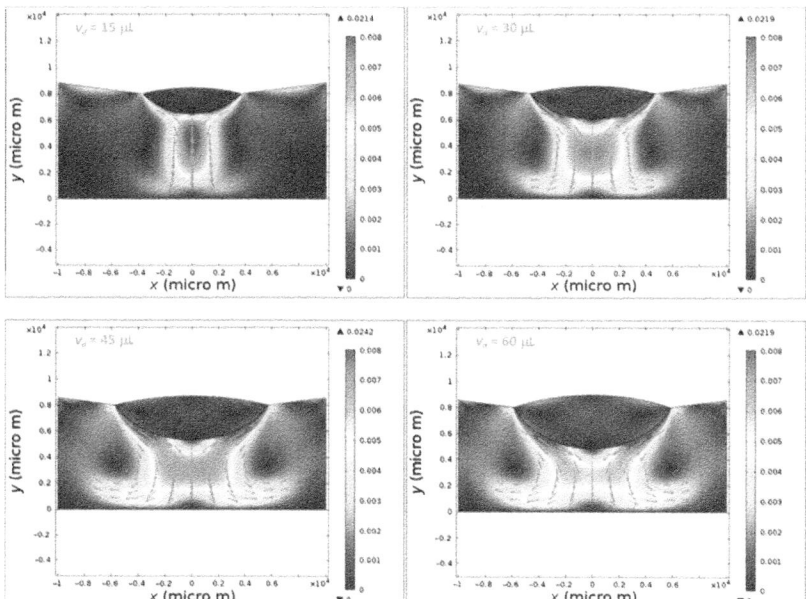

FIGURE 5.93 Velocity contours in droplet and water for various droplet volumes. Velocity is m/s [8].

droplet volume results in increased normalized droplet height than normalized droplet diameter at the three-phase contact line. This situation can also be seen from Fig. 5.92B, in which droplet height over droplet diameter at the three-phase contact line (h/D) is shown. This behavior is associated with the Laplace pressure inside the droplet, which gives rise to more circular droplet as the volume increases. Since the density of the droplet fluid (875 kg/m^3 at 300K) is lower than density of water (996.5 kg/m^3 at 300K), the droplet floats on the water surface and the weight of the fluid displaced due to the droplet immersion is the same as the buoyancy force acting on the droplet. In this case, the hydrostatic force acting on the interfacial curved surface (between droplet liquid and water) of the immersed droplet remains same, which influences the Laplace pressure inside the droplet. The vertical force balance inside the droplet alters the horizontal cross-section of the droplet and the Laplace pressure ($\Delta P = (6h_s^2\gamma_f)/4S_{oil}r_o^3 + h_s^3$, where γ_f is the surface tension of the droplet fluid, S_{oil} is the specific gravity of the droplet fluid, h_s is the droplet height immersed in water, and r_o is the radius of the droplet at three-phase contact line). The pressure formulation is given in Eq. (5.75).

Fig. 5.93 shows the velocity contours in water and inside the droplet while Fig. 5.8 shows velocity contours inside the droplet together with the high-speed camera droplet images. Since velocity magnitude remains

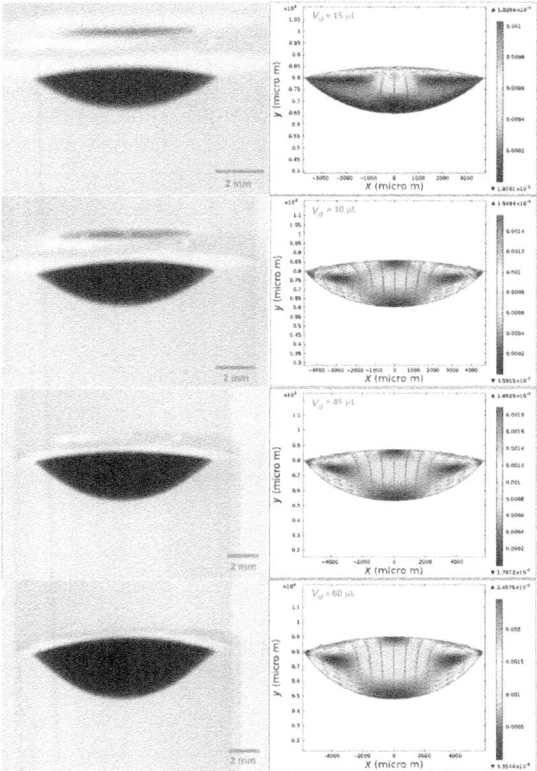

FIGURE 5.94 Velocity contours inside the droplet and droplet optical image in water. Velocity is in m/s [8].

significantly smaller inside the droplet as compared to that of water, Fig. 5.94 provides flow contours and patterns, which are not visible in Fig. 5.93. Flow inside water attains higher velocity than that of inside the droplet. Because of the Marangoni and buoyancy currents in water, two counter-rotating circulations cells develop inside the water. The size of circulation cells increases with increasing the droplet volume. In this case, the droplet bottom, which is immersed inside water, acts like an obstacle for the flow while causing streamline curvature in the vicinity of the droplet bottom.

The maximum velocity inside the water remains almost the same for all the cases considered in the present study. Because of the vertical force balance, the free surface of water does not remain horizontal, but makes a small inclination angle from the horizontal line. This situation can also be seen from the high-speed camera images in Fig. 5.94. In the case of flow field inside the droplet, two counter-rotating circulation cells are formed inside the droplet (Fig. 5.94). This behavior is related to the Marangoni and

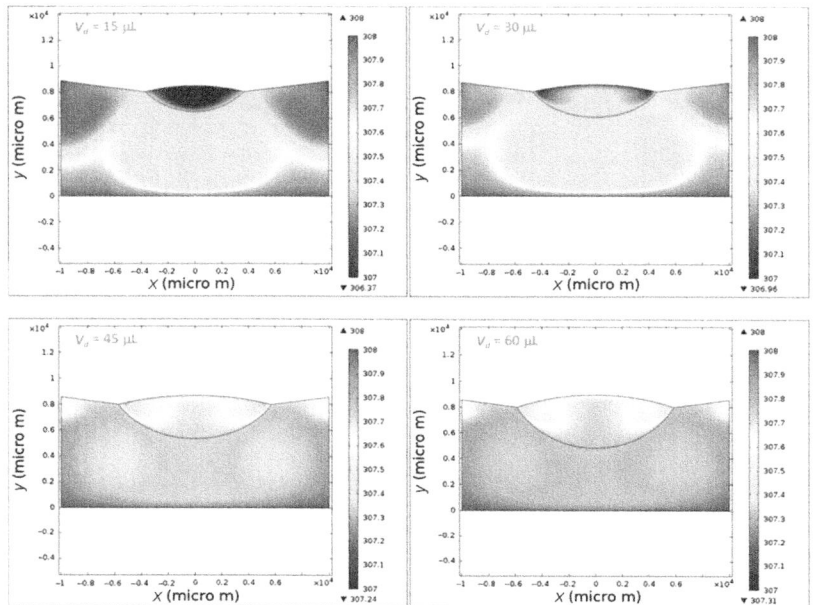

FIGURE 5.95 Temperature contours in droplet and water for various droplet volumes. Temperature is in K.

the buoyancy forces developed due to temperature variation on the droplet surface and the droplet interior. The maximum flow velocity inside the droplet increases with increasing droplet volume; in which case, the maximum flow velocity is in the order of 1.0256×10^{-3} m/s for 15 µL droplet and it is 2.4575×10^{-3} m/s for 60 µL droplet. Increased flow velocity is associated with: (1) enhancement of $d\gamma_f/dT$ because of increased droplet surface arc length, which in turn enhances the Marangoni current; and (2) increased droplet height (h), which enhances the temperature gradient in the vicinity of droplet bottom while contributing to the buoyancy current enhancement inside the droplet. The interfacial shear acting on the curved surface of the droplet, which is immersed in water, generates the frictional force, is related to the formation of two counter-rotating cells inside the droplet, which gives rise to almost zero net frictional force acting on the droplet curved surface. Consequently, the pinning force is mainly governed by the lateral component of the surface tension force, which is $\sim F_\gamma \cos(\theta_c + \alpha)$. Hence, the droplet remains stationary on the water surface during the heating period, which is consistent with the experimental findings.

Fig. 5.95 shows temperature contours inside the water and the droplet for various droplet volumes while Fig. 5.96 shows the temperature contours inside the droplets after 500 seconds of heating. It should be noted

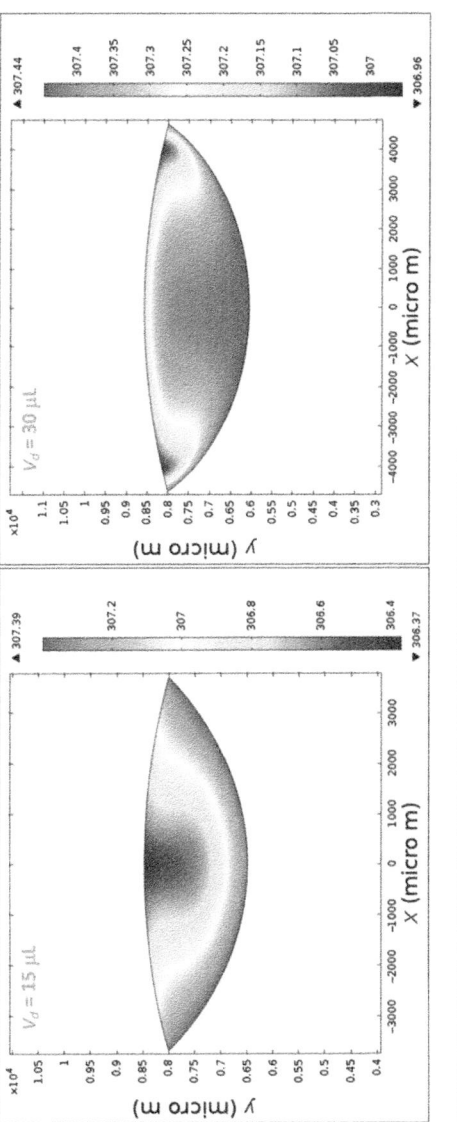
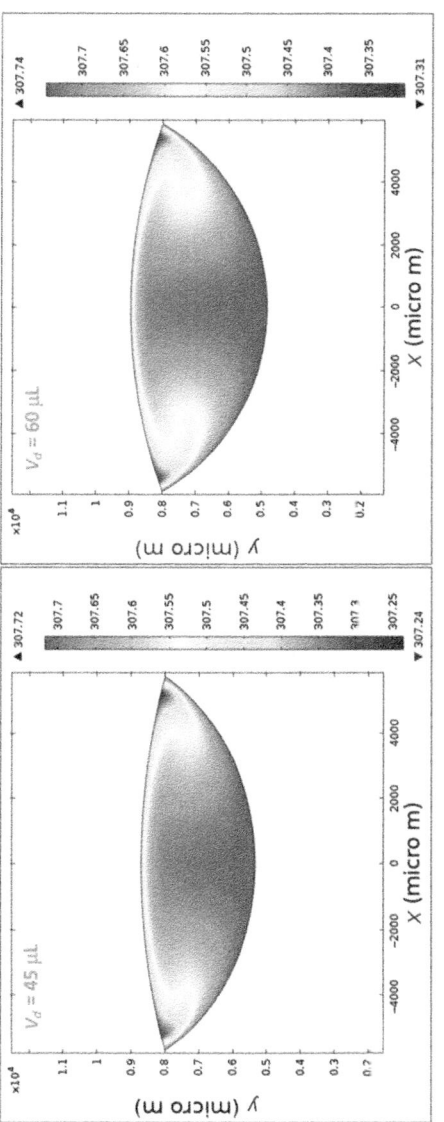

FIGURE 5.96 Temperature contours in droplet for various droplet volumes. Temperature is in K [8].

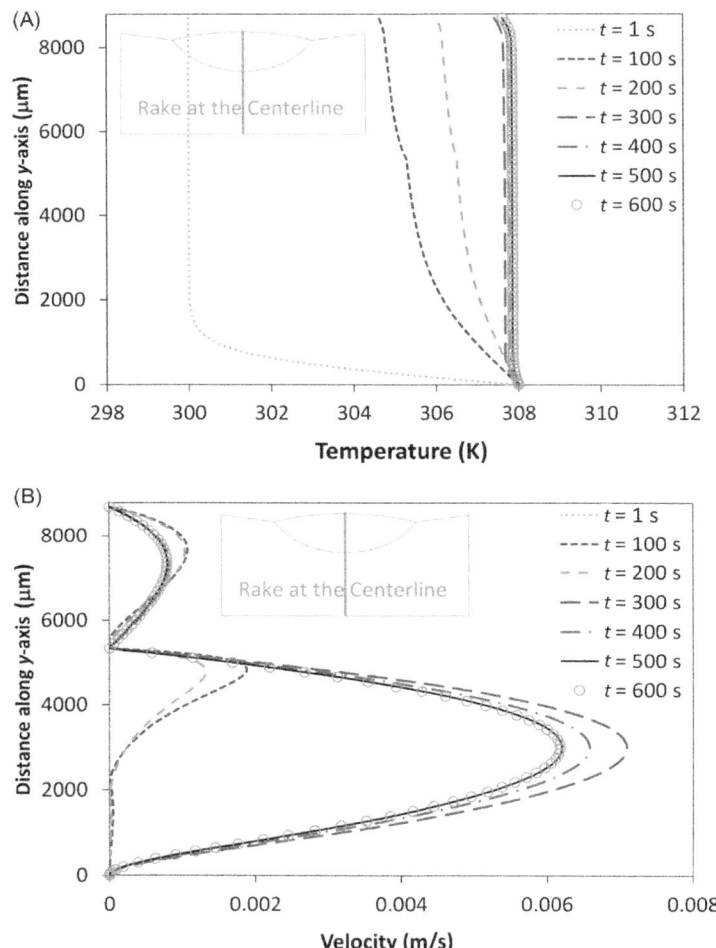

FIGURE 5.97 Temperature and velocity distribution along vertical rake for various heating durations: (A) temperature distribution, (B) velocity distribution [8].

that heating of droplet fluid results in quasisteady temperature increase inside the droplet after 500 seconds of heating. This situation is seen from Fig. 5.97, in which temperature and velocity variation along the central line of the droplet is shown for various heating durations. Temperature profiles become almost self-similar along the vertical line after 500 seconds of heating (Fig. 5.97A), which is also true for the velocity distribution along the centerline in the droplet (Fig. 5.97B). Moreover, temperature remains low in the close region of the circulation cells inside water. Since no flow crosses the circulation cell, the heated fluid cannot be carried by the convection current toward the cell center. However, the

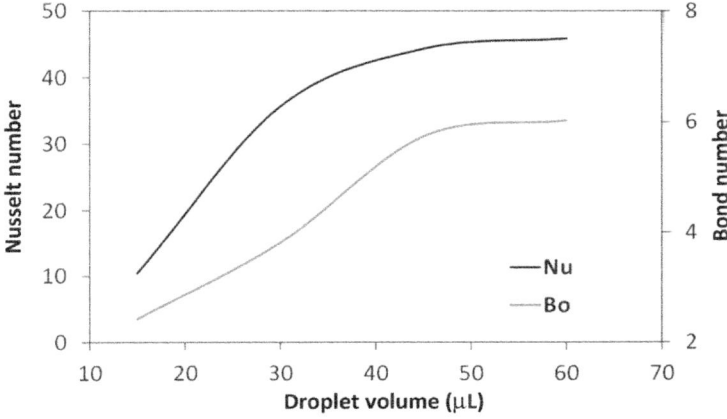

FIGURE 5.98   Nusselt and Bond number variation with droplet volume [8].

heat diffusion from the cell outer boundary to the cell center alters the temperature inside the circulation cells. In addition, the buoyancy current carries the heated fluid from the droplet bottom toward the droplet tip in the region of the circulation cell outer boundaries, which is more pronounced for the large volume droplets. Since the flow velocity in the outer region of the circulation cell is higher for the large volume droplet, the convection current and diffusional heat transfer are responsible for the high-temperature zone along the vertical centerline of the droplet for the large droplet volume. The heated fluid extends from the droplet bottom region toward droplet upper region, which is particularly true for small volume droplet (15 µL); consequently, temperature behavior inside the droplet fluid is significantly influenced by the droplet volume. In addition, because of the natural convection boundary condition at the free surface of the droplet, temperature remains lower in the surface vicinity than that of the droplet interior.

Fig. 5.98 shows the Nusselt and the Bond numbers with the droplet volume. The bond number remains greater than one for all the volumes of droplet incorporated in the present study. Since the Bond number is associated with the ratio of the buoyancy force over the Marangoni force, the attainment of the Bond number greater than unity demonstrates that the buoyancy current mainly dominates the flow field inside the droplet. This situation is more pronounced for large volumes of the droplet; in which case, the Bond number increases with increasing the droplet volume. This increase becomes gradual after the droplet volume ≥ 45 µL. This indicates that further increasing of droplet volume beyond 45 µL does not alter the ratio of buoyancy over the Marangoni currents inside the droplet. The Nusselt number increases with increasing the droplet volume; however, this increase becomes gradual for the droplet volume ≥ 45 µL, similar to the case observed for the

Bond number. In this case, the buoyancy current contributes to the heat transfer from the droplet bottom to the droplet interior in addition to the heat diffusion from the droplet bottom toward the droplet interior.

REFERENCES

[1] A. Al-Sharafi, B.S. Yilbas, H. Ali, Adhesion of a water droplet on inclined hydrophilic surface and internal fluidity, Int. J. Adhesion Adhesives (2018). in press.

[2] A. Al-Sharafi, B.S. Yilbas, H. Ali, N. AlAqeeli, A. Water, Droplet pinning and heat transfer characteristics on an inclined hydrophobic surface, Sci. Rep. 8 (1) (2018) 3061.

[3] B.S. Yilbas, A. Al-Sharafi, H. Ali, N. Al-Aqeeli, Dynamics of a water droplet on a hydrophobic inclined surface: influence of droplet size and surface inclination angle on droplet rolling, RSC Adv. 7 (77) (2017) 48806–48818.

[4] A. Al-Sharafi, B.S. Yilbas, H. Ali, Water droplet mobility on a hydrophobic surface under a thermal radiative heating, Appl. Therm. Eng. 128 (2018) 92–106.

[5] A. Al-Sharafi, B.S. Yilbas, A.Z. Sahin, H. Ali, H. Al-Qahtani, Heat transfer characteristics and internal fluidity of a sessile droplet on hydrophilic and hydrophobic surfaces, Appl. Therm. Eng. 108 (2016) 628–640.

[6] B.S. Yilbas, H. Ali, A. Al-Sharafi, N. Al-Aqeeli, Droplet dynamics on a hydrophobic surface coated with N-octadecane phase change material, Colloids Surf. A: Physicochem. Eng. Aspects 546 (2018) 28–39.

[7] A. Al-Sharafi, B.S. Yilbas, H. Ali, Droplet heat transfer on micro-post arrays: effect of droplet size on droplet thermal characteristics, Int. J. Heat Fluid Flow 68 (2017) 62–78.

[8] A. Al-Sharafi, B. Yilbas, A. Al-Zahrani, Ferro-liquid droplet heat transfer on water surface: effect of droplet volume on droplet fluidity, J. Thermophys. Heat Transfer (2018) 1–16.

[9] D. Ebert, B. Bhushan, Wear-resistant rose petal-effect surfaces with superhydrophobicity and high droplet adhesion using hydrophobic and hydrophilic nanoparticles, J. Colloid Interface Sci. 384 (1) (2012) 182–188.

[10] Y.H. Kwon, H.K. Myong, Numerical analysis for actual droplet movement on hydrophilic/hydrophobic surfaces, in: HACRA 2014 – Proceedings of the 7th Asian Conference on Refrigeration and Air Conditioning, 2014.

[11] X. Yang, X. Liu, Y. Lu, J. Song, S. Huang, S. Zhou, et al., Controllable water adhesion and anisotropic sliding on patterned superhydrophobic surface for droplet manipulation, J. Phys. Chem. C 120 (13) (2016) 7233–7240.

[12] Y. Shi, G. Tang, H. Xia, Investigation of coalescence-induced droplet jumping on superhydrophobic surfaces and liquid condensate adhesion on slit and plain fins, Int. J. Heat Mass Transfer 88 (2015) 445–455.

[13] J. Ou, Q. Shi, Z. Wang, F. Wang, M. Xue, W. Li, et al., Sessile droplet freezing and ice adhesion on aluminum with different surface wettability and surface temperature, Sci. China Phys. Mech. Astron. 58 (7) (2015) 1–8.

[14] B. Bhushan, Y.C. Jung, Wetting, adhesion and friction of superhydrophobic and hydrophilic leaves and fabricated micro/nanopatterned surfaces, J. Phys. Condensed Matter 20 (22) (2008) 225010.

[15] N. Moronuki, H. Tachi, Y. Suzuki, Hydrophilic/hydrophobic surface pattern design for oil repellent function in water, Int. J. Nanomanuf. 11 (1–2) (2015) 46–55.

[16] Z. Zhang, J. Yang, X. Men, X. Xu, X. Zhu, Reversible switching of surface wettability and water adhesion on a polymer nanocomposite coating, J. Adhesion Sci. Technol. 26 (8–9) (2012) 1083–1091.

[17] Q. Chang, J. Alexander, Analysis of single droplet dynamics on striped surface domains using a lattice Boltzmann method, Microfluid. Nanofluid. 2 (4) (2006) 309−326.

[18] H. Chen, E. Dong, J. Li, H.A. Stone, Adhesion of moving droplets in microchannels, Appl. Phys. Lett. 103 (13) (2013) 131605.

[19] X. Liu, Z. Liu, Y. Liang, F. Zhou, In situ surface reaction induced adhesion force change for mobility control, droplet sorting and bio-detection, Soft Matter 8 (40) (2012) 10370−10377.

[20] E. Gogolides, K. Ellinas, A. Tserepi, Hierarchical micro and nano structured, hydrophilic, superhydrophobic and superoleophobic surfaces incorporated in microfluidics, microarrays and lab on chip microsystems, Microelectr. Eng. 132 (2015) 135−155.

[21] D.-G. Lee, C.-K. Oh, S.-H. Yang, S.-J. Han, O.-C. Jeong, Fabrication of hydrophilic poly (dimethylsiloxane) with periodic wrinkling surface and its application, Trans. Korean Inst. Electr. Eng. 63 (5) (2014) 671−675.

[22] A. ElSherbini, A. Jacobi, Retention forces and contact angles for critical liquid drops on non-horizontal surfaces, J. Colloid Interface Sci. 299 (2) (2006) 841−849.

[23] D. Pilat, P. Papadopoulos, D. Schaffel, D. Vollmer, R. Berger, H.-J. Butt, Dynamic measurement of the force required to move a liquid drop on a solid surface, Langmuir 28 (49) (2012) 16812−16820.

[24] A.H. Ayyad, Thermodynamic derivation of the Young−Dupré form equations for the case of two immiscible liquid drops resting on a solid substrate, J. Colloid Interface Sci. 346 (2) (2010) 483−485.

[25] T. Kajiya, F. Schellenberger, P. Papadopoulos, D. Vollmer, H.-J. Butt, 3D Imaging of water-drop condensation on hydrophobic and hydrophilic lubricant-impregnated surfaces, Sci. Rep. 6 (2016) 23687.

[26] S. Dash, S.V. Garimella, Droplet evaporation on heated hydrophobic and superhydrophobic surfaces, Phys. Rev. E 89 (4) (2014) 042402.

[27] http://www.comsol.com/comsol-multiphysics, 2018.

[28] A. Al-Sharafi, H. Ali, B.S. Yilbas, A.Z. Sahin, M. Khaled, N. Al-Aqeeli, et al , Influence of thermalcapillary and buoyant forces on flow characteristics in a droplet on hydrophobic surface, Int. J. Therm. Sci. 102 (2016) 239−253.

[29] Q. He, D. Jiao, Explicit and unconditionally stable time-domain finite-element method with a more than "optimal" speedup, Electromagnetics 34 (3−4) (2014) 199−209.

[30] B. Yilbas, H. Ali, N. Al-Aqeeli, M. Khaled, N. Abu-Dheir, K. Varanasi, Solvent-induced crystallization of a polycarbonate surface and texture copying by polydimethylsiloxane for improved surface hydrophobicity, J. Appl. Polym. Sci. 133 (22) (2016) 43467.

[31] F. Heib, R. Hempelmann, W. Munief, S. Ingebrandt, F. Fug, W. Possart, et al., High-precision drop shape analysis (HPDSA) of quasistatic contact angles on silanized silicon wafers with different surface topographies during inclining-plate measurements: influence of the surface roughness on the contact line dynamics, Appl. Surf. Sci. 342 (2015) 11−25.

[32] A. Al-Sharafi, A.Z. Sahin, B.S. Yilbas, S. Shuja, Marangoni convection flow and heat transfer characteristics of water−CNT nanofluid droplets, Numerical Heat Transfer, Part A: Applicat. 69 (7) (2016) 763−780.

[33] J. Li, L. Shi, Y. Chen, Y. Zhang, Z. Guo, B.-L. Su, et al., Stable superhydrophobic coatings from thiol-ligand nanocrystals and their application in oil/water separation, J. Mater. Chem. 22 (19) (2012) 9774−9781.

[34] Y.C. Jung, B. Bhushan, Wetting transition of water droplets on superhydrophobic patterned surfaces, Scripta Mater. 57 (12) (2007) 1057−1060.

[35] T. Deng, K.K. Varanasi, M. Hsu, N. Bhate, C. Keimel, J. Stein, et al., Nonwetting of impinging droplets on textured surfaces, Appl. Phys. Lett. 94 (13) (2009) 133109.

[36] D. Tam, V. von ARNIM, G. McKinley, A. Hosoi, Marangoni convection in droplets on superhydrophobic surfaces, J. Fluid. Mech. 624 (2009) 101−123.

[37] C. Antonini, F. Carmona, E. Pierce, M. Marengo, A. Amirfazli, General methodology for evaluating the adhesion force of drops and bubbles on solid surfaces, Langmuir 25 (11) (2009) 6143−6154.

[38] E. Tuck, L. Schwartz, Thin static drops with a free attachment boundary, J. Fluid. Mech. 223 (1991) 313−324.

[39] P. Aussillous, D. Quéré, Shapes of rolling liquid drops, J. Fluid. Mech. 512 (2004) 133−151.

[40] F. Brochard-Wyart, H. Hervet, C. Redon, F. Rondelez, Spreading of "heavy" droplets: I. Theory, J. Colloid Interface Sci. 142 (2) (1991) 518−527.

[41] W.Y.D. Yong, Z. Zhang, G. Cristobal, W.S. Chin, One-pot synthesis of surface functionalized spherical silica particles, Colloids Surf. A: Physicochem. Eng. Aspects 460 (2014) 151−157.

[42] A. Al-Sharafi, A.Z. Sahin, B.S. Yilbas, Measurement of thermal and electrical properties of multiwalled carbon nanotubes−water nanofluid, J. Heat Transfer 138 (7) (2016) 072401.

[43] B. Erzincanli, M. Sahin, An arbitrary Lagrangian−Eulerian formulation for solving moving boundary problems with large displacements and rotations, J. Comput. Phys. 255 (2013) 660−679.

[44] L. Mahadevan, Y. Pomeau, Rolling droplets, Phys. Fluids 11 (9) (1999) 2449−2453.

[45] L.T. Elkins-Tanton, P. Aussillous, J. Bico, D. Quere, J.W. Bush, A laboratory model of splash-form tektites, Meteorit. Planet. Sci. 38 (9) (2003) 1331−1340.

[46] R. Brown, L. Scriven, The shape and stability of rotating liquid drops, Proc. R. Soc. Lond. A 371 (1746) (1980) 331−357.

[47] P. Aussillous, D. Quéré, Liquid marbles, Nature 411 (6840) (2001) 924−926.

[48] Z. Xu, Z. Ao, D. Chu, A. Younis, C.M. Li, S. Li, Reversible hydrophobic to hydrophilic transition in graphene via water splitting induced by UV irradiation, Sci. Rep. 4 (2014) 6450.

[49] M. Sha, D. Niu, Q. Dou, G. Wu, H. Fang, J. Hu, Reversible tuning of the hydrophobic−hydrophilic transition of hydrophobic ionic liquids by means of an electric field, Soft Matter 7 (9) (2011) 4228−4233.

[50] B.S. Yilbas, B. Salhi, M.R. Yousaf, F. Al-Sulaiman, H. Ali, N. Al-Aqeeli, Surface characteristics of silicon nanowires/nanowalls subjected to octadecyltrichlorosilane deposition and n-octadecane coating, Sci. Rep. 6 (2016) 38678.

[51] B. Yilbas, H. Ali, N. Al-Aqeeli, N. Abu-Dheir, M. Khaled, Influence of mud residues on solvent induced crystalized polycarbonate surface used as PV protective cover, Solar Energy 125 (2016) 282−293.

[52] F. Heib, M. Schmitt, Statistical contact angle analyses with the high-precision drop shape analysis (HPDSA) approach: basic principles and applications, Coatings 6 (4) (2016) 57−74.

[53] T.M. Koller, T. Klein, C.D. Giraudet, J. Chen, A. Kalantar, G.P. van der Laan, et al., Liquid viscosity and surface tension of n-dodecane, n-octacosane, their mixtures, and a wax between 323 and 573 K by surface light scattering, J. Chem. Eng. Data 62 (10) (2017) 3319−3333.

[54] D. Richard, D. Quéré, Viscous drops rolling on a tilted non-wettable solid, EPL (Europhys. Lett.) 48 (3) (1999) 286.

[55] B.W. McCormick, Aerodynamics, Aeronautics, and Flight Mechanics, Wiley, New York, 1995.

[56] A. Al-Sharafi, B.S. Yilbas, H. Ali, Water droplet adhesion on hydrophobic surfaces: influence of droplet size and inclination angle of surface on adhesion force, J. Fluids Eng. 139 (8) (2017) 081302.

[57] M. Schmitt, K. Groß, J. Grub, F. Heib, Detailed statistical contact angle analyses; "slow moving" drops on inclining silicon-oxide surfaces, J. Colloid Interface Sci. 447 (2015) 229–239.

[58] J.D. Smith, R. Dhiman, S. Anand, E. Reza-Garduno, R.E. Cohen, G.H. McKinley, et al., Droplet mobility on lubricant-impregnated surfaces, Soft Matter 9 (6) (2013) 1772–1780.

[59] Z. Pan, S. Dash, J.A. Weibel, S.V. Garimella, Assessment of water droplet evaporation mechanisms on hydrophobic and superhydrophobic substrates, Langmuir 29 (51) (2013) 15831–15841.

[60] H. Kim, H.-C. Lim, Mode pattern of internal flow in a water droplet on a vibrating hydrophobic surface, J. Phys. Chem. B 119 (22) (2015) 6740–6746.

[61] M. He, H. Qiu, Internal flow patterns of an evaporating multicomponent droplet on a flat surface, Int. J. Therm. Sci. 100 (2016) 10–19.

[62] M. Sakai, J.-H. Song, N. Yoshida, S. Suzuki, Y. Kameshima, A. Nakajima, Direct observation of internal fluidity in a water droplet during sliding on hydrophobic surfaces, Langmuir 22 (11) (2006) 4906–4909.

[63] L. Hao, P. Cheng, An analytical model for micro-droplet steady movement on the hydrophobic wall of a micro-channel, Int. J. Heat Mass Transfer 53 (5–6) (2010) 1243–1246.

[64] T.K. Pradhan, P.K. Panigrahi, Thermocapillary convection inside a stationary sessile water droplet on a horizontal surface with an imposed temperature gradient, Exp. Fluids 56 (9) (2015) 178.

[65] B.S. Sikarwar, S. Khandekar, K. Muralidhar, Simulation of flow and heat transfer in a liquid drop sliding underneath a hydrophobic surface, Int. J. Heat Mass Transfer 57 (2) (2013) 786–811.

[66] G. Rosengarten, R. Tschaut, Effect of superhydrophobicity on impinging droplet heat transfer, in: 2010 14th International Heat Transfer Conference, American Society of Mechanical Engineers, 2010, pp. 741–746.

[67] J.J. Thalakkottor, K. Mohseni, Effect of slip on circulation inside a droplet, J. Fluids Eng. 137 (12) (2015) 121201.

[68] D. Chatzikyriakou, S. Walker, G. Hewitt, C. Narayanan, D. Lakehal, Comparison of measured and modelled droplet–hot wall interactions, Appl. Therm. Eng. 29 (7) (2009) 1398–1405.

[69] G. Azimi, R. Dhiman, H.-M. Kwon, A.T. Paxson, K.K. Varanasi, Hydrophobicity of rare-earth oxide ceramics, Nat. Mater. 12 (4) (2013) 315–320.

[70] T.L. Bergman, F.P. Incropera, D.P. DeWitt, A.S. Lavine, Fundamentals of Heat and Mass Transfer, John Wiley & Sons, 2011.

[71] J. Mackenzie, W. Mekwi, An unconditionally stable second-order accurate ALE–FEM scheme for two-dimensional convection–diffusion problems, IMA J. Numer. Anal. 32 (3) (2011) 888–905.

[72] A. Al-Sharafi, B.S. Yilbas, H. Ali, Heat transfer and fluid flow characteristics in a sessile droplet on oil-impregnated surface under thermal disturbance, J. Heat Transfer 139 (9) (2017) 092004.

[73] H.-J. Choi, S. Choo, J.-H. Shin, K.-I. Kim, H. Lee, Fabrication of superhydrophobic and oleophobic surfaces with overhang structure by reverse nanoimprint lithography, J. Phys. Chem. C 117 (46) (2013) 24354–24359.

[74] H. Fujimoto, W. Obana, M. Ashida, T. Hama, H. Takuda, Hydrodynamics and heat transfer characteristics of oil-in-water emulsion droplets impinging on hot stainless steel foil, Exp. Therm. Fluid Sci. 85 (2017) 201−212.

[75] D. Halliday, J. Walker, Fundamentals of Physics, John Wiley & Sons, 2005.

[76] A.I. Zografos, W.A. Martin, J.E. Sunderland, Equations of properties as a function of temperature for seven fluids, Comp. Methods Appl. Mech. Eng. 61 (2) (1987) 177−187.

[77] http://www.surface-tension.de/solid-surface-energy.htm, 2018.

[78] https://www.osti.gov/scitech/servlets/purl/1264508, 2018.

[79] B. Bhushan, M. Nosonovsky, The rose petal effect and the modes of superhydrophobicity, Philos. Trans. Royal Soc. London A: Math. Phys. Eng. Sci. 368 (1929) (2010) 4713−4728.

[80] G. Lu, Y.-Y. Duan, X.-D. Wang, D.-J. Lee, Internal flow in evaporating droplet on heated solid surface, Int. J. Heat Mass Transfer 54 (19−20) (2011) 4437−4447.

[81] C. Yang, Y. Ning, X. Ku, G. Zhuang, G. Li, Automatic magnetic manipulation of droplets on an open surface using a superhydrophobic electromagnet needle, Sensors Actuat. B: Chem. 257 (2018) 409−418.

[82] K. Zhao, D. Li, Manipulation and separation of oil droplets by using asymmetric nano-orifice induced DC dielectrophoretic method, J. Colloid Interface Sci. 512 (2018) 389−397.

[83] M. Wegener, N. Paul, M. Kraume, Fluid dynamics and mass transfer at single droplets in liquid/liquid systems, Int. J. Heat Mass Transfer 71 (2014) 475−495.

[84] Q. Zhang, H. Li, C. Zhu, T. Fu, Y. Ma, H.Z. Li, Micro-magnetofluidics of ferrofluid droplet formation in a T-junction, Colloids Surf. A: Physicochem. Eng. Aspects 537 (2018) 572−579.

[85] U. Sen, S. Chatterjee, S. Sen, M.K. Tiwari, A. Mukhopadhyay, R. Ganguly, Dynamics of magnetic modulation of ferrofluid droplets for digital microfluidic applications, J. Magn. Magn. Mater. 421 (2017) 165−176.

[86] R. Cimbala, J. Kurimský, M. Rajňák, K. Paulovičová, M. Timko, P. Kopčanský, et al., Magnetic fluid droplet deformation in electrostatic field, J. Electrost. 88 (2017) 55−59.

[87] M. Rajnak, M. Timko, P. Kopcansky, K. Paulovicova, J. Tothova, J. Kurimsky, et al., Structure and viscosity of a transformer oil-based ferrofluid under an external electric field, J. Magn. Magn. Mater. 431 (2017) 99−102.

[88] A. Carré, J.-C. Gastel, M.E. Shanahan, Viscoelastic effects in the spreading of liquids, Nature 379 (6564) (1996) 432.

[89] http://www.etc-cte.ec.gc.ca/databases/Oilproperties/default.aspx, 2018.

Chapter 6

Dust Effects on Surfaces in Humid Environment and Applications

Chapter Outline

Self-Cleaning of Surfaces and Water Droplet Mobility. DOI: https://doi.org/10.1016/B978-0-12-814776-4.00006-9

6.1 INTRODUCTION

The adhesion between dust particles and a settled surface, in some cases, remains large and external efforts are needed to remove them from the settled surfaces. In general, adhesion force between dust particles and a solid surface is influenced by the free energies of the solid surface and the dust particles, and the interfacial energy between the dust particles and the solid surface. The surface free energy is high for metallic surfaces, which makes it difficult to remove the dust particles settled from metallic surfaces. In addition, water molecules condensate on the settled dust particles in ambient humid air conditions. This, in turn, modifies the physical and chemical properties of the dust particles. In addition, alkaline and alkaline earth metals in dust particles dissolve in water condensate and form a chemically active liquid solution. Due to gravity, the liquid solution flows through the solid surface and forms an interlayer between the dust particles and solid surface. Once the liquid solution layer dries out, it forms crystalline structures at the interface of the dust particles and the solid surface. This further increases adhesion between the dust particles and the solid surface. Consequently, efforts required to remove the dust particles from the solid surface become significant and costly. Moreover, the liquid solution at the interface has basic ions, with an adverse effect on the solid surface integrity; in which case, the liquid solution undergoes a chemical attack on the solid surface giving rise to the corrosion effect on the surface prior to drying. Once the surface integrity of aluminum is damaged due to the chemical modification by the liquid solution, the optical properties of the surface, such as reflectivity and absorptivity, can change even the dried mud solution is removed from the surface. On the other hand, aluminum is widely used in solar-energy harvesting systems as reflecting surfaces such as a trough, which improves solar concentration via focusing of the solar radiation onto a thermal harvesting unit such as a solar-heat exchanger. However, environmental dust settlements and mud formation on the aluminum surface modify the optical characteristics of the aluminum trough surface while lowering concentrated solar radiation in a solar-harvesting system.

The dust adhesion on surfaces is critically important in terms of their removal from the surface and their aftereffects related to optical properties of the surface. This issue is critical for the surface of energy harvesting devices. In this chapter, dust and mud effects on metallic and nonmetallic surfaces are presented in light of previous studies [1−7].

6.2 DUST AND MUD EFFECTS ON GLASS SURFACES

Dust particles are composed of compounds containing alkali (NaOH) and alkaline earth metals ($CaCO_3$) that dissolve in condensed water vapor and increase the pH of the water. After mud forms from dust particles and water, some of the components of the dust, such as the alkali/alkaline earth compounds, dissolve into the water and form a chemically active solution. Because mud is composed of porous structures, the solution (water with dissolved ionic compounds) forms sediments at the interface between the substrate and the mud. The solution dries and forms a crystalline layer between the dried mud and the substrate. This process modifies the surface chemistry (i.e., the adhesive and cohesive forces) and increases the force required to remove the mud from the surface because of the additional covalent bonding. Because the mud formation and removal are interrelated, these processes are complex and require a thorough investigation of the aftereffects of mud deposition, including the chemical, optical, morphological, and mechanical (adhesion, friction, hardness, etc.) effects on the substrate surfaces. Consequently, the characteristics of the mud formed from dust particles on glass surfaces and its aftereffects on the surface characteristics are presented in this section in light of the previous study [1].

6.2.1 Experimental

Photovoltaic (PV)-protective glass samples with dimensions of 30 mm × 30 mm × 2 mm (width × length × thickness) were used as workpieces. The chemical composition of the glass was 76.5% SiO_2, 9.9% CaO, 1.2 MgO, and 12.4% Na_2O. The dust was collected from PV modules in the area of Dhahran in Saudi Arabia after a dust storm in 2014. Characterization of the dust was performed using SEM, energy dispersive spectroscopy (EDS), and X-ray diffractogram (XRD). A JEOL 6460 SEM was used for the SEM and EDS examinations, and a Bruker D8 Advanced diffractometer with a CuKα radiation source was used for XRD analysis. The typical settings of the XRD instrument were as follows: 40 kV and 30 mA for the X-ray source and a scanning angle (2θ) range of 20−80 degrees. Roughness measurements and surface profile characterization were performed using a 5100 atomic force microscope/scanning force microscope (AFM/SPM) microscope by Agilent in contact mode. The probe tip was made of silicon nitride ($r = 20-60$ nm) with a manufacturer-specified force constant, k, of 0.12 N/m.

A MicroPhotonics digital microhardness tester (MP-100TC) was used for the surface microhardness measurements. The standard test method for the Vickers indentation hardness of advanced ceramics (ASTM C1327-99) was adopted. The measurements were repeated five times at each location to ensure the consistency of the results. A linear microscratch tester (MCTX-S/N: 01-04300) was used to determine the friction coefficient of the glass surfaces. The contact load was set at 0.03 N, and the end load was set at 2.5 N. The scanning speed was 5 mm/min, and the loading rate was 0.01 N/s. The total length for the scratch tests was 0.5 mm.

The optical transmittance was measured using a UV spectrometer (Jenway—67 Series spectrophotometer), and Fourier transform infrared

(FTIR) spectroscopy (Bruker—VERTEX70) was performed to collect the infrared absorption spectrum of the glass.

To investigate the effects of dust and mud on the surface characteristics of the glass, actual dust accumulation and mud formation were simulated in a laboratory. In actual environments, the mud was formed from accumulated dust particles due to the condensation of water vapor onto the particles. The accumulated dust thickness was measured over a period of 2 weeks during a dust storm in Saudi Arabia in 2014. This accumulation was in the order of 300 μm. To simulate dust accumulation in the laboratory, 300-μm layers composed of dust particles collected from the local environment were formed on the cleaned glass surfaces. Desalinated water, which was equal to the amount of water vapor that condensed on the same volume of the dust in the open environment, was dispensed gradually onto the dust layer. The initial condensation tests were performed in ambient humid air to estimate the amount of condensate that accumulated over time. Moreover, the dispensed water was left on the surface of the dust layer without mechanical mixing to resemble water condensation from humid air. Therefore the simulated formation of mud on the glass surfaces was similar to the deposition that occurred naturally in the open environment. Next, the glasses were kept in ambient air at room temperature for 3 days to dry. Scratch tests were performed to measure the tangential force required to remove the mud from the glass surfaces. The tangential force provided information regarding the adhesion, cohesion, and frictional work during the dry mud removal. To examine the aftereffects of the mud on the glass surfaces, the dry mud was removed from the workpiece surfaces using a desalinated waterjet that was 2 mm in diameter with a velocity of 2 m/s. The cleaning process was applied for 15 minutes to each glass surface. Finally, the morphology, optical transmittance, molecular characteristics, and microhardness of the mud-removed glass surfaces were analyzed using the analytical tools.

The microhardness, friction, and adhesion tests were repeated 12 times to secure confidence levels for the experimental uncertainty assessments. Based on the distribution of the experimental data, a confidence level of 95% resulted; in which case, the mean (μ) of the data distribution was within ± 1.75 of the standard deviation of the distribution of a single measurement from that distribution. The experimental uncertainty analysis revealed that an uncertainty less than 2% resulted for the microhardness measurements while an uncertainty of about 3% was obtained for the friction and adhesion tests.

6.2.2 Characteristics of Dry Mud and Its Influence on Glass Surfaces

Fig. 6.1 shows SEM micrographs of the top surface and a cross section of the mud formed on the glass surface. The mud is formed using the collected dust and the desalinated water application that mimics the condensation of

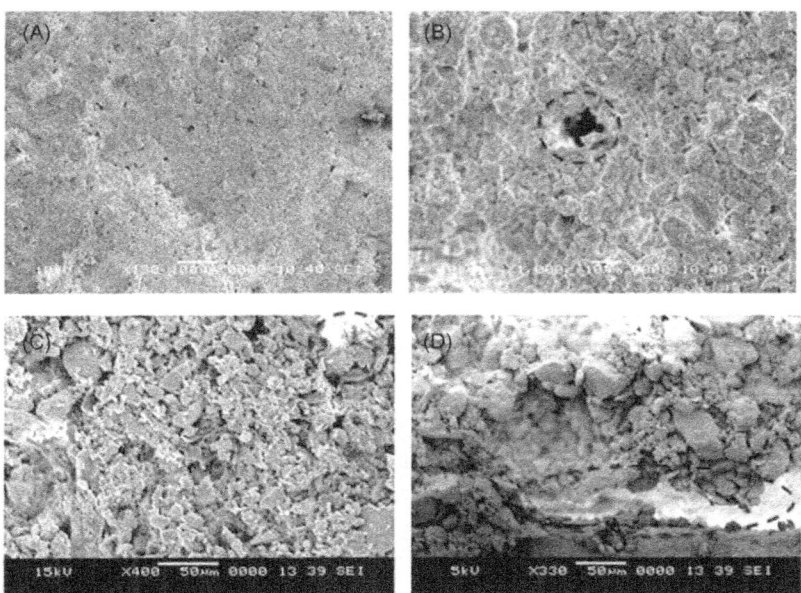

FIGURE 6.1 SEM micrographs of the dry mud on the glass surface: (A) top view of the mud at the glass surface; (B) cavity formed due to agglomeration of large particles in the mud (highlighted by the circle); (C) dry mud cross section (white color represents the dried solution, as indicated by the circle); and (D) dry mud glass interface (white color is dry mud solution at the interface, as highlighted by the ellipse) [1].

water from humid air. During the formation of the mud, no mechanical mixing was used, and the mud was left to dry at room temperature for 2 days prior to analysis. The mud surface consisted of closely adhered fine dust particles and cavities in-between the large particles (Fig. 6.1). Closely packed particles resulted in cohesive forces in the mud while improving the microhardness in this region (Table 6.1). Because no strong bonding occurred in-between the large particles because of the cavities formed in-between them, the microhardness was small in this region. Note that the microhardness of the individual large dust particles (within the range of 20 µm) is in the order of 26.3 HV.

Close examination of the surface (Fig. 6.1) reveals the presence of a residue from the dried liquid solution in-between the small particles. The alkali and alkaline earth metals (Na and Ca) in the particles dissolve in the water during the mud formation, forming a liquid mud solution. Therefore when the liquid solution dries, some of the dissolved alkali and alkaline earth metal compounds remain in-between the fine-sized dust particles. These regions appear as a bright color in the SEM micrographs. The presence of alkali metals is also evident from the EDS data obtained from the top surface of the dry mud (Table 6.2).

TABLE 6.1 Microhardnesses of the As-Received Glass, Dust, Dry Mud, and Glass Surface After Mud Removal [1]

	Hardness (HV)
As-received glass surface	175 (+10/−10)
Dust particle	26.3 (+1/−1)
Dry mud surface	20.2 (+1/−1)
Glass surface after mud removal	210 (+10/−10)

TABLE 6.2 EDS Data From the Surface of the Mud, the Glass Surface After Mud Removal, and the Dried Solution at the Interface [1]

	Si	Ca	Na	S	Mg	K	Fe	Cl	O
Dried mud surface	16.3	7.2	0.4	2.2	1.7	0.2	0.4	0.1	Balance
Mud-removed glass surface	11.7	6.3	3.8	1.4	2.7	1.1	0.5	−	Balance
Dried solution at the interface	−	7.1	2.1	1.1	1.8	1.2	0.4	0.6	Balance

The presence of undissolved large dust particles results in the formation of pores throughout the dry mud. The pores appear to be randomly scattered throughout the mud cross section. The water containing dissolved mud compounds flows in-between the large particles through the cavities and eventually reaches the surface of the glass due to the effects of gravity. However, some of the mud solution is retained in the cavities in the mud forming a dense structure with a white color in the cavity upon drying. The EDS analysis reveals that the dried mud solution in the cavities exhibits similar characteristics as at the interface between the glass and the mud. Chlorine, sodium, potassium, and sulfur are observed in the EDS data (Table 6.2). The dust and water mixture is ultrasonically shaken in a tube for 1 hour, and then the solution is extracted; the pH of the mud solution is 8.4, which increases the concentration of OH^- ions in the solution. As the solution dries, it forms crystal structures at the interface and in the cavities, which appear white (Fig. 6.1).

To assess the morphology and the effects of the dried mud solution at the interface, the dried mud is removed from the glass surface using a pressurized desalinated waterjet with a diameter of 2 mm and a jet velocity of 2 m/s. Fig. 6.2 shows the SEM micrographs of the glass surface after the

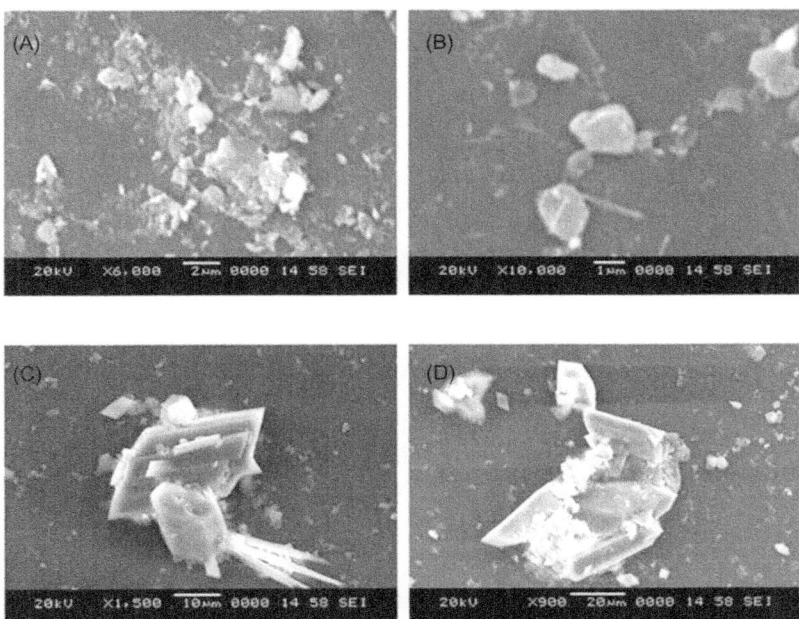

FIGURE 6.2 SEM micrographs of mud residue after mud removal and the dried solution: (A) large mud residue strongly attached to the glass surface; (B) fine-sized mud residue and (C and D) crystals formed due to the dried solution at the glass surface [1].

mud removal. Mud residue of different sizes is visible in the micrographs (Fig. 6.2). Close examination of the glass surface reveals that fine-sized crystals are formed on the surface (Fig. 6.2). These particles are associated with the residue from the mud solution after the removal of the dry mud. This situation is observed in the AFM images shown in Fig. 6.2. Moreover, due to the presence of OH^- ions in the mud solution (pH = 8.4), small cavities are formed in the glass surface because of surface etching. In this case, KOH ions in the mud solution are responsible for the local etching of the glass surface.

The local etching increases the surface texture of the glass after the mud removal, as indicated by the line profile shown in Fig. 6.3. The possible ion-exchange mechanism of the glass—alkaline ions at the interface between the glass surface and the mud in the presence of the mud solution can be expressed as [8]: $Si - O^- \cdots K^+(OH)^-$ water $\rightarrow \equiv Si - OH + K^+ + OH^-$. The breakdown of the glass network through alkaline attack due to OH^- ions also generates the reaction $Si - O - Si \equiv + OH^- \equiv Si - OH + {}^-O - Si \equiv$. Note that KOH attack causes the formation of silanol groups ($\equiv Si - OH$) on the glass surface, which produces cracks in the glass. Moreover, the OH^- ions destroy siloxane bonds ($\equiv Si - O - Si \equiv$) at the glass surface during the

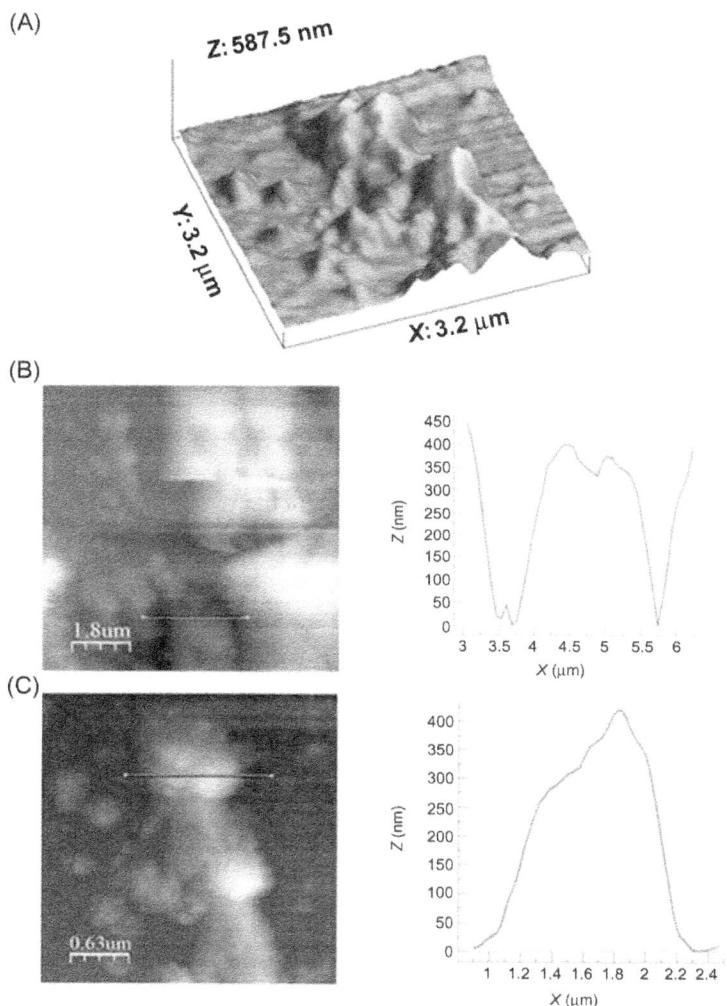

FIGURE 6.3 ATM micrographs: (A) fine-sized mud residue attached to the glass surface; (B) small cavities formed on the glass surface; and (C) horizontal profile of the mud residue [1].

alkaline attack, which favors the penetration of water molecules (H^+ and OH^- ions) into the glass. This process results in the formation of small cavities in the surface of the glass (Fig. 6.3). During the KOH attack, the diffusion of potassium ions into the surface causes a volume change because of the large diameter of potassium. This volume change results in chemical toughening of the glass surface and an increase in the microhardness of the surface after the mud removal (Table 6.1).

To examine the effects of the dry mud solution on the molecular characteristics of the glass, FTIR spectroscopy was performed on the as-received glass and the glass after the removal of the dry mud. Fig. 6.4 shows the FTIR data for the as-received glass and the glass after mud removal. In the case of the as-received surface, the $550-600$ cm^{-1} region corresponds to the Si$-$O$-$Si bending vibration [9], and the stretching vibration of Si$-$O$-$Si occurs at approximately 910 and 990 cm^{-1}. In the case of the glass surface after the mud removal, the absorbance peaks exhibit a slightly different behavior than those of the as-received glass. This difference is due to the stress induced in the region close to the glass surface due to the attack by the alkali and alkaline earth hydroxides and the diffusion of potassium into this region [9]. The stretching vibration of nonbridging oxygens (Si$-$O$^-$) is observed at 920 cm^{-1}. This peak appears in the glass after dust removal, but it is not observed in the as-received glass. The peak is related to the ionic bonding between the nonbridging oxygen and the network modifiers. In this case, the dissolution of the glass network and the formation of a new SiO$_2$-enriched network due to surface reconstruction are responsible for the stretching vibration of the nonbridging oxygen [10].

Fig. 6.5 shows the data obtained from the transmittance measurements for the as-received and the mud-removed glasses. The transmittance is reduced by nearly 35% (on average) for the mud-removed glass surface. The reduction in the transmittance is associated with (1) the mud residue, which blocks the

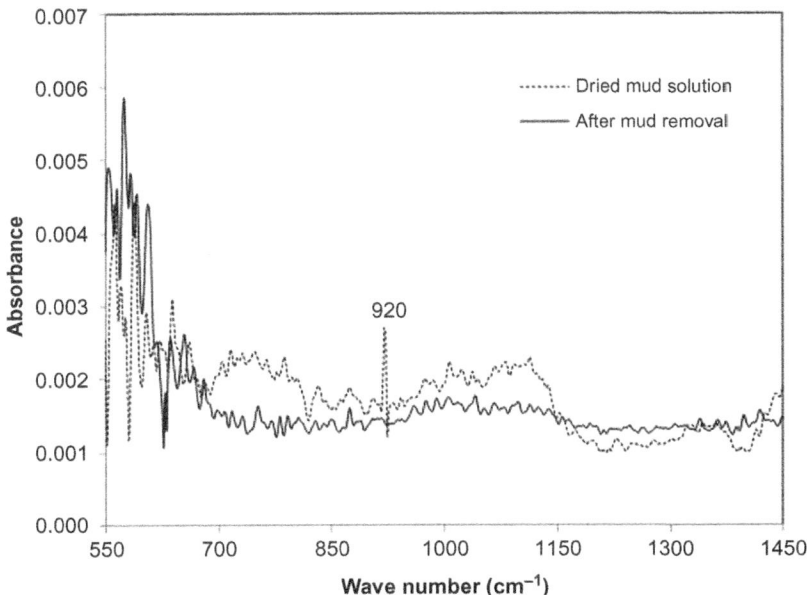

FIGURE 6.4 FTIR data for the as-received glass and the glass after dust removal [1].

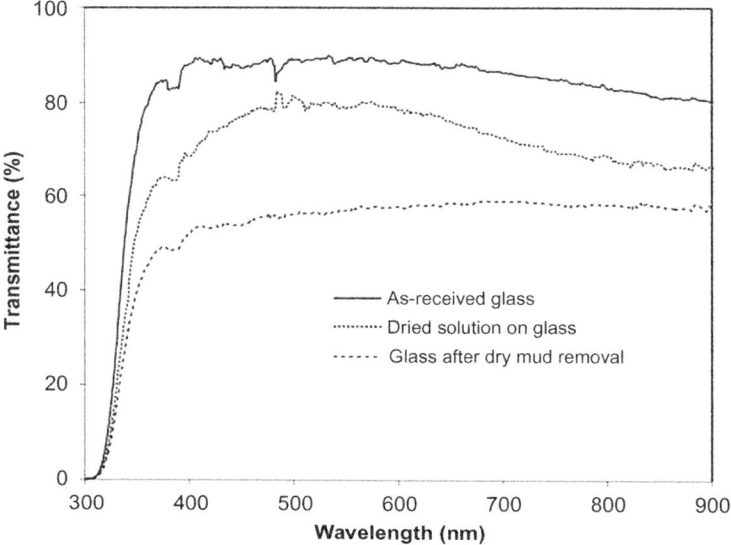

FIGURE 6.5 Transmittance of the as-received glass, glass after dry mud removal, and glass with the dried solution [1].

incident radiation; and (2) the molecular changes that occur in the surface region of the glass due to the alkali and alkaline earth hydroxide attack.

The mud residue forms strong bonds on the glass surface, which prevents the pressurized desalinated waterjet from removing it (Fig. 6.2). The total area covered by the mud residue is estimated to be 3%. This result indicates the strong adhesion of the mud residue remaining on the glass surface. The XRD of the dust residue on the glass surface after the dry mud removal reveals $CaCO_3$, NaCl, and MgO peaks (Fig. 6.6). Similar peaks are also observed in the XRD of the dry mud surface (Fig. 6.6).

Because adhesion of the dry mud to the glass surface is strong and mud residue remains after the surface is cleaned with a jet of desalinated water, the adhesion, cohesion, and frictional work required to remove the dry mud from the glass surface were determined from the tangential force analysis. Several forces can contribute to the adhesion of the mud to the glass surface, including van der Waals and electrostatic forces [11]. In the case of the dry mud on the glass surface, the dried solution, which is composed of dissolved alkali metals, alkaline earth metals, and other ions, forms a fine layer at the interface between the dry mud and the glass surface. Consequently, considering only strong van der Waals forces may not correctly describe the adhesion force because of the covalent bonds formed by the dried solution at the interface.

The tangential force measured by the microtribometer when removing the dry mud from the glass surface is the combination of the adhesive and cohesive forces. The tangential force due to the as-received glass surface is

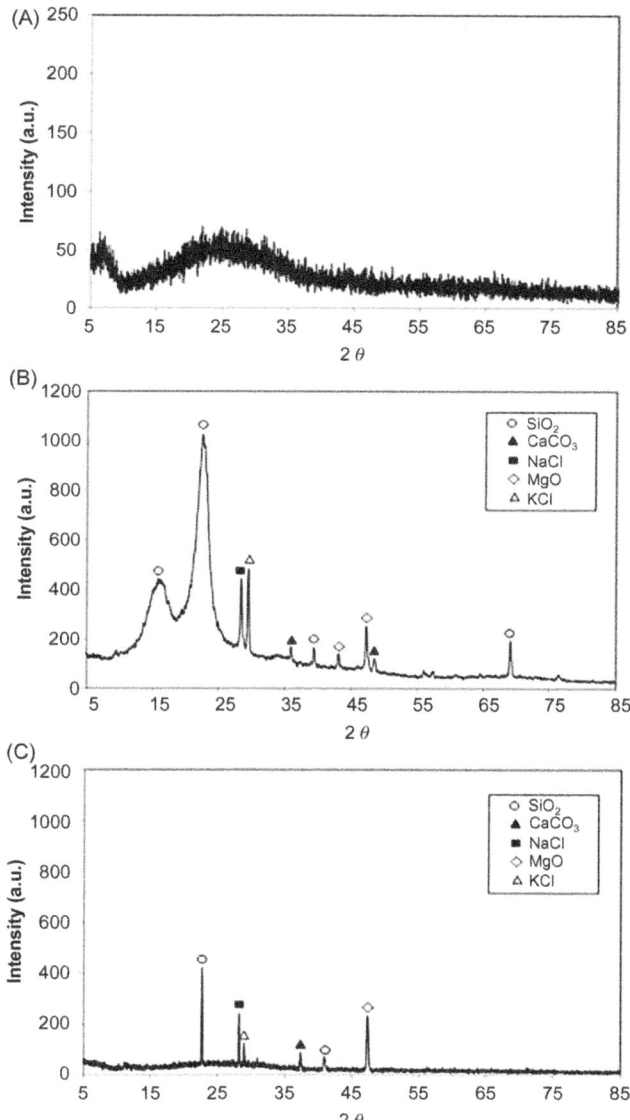

FIGURE 6.6 XRD of the different types of glass: (A) as-received glass; (B) dry mud on the glass surface; and (C) after removal of the dry mud from the glass surface [1].

associated with the frictional force; however, the tangential force due to the mud on the glass surface includes the frictional, adhesion, and cohesive forces. Therefore integration of the tangential force over the distance traveled by the microtribometer gives the work done against friction for the

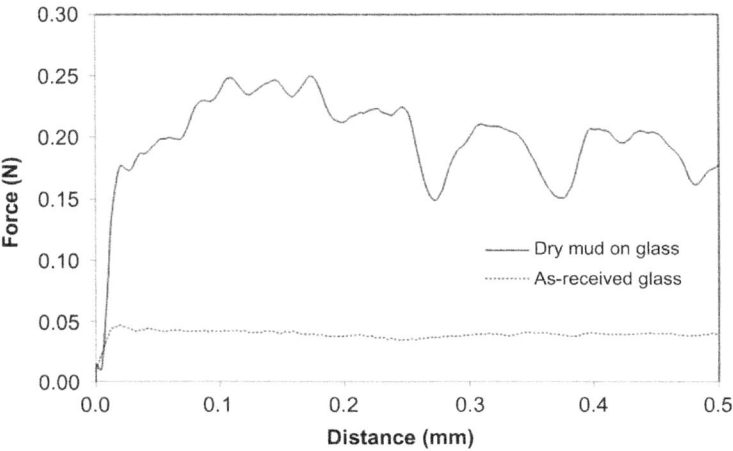

FIGURE 6.7 Frictional force for the as-received glass surface and the tangential force required to remove the mud from the surface [1].

TABLE 6.3 Frictional and Adhesion Work Obtained From the Tangential Force [1]

	Frictional Work (mJ)	Adhesion and Cohesion Work (mJ)
As-received surface	0.000403	–
As-received surface with mud	–	0.9213
Surface with mud residue	0.0293	–

as-received glass surface, while it provides the frictional, cohesive, and adhesion work for the dry mud on the glass surface. The cohesion and adhesion work can be obtained after subtracting the frictional work from the work performed due to friction, cohesion, and adhesion. Fig. 6.7 shows the tangential force variation with distance for (1) the as-received glass and (2) the dry mud on the glass surface. Table 6.3 presents the work performed against friction and the combination of the cohesion and adhesion work. Table 6.3 shows that the frictional work is considerably smaller than the work corresponding to the combination of cohesion and adhesion. In addition, the coefficient of friction for the mud-removed glass surface is larger than that of the as-received glass surface. This difference is associated with the mud residue remaining on the surface and the fine-sized cavities formed due to the KOH attack. The tangential force measurements are repeated five times, and the experimental error is estimated to be approximately 6%.

6.3 MUD CHARACTERISTICS ON BISPHENOL-A POLYCARBONATE

Bisphenol-A polycarbonate (PC) sheets are used as a protective cover for PV panels because of their mechanical flexibility, high fracture toughness, and low density. However, dust accumulation on PC surfaces is problematic and adversely affects the PV efficiency due to the lowered transmittance of the solar radiation to the active area of the PV panel. In addition, the liquid solution that results from the mud formed in humid environments is chemically active because of the dissolved alkaline materials and alkaline earth metal ions and sediments at the PC surface and alters its physical and chemical characteristics. When the solution at the mud−PC interface dries, dissolved minerals crystallize on the surface and the force required to remove the dry dust from the surface increases significantly. The dry mud residue and the crystals of the dried solution degrade the optical transmittance of the PC, thus lowering the PV efficiency.

While airborne dust and its characteristics have been thoroughly examined, studies regarding the effects of environmental dust on surfaces have not been extensively reported [12]. Characterization of atmospheric airborne dust during the wet seasons in East Africa was investigated by Mkoma et al. [13]. They demonstrated that common crystal and sea-salt elements including Na, Mg, Al, Si, Cl, Ca, Ti, Mn, Fe, Sr, NO_3^-, and P (and to a lesser extent Cu and Zn) tended to be coarse particles. In addition, the aerosol chemical mass content of the aerosol was determined to consist of 48% organic matter, 44% crustal matter, 4% sea-salt, and 2% elemental carbon was observed. A characterization of atmospheric aerosols was also carried out by Maenhaut et al. [14]. They showed that most of the Ca was water soluble; the mineral dust Ca was presumably mostly present as $CaCO_3$, and perhaps also in part as gypsum. In contrast, only half of the K content was water soluble, indicating that it was to a large extent associated with insoluble mineral dust. Patterns of dust retention on urban trees in oasis cities were examined by Baidourela and Zhayimuj [15]. The findings revealed that dust that had accumulated on tree leaves was mainly of local urban origin, and the heavy metal concentrations at different sites varied significantly. The morphology of atmospheric particles in a semiarid region of India was studied by Mishra et al. [16]. They demonstrated that the influence of the dust aspect ratio on dust scattering was significant for dust with high hematite content. Petruk and Skinner [17] characterized particles in airborne dust and showed that small particles had low aspect ratios.

Although the relationship between dust accumulation and performance of PV panels was investigated previously [18], the aftereffects of the mud formed on the PC surface in terms of the surface characteristics have not been addressed. Therefore in this study, dust characteristics and the aftereffects of the mud that forms due to dust particle accumulation on the surface properties of PC were examined. The surface characteristics investigated include microhardness, surface energy, topology, elemental compositions and compounds, and the molecular states of the surface region. Analytical tools, including optical, electron scanning, and atomic force microscopies, X-ray diffraction, energy-dispersive X-ray spectroscopy, microtribometry, UV transmission spectroscopy, and FTIR spectroscopy, were used to characterize the mud aftereffects on the PC surface. In addition, the elemental composition of the mud solution was assessed using quadrupole inductively coupled plasma mass spectrometry. The analysis and the findings are presented in light of the previous study [2].

6.3.1 Mud Formed From Dust Particles

Figs. 6.8 and 6.9 show the SEM images of the top surface and cross section of the dry mud formed on the PC surface, respectively. The dry mud surface was composed of closely packed dust particles (Fig. 6.9A), microvoids (Fig. 6.9B), and adhered large dust particles (Fig. 6.9C). Furthermore, the morphology of the dry mud surface exhibited an irregular topology with an average surface roughness in the order of 2.6 μm. Close examination of the SEM image in Fig. 6.9D revealed the presence of locally scattered white dense regions on the surface, which is likely related to the dried mud solution that is

FIGURE 6.8 SEM images of the surface of the mud formed from the dust particles: (A) fine particles; (B) voids between the dust particles; (C) large particles and surrounding voids as indicated by the circle; and (D) dried mud solution at the mud surface [2].

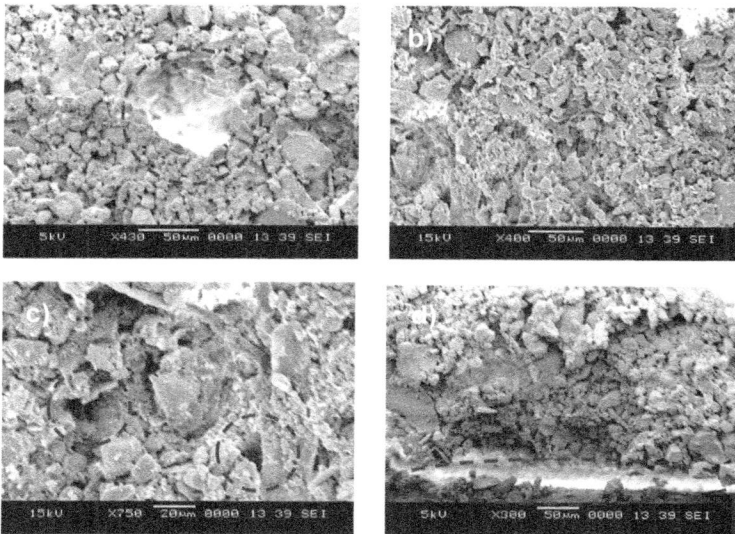

FIGURE 6.9 SEM images of the dry mud cross section: (A) small dust particles forming a dense structure and dried solution trapped in the cavity as indicated by the circle; (B) large particles as indicated by the circle; (C) voids in the dry mud as indicated by the circles, and (D) dried mud solution at the interface as indicated by the ellipse [2].

primarily composed of Na, K, and Ca (Table 6.4). The alkaline materials (e.g., Na and K) and the alkaline earth metallic compounds dissolved in the water, thus creating a mud solution. Some of the mud solution sediments at the mud—PC interface were obtained upon gravitational settling; however, a small amount remained at the surface where dust particles were small, and adhered strongly and formed a white residue after drying.

To determine the pH of the mud solution, the dust particles were mixed with ultraclean desalinated water at a ratio of 1:4 and placed in an ultrasonic shaker for 15 minutes. The pH of the mud solution was recorded over a given time period. The temporal variations in the mud solution pH are shown in Fig. 6.10. As observed, the pH increased with time, and the solution remained basic at pH = 8.4, which is associated with the presence of OH^- ions in the solution. In this case, the dissolution of the alkaline and alkaline earth metallic compounds in the water was responsible for the formation of OH^- ions. The data obtained from the quadrupole inductively coupled plasma mass spectrometry analysis revealed that the mud solution contained alkaline and alkaline earth metals. Consequently, the sediment mud solution contained alkaline and alkaline earth metals after drying, as evidenced in the EDS data (Table 6.4).

The surface microhardness data, given in Table 6.5, were obtained at different locations on the dry mud surface. The microhardness increased in

TABLE 6.4 Elemental Composition of the Dried Mud Solution on the PC Surface Assessed by EDS

Dried Solution at Interface	Ca	Na	S	Mg	K	Fe	Cl	O
Spectrum 1	7.2	0.4	2.2	1.7	0.2	0.4	0.1	Balance
Spectrum 2	6.3	3.8	1.4	2.7	1.1	0.5	–	Balance
Spectrum 3	7.1	2.1	1.1	1.8	1.2	0.4	0.6	Balance
Spectrum 4	5.2	1.3	0.9	1.2	0.9	0.4	0.3	Balance
Spectrum 5	4.5	2.6	0.7	3.1	1.2	0.3	0.2	Balance
Spectrum 6	6.3	0.9	1.2	2.5	1.1	0.1	0.2	Balance
Spectrum 7	6.8	2.4	1.1	1.9	1.2	0.4	0.3	Balance

Each spectrum corresponds to a different location on the surface as indicated in the SEM image [2].

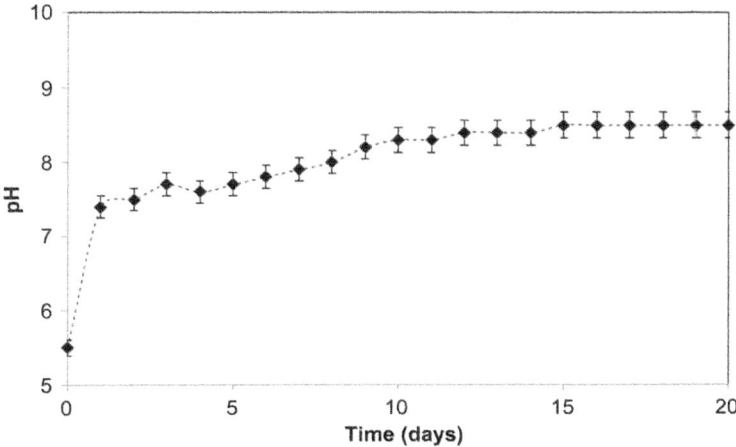

FIGURE 6.10 Variation in the pH of the liquid solution consisting of dust particles and water as a function of time [2].

TABLE 6.5 Microhardness of the As-Received PC Sheet, Dust Particles, Dry Mud Surface, PC Surface After Mud Removal, and Dried Mud, NaOH, and KOH Solutions

	Hardness (HV)
As-received PC surface	8.2 ± 1
Dust particle (size ≥ 10 µm)	26.3 ± 1
Dry mud surface	20.2 ± 1
PC surface after mud removal	15.1 ± 1
Dried mud solution	13.6 ± 1
Dried NaOH solution	11.7 ± 1
Dried KOH solution	10.4 ± 1

The errors are based on several measurements [2].

regions where the small dust particles were in close proximity. This result can be attributed to the strong cohesive forces among the small particles and the presence of the dry solution in this region, which acts as a binding agent among the particles. In contrast, the microhardness was lower in regions containing large particles, which is related to the weak bonding among the large particles due to the voids present among them. The undissolved oddly shaped, large dust particles were responsible for the nearby fine void formations in the dry mud; however, these voids appeared to be randomly scattered on the mud surface. The microhardness values of individual dust particles that were $\geq 10\ \mu m$ in size were also measured. The findings showed that the microhardness values varied significantly within the range of 6.7–26.3 HV. In the mud cross-sectional images (Fig. 6.9A–D), porous structures were found to cover a large area of the cross-sectional surface that were likely due to the naturally formed dust layer from the irregularly shaped dust particles. It should be noted that mechanical compaction was not applied to the dust layer on the PC surface prior to mud formation. The liquid solution flowed within the porous structures and hardened at the interface, forming a thin layer between the mud and the PC surface. However, some of the liquid solution remained in the cavities formed between the undissolved irregularly shaped dust particles in the mud. The film formed from the dried solution at the interface contained alkaline elements (e.g., Na, K), alkaline earth metals (e.g., Ca), and Fe, Mg, and Cl, as also shown in the EDS data in Table 6.4.

6.3.2 Analysis of Mud Residue

To assess the mud that remained on the surface, the PC surface interface with the dry mud was cleaned using a jet of water. A pressurized waterjet that was 2 mm in diameter was sprayed at a velocity of 2 m/s and directed normally to the mud surface during the cleaning process for 20 minutes.

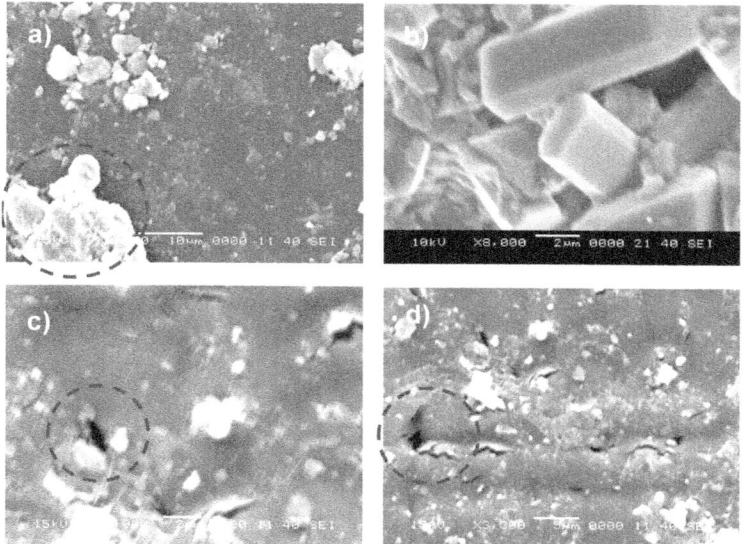

FIGURE 6.11 Mud residue and cavities formed at the PC surface after mud removal: (A) large and small mud residues—the large residues are indicated by the circle; (B) dried solution crystals; and (C and D) cavities of various sizes formed at the surface [2].

After drying the surface naturally in an air, the PC surface was examined. Figs. 6.11A−D show SEM images of the PC surface after mud removal. A few locally scattered mud residues were observed on the PC surface (Fig. 6.11A). Additionally, crystal structures were found on the surface (Fig. 6.11B). Mud residue was also observed in the AFM micrographs in Fig. 6.12.

The presence of mud residue after cleaning with pressurized water indicates the presence of strong adhesion between the dry mud and the PC surface. Due to the mud solution sediment at the interface of the mud and PC, which contains alkaline hydroxides (e.g., NaOH and KOH) due to the high pH (8.4), nucleophile (OH^-) attacks take place at the PC surface depending on the local concentration of NaOH and KOH in the mud solution. Thus

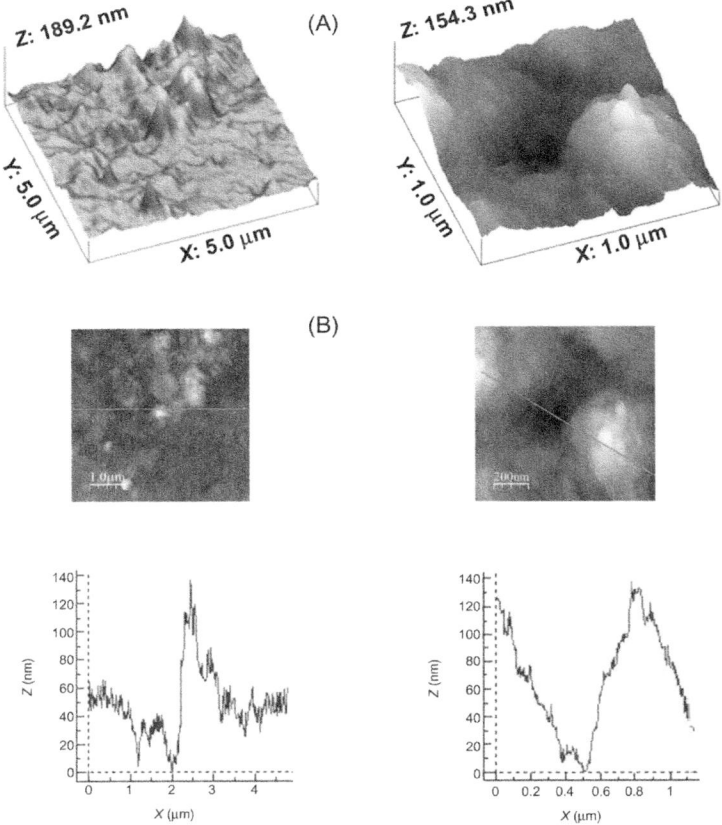

FIGURE 6.12   AFM images of the PC surface after mud removal: (A) mud residue and the cavities formed in the region near the mud residue and (B) line scans of the mud residue and associated profiles of the mud residue and the cavity [2].

micromolecules with carbon bonded to three oxygen molecules of the PC (i.e., a carbonate link) at the surface are attacked by OH^- due to differences in electronegativity. An anion returns to the aqueous phase, and the reaction continues until complete depolymerization of the PC at the surface. A schematic view of this process is shown in Fig. 6.13, in which NaOH attack of the PC surface (Fig. 6.13A) and OH^- attachment on PC surface (Fig. 6.13B) are demonstrated. Similar scenarios apply to KOH. The degradation of carbonate groups leads to a series of reactions due to the elimination of CO_2 and CO in the absence of free radical reactions (i.e., first-order reaction [19]). Consequently, local degradation creates small cavities at the surface, particularly in the region near the mud residue (Fig. 6.12A) where the mud solution sediments locally at the surface. The dried mud solution at the interface of the dry mud and the PC surface increases adhesion between the mud residue and the PC surface despite the use of the high-pressure waterjet during surface cleaning. Furthermore, the presence of Cl in the dried mud solution at the dry mud—PC interface

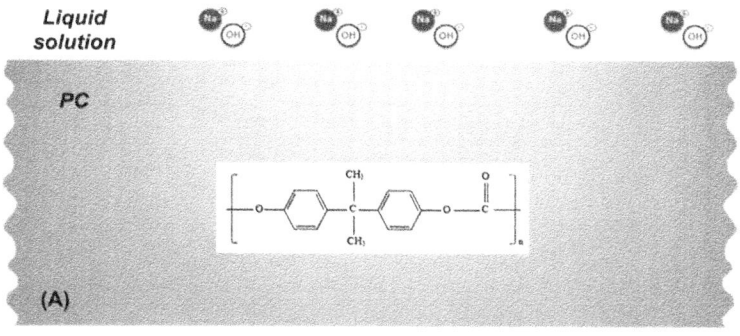

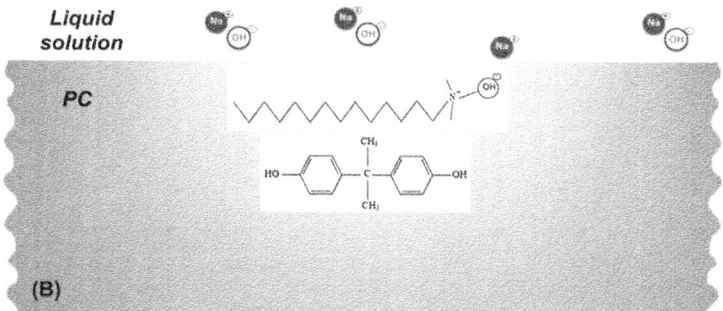

FIGURE 6.13 Schematic of NaOH attack of the PC surface: (A) the PC surface and the liquid solution and (B) PC surface and OH^- attachment [2].

(Table 6.4) suggests that the simultaneous presence of bisphenate anions and chlorine could lead to chlorination of the aromatic rings with the formation of polychlorinated bisphenates. However, under basic conditions (e.g., pH = 8.4), polychlorinated bisphenates are known to oxidize to form radicals [20], in which quinones form depending on the oxidative process that occurs [21]. This process contributes to the degradation of the PC surface, thereby enhancing the formation of small cavities on the PC surface (Figs. 6.11C and 6.12A).

6.3.3 Analysis of Mud Removal

Fig. 6.14 shows the FTIR spectroscopy data of the as-received PC sheet and PC sheet following removal of mud with a pressurized waterjet. The as-received PC sheet exhibited an absorption spectrum typical of PC glass [22]. The absorption bands, corresponding to C−H bond stretching vibration, were observed at $2874-2969 \text{ cm}^{-1}$. The absorption band at $860-680 \text{ cm}^{-1}$ corresponded to the bending vibration of the C−H bond, and the band at 1496 cm^{-1} was attributed to the C−H bending vibrations of methylene groups. Characteristic absorption bands of aromatic C−H bending vibration were observed at $860-680 \text{ cm}^{-1}$ and those of aromatic C = C bending vibrations were observed at $1700-1500 \text{ cm}^{-1}$. In addition, the absorption peak at 1770 cm^{-1}, which is typically associated with the C = O stretching vibration band of ethers, was observed. The FTIR spectrum of the PC sheet after mud removal displayed a peak in the range of $3070-3580 \text{ cm}^{-1}$ that

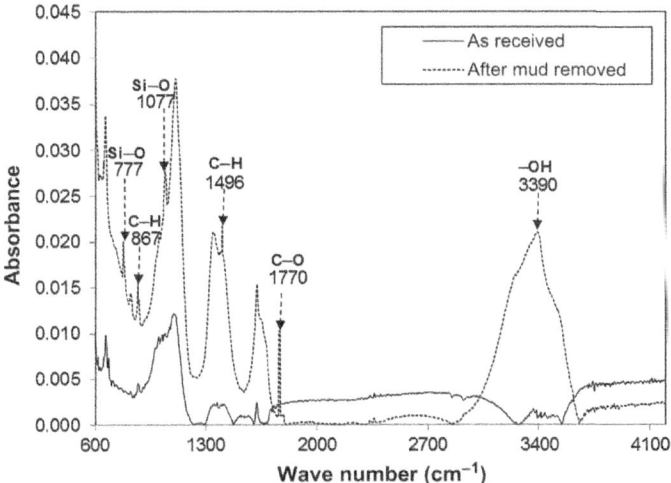

FIGURE 6.14 FTIR spectra of the as-received PC sheet and the PC sheet following mud removal [2].

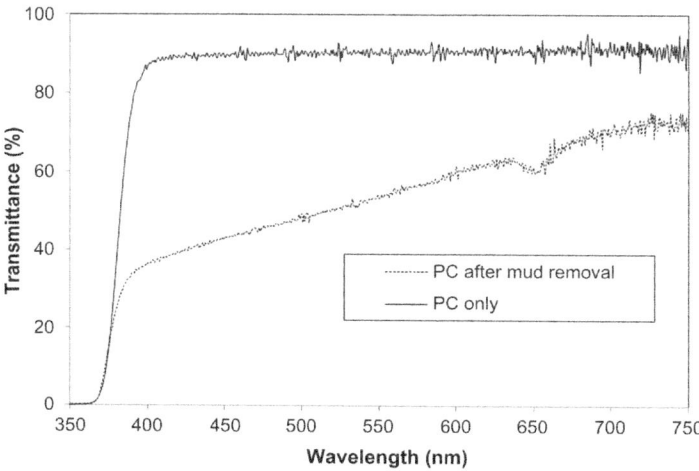

FIGURE 6.15 UV—visible transmittance spectra of the as-received PC sheet and the PC sheet following removal of the dry mud [2].

corresponded to hydroxyl groups (—OH) due to the formation of bisphenol-A carbonate monomers or other substituted phenols. The presence of Si—O bonds was evidenced by the 777 and 1077 cm^{-1} peaks, corresponding to the stretching associated with SiO_2 [23], which is remains as a residue on the surface after mud removal. In addition, the peaks at 867 and 1431 cm^{-1} were likely caused by the stretching vibrations of CO_3^{-2} [24], which are related to residue that remains after mud removal from the surface.

Fig. 6.15 shows the data obtained from the UV—visible transmittance measurements of the as-received PC sheet and PC sheets following removal of mud. As observed, the transmittance decreased by nearly 45% on average for PC surface following mud removal. This reduction in transmittance was attributed to (1) the presence of mud residue, which blocks incident light at the surface and (2) molecular changes that occur at the surface of PC due to the alkaline and alkaline earth metal hydroxides attacks.

Fig. 6.16 shows the variation in the tangential force along the distance at the PC surface with and without dry mud present. The tangential force was recorded using a microtribometer during the tests for the PC surface with the dried mud and the frictional force variation on the as-received PC surface. The area under the tangential force profile describes the adhesion and frictional work required to remove the dry mud from the surface, and the area under the frictional force describes the frictional work performed [25]. Therefore subtraction of the frictional work from the adhesion work yields the adhesion work required to remove the dry mud from the PC surface. The tangential force measurements were repeated five times, and the estimated experimental error was approximately 6%.

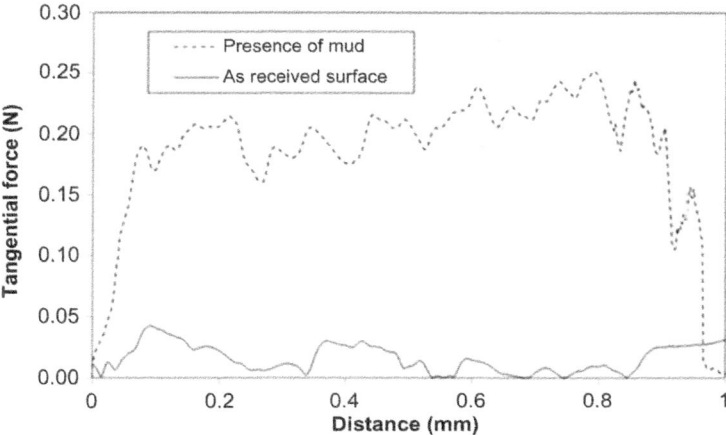

FIGURE 6.16   Tangential force due to friction on the as-received PC surface and on the PC surface containing mud [2].

6.3.4   Effect of Dried Solution

To assess the effect of the dried solution between the dry mud and the PC surface, the liquid solution extracted from the dust was dispensed on the as-received PC surface. The adhesion force measurements were repeated for the dried mud solution only at the PC surface. Because the liquid solution extracted from the dust contained dissolved alkaline and alkaline earth metal hydroxides, the effects of the individual NaOH and KOH solutions on the adhesion force were also tested. In these tests, 10% solutions of NaOH and KOH were individually dispensed onto the PC surface, and the adhesion tests were performed after the solutions had dried. Fig. 6.17 shows the tangential force versus the distance along the sample, as obtained from the microtribometer data for the dried solutions. The findings show that the tangential force remained high for the KOH and NaOH dried solutions. In contrast, the tangential force due to the dried mud solution was lower than that of the NaOH and KOH dried solutions. This result can be attributed to (1) the concentration of KOH and NaOH in the solution extracted from the mud prior to drying, which was less than that of the individually dispensed NaOH and KOH solutions at the PC surface and (2) the dried mud solution containing other elements such as Fe, Cl, and Mg (Table 6.4). Therefore the combined presence of the compounds formed from these elements and from NaOH and KOH in the dried mud solution lowered the tangential force required for mud removal.

Table 6.6 shows the adhesion work required to remove the dry mud, dry mud solution, and dry NaOH and KOH solutions from the PC surface. The adhesion work required to remove the dry mud was higher than that required

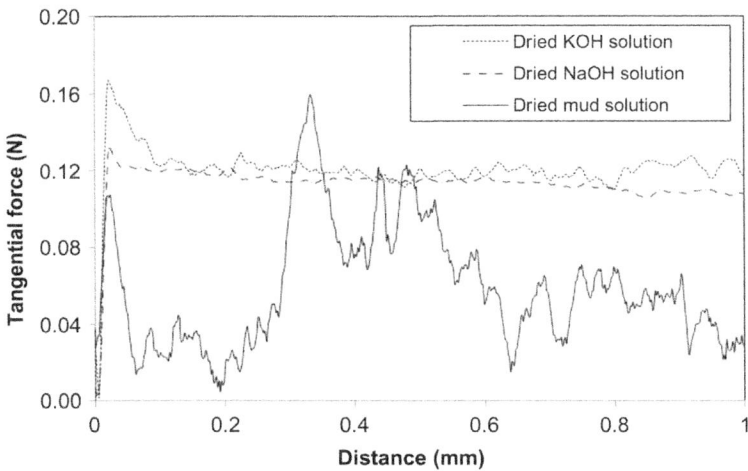

FIGURE 6.17 Variation in tangential force as a function of distance due to the presence of dried KOH, NaOH, and mud solutions on the PC surface [2].

TABLE 6.6 Measured Adhesion Work Required for the Removal of Dry Mud, Dried Mud Solution, Dried NaOH Solution, and Dried KOH Solution From the PC Surface

	Adhesion Work (mJ)
Dry mud	0.15558
Dried mud solution	0.03664
Dried NaOH solution	0.08396
Dried KOH solution	0.09167

The friction of the as-received PC surface was 0.000603 mJ [2].

to remove the remaining dry solutions from the PC surface. The van der Waals forces and the covalent bonds that formed at the PC surface were responsible for the high adhesion work required to remove the dry mud from the PC surface. To further explore the adhesion work required for the removal of the dry mud, the surface energy of the dry mud was measured using the sessile drop technique [26]. Table 6.7 shows the data used for the surface energy measurements of the dry mud. The surface energy of the dry mud was 55.39 mJ/m² (Table 6.8), which was higher than that of the PC surface (i.e., 34.5 mJ/m²) [27]. Hence, the adhesion work required to remove the dry mud removal was high due to the high surface energy of the dry mud and presence of the dry mud solution at the interface, in which case the

TABLE 6.7 Lifshitz−van der Waals Components and Electron-Donor Parameters Used in the Simulation [22]

	γ_L (mJ/m^2)	γ_L^L (mJ/m^2)	γ_L^+ (mJ/m^2)	γ_L^- (mJ/m^2)
Water	72.8	21.8	25.5	25.5
Glycerol	64	34	3.92	57.4
Ethylene glycol	48	19	0.41	1.28

L indicates the liquid phase; γ_L is the liquid surface tension; γ^L is the apolar component due to the Lifshitz−van der Waals intermolecular interactions; and γ^+ and γ^- are the electron-acceptor and electron-donor parameters, respectively, of the acid−base component of the solid and liquid surface free energy.

TABLE 6.8 Lifshitz−van der Waals Components and Electron-Donor Parameters of Smooth Surface

γ_S (mJ/m^2)	γ_S^+ (mJ/m^2)	γ_S^- (mJ/m^2)	γ^P (mJ/m^2)	γ_S^L (mJ/m^2)
55.39	0.2688	48.17	7.197	48.19

S indicates the solid phase; γ^+ and γ^- are the electron-acceptor and electron-donor parameters, respectively, of the acid−base component of the solid and liquid surface free energy; γ^P is due to electron-acceptor and electron-donor intermolecular interactions; and γ_S^L is the interfacial solid−liquid free energy.

formation of hydroxyl groups (−OH) (Fig. 6.13) enhanced chemical bonding between the dry mud and the PC surface, thereby considerably enhancing the adhesion work.

Fig. 6.18 shows the SEM images of the crystals formed at the PC surface due to the dried NaOH, KOH, and mud solutions. Close examination of the images revealed that fine crystals formed on the surface (Fig. 6.18A−D), which can be associated with the crystallization of the dried solution at the PC surface. Because calcium was present in the dried mud solution (Table 6.4) at the interface of the dry mud and the PC surface, $CaCO_3$ crystals formed at the PC surface, as shown in the XRD of the PC surface after mud removal in Fig. 6.19. The peak corresponding to $CaCO_3$ in the diffractogram indicates the presence of $CaCO_3$ crystals at the PC surface. The SEM images of the crystals showed that calcite was most likely connected to polycrystalline spherulites at the PC surface, which thus contributed to the bonding between the mud and the PC surface and, in turn, increased the adhesion work. However, the presence of the NaOH and KOH solutions prior to drying resulted in shallow depth of etching which produced submicrometer-sized cavities at the surface, particularly in regions near the fine crystals (Fig. 6.18A and B). This phenomenon was more pronounced for the KOH

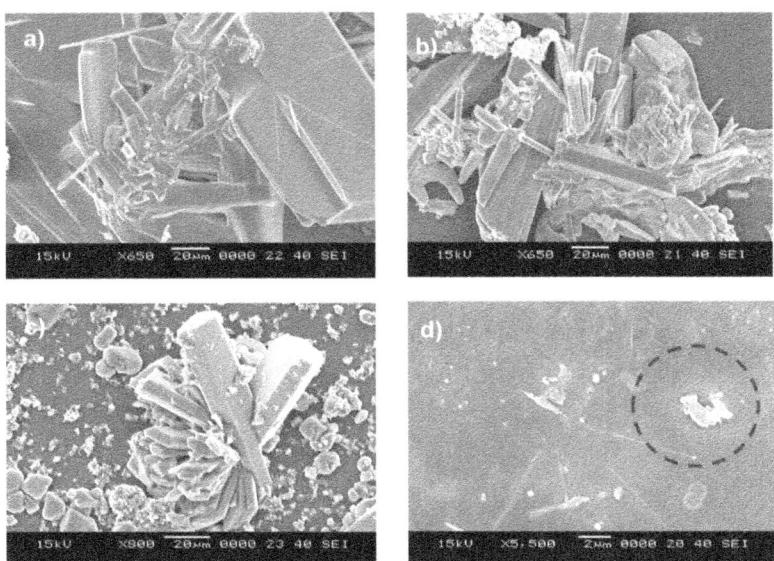

FIGURE 6.18 SEM images of the dried solution crystals on the PC surface: (A) dried NaOH solution; (B) dried KOH solution; (C) dried mud solution; and (D) dried $CaCO_3$ solution ($CaCO_3$ residue and PC crystals are indicated by the circle) [2].

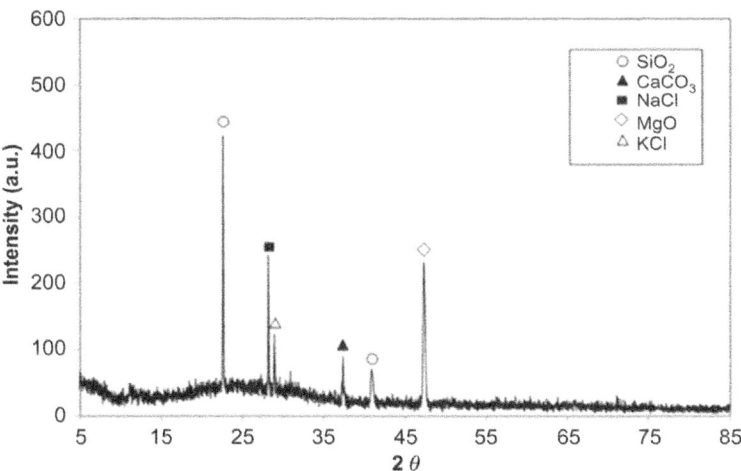

FIGURE 6.19 XRD of the PC surface after removal of mud [2].

dried solution. Local etching increased the surface roughness of the PC sheet, as shown in the line profile in Fig. 6.12. The microhardness data presented in Table 6.5 showed that the PC after dry mud removal achieved the highest microhardness among the tested samples, followed by the dried mud solution, dried NaOH solution, and dried KOH solution. Increase in micro-hardness is typically associated with (1) the presence of mud residuals, such as $CaCO_3$ and silica, at the PC surface and (2) chemical modification of the PC surface by the mud solution during NaOH and KOH attack. The micro-hardness data were obtained for the dried solutions individually dispensed onto the PC surface.

6.4 MUD EFFECT ON INCONEL 718 SURFACE

Inconel 718 alloy is a nickel-based alloy with chromium—molybdenum with superior properties in corrosive environments. In addition, it displays high tensile and creep—rupture properties. The addition of niobium in the alloy constitute allows age hardening, annealing, and welding without undergoing spontaneous hardening during the cooling and the heating cycles. The alloy finds wide applications in various industries some of which include aerospace, chemical and marine engineering, and nuclear reactors. Although Inconel 718 alloy has several applications because of its superior thermal properties and corrosion resistance, and the surface properties can be improved further for tribological applications. One of the methods to improve the surface properties is to treat the surface via using a high-power laser beam. The laser surface treatment of engineering alloys has several advantages, some of which include precision of operation, short processing time, localized heating, and low cost. On the other hand, laser surface treatment involves high-temperature thermal processing, which is critical for the end product quality. This is because of thermal erosion and surface asperities, such as microcracks and large cavities, which occur at high temperatures, while limiting the practical applications of laser high-temperature processing. However, the process optimization and proper selection of laser-processing parameters minimized these defects and provides satisfying end product quality with desired surface properties. Laser surface texturing involves with local ablation of surface with limited melting in the irradiated region. Since laser intensity distribution is Gaussian in nature, peak power intensity occurs at the irradiated spot center while intensity reduces toward the spot edges. This, in turn, allows surface evaporation in the close region of the laser-irradiated spot center while melting takes place in the edge region of the irradiated spot. The melt flow from the region of the irradiated spot edges toward the irradiated spot center modifies the surface texture. In this case, surface texture with micro/nano-pillars is produced while altering the wetting state of the surface. The third particle adhesion, such as dust particles, on the surface becomes weak because of the micro/nanopillars formed at the surface. Hence the adhesion force between particles and textured surface remains weak and efforts required for removable of those particles from the surface become small. Since the particle adhesion on the laser-textured surfaces remains critical in terms of surface self-cleaning, the wetting state of the surface plays an important role toward reduction of the adhesion force, which is particularly true for the environmental dust adhesion on laser-textured surfaces. In this section, a laser gas-assisted texturing of Inconel 718 alloy is carried out. The resulting surface texture morphology and wetting state are assessed using the analytical tools. Surface energy of the laser-treated layer is determined from the droplet contact angle method [28]. The dust particles are mixed with the desalinated water resembling the water condensation of the dust particles in humid air ambient. The liquid solution extracted from the water−dust particles mixture is analyzed using the analytical tools. The liquid solution is dispensed on the laser-textured surface and the spreading rate of liquid solution and its effects on the surface care examined in line with the previous study [3].

6.4.1 Experimental Analysis

A CO_2 laser (LC-ALPHA III) providing a nominal output power of 2 kW in repetitive pulses of 1500 Hz frequency was used to irradiate the workpiece surface. The laser beam diameter at the irradiated spot surface was in the order of 0.2 mm. During the laser texturing process, high-pressure nitrogen-assisting gas was used. Inconel 718 samples were cut in the size 15 mm × 13 mm × 3 mm and they were used in the experiments. Several tests were carried out to optimize the laser treatment parameters. The laser treatment parameters resulting in defect-free surfaces and texture characteristics with micro/nanopillars are selected. The laser treatment parameters are given in Table 6.9.

A JEOL JDX-3530 was used to obtain SEM micrographs related to the morphological changes of the laser-textured surfaces. Elemental composition of the laser-treated surface was determined via EDS analysis, which was carried out at textured surface at different locations. AFM/SPM, by Agilent, in contact mode was used to analyze the surface texture characteristics. The AFM tip was made of silicon nitride probes ($r = 20−60$ nm) with a manufacturer-specified force constant, k, of 0.12 N/m. A Bruker D8 Advance having CuKα radiation was incorporated for XRD analysis. Typical XRD equipment settings were 40 kV and 30 mA.

TABLE 6.9 Laser Heating Conditions Used in the Experiment [3]

Scanning Speed (cm/s) (mm/min)	Power (W)	Frequency (Hz)	Nozzle Gap (mm)	Focus Setting (mm)	N_2 Pressure (kPa)
10	105	1500	1.5	127	600

A Microphotonics digital microhardness tester (MP-100TC) was used for the microhardness measurements at the laser-treated surface. The standard test method for Vickers indentation hardness of advanced ceramics (ASTM C1327-99) was adopted and a 300 g load was used during the tests. The measurements were repeated five times at each location at the surface and the error estimated is in the order of 5%. A linear microtribometer (MCTX-S/N: 01-04300) was used to determine the tangential force friction coefficient of the dry mud formed TiN-coating surface. The equipment was set at the contact load of 0.03 N and end load of 5 N. The scanning speed was 5 mm/min and loading rate was 1 N/s.

The wetting experiment was performed using Kyowa (model—DM 501) contact angle goniometer. A static sessile drop method was considered for the contact angle measurement. Droplet volume was controlled with an automatic dispensing system having a volume step resolution of 0.1 μL. Still images were captured, and contact angle measurements were performed after one second of deposition of a water droplet on the surface.

The environmental dust particles were collected and water condensation on the dust particles were simulated while mimicking humid air outdoor ambient. The aftereffects of the dust particles on the laser and as-received surfaces analyzed. The dust particles were collected using the soft brushes from the outdoor PV panel surfaces in Dhahran area of Saudi Arabia. The dust layer thickness on PV panel surfaces was measured over the period of 4 weeks and it was found to be in the order of 300 μm. Therefore a dust layer of 300 μm thickness was incorporated in the experiments resembling outdoor dust accumulation on the surfaces. The initially condensation tests were carried out on the dust particles in local outdoor humid air ambient. The tests provided the amount of water condensate accumulating on the dust particles over a 6-hour period. The findings revealed that the amount of water condensate had the same volume of the dust particles over the 6-hour period. Consequently, the desalinated water with the same volume of the dust particles was mixed in a small sealed container resembling the outdoor environment. The mixture of desalinated water and the dust particles was left any mechanical mixing in sealed container without mechanical mixing. This results in a natural formation of liquid solution in the container. The liquid solution was extracted from the container and dispensed onto the laser-treated and as-received surfaces. The workpieces were, then, kept in a local standard ambient air for 1 day to dry. The laser-treated and as-received surfaces with the presence of the dried liquid solution were tested for adhesion force measurements using the microtribometer. The dry liquid solution was, later, removed from the laser-treated and as-received surfaces with a desalinated waterjet of 2 mm diameter and 1.5 m/s velocity. The cleaning process was continued for 25 minutes for each workpiece surfaces. Aftereffects of the liquid solution on the laser-textured and as-received surfaces were analyzed using the analytical tools.

6.4.2 Characteristics of Laser-Treated Surface

Fig. 6.20 shows SEM micrographs of laser-textured Inconel 718 surface. The laser-textured surface consists of regular laser-scanning tracks and the spacing between the scanning tracks is in the order of 200 μm (Fig. 6.20A). The formation of laser-scanning tracks is attributed to the high-frequency laser repetitive pulses, which result in overlapping of irradiated spots on the surface along the scanning direction. It should be noted that the laser repetitive pulse frequency is kept at 1500 Hz in the experiments and the overlapping of

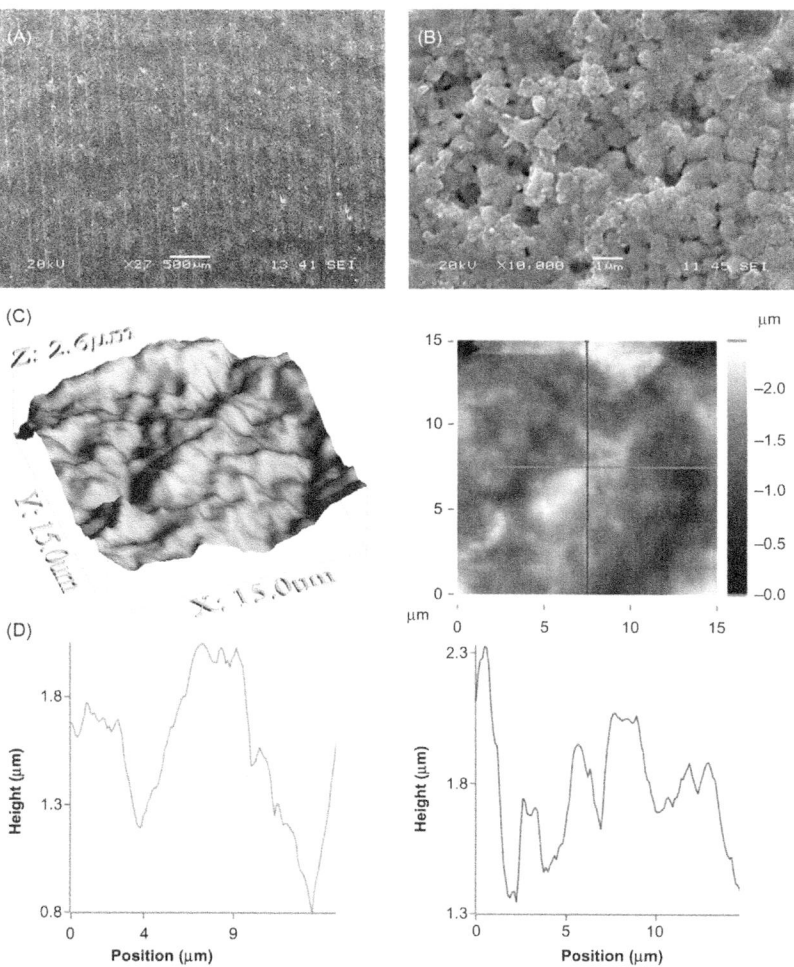

FIGURE 6.20 SEM micrographs and AFM images of laser-textured surface: (A) laser-scanning tracks; (B) micro/nanosize pillars formed at surface; (C) 3D image of surface; and (D) line scan along horizontal and vertical limes across surface [3].

the irradiated spots gives rise to laser-scanning tracks (Fig. 6.20A). The over-lapping ratio of the irradiated spots on the surface is in the order of 70%. The close examination of surface reveals that the irradiated surface is free from cracks, voids, and large size cavities. The high cooling rates take place in the surface region results in high thermal stresses in this region [3]. However, the closely spaced laser-scanning tracks modify the cooling rates in the laser-treated layer through the self-annealing effect. It should be noted that recently formed closely spaced laser-scanning tracks act like a heat source for the neighboring previously formed laser-scanning tracks. In this case, heat diffusion from the recently formed laser-scanning tracks toward the previously formed-scanning tracks modifies the cooling rates in the laser-treated layer. Moreover, the spatial distribution of the laser pulse intensity is Gaussian with the peak intensity occurring at the irradiated spot center. The irradiated surface undergoes evaporation at the spot center while melting takes place at the irradiated spot edges. Therefore the melt flow from the irradiated spot edges toward the irradiated spot center modifies the cavity size, in terms of depth and width, which is formed initially by the surface evaporation due to the high-beam intensity at the irradiated center. Consequently, the combination of evaporation and melting at the surface results in surface texture composed of micro/nanopillars (Fig. 6.20B). In order to assess the surface roughness and the roughness parameter, which is defined through the ratio of area covered by pillars over the projected area at the surface, AFM imaging is carried out at the laser-treated surface. Fig. 6.20C shows a 3D image of the surface while Fig. 6.20D shows the line scan across the surface. The presence of micro/nanopillars-like structures at the surface is evident. The line scan of the surface (Fig. 6.20D) demonstrates the peaks and valleys of the surface texture. In addition, some small spikes-like textures correspond to the submicrometer whisker-like structures formed at the surface. The roughness parameter (r) is determined as 0.54 and the average surface roughness is in the order of 1.75 μm. The surface microhard-ness increases after the laser treatment process. The microhardness increase is associated with the nitride species ($\gamma''N$ and $\gamma'N$) formation in the surface region. This can be observed from Fig. 6.21, which shows the XRD of the laser-treated surface; in which case, the peaks of $\gamma''N$ and $\gamma'N$ nitride species are observed, which are formed in the surface region. Table 6.10 gives the EDS data for the laser-textured surface. The elemental composition of the laser-treated surface remains similar to the untreated surface; however, nitro-gen is observed at the laser-textured surface, which is attributed to the nitride species formed during the laser texturing of the surface under the high-pressure nitrogen-assisting gas.

Although light elements, like nitrogen, are not accurately quantified from the energy-dispersive spectrum, its presence at the surface is evident. In addition, the dissolution of Laves phase in the surface region contributes to the microhardness increase; in which case, the amount of Nb available for γ''

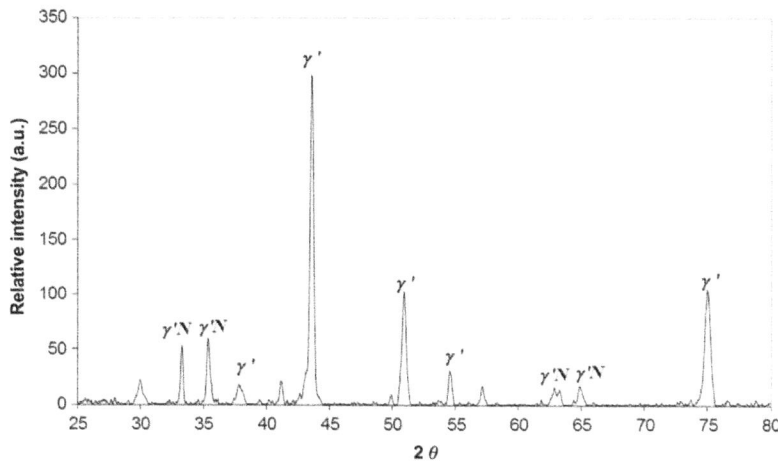

FIGURE 6.21   XRD of laser-textured surface [3].

TABLE 6.10 Elemental Composition (wt.%) of As-Received and Laser-Treated Inconel 718 Alloy [3]

	N	Al	Ti	Cr	Fe	Nb	Ni
As-received	0	0.7	1.1	18.1	17.2	5.1	Balance
Laser-treated	6	0.5	0.9	17.4	17.1	5.1	Balance

TABLE 6.11 Microhardness Results Prior to and After the Laser Treatment Process [3]

	Microhardness (HV)
Laser-treated	460 ± 20
Untreated	310 ± 10

precipitation increases. Table 6.11 gives the microhardness measurement results at the workpiece surface.

The precipitation of niobium-rich phases [δ-phase (Ni$_3$Nb)] in the surface region can cause grain-boundary cracking [29]. However, this situation is not observed from the micrographs of the surface (Fig. 6.20B). Therefore the δ-phase of niobium-rich compound in the surface region remains less in amount. The fine discrete particles occurring at the surface are related to the Laves morphology. Hence, the discrete Laves phase demonstrates the presence of the precipitation of strengthening phases γ'' (Ni_3Nb) and γ' (Ni_3 (Al, Ti)) in the surface region. However, the highly interconnected coarse Laves phase can trigger the early initiation of cracks and gives rise to the low energy fracture path [30]. Nevertheless, this structure is not evident in the surface region of the irradiated layer. Consequently, the self-annealing effect generated in the laser-treated region via closely spaced laser-scanning tracks minimizes the stress levels and reforms the metallurgical changes toward preventing the crack initiation and propagation.

6.4.3 Surface Hydrophobicity and Surface Free Energy

The wetting state of the surface can be assessed by the water-droplet contact angle measurements. The Young's equation for a sessile water droplet on perfectly smooth and chemically homogeneous surface can be used to formulate the water-droplet contact angle; in which case, it yields [31]: $cos\theta = (\gamma_{sv} - \gamma_{sl})/\gamma_{lv}$, where θ is the water-droplet contact angle, γ_{sv} is the interfacial tensions of solid−vapor, γ_{sl} is the interfacial tensions of solid−liquid, and γ_{lv} is the interfacial tensions of liquid−vapor. Since the laser-textured surface has the roughness and the Young's equation for the contact angle calculation needs to be modified while introducing the roughness parameter (r), the arrangement yields [32]: $cos\theta_w = r(\gamma_{sv} - \gamma_{sl})/\gamma_{lv}$, where θ_w is the water-droplet contact angle on the rough surface, and r is the roughness parameter, which was defined earlier as the ratio of area covered by the pillars over the total projected area of the surface. In this case, $r = 1$ corresponds to the perfectly smooth surface and $r < 1$ represents the rough surface. The modified Young's equation can be rearranged to include the influence of the surface and interfacial tension components between the liquid−solid and the liquid−vapor interfaces. Hence, the resulting equation becomes [33]: $cos\theta_c = f_1cos\theta_1 + f_2cos\theta_2$, where θ_c is the apparent contact angle; f_1 is the surface fraction of the liquid−solid interface; f_2 is the surface fraction of the liquid−vapor interface; θ_1 is the contact angle for the liquid−solid interface; and θ_1 is the contact angle for the liquid−vapor interface. However, for the air−liquid interface, f_1 can be written in terms of the solid fraction (f_1) and air fraction (f_2), which is ($1-f_2$). Here, f_2 ranges within 0 to 1. Hence, $f_2 = 0$ corresponds to the liquid droplet that is not contacting the surface. Alternatively, $f_2 = 1$ represents the surface that is completely wetted. In the laser-textured surface, the Cassie−Baxter and Wenzel states could coexist on the textured surfaces; in which case, if a liquid−air interface can remain pinned at the micro/nanopillars tops, the transition from the Cassie and Baxter to the Wenzel states becomes possible [34]. The laser-textured surface is composed of varying pillars height across the surface; consequently, air gaps with varying sizes are in-between the pillars. This alters the wetting state of the laser-textured surface from Cassie and Baxter to the Wenzel state. The water-droplet contact angle varies in-between

87 ± 3 degrees and 130 ± 3 degrees on the textured surface. However, the area coverage for the Wenzel state on the textured surface ($\theta_C = 87 \pm 3$ degrees, where θ_C is the water-droplet contact angle) is in the order of 6%, which is considerably small. Moreover, the contact angle hysteresis varies within $11-32$ degrees on the laser-textured surface. In this case, the large air gaps on the textured surface are not sufficient to generate the Lotus effect and give rise to the Wenzel state at the textured surface. However, whisker-like texture generates the lotus effect while reducing the contact angle hysteresis and enables increasing droplet mobility on the textured surface. Moreover, the free energy of the laser-treated surface is determined incorporating the droplet contact angle method [28]. In the assessment of the surface energy of the laser-treated surface, the liquid droplets including glycerol, and diiodomethane are incorporated beside the water droplet. The brief analysis of the surface energy formulation is presented in accordance with the previous study [28,35]. The surface energy formulation for solid surfaces was presented by van Oss et al. [35]. The resulting relation is in the form of $\gamma_L(\cos\theta + 1) = 2\sqrt{\gamma_S^L \cdot \gamma_L^L} + 2\sqrt{\gamma_S^+ \cdot \gamma_L^-} + 2\sqrt{\gamma_S^- \cdot \gamma_L^+}$, where γ_S is the solid surface free energy, γ_{SL} is the interfacial solid–liquid free energy, γ_L is the liquid surface tension, θ is the contact angle, γ^L is the apolar component due to Lifshitz–van der Waals intermolecular interactions, and γ^P is due to the electron-acceptor and electron-donor intermolecular interactions, and γ^+ and γ^- are the electron-acceptor and electron-donor parameters of the acid–base component of the solid and liquid surface free energy, respectively. S and L represent the solid and liquid phases, respectively.

The contact angle relation could be incorporated determining the values of γ_S^L, γ_S^+, and γ_S^- while using the contact angle data and γ_L^L, γ_L^+, and γ_L^-. The data related to the values of γ_L^L, γ_L^+, and γ_L^- can be found in the open literature for water, glycerol, and diiodomethane and are given in Table 6.12 [26,28]. To ensure the repeatability of the measurements, the contact angle for each fluid was measured 10 times. Based on the experimental repeatability, the measurement error is in the order of 2%, which is considerably small.

Using the contact angle relation ($\gamma_L(\cos\theta + 1) = 2\sqrt{\gamma_S^L \cdot \gamma_L^L} + 2\sqrt{\gamma_S^+ \cdot \gamma_L^-} + 2\sqrt{\gamma_S^- \cdot \gamma_L^+}$), the surface free energy of the laser-treated Inconel

TABLE 6.12 Lifshitz–van der Walls Components and Electron-Donor Parameters Used in the Simulation [26,28]

	γ_L (mJ/m^2)	γ_L^L (mJ/m^2)	γ_L^+ (mJ/m^2)	γ_L^- (mJ/m^2)
Water	72.8	21.8	25.5	25.5
Glycerol	64	34	3.92	57.4
Diiodomethane	50.8	50.8	0.72	0

718 was found to be in the order of 61.7 mJ/m². The surface free energy of as-received surface was in the order of 117 mJ/m², which is slightly smaller than pure nickel (120 mJ/m² [36]). Consequently, the surface texture, with micro/nanosize pillars and cavities, and reduced surface free energy contribute to the hydrophobic wetting state of the laser-textured surface.

6.4.4   Dried Liquid Solution on Laser-Textured Surface

The alkaline (Na, K) and alkaline earth (Ca) metal compounds of dust particles are dissolvable in water condensate in ambient humid air while forming a liquid solution on the dust particles. Therefore further studies related the influence of the dissolved compounds on the laser-textured surface is necessary to mimic water condensation on dust particles. Thus the dust particles were mixed with desalinated water to resemble the ratio that occurs in the environmental conditions of Saudi Arabia on a humid day. The volume ratio of the dust particles to water condensate was found to be unity; hence, this ratio is incorporated for the dust particles and the desalinated water mixture. After mixing the dust particles and the desalinated water, the liquid solution was extracted and analyzed incorporating the inductive coupled plasma (ICP). The ICP data for the liquid solution are given in Table 6.13. The presence of Ca, Na, K, Cl, and Mg in the liquid solution is observed from Table 6.13. In addition, the alkalinity of the liquid solution was measured incorporating pH meter and the findings reveal that the pH of the liquid solution is in the order of 8.4, which shows that the liquid solution is basic. Hence, the water condensation in ambient humid air causes the dissolution of compounds from the dust particles while forming the liquid solution around the dust particles. The liquid solution then flows on the laser-textured surface under the gravitational influence.

The liquid solution may form the continuous film onto the laser-textured surface. To assess if the liquid solution forms a continuous film on the surface or remains as the liquid islands, the spreading coefficient of the liquid solution on the laser-textured surface is analyzed. However, the spreading of the liquid solution on the laser-textured surface depends on the spreading coefficient of the liquid on the surface. Consequently, the negative value of the spreading coefficient indicates the complete spreading of the liquid solution while forming the liquid film on the textured surface. On the other hand,

TABLE 6.13 ICP Data (ppm) for the Mud Solution After 8 h Dissolution Time of Dust Particles in Desalinated Water [3]

Ca	Na	Mg	K	Fe	Cl
315,000	48,000	67,000	28,000	2100	41,000

the spreading coefficient of the liquid solution is associated with the interfacial tension among the laser-textured surface, the liquid solution, and air. The spreading coefficient can be formulated as $S_{sl(a)} = \gamma_{la} - \gamma_{sa} - \gamma_{sl}$ [26], where γ_{la} is the surface tension of the liquid solution in air, γ_{sa} is the surface free energy of the laser-treated surface in air, and γ_{sl} is the interfacial tension between the liquid solution and the treated surface. The surface tension of the liquid solution is measured using the capillary tube method [37]. The surface tension of the liquid solution in air is measured in the order of 0.085 N/m, which is close to the surface tension of water (0.072 N/m). Therefore the interfacial tension between the laser-textured surface and a liquid solution is assumed to be of same order as the interfacial tension between the laser-textured surface and the desalinated water. Moreover, the interfacial tension between the laser-treated surface and the desalinated water can be determined from the modified Young's equation, which is $\gamma_{sl} = \gamma_{sv} - (1/r)\gamma_{lv}cos\theta_w$ and results in the interfacial tension in the order of 59.83 mJ/m². However, the free energy of the laser-treated surface measured from the contact angle method is 61.7 mJ/m², and the spreading coefficient of liquid solution ($S_{Liquid(a)}$) on the laser-treated surface becomes in the order of $S_{Liquid} = -$ 49.53 mJ/m². Consequently, the spreading rate $S_{Liquid(a)} < 0$ demonstrates that the liquid solution forms a film on the laser-treated surface. The dried liquid solution forms crystal structures at the surface. This situation can be seen from Fig. 6.22A and B, which shows SEM micrographs of dried liquid solution crystals. The crystals formed are in various shapes and sizes and EDS analysis reveals that it contains Ca, Si, K, Na, Mg, and Cl (Table 6.14). Consequently, the dissolved alkaline and alkaline earth metals in desalinated water are mainly responsible for the formation of the liquid solution. In order to investigate the effect of dried liquid solution on the morphology of the laser-textured and as-received surfaces, the dried liquid solution is removed by a waterjet. Fig. 6.22C and D show SEM micrographs of the laser-textured and as-received surfaces after the dried mud is removed by a waterjet. Although the dried liquid solution contains alkaline and alkaline earth metals together with chlorine, the liquid solution did not leave any corrosion damage site, such as pit sites, on the surface. This is attributed to one or all of the following: (1) the time required for the liquid solution film to dry out on the solid surface is short so the corrosion effect is not observed and (2) Inconel alloy is resistant to corrosion and no pit sites are formed at the surface. Nevertheless, the liquid solution formed by the water condensate and the dissolution of some dust particles compounds has a physical consequence such as increasing adhesion of the dust particles on the surface.

In order to assess the adhesion of the dried liquid solution on the laser-textured and as-received surfaces, scratch tests were carried out and the tangential force required to remove the dried liquid solution from the laser-textured and as-received surfaces are presented along the scratch length.

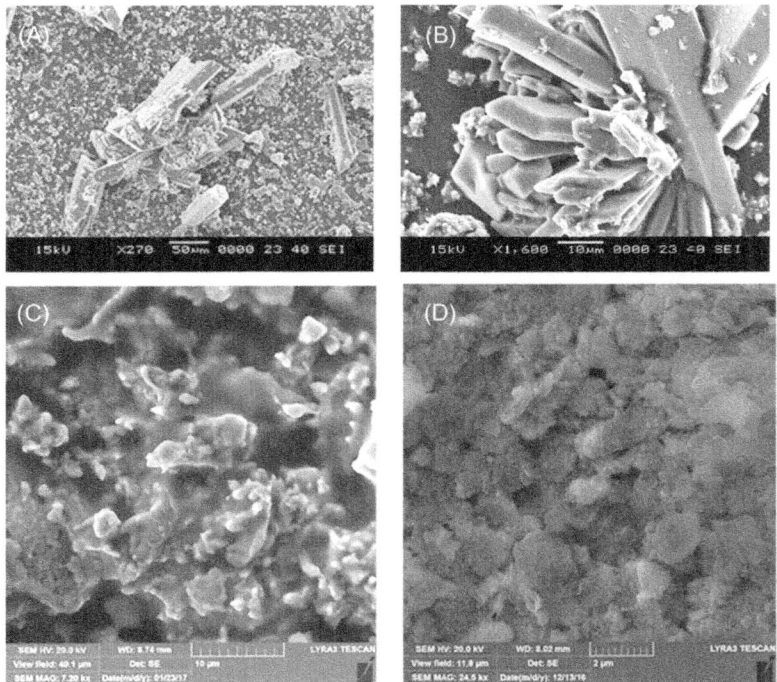

FIGURE 6.22 SEM micrographs of dried liquid solution: (A) crystals formed on laser-textured surface; (B) various size crystals due to drying of liquid solution; (C) laser-textured surface after dried liquid solution removed from surface; and (D) as-received surface after dried liquid solution removed from surface. The residue of dried mud solution is evident on as-received surface [3].

TABLE 6.14 Elemental Composition of Dust (wt.%) [3]

	Si	Ca	Na	S	Mg	K	Fe	O
Dust particles	12.1	8.3	2.5	2.1	1.9	1.1	1.2	Balance
Dried mud solution	–	7.5	2.1	–	1.5	0.9	–	Balance

Fig. 6.23 shows the tangential force obtained from the microtribometer for laser-textured and as-received surfaces. In addition, the friction force is also shown for as-received and laser-textured surfaces without the presence of the dried liquid solution for comparison. It should be noted that the tangential force corresponds to the force required to remove the dried solution from the surface. The tangential force attains considerably larger values than the frictional force. This shows the strong adhesion between the surface and the dried solution. In the case of the laser-textured surface, the tangential force

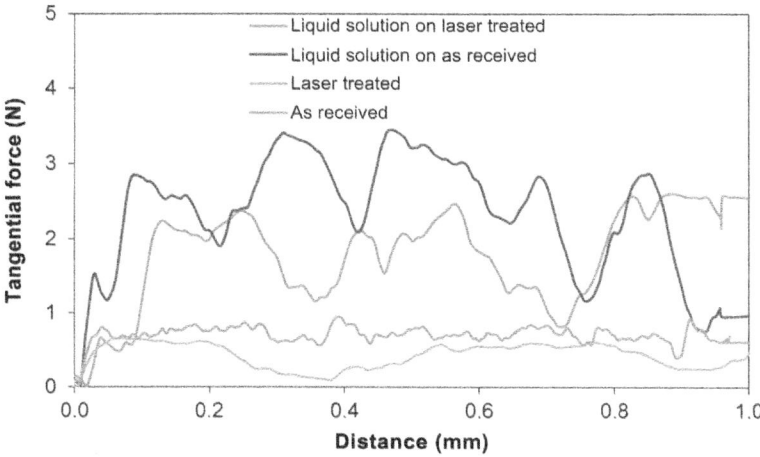

FIGURE 6.23 Tangential force along the distance on as-received and laser-textured surfaces with and without presence of dried liquid solution [3].

remains higher for the as-received surface than that of the laser-textured surface. Although the liquid solution forms a film on both surfaces prior to drying, the air gaps in the textured surface results in uncovered local regions by the dried liquid solution. This contributes significantly to the tangential force reduction for the laser-textured surface. In addition, the formation of nitride compounds at the surface lowers the surface energy of the laser-textured surface. Therefore the low interfacial resistance, which gives rise to low adhesion of the dried liquid solution, results in reduced tangential force for the laser-textured surface. Consequently, when the liquid solution is formed in-between the solid surface and the dust particles, the adhesion of the dust particles remains strong once the liquid solution dries. The effort required to remove the dust particles on the dried liquid solution becomes significant. In addition, laser-textured surface results in smaller frictional force as compared to as-received sample.

6.5 DRY MUD EFFECT OF TITANIUM NITRIDE–COATED SURFACE

Titanium nitride (TiN) coating has wide applications in industry because of its superior resistance toward corrosion and wear. TiN-coated surfaces are also proposed as a solar-selective absorber for high-temperature air-stable solar receivers [38]. High-temperature air-stable solar-selective absorbers, in general, operate in air ambient with temperatures over 400°C to provide cost-effective energy harvesting and efficiency improvement in solar-thermal power plants and solar-cooling applications [39]. Research on utilization of air-stable solar-thermal selective surfaces is in progress, particularly for high-temperature applications [40]. Several research studies have been carried out to examine the surface characteristics of TiN coating in relation to solar-energy applications. The optical properties of TiN-based spectrally selective solar absorbers were studied by Gao et al. [41]. They showed that the SS/TiN/Al_2O_3 coating exhibited good thermal stability in a vacuum at 500°C for 5 hours with enhanced absorptance and reduced emittance. The characterization and performance evaluation of Ti/AlTiN/AlTiON/AlTiO high-temperature spectrally selective coatings for solar-thermal power applications was investigated by Barshilia [42]. He demonstrated that the absorber coating displayed improved adhesion, UV stability, corrosion resistance, and thermal stability in air and vacuum with high absorptance and low emittance. A review on physical vapor deposited (PVD) spectrally selective coatings for mid- and high-temperature solar-thermal applications was presented by Selvakumar and Barshilia [43]. They indicated that solar-selective coatings based on transition metal nitrides, oxides, and oxynitrides hold great potential for high-temperature applications because of their excellent mechanical and optical properties, which were yet to be commercialized. Investigation of optical properties of solar absorber based on cermet of TiN in SiO_2 deposited on lanthanum aluminate was studied by Cao et al. [44]. The findings revealed that the optimized cermet contains TiN with a volume fraction of 60% and 65% demonstrated an absorptance higher than those of approximately 95% before annealing and approximately 94% after annealing at 700°C, which appeared to be useful for especially concentrated solar applications. A refractory selective solar-absorber incorporation TiN thin coating was examined by Jiang et al. [45] for a high-performance thermochemical steam-reforming process. They showed that the selective surface of TiN coating resulted in superior performance for steam reforming and hydrogen production rate at high surface temperatures. A study of porous TiN microspheres in relation to dye-sensitized solar cells was carried out by Wang and Liu [46]. They indicated that the dye-sensitized solar cell with the unique structure resulted in a high conversion efficiency of 6.8%, which was favorably comparable to the cell based on a conventional platinum counter electrode. Solar-selective absorbing coatings TiN/TiSiN/SiN prepared on stainless steel substrates were investigated by Feng et al. [47]. They demonstrated that the solar-selective absorbing performance of the Cu/TiN/TiSiN/SiN did not show significant changes after they were heat-treated up to 700°C in vacuum. An optimization study of TiAlN/TiAlON/Si_3N_4 coatings for solar-absorber applications was carried out by An et al. [48]. They analyzed the reflectance spectrum of the coatings and indicated that the optimized coating resulted in a solar absorptance and thermal emittance of 94.6% and 5.2% at 400 K, respectively. A new TiN coating combining broadband visible transparency was introduced by Smith et al. [38]. They showed that depositing thin films resulted in surface reflection in the near infrared region. This, in turn, made it possible to produce films that transmit daylight neutrally at reasonably high levels, while still maintaining low emittance and solar control in the region of the near infrared radiation. A new structure of durable metallic thin film contacts for solar cells was investigated by Matenoglou et al. [49]. They used stoichiometric TiN with a thin nonstoichiometric titanium nitride (TiNx) buffer layer giving the desirable mechanical properties and metallic behavior. They showed that the metallic contacts remained cohered during the scratch test and coating strongly adhered to the substrate surface.

The environmental dust and mud effect on selective surfaces for solar-energy harvesting was left for future study. In solar-energy harvesting system, surfaces are exposed to outdoor conditions and suffer from dust accumulation and mud formation in harsh environments with dusty and humid ambient air. Since the environmental dust particles possess various elements, their effects on the selective surface properties are critically important for the sustainability of solar-harvesting system performance. Consequently, investigation of the dust and mud effects on selective surfaces is essential. In this section, environmental dust settlement and mud formation on TiN coating are presented in line with the previous study [4]. The dust particles and dry mud formed on TiN coating are characterized. The influence of mud formed on TiN surfaces, due to water condensate on the dust particles, is examined using the analytical tools including electron scanning microscope and AFM, FTIR, XRD, and EDS. The tangential force required to remove the dry mud from the surface of TiN coating is measured incorporating the microtribometer. Surface energy and hydrophobic characteristics prior and after the dry mud removal were determined using the goniometer and water-droplet technique.

6.5.1 Laser-Treated Surface and Mud Characteristics

TiN coating was deposited onto 3 mm thick Ti6Al4V alloy surface using the PVD coating unit as described in the previous study [50]. The PVD coating results in a uniform thickness of TiN coating, which can be seen from Fig. 6.24A and B, in which SEM micrographs of a cross section of the coating are shown. The coating surface was examined thoroughly using the optical microscope and SEM. The findings revealed the coating surface was free from coating defects such as pinholes, voids, and microcracks.

In order to assess the amount of dissolved alkaline (Na, K) and alkaline earth (Ca) metals in water condensate, dust particles were mixed with the desalinated water mimicking the water condensation on the dust particles in humid air conditions. The liquid solution from the mixture of desalinated water and the dust particles was extracted. Inductively coupled plasma (ICP) data carried out in the liquid solution showed that Ca, Mg, Na, and K are

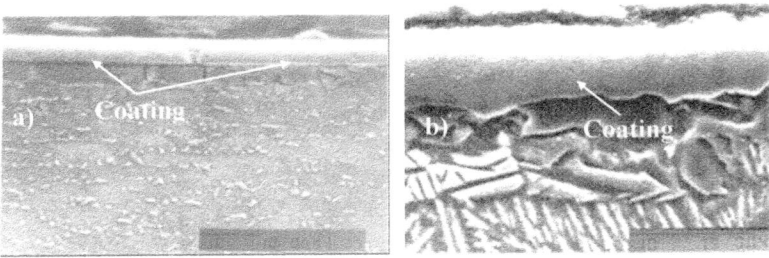

FIGURE 6.24 SEM micrographs of cross section of TiN-coated and laser-treated workpieces: (A) TiN-coated Ti6Al4V alloy and (B) laser-treated and TiN-coated Ti6Al4V alloy study [4].

FIGURE 6.25 SEM micrographs of crystals formed during drying of liquids solution: (A) crystal formed on TiN coating. Since spreading coefficient allows island of liquid solution on the TiN-coating surface, locally scattered crystals are formed at the surface, and (B) two crystals with different sizes on TiN-coating surface study [4].

present in the extracted liquid solution. The alkalinity of the liquid solution was also measured via pH meter. The results show that the liquid solution is basic with pH = 8.6. Consequently, alkaline and alkaline earth metal compounds dissolve in water condensate and form a liquid with a basic (not acidic) characteristic, which then flows toward the TiN-coating surface under the gravitational force. To examine the effect of liquid solution on the coating surface, the liquid solution is deposited onto the coating surface and left to dry for 6 hours. Fig. 6.24 shows SEM micrographs of the dried liquid solution on TiN-coating surface. The crystals with varying sizes are observed from the micrographs (Fig. 6.25A and B). The variation of crystal sizes is associated with the concentration of dissolved compounds and temperature variation at the surface. The dissolved alkaline and alkaline earth metal compounds are responsible for the formation of the crystal structures on the surface. Consequently, these structures increase the adhesion of the dust particles to the coating surface when the liquid solution dries out in-between the dust particles and the coating surface.

Fig. 6.26 shows FTIR data for dried mud solution on the surface of the TiN coating. The strong peak at $670 \, \text{cm}^{-1}$ corresponds to Ti−N bonding. Because of the residues of mud in the mud solution, silica particles appear on the dried liquid solution (Fig. 6.25A and B). In this case, the peaks at 777 and $1077 \, \text{cm}^{-1}$ correspond to the Si−O bonds, which is associated with stretching of SiO_2 [23]. The peaks at 867 and $1431 \, \text{cm}^{-1}$ are attributed to the stretching vibrations of CO_3^{-2} [24]. These peaks are attributed to $CaCO_3$ residues in the dried liquid solution on the TiN-coating surface.

To resemble the mud formation on the TiN-coating surface, a film of dust particles of 300 μm was formed and desalinated water dispensed onto the dust particles with an amount obtained to mimic the water condensate onto the dust particles in ambient humid air. The mixture of dust particles and desalinated water was left to dry over 6 hours. Fig. 6.27A and B shows SEM micrographs of the dry mud surface. In general, the dry mud surface is composed of dust particles and condensed cement-like structures formed in-between the particles. In addition, some small void-like textures are also observed on the surface. The void-like textures are formed by the dissolved dust compounds. These compounds form a liquid solution, which flows through the dust particles toward the interface between the dust particles and the coating surface while leaving the void-like texture on the dry mud surface.

On the other hand, some of the liquid solution is captured in the cavity-like structures in the mud cross section and upon drying it appears like a

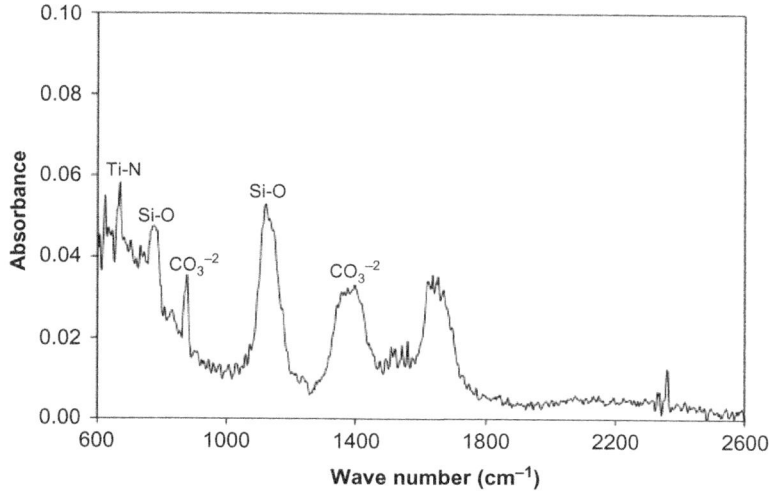

FIGURE 6.26 FTIR data for dried mud solution on the coating surface study [4].

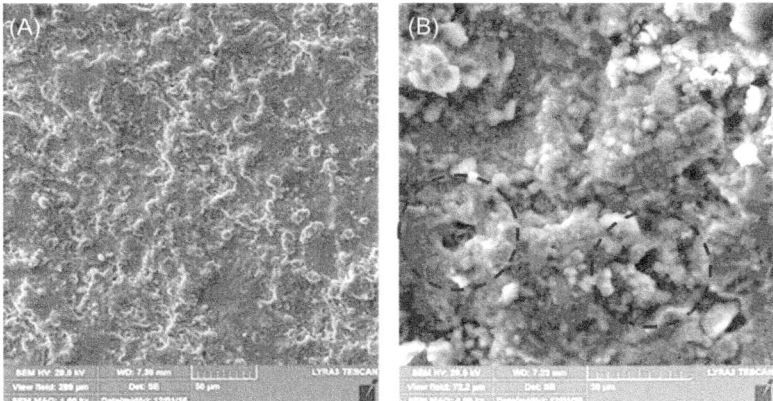

FIGURE 6.27   SEM micrographs of dry mud surface: (A) dust particles and cement-like texture at the surface and (B) porous-like structures, marked by dotted circles, on the dry mud surface because of mud liquid flowing toward TiN-coating surface under gravitational force study [4].

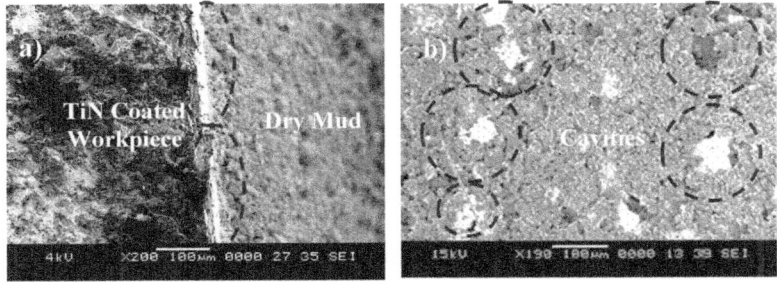

FIGURE 6.28   SEM micrographs of dry mud cross section: (A) island of dried liquid solution at the interface of the dry mud and TiN-coating surface, marked by dotted circles and (B) cavities across the dry mud cross section, which are marked in dotted circle. The dried liquid solution, appearing as a bright color, is captured in some cavities [4].

bright region across the dry mud layer (Fig. 6.28A). The liquid solution spreads on the coating surface while forming a film at the interface between the coating surface and the mud. However, the continuity and the uniformity of the liquid film depend on the spreading coefficient over the coating surface. In general, the spreading coefficient of liquid depends on the interfacial tension between TiN-coating liquid and air according to $S_{sl(a)} = \gamma_{la} - \gamma_{sa} - \gamma_{sl}$, where γ_{la} is the surface tension of liquid solution in air; and γ_{sa} is the surface free energy of TiN coating in air; and γ_{sl} is the interfacial tension between liquid solution and TiN coating. The surface energy of TiN coating is determined by using the liquid contact angle measurements method [28]. In this case, water, glycerol, and ethylene glycol are used to measure the droplet contact angle of each fluid.

6.5.2 Assessment of Surface Free Energy and Mud Adhesion

The surface energy of the TiN-coated surface is obtained, which is given in Table 6.15. In addition, the surface tension of the liquid solution is measured using the capillary tube method [37]. The measurement reveals that the surface tension of the liquid solution is in the order of 0.085 N/m. The interfacial tension between the TiN coating and a liquid solution is taken to be the same as the interfacial tension between the TiN coating and water, which is in the order of 80.41 mJ/m^2 (Table 6.15). It should be noted that the liquid solution has a similar surface tension as water; therefore the consideration for interfacial tension between the liquid solution and TiN-coating surface is justified. Since the surface energy of TiN coating is determined to be in the order of 120 mJ/m^2 (Table 6.15) and the surface tension of liquid solution is 0.085 mJ/m^2, the spreading coefficient of liquid solution on the TiN surface ($S_{sl(a)}$) becomes in the order of -200 mJ/m^2, which is $S_{sl(a)} < 0$. Therefore the liquid solution partially wets the surface rather forming a liquid film between the dust particles and the TiN-coating surface.

The partial wetting of the liquid solution can be observed from the SEM micrographs of dry mud cross section (Fig. 6.28B). In this case, the bright region at the interface of the dry mud and TiN-coating surface represents the dried liquid solution. However, the dried liquid solution does not completely cover the entire interface, but it is present locally. The local appearance of the dried mud solution demonstrates the nonwetting characteristics of the liquid solution at the interface prior to drying. In order to assess the aftereffects of the dry mud on the surface of TiN coating, the dry mud is removed with a desalinated waterjet of 2 mm diameter and 2 m/s jet velocity.

Fig. 6.29A and B shows SEM micrographs of the dry mud-removed surface. A cluster of dry mud residue is evident on the coating surface. The close view of the dry mud residue demonstrates that it was formed from the small particles, which are strongly attached together like bonded structures. This can be attributed to the bonding of small dust particles with the liquid solution prior to drying on the coating surface. These structures strongly attach to the coating surface after drying and become difficult to remove from the surface via waterjet. The elemental composition of the dry mud residue on the coating surface is given in Table 6.16. The presence of Ca, Si, Na, Mg, Cl, and O indicates that the dry mud residue is most likely

TABLE 6.15 Surface Energy and Lifshitz–van der Walls Components and Electron-Donor Parameters Determined for TiN-Coating Surface Study [4]

γ_S (mJ/m^2)	γ_S^+ (mJ/m^2)	γ_S^- (mJ/m^2)	γ^p (mJ/m^2)	γ_S^l (mJ/m^2)
120	35.93	10.93	39.63	80.41

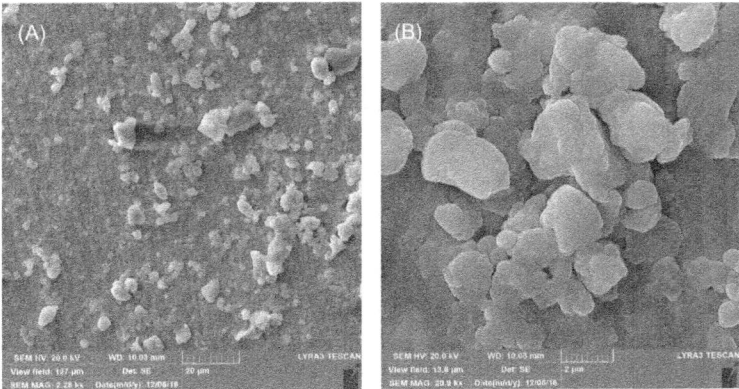

FIGURE 6.29 SEM micrographs of dry mud-removed surface: (A) mud residue on TiN-coating surface and (B) clustered small mud residue on TiN-coating surface study [4].

TABLE 6.16 Elemental Composition of the Dust Particles (wt.%) [4]

	Si	Ca	Na	S	Mg	K	Fe	Cl	O
Dust particle size > 2 μm	11.4	7.6	2.8	1.8	2.2	1.2	1.1	0.9	Balance
Dust particle size < 2 μm	11.5	7.1	5.1	1.1	2.9	2.1	0.9	2.1	Balance
Mud residues	10.5	6.3	3.4	–	2.1	2.1	–	1.8	Balance

composed of $CaCO_3$, NaCl, and MgO. Moreover, Fig. 6.30 shows the AFM images of the coating surface after removal of the dry mud by a waterjet. The dry mud residue forms a texture on the coating surface (Fig. 6.30A) and the residue of the small dust particles attaches to the coating surface. The texture height is observed to be varies on the surface (Fig. 6.30B), which results in the average surface roughness, due to presence of the dry mud residue, in the order of 1.15 μm.

In order to assess the dry mud adhesion on the TiN-coating surface, the tangential force required to remove the dry mud from the surface was measured using the microtribometer. Fig. 6.31 shows the tangential force along the scanning (scratch) distance on the TiN coating. The friction force on the plain TiN coating is also shown in Fig. 6.31. The tangential force required to remove the dry mud from the coating surface is considerably larger than that of the frictional force on the as-received coating surface. This behavior is related to the adhesion of the dry mud on the coating surface. In addition, the friction force on the coating surface also contributes to the tangential force. Some small peaks are observed along the tangential force, which is

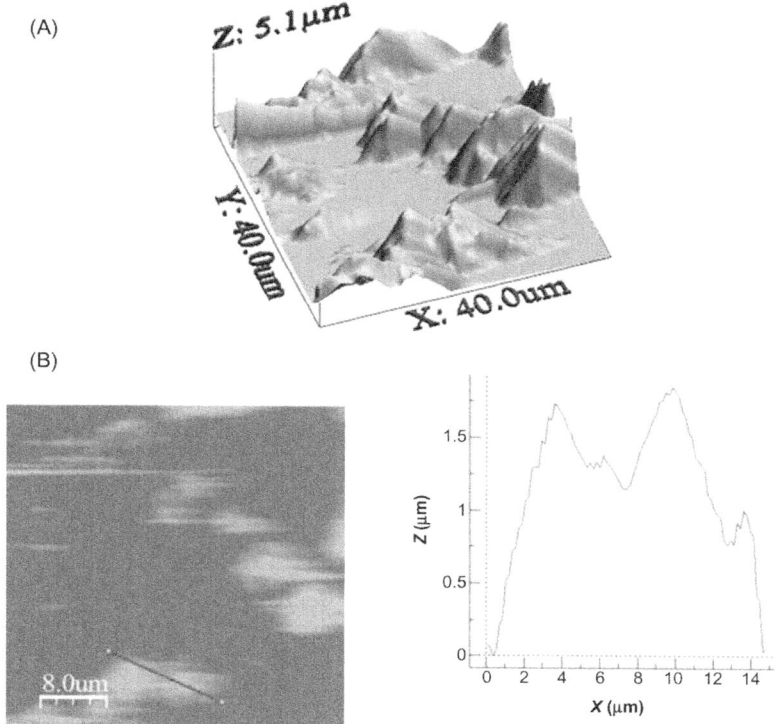

FIGURE 6.30   AFM images of the dry mud-removed surface: (A) 3D micrograph of dry mud-removed surface and (B) line scan of mud-removed surface study [4].

associated with the strong adhesion of the dust particles on the coating surface. Since the liquid solution partially wets the surface, upon drying small islands of dried solution film occurs at the interface between the dust particles and the coating surface. These islands of dried liquid solution at the interface are responsible increasing the tangential force. Moreover, the ratio of the maximum tangential force to the average value of the tangential force is in the order of 1.5, which demonstrates that the adhesion increases almost 30% more as the liquid solution forms locally dried regions at an interface between the dust particles and the coating surface. In order to assess the tangential force required to remove the dry dust particles from the surface, the tangential force measured is repeated for the removal of the dust particles from the coating surface. The tangential force findings for the dust particles are also shown in Fig. 6.31 for comparison. It is evident that the average tangential force required to remove the dry dust particles is almost 0.3 of the tangential force required to remove the dry mud from the coating surface. On the other hand, the tangential work for removing the dry mud from the surface can be determined through the integration of tangential force over

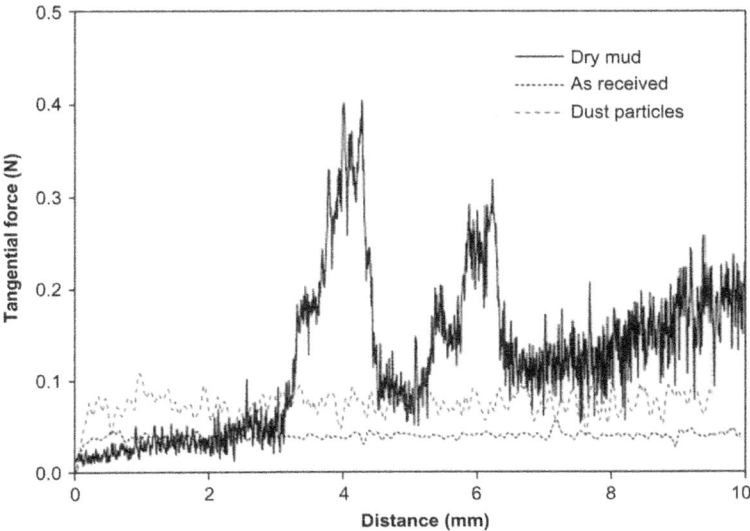

FIGURE 6.31 Tangential force measured to remove the dry mud and dust particles from TiN-coating surface. Frictional force on the TiN-coating surface is also included for comparison [4].

TABLE 6.17 Adhesion Work Obtained From the Tangential Force Study [4]

TiN Coating	Adhesion Work (mJ)
As-received	0.119
Dust particles	0.387
Dry mud	1.244

the scanning length in Fig. 6.31. The friction work can also be calculated in a similar manner. Therefore subtraction of friction work from the tangential work gives the adhesion work. The adhesion work was determined from the integration and subtraction of the tangential and the friction forces. The adhesion work for removing the dry mud when liquid solution totally wets the coating surface while forming a thin film at the interface is almost twice the adhesion work required to remove dry mud from the surface. In addition, the adhesion work required to remove the dry dust particles is almost 0.3 of the adhesion work of the dry mud removal from the coating surface. Table 6.17 outlines the adhesion work required for the removal of the dry dust particles, the dry mud, and the dry mud with the dried liquid solution at the interface. Consequently, the dust particles removal from TiN-coating surface in the dry air ambient is favorable in terms of the efforts required for surface cleaning.

6.6 MUD EFFECTS ON ALUMINUM SURFACES

Aluminum and aluminum alloys are being used extensively in various solar-energy harvesting systems such as concentrated solar-thermal and PV applications. Aluminum has unique characteristics such as high-strength, lightweight, high surface reflectivity, and good thermal and electrical conductivities. These properties enable aluminum to become an inseparable component of the solar-energy harvesting systems. In general, aluminum troughs or dishes are used in concentrated solar-power systems because of high reflectance and low absorption of aluminum in solar-wavelength range. However, environmental dust accumulation alters the optical characteristics of aluminum troughs or dishes while lowering the solar-concentrated power in the energy harvesting system. In humid air ambient, condensation of water on to the dust particles results in dissolution of alkaline and earth alkaline compounds of the dust while forming a chemically active liquid with high pH on the through or dish surfaces [1]. Since aluminum surface is susceptive to high pH liquids, formation of hydroxyl groups further modifies the optical characteristics of aluminum surface. Exposure of surfaces to high-temperature environments, such as high-temperature harsh environments, the mud solution formed in the humid ambient can dry out and this can form an interfacial layer between aluminum surface and the dust particles. This enhances the adhesion between the dust particles and the aluminum surface while increasing the efforts required for the dust removal from the surface. Thus investigation of the adhesion of the dry mud solution on the aluminum surface at different mud solution drying temperatures is essential.

Dust accumulated of solar concentrators and solar collectors ranked with climate conditions (humidity, temperature) and irradiance (spectrum, intensity, and uniformity) as the major concerns for system components and reliability [51]. Most solar-concentration applications have been established in deserts due to the availability of direct light beams. Deserts also have hot environments with high dust accumulation rates, and small dust deposition increases reflection losses, which significantly affect solar-system performance [52]. Tsuyoshi et al. [53] reported that concentrating solar systems use direct sunlight beams only and some of the light is scattered by soiled collector surfaces. Moreover, PV flat panels are less sensitive to soiling compared to concentrating systems due to the use of both direct and indirect beams. In addition to dust foiling, solar concentrators are operated at relatively high temperatures compared to flat plate collectors due to the smaller area required for heat loss [54]. Therefore in a humid environment, water vapor absorbed by dust particles forms mud on the aluminum surface. The formed mud dries and adheres to the aluminum surface due to high temperature and becomes difficult to remove [55]. Thus solar-energy harvesting surfaces require frequent cleaning in order to maintain the required efficiency. Guan et al. [56] installed a dust monitor to collect weather and dust data for a solar-thermal power plant in Australia. The results revealed that the average size of the dust particles was 20 μg, which reduced the solar mirror reflectivity from 92% to 70% in 1 month.

The influence of the dry mud solution on the surface characteristics of aluminum at different environmental temperatures is critical to understand for practical applications of solar troughs. Dust and dry mud adhesion on energy harvesting device surfaces is also critical to understand in terms of device performance and the efforts required to clean them. Consequently, in the present section, adhesion of dry mud solution on aluminum surface is investigated and the effect of dry mud solution on the surface characteristics of aluminum sheets is presented in light of the previous study [5].

6.6.1 Experimental Analysis

Mirror finished aluminum sheet with 1 mm thickness was used in the experiments. The aluminum surface was ultrasonically cleaned prior to mud solution formation from environmental dust and water mimicking water condensation on dust particles. It was reported that the chemical composition of dust particles does not change among different locations in Saudi Arabia but the deposition rate varies with seasons [57]. The dust particles were collected from the local environment in the Dhahran area of Saudi Arabia. The dust layer thickness on the surfaces in open environment was measured after the dust storms and the dust layer thickness was estimated to be 300 μm. Therefore a dust layer with 300 μm thickness was formed on the aluminum surfaces in the laboratory environment to resemble the real dust accumulation in the open environment. In addition, the amount of water condensate on the dust particles was measured after 6 hours of exposure in the open environment of a humid day. The measurement of condensate revealed that the weight ratio of dust particles to water condensate was in the order of 0.2. Hence, this ratio was adopted in the laboratory environment mixing the dust particles with desalinated water to form the mud liquid. It should be noted that the mixture of the dust particles and water formed a mud and gradually the mud liquid separated from the mud while accumulating at the bottom of the container under the gravity. The mud liquid was then extracted and the pH of the solution measured. It was observed that the pH of the mud solution reached 7.6, which was attributed to dissolution of alkaline and alkaline earth compounds of the dust particles within water during the mixing. The mud solution was deposited at the aluminum workpiece surfaces. Since the temperature of the environment changes from 30°C to 70°C during a local humid day, aluminum workpieces with the mud solution were dried in a furnace at different temperatures within the range of 30°C−70°C to mimic actual drying conditions.

Once the mud solution was dried under different temperatures, the aluminum surfaces with the dried mud solution were used to determine the adhesion work required to remove the dried mud solution from the aluminum surface. In this case, the scratch resistance, friction coefficient, and tangential force for the adhesion work were determined from the data obtained from a linear microscratch tester (MCTX-S/N: 01-04300 type Rockwel 1−176). During the experiments, the equipment tip (made of diamond with $100 \mu m$ radios) was set at a contact load of 0.03 N and an end load of 2 N. The total length for the scratch groove was 10 mm, and the scanning speed was maintained constant at 6 mm/min with a loading rate of 5 N/s. The scratch hardness (Hs) according to the ASTM D7027-05 standard can be given as the following expression (ASTM D7027-13, 20113): $Hs = (4qP/\pi w^2)$, where P is the normal constant load used during the scratch tests in Newton, w is the scratch groove width in millimeters, and q is the dimensionless parameter, which depends on the elastic recovery extend of the material during scratching test. In the case of no recovery $q = 2$ and for full elastic recovery $q = 1$. The aluminum workpiece surfaces were cleaned from the dried mud solution by using a desalinated waterjet of 2 m/s velocity. The cleaning process of each workpiece surface continued for 15 minutes, and process provided information on the friction characteristics of the aluminum surfaces after dried mud removal.

6.6.2 Dry Mud Characterization on Aluminum Surface

The dry mud solution contained some alkali and alkali earth metal compounds as given in Table 6.18. These compounds were dissolved in water, forming a chemically active mud solution, which segregates from the mud and accumulates at the mud bottom due to gravity. The pH increased to reach 7.6. This behavior is attributed to the dissolution of the alkaline and alkaline earth metal compounds in water and is mainly due to the presence of −OH ions from the dissolution of alkaline earth metal compounds like Ca and alkaline (Na, K) in the mud solution consistent with the previous study [58]. The increase of pH of the mud liquid in a short time demonstrates that it can potentially alter the surface characteristics of aluminum sheets. In order to assess the effect of the mud drying temperature on the elemental

TABLE 6.18 Elemental Composition of the Dried Mud Solution at Different Drying Temperatures (wt.%) [58]

Temperature (°C)	Na	Mg	K	Cl	Ca	O
70	15.3	6.3	3.9	15.2	15.8	Balance
50	9.9	5.1	4.0	12.1	16.5	Balance
40	9.4	7.8	2.9	13.9	15.9	Balance

composition of the dry mud, EDS data for the dry mud solution are provided for different mud solution drying temperatures and given in Table 6.18. Some small changes of elemental distribution are observed in the dried mud solution. However, this variation is attributed to the composition of the mud liquid prior to drying. Nevertheless, the variation of elemental composition is small. However, in the dried mud solution, no Si is observed. This is related to the undissolved silicon compounds in the desalinated water. Moreover, ICP mass spectrometry was carried out to determine presence of the elements in the mud solution prior to the drying. The data obtained from the ICP are reported in Table 6.19. It is evident from the ICP data that the elements observed in the EDS data are also present in the ICP data. Therefore the dissolution of alkaline and alkaline earth metal compounds in the dust particles modifies the elemental composition of condensate water.

On the other hand, partial oxidation of an aluminum surface results in undissociated hydroxyl group formation on the surface under the aqueous mud solution. The pH of the mud solution (pH = 7.6) gives rise to a positive charge at the surface, that is, $-OH_{surf} + OH^- \rightleftarrows -C_{surf}^- + H_2O$ [59]. In this case, the liquid mud solution reacts with the aluminum surface to form hydroxyl groups due to increasing of OH^- during the surface exposure to the liquid mud solution. Consequently, locally distributed hydroxylated aluminum oxide (AlOOH) regions are formed at the surface [60]. This can also be seen from XRD of aluminum surface after dry mud removal (Fig. 6.32). The AlOOH regions have a protruding appearance at the surface, which can be seen from SEM micrographs of the surface as shown in Fig. 6.33.

Fig. 6.34 shows FTIR spectroscopy data for the aluminum surface after the dry mud removal. Hydroxylated aluminum on the dry mud-removed surface is evident from the data. The band at 1091 cm^{-1} is related to Al-OH hydroxyl bending mode in aluminum oxy-hydroxides [61,62]. The large band seen within $3260-3680 \text{ cm}^{-1}$ is attributed to the stretching vibration of OH of the hydroxyl groups. The intensity peaks at 3372 and 3112 cm^{-1} correspond to the (Al)O-H stretching vibrations [63]. The band at the shoulder 1173 cm^{-1} is associated with the Al-O-H mode of γ-AlOOH. The intensity peak at 743 and 607 cm^{-1} is related to the vibration mode of Al-O. In addition, the absorption peaks 711 and 875 cm^{-1} correspond to the in-plane bending and out-of-plane bending vibration of carbonate ions, respectively.

TABLE 6.19 ICP Data (ppb) for the Mud Solution After the Dust Particles Were Dissoluted in Desalinated Water for 6 h [58]

Ca	Na	Mg	K	Fe	Cl
309,800	44,600	69,950	33,400	1830	37,600

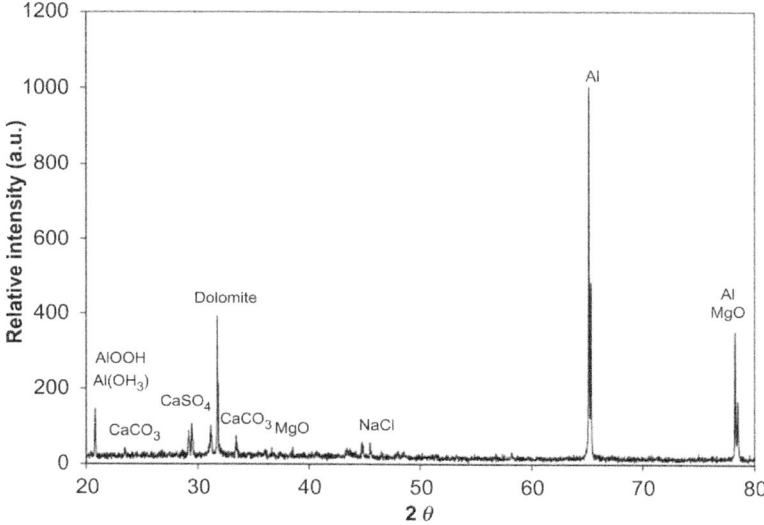

FIGURE 6.32 XRD of dry mud solution-removed surface of aluminum by a waterjet [58].

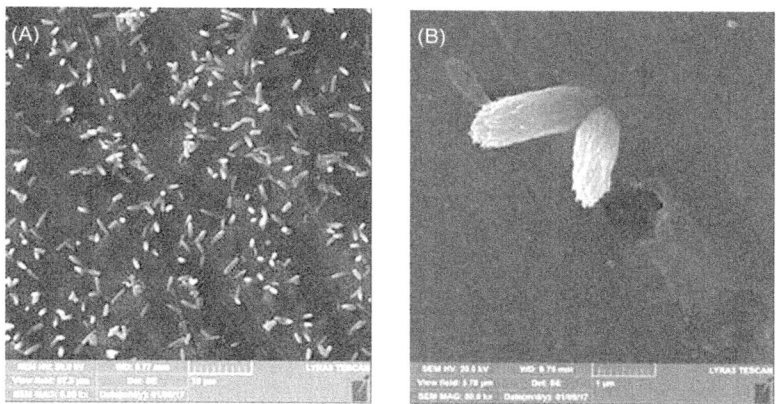

FIGURE 6.33 SEM micrographs of mud solution-removed aluminum surface: (A) locally distributed aluminum oxidehydride regions and (B) close view of protruded oxidehydride [58].

6.6.3 Mechanical Properties of Dry Mud Solution on Aluminum

Scratch tests for the removal of dry mud solution were carried out and adhesion force obtained from the tangential force measurements during the scratch tests. It should be noted that the difference between the tangential force from the removal of the dry mud solution and the friction force due to the as-received aluminum surface is considered as the adhesion force for the dry mud solution on the aluminum surface [2]. The adhesion force,

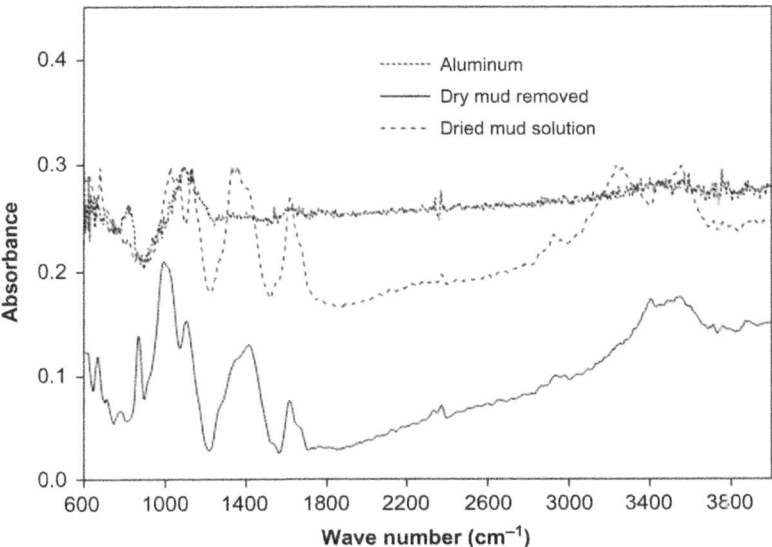

FIGURE 6.34 FTIR data for aluminum, dry mud solution, and dry mud solution-removed surface by a waterjet [58].

in-between the dried mud solution and aluminum surface, increases with increasing mud drying temperature. The explanation for the adhesion enhancement phenomena is related to the evaporation rate of water from the mud solution during the drying process. The water evaporation at high temperature causes the high rate of volume shrinkage of the mud solution, which in turns increases the formation of stress fields in dry mud solution resulting in adhesion amelioration. Crystals are formed after drying of the mud solution, as shown in the SEM images in Fig. 6.35. The formation of crystals at different sizes is evident from the SEM micrographs. Moreover, the high rate of drying of the mud solution on the aluminum surface, which occurs at high drying temperatures, leads to small crystals at the surface. Since the adhesion force increases with the drying mud liquid temperature, the small crystals increase the ionic bonding at the interface between the dry mud solution and the aluminum surface.

The total adhesion work is represented by integration of the tangential force along a specified distance (10 mm). The adhesion work data reveal that the adhesion work is in the order of 0.54 mJ (N.mm) at 30°C drying temperature, and it increases to reach maximum adhesion work of 1.3 mJ at 70°C drying temperature. The general trend of adhesion work variation with temperature is shown in Fig. 6.36. The increased of the adhesion work resulted in dried mud residue along the scratch mark on the aluminum surface. This situation can be seen from Fig. 6.37, which shows the scratch marks. For further assessment of the residue along the scratch marks, SEM was carried out

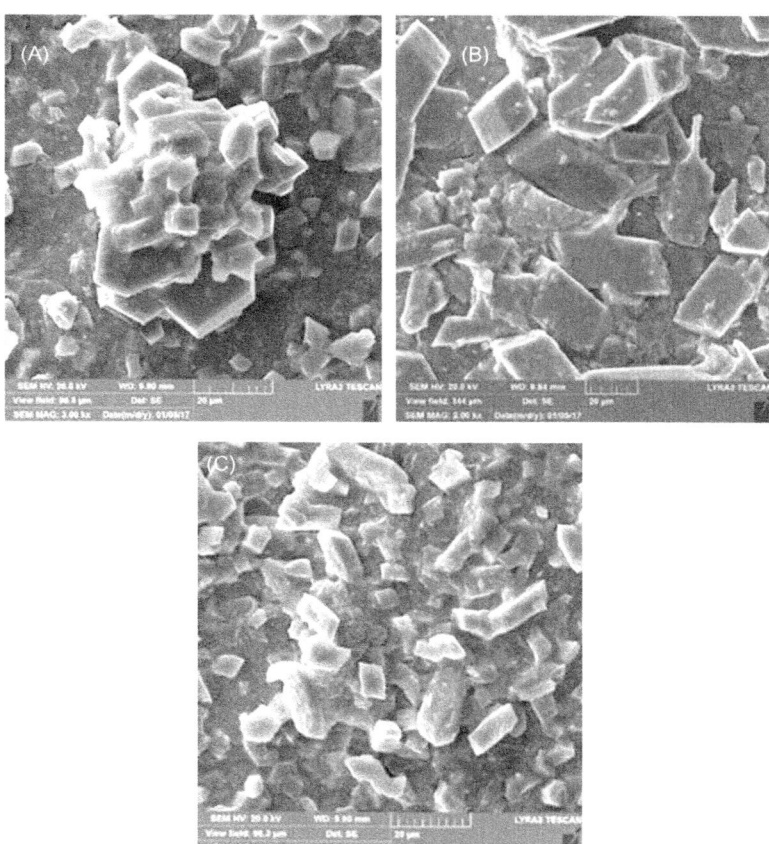

FIGURE 6.35   Dry mud solution crystals formed at the aluminum surface at different drying temperatures: (A) 30°C mud solution drying temperature; (B) 50°C mud solution drying temperature; and (C) 70°C mud solution drying temperature [58].

after scratch tests. The SEM micrographs are shown in Fig. 6.38 together with the EDS line scan. The presence of Al in the EDS data (Table 6.20) revealed that the dry mud solution is removed from the surface, provided that some mud solutions are present along the scratch length, as can be observed in Fig. 6.38.

The friction coefficient of the dry mud solution surface is shown in Fig. 6.39. The dry mud solution surface results in lower friction coefficient compared to as-received aluminum surface regardless of the drying temperature. Therefore the dry mud solution alters the surface characteristics by reducing the friction coefficient. This behavior is attributed to the hardness increase at the dry mud solution with temperature (Table 6.21). However, the friction coefficient further reduces with increasing drying mud temperature.

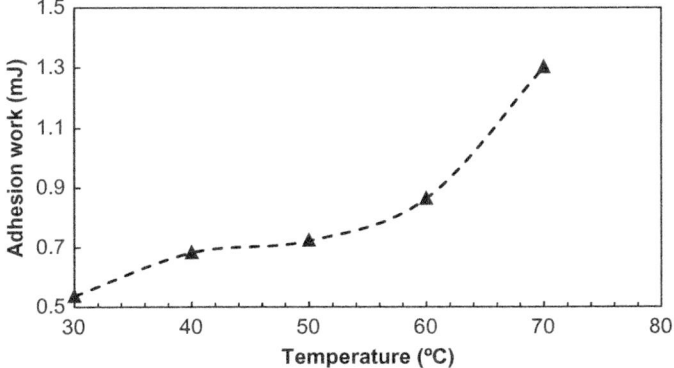

FIGURE 6.36 The adhesion work variation with mud solution drying temperatures [58].

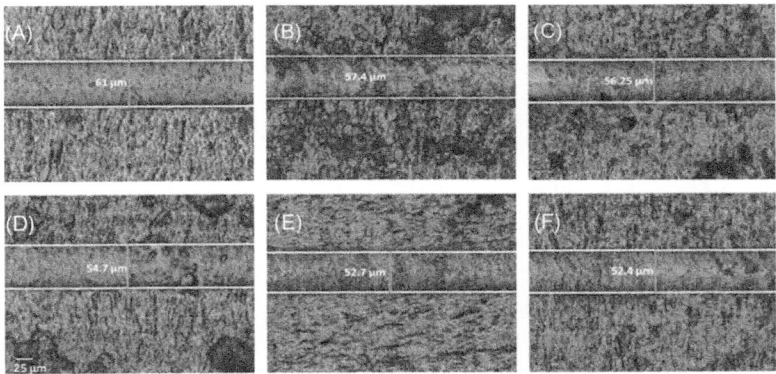

FIGURE 6.37 Optical images of scratch marks for different mud solution drying temperatures: (A) as-received, (B) 30°C, (C) 40°C, (D) 50°C, (E) 60°C, and (F) 70°C drying temperatures. Scratching width is in μm [58].

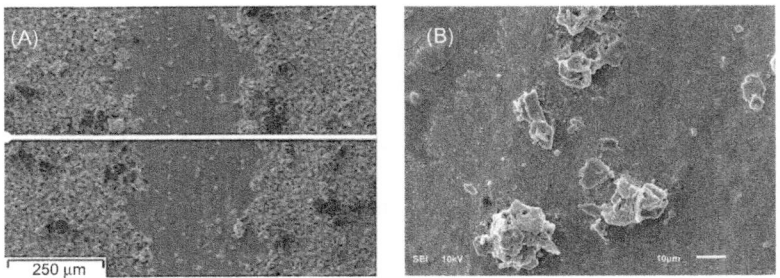

FIGURE 6.38 SEM micrographs of scratch mark and line scan for EDS data analysis: (A) scratch groove line scan for EDS data and (B) residue of dry mud solution after scratch test [58].

TABLE 6.20 EDS Data Along the Line Scan SEM Image (Fig. 6.38) [58]

wt.%	Na	K	Al	Cl	Ca	O
	8.1	5.05	54	12	8.9	Balance

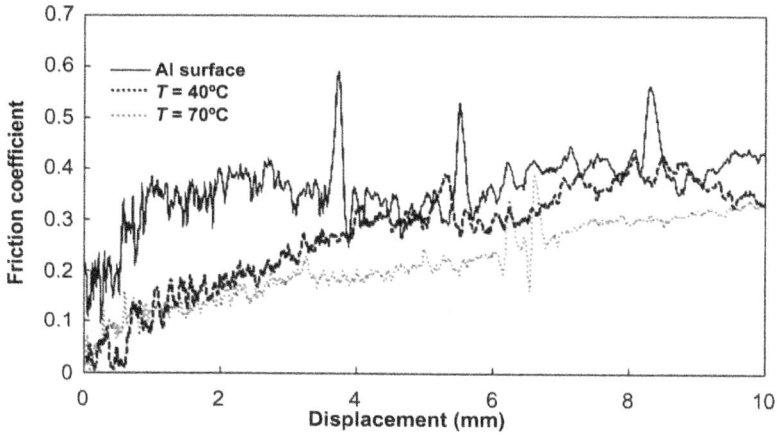

FIGURE 6.39 Friction coefficient of aluminum surface and dry mud solution formed at different temperatures [58].

TABLE 6.21 Scratch Hardness of Aluminum Surface at Different Drying Temperatures of Mud Solution [58]

Temperature	Scratch Groove Width (μm)	Scratch Hardness (HV)
As-received	61.02	34.88
30 (°C)	57.35	39.48
40 (°C)	56.24	41.06
50 (°C)	54.70	43.41
60 (°C)	52.73	46.70
70 (°C)	52.39	47.32

In addition, the scratch hardness of the dry mud solution enhances with increasing drying mud temperature as given in Table 6.20. In this case, the scratch hardnesses of the dried mud solution are 39.5, 43.4, and 47.3 HV for 30°C, 50°C, and 70°C drying mud solution temperature, respectively. The scratch hardness is inversely proportional to the scratch groove width as listed in Table 6.21. The variation of the width of the scratch marks is shown in Fig. 6.37 for as-received and for different mud solution drying temperatures.

6.7 MUD EFFECT ON LASER-TREATED ALUMINA SURFACE

Alumina tiles have high resistance to corrosion and wear and have various applications in industry [64]. Although alumina tiles can be used efficiently in harsh environments, their thermal and electrical conductivities are considerably low. In addition, they are generally produced from alumina powders through sintering; in which case, micro/nanopores are formed in the tile surface. The practical applications of aluminum tiles are limited due to low conductivities and the presence of pores in the surface region. However, aluminum nitride has high thermal and electrical conductivities and demonstrates piezoelectric properties [65]; therefore forming an aluminum nitride layer at the alumina tile surface extends its application in electric and electronic industries. Several techniques can be used to form an aluminum nitride layer on the alumina tile surface, such as vacuum coating [66], powder sintering [67], atomic layer deposition (ALD) [68], and PVD coating [69]. Some of these deposition/coating processes have disadvantages such as limited coating layer thickness (physical vapor deposition and atomic layer deposition) or high vacuum conditions (vacuum coating). On the other hand, the gas-assisted laser nitriding process offers considerable advantages over the other techniques because of high processing speed, precision of operation, and low cost. Laser gas-assisted nitriding involves high-temperature processing with high cooling rates. This, in turn, increases thermal stress levels due to the presence of the high-temperature gradients in the laser-irradiated region. Once the stress levels exceed the elastic limit of the substrate material, cracks and crack networks can be formed at the laser-treated surface, which in turn limits the practical applications of the treated surface. However, forming regular laser-scanning tracks during the treatment process alters the cooling rates while lowering stress levels in the laser-treated region [70]. The combination of laser melting and ablation at the surface forms the texture composed of micro/nanopillars and whiskers, which can generate hydrophobic features at the surface. In addition, the aluminum nitride layer formed on the tile surface during the laser nitrogen gas-assisted processing lowers the surface energy [71], which contributes significantly to the hydrophobicity of the laser-treated surface.

The response of the laser-treated alumina surface to dust accumulation and mud aftereffects is critical to understand for practical applications. Consequently, in the present section, laser gas-assisted texturing of alumina surfaces is conducted to improve hydrophobicity. Morphological and metallurgical changes as well as contact angle of at the resulting surface were carried out using the analytical tools. Aftereffects of the mud, which is formed from the environmental dust at the laser-textured surface, on the surface hydrophobicity are examined. In addition, the mud adhesion at the laser-treated surface is investigated and adhesion force required to remove the mud from the surface is analyzed. The analysis and findings are presented in light of the previous study [6].

6.7.1 Experimental Study

A CO_2 laser (LC-ALPHAIII) delivering nominal output power of 2 kW was used to irradiate the workpiece surface. The focusing lens was used to achieve the irradiated spot diameter of 200 μm at the workpiece surface. Nitrogen-assisting gas emerging from the conical nozzle and coaxially with the laser beam was used during the laser treatment process. The frequency of the laser pulses was 1500 Hz, which resulted in a 70% overlapping ratio of the irradiated spots at the workpiece surface. The laser treatment parameters were selected according to initial tests; several tests were carried out by incorporating different laser treatment parameters. Consequently, the laser parameters giving rise to minimum surface defects including small cavities and without crack networks were selected in light of the previous studies [72,73]. It was noted that increasing laser power at the workpiece surface by 10% and keeping the laser-scanning speed the same as resulted in large cavity formation at the surface. In addition, similar behavior was also observed when laser-scanning speed was reduced by 10% and keeping the laser output power the same. Therefore increasing laser-scanning speed or lowering laser output power did not result in a surface texture with sufficiently high contact angles. Since the laser-power intensity distribution at the irradiated surface is Gaussian, the peak power intensity occurs at the irradiated spot center. Therefore temperature at the irradiated spot center reaches the evaporation temperature of the substrate material while at a small distance away from the irradiated spot center melting takes place at the surface due to low power intensity in this region. Since the laser-power intensity was controlled during the process, the excessive melt flow during the melting was avoided across the laser-scanning tracks. The melted material in the vicinity of the irradiated spot center flows into the cavity while reducing the cavity size. Consequently, through controlling the laser-power settings, beam intensity distribution, pulse repetition rate, spot size, and scanning speed, the desired texture could be achieved at the workpiece surface. The laser treatment conditions are given in Table 6.22.

Alumina (Al_2O_3) tiles (Ceram Tec-ETEC, 2010) 3 mm thick were used as workpieces. Material characterization of the laser-treated surfaces was carried out using SEM, EDS, and XRD. Jeol 6460 electron microscopy was used for SEM and EDS examinations and Bruker D8 Advanced having CuKα radiation was used for XRD analysis. A typical setting of XRD was 40 kV and 30 mA and scanning angle (2θ) ranged from 20 to 80 degrees. Surface roughness measurement of the laser-melted surfaces was performed using a 5100 AFM/SPM microscope by Agilent in contact mode. The tip

TABLE 6.22 Laser Treatment Conditions Used in the Experiment

Scanning Speed (cm/s) (mm/min)	Power (W)	Frequency (Hz)	Nozzle Gap (mm)	Focus Setting (mm)	N₂ Pressure (kPa)
10	90	1500	1.5	127	600

was made of silicon nitride probes ($r = 20-60$ nm) with a manufacturer-specified force constant, k, of 0.12 N/m.

A Microphotonics digital microhardness tester (MP-100TC) was used for the surface microhardness measurements. The standard test method for Vickers indentation hardness of advanced ceramics (ASTM C1327-99) was adopted. The measurements were repeated five times at each location for consistency of results.

The XRD technique was used to measure the residual stresses and the XRD data were obtained in the surface region of the laser-treated layer due to the penetration depth of CuKα radiation, which was in the order of 5 μm. The position of the diffraction peak exhibited a shift as the laser-treated workpiece was rotated by an angle ψ. The relationship between the peak shift and the residual stress (σ) can be written as [74]: $\sigma = [E/(1 + v)sin^2\psi][(d_n - d_o)/d_o]$, where E is Young's modulus, v is Poisson's ratio, ψ is the tilt angle, and d_n and d_o are the d spacing measured at each tilt angle. It should be noted that the d spacing changes linearly with $sin^2\psi$ when there is no shear strain present in the surface region of the workpiece. Moreover, the γ-Al₂O₃ peak (ICDD 29-1486) takes place at 37.5 degrees, which corresponds to the (311) plane with the interplaner spacing of 0.239 nm. The linear dependence of d(311) results in the slope of -1.4378×10^{-12} m/degrees and the intercept of 0.239 nm. The residual stress determined from the XRD technique at the surface vicinity is in the order of -1.8 ($+0.07/-0.07$) GPa. XRD measurements were repeated three times and the error related to the measurements was in the order of 4%.

A linear microscratch tester (MCTX-S/N: 01-04300) was used to determine the friction coefficient of the laser-treated and untreated surfaces. The equipment was set at the contact load of 0.03 N and end load of 5 N. The scanning speed was 5 mm/min and loading rate was 1 N/s. The total length for the scratch tests was 1 mm.

The wetting experiment was performed using a Kyowa (model—DM 501) contact angle goniometer. A static sessile drop method was considered for the contact angle measurement. The water contact angle between the water droplet and the heat-treated surface was measured with the fluid medium as deionized water. Droplet volume was controlled with an automatic dispensing system with a volume step resolution of 0.1 μL. Still images were captured, and contact angle measurements performed after 1 second of deposition of water droplets on the surface.

To investigate the effect of mud on the surface characteristics of the laser-treated and as-received workpieces, actual dust accumulation and mud formation was simulated in the laboratory. In reality, mud s formed from the accumulated dust particles in humid air because of the condensation of water vapor on to the dust particles. To resemble the dust accumulation, the dust thickness was measured over a period of 2 weeks during a dust storm in Saudi Arabia. Therefore a layer 300 μm thick composed of dust particles collected from the local environment was formed at the laser-treated and as-received alumina workpiece surfaces. The desalinated water, which was equal to water vapor condensate with same volume of dust in the open environment, was dispensed gradually on to the dust layer. It should be noted that the initial condensation tests were carried out in the local humid air to estimate the amount of condensate accumulated over time. It was found that the amount of condensate had almost the same volume of dust. In order to resemble the water condensation in the humid air, dispensed water was left on the surface of the dust layer without mechanical mixing. This resulted in formation of mud at the workpiece surface similar to that occurring naturally in the open environment. The workpieces were kept in ambient air at a local room temperature for 3 days to dry. Scratch tests were carried out to measure the adhesion force required to remove the mud from the laser-treated and as-received workpiece surfaces. To investigate the mud aftereffects on the laser-treated surface, the dry mud was removed from the workpiece surfaces with a desalinated waterjet with 2 mm diameter and 1.5 m/s velocity. The cleaning process was continued for 15 minutes for each workpiece surface. Finally, the morphology, microhardness, fracture toughness, residual stress, and hydrophobicity of the mud-removed surfaces were analyzed using the analytical tools.

6.7.2 Morphological and Metallurgical Aspects of Surface Prior to and After Mud Forming

Fig. 6.40 shows SEM micrographs of laser-treated surfaces without mud and after mud removal. The laser-treated surface is composed of regular laser-scanning patterns (Fig. 6.40A) with overlapping laser-irradiated spots (Fig. 6.40B). Laser pulse repletion results in overlapping of irradiated spots at the surface during laser scanning. Since the pulse repetition rate was 1500 Hz, the overlapping ratio of laser-irradiated spots at the surface was in the order of 70%, which in turn resulted in almost continuous regular scanning patterns. Laser beam intensity remained high at the irradiated spot center and reduced toward the irradiated spot edges due to Gaussian intensity distribution. Thus temperature exceeds the evaporation temperature at the irradiated spot center while forming a small cavity via surface evaporation.

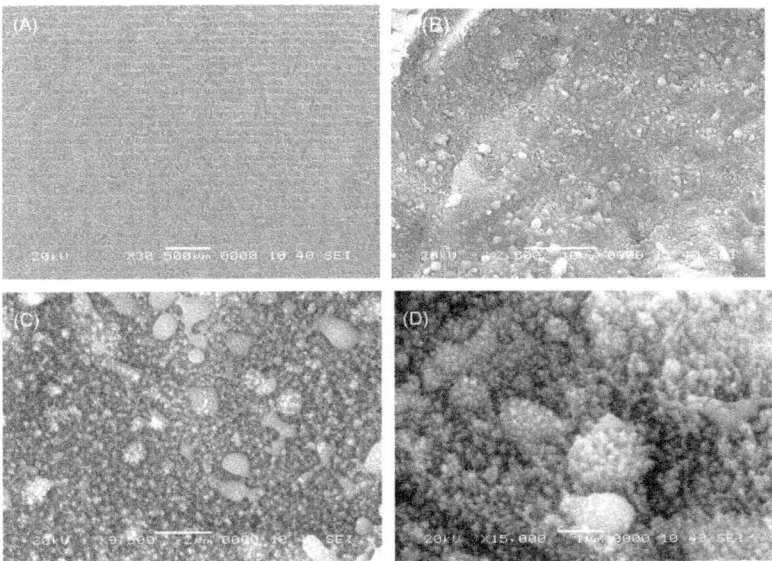

FIGURE 6.40 Laser-textured surface: (A) regular laser-scanning tracks; (B) overlapping of laser-irradiated spots; and (C and D) micro/nanopillars at the laser-treated surface.

In the region close to the edge of the irradiated surface, the temperature remains at the melting temperature, which causes melting in this region. Consequently, melt flow from this region toward the cavity center reduces the cavity size while lowering the surface roughness. This situation can be observed from the AFM images shown in Fig. 6.41A. The laser-treated surface is free from cracks despite the high cooling rates in the surface region. Close examination of the SEM micrographs reveals that the laser-treated surface is composed of fine micro/nanosize structures (Fig. 6.40C and D) without a regular pattern. Moreover, the use of high-pressure nitrogen-assisting gas enhances the cooling rates in the laser-treated region. However, heat conduction from the recently formed laser-scanning tracks toward early formed tracks modifies the cooling rates. This, in turn, reduces the temperature gradients and thermal stress levels in the laser-treated surface. Consequently, formation of surface cracks and crack networks are avoided through laser scanning at the surface. Fig. 6.41 shows AFM images of the laser-treated surface. The formation of cavities and melt flow at the surface contributes to the surface roughness. The presence of micro/nanopoles at the treated surface is evident from the AFM images (Fig. 6.41B). The surface roughness is in the order of 0.6 μm (Fig. 6.41B). However, the presence of mud residue at the surface alters the roughness of the laser-treated surface, which can be seen from the AFM images in Fig. 6.41C and D.

In the case of the laser-treated surfaces after mud removal, the presence of mud residue is evident from the SEM micrographs (Fig. 6.42A). This is

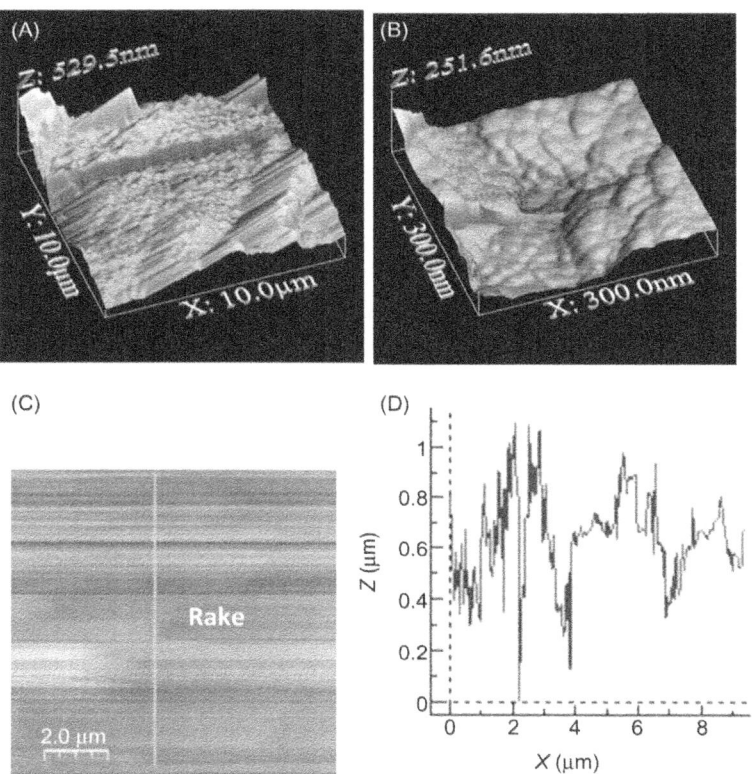

FIGURE 6.41 AFM images of laser-textured surface: (A) micro/nanopillars; (B) nanosize cavity due to laser ablation; and (C and D) surface roughness along the rake shown in the image.

attributed to adhesion between the mud residue and the laser-treated surface. It should be noted that mud residues composes of silica, metal oxides, and alkaline and alkaline earth metals. However, alkaline and alkaline earth metals can dissolve in water during the mud formation. Since the mud cross section has porous structure (Fig. 6.42B), the liquid solution containing alkaline and alkaline earth metals accumulates and wets the laser-treated surface under the gravity. Once the mud dries by time, the liquid solution also dries and forms the crystal structures at the laser-treated surface (Fig. 6.42C). This, in turn, increases the adhesion between the mud and the laser-treated surface. Therefore the mud residue remains at the surface after cleaning by a waterjet. The SEM image for the laser-treated and mud-removed surfaces shows that the micro/nanostructures formed at the surface are partially filled with the mud residue (Fig. 6.42D), which is scattered locally at the surface.

The use of high-pressure nitrogen as an assisting gas results in the formation of nitride compounds at the surface. In this case, the AlN compound is

FIGURE 6.42 SEM micrographs of laser-textured surface after the mud removal: (A) laser-scanning tracks and textured surface are partially filled by mud residue after mud removal; (B) dry mud cross section; (C) crystalized mud solution after drying at the textured surface; and (D) mud residue fill the textured surface.

formed during the laser treatment process, which is evident from Fig. 6.43, in which XRD for laser-treated and as-received surfaces is shown. Aluminum nitride formation occurs into two steps such that $Al_2O_3 + 2C \rightarrow Al_2O + 2CO$ is formed in the first step and in the second step high-pressure nitrogen results in the reaction $Al_2O + CO + N_2 \rightarrow 2AlN + CO_2$ at the surface. However, the carbonic gas (carbon dioxide) formed during the nitriding process escapes from the surface. Moreover, the AlON peak is present in the XRD. Although the peak intensity is low in the diffractogram, the formation of AlON is associated with high-temperature processing at the surface. During the melting of the surface, $Al_2O_3 \rightarrow 2AlO + \frac{1}{2}O_2$ can be formed; in which case, oxygen atom can remain in the alumina structure [75]. Since the gas pressure remains high at the surface because of the stagnation flow of the impinging N_2-assisting gas, the $2AlO + N_2 \rightarrow 2AlON$ reaction can take place at the surface while forming the AlON phase. However, transformation of γ-Al_2O_3 phase into thermodynamically stable α-Al_2O_3 can also take place, which can be observed after comparing the diffractograms of the laser-treated and as-received workpiece surfaces. In the case of the laser-treated surface after the mud removal, the XRD shows the presence of $CaCO_3$, NaCl, and MgO peaks, which are associated with the mud residue at the surface. In addition, the small peak for AlO(OH) is also observed from the

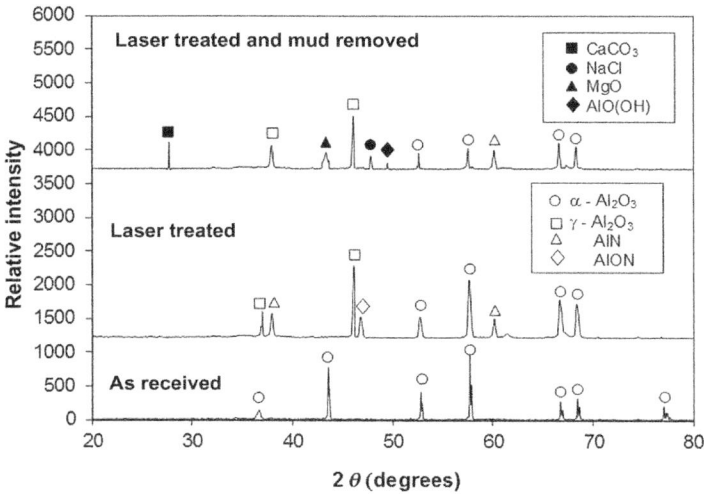

FIGURE 6.43 XRD of as-received, laser-treated, and laser-treated and mud-removed surfaces.

TABLE 6.23 Elemental Composition of As-Received, Laser-Treated, and Laser-Treated and Mud-Removed Surfaces Obtained From EDS Data (wt.%)

	N	O	Ca	Na	Mg	Al
As-received	–	48.2	–	–	–	Balance
Laser-treated	8.3	39.4	–	–	–	Balance
Laser-treated and Mud-removed	3.2	42.4	3.1	1.8	1.2	Balance

XRD. This is related to the hydrolysis effect of the mud solution on the laser-treated surface prior to drying. Since the mud solution has high pH (8.4 base), it increases the activation energy of the phase transition to aluminum oxyhydroxide [76], due to high OH^- ions in the solution. The elemental composition of laser-treated surfaces without mud and after mud removal and as-received surface is given in Table 6.23. It is evident that elemental composition remains almost the same after the laser treatment process; however, nitrogen is observed for the laser-treated surface despite the fact that the quantification of the light elements, such as nitrogen, involves errors in EDS data. Nevertheless, nitrogen presence is attributed to nitride species formation at the surface. In the case of laser-treated and mud-removed surface, Ca, Na, Mg, and Si are observed in addition to the elemental composition of the laser-treated surface. The presence of alkaline metal (Na) and alkaline earth metal (Ca) is due to mud residue at the laser-treated surface.

6.7.3 Properties and Adhesion Work for Dry Mud Removal

Table 6.24 lists the surface microhardness data for as-received workpieces and laser-treated without mud and after mud removal ones. The laser treatment results in microhardness enhancement at the surface. Surface microhardness enhancement is associated with the following: (1) high cooling rates at the surface results in grain refinement, which in turn increases microhardness and (2) formation of nitride compounds contributes to the volume shrinkage at the surface while giving rise to microhardness increase. Microhardness of the laser-treated surface after mud removal is slightly higher than that of the laser-treated surface. This is attributed to the presence of dried mud solution, which forms crystals at the surface after the mud removal (Fig. 6.42C). Therefore their effects on the surface characteristics are responsible for the microhardness increase after the mud removal. Table 6.24 gives the residual stress determined by the XRD technique for laser-treated surface without mud and after the mud removal. Residual stress is compressive and is in the order of -1.8 GPa. The formation of high residual stress in the surface region is related to grain refinement under the high cooling rates and volume shrinkage due to AlN phase formation. However, the residual stress for the laser-treated workpiece surface after the mud removal is slightly higher compared to that of the laser-treated surface. This is attributed to the dried solution at the surface, which could cause the hydrolysis of the laser-treated surface by the wet solution in the mud prior to drying. In this case, the high pH of the mud solution (pH = 8.4 base) increases the concentration of OH^- ions in the solution while enhancing activation energy for phase transition to aluminum oxyhydroxide in the surface region [76]. This situation can be observed from the small peak of AlO(OH) occurring in the XRD for the laser-treated surface after the mud removal. Nevertheless, the residual stress increase after the mud removal is small, in the order of 400 MPa.

Fig. 6.44 shows FTIR data obtained for the laser-treated workpiece surface without mud and after the mud removal. Absorption broadening band of Al-O vibration occurs in-between 1000 and 400 cm^{-1}. AlN vibrations occur at peaks 1695 and 670 cm^{-1}. In addition, the stretching vibration of Al-OH

TABLE 6.24 Microhardness and Residual Stress at the Workpiece Surface Prior to and After the Laser Treatment

	Hardness (HV)	Residual Stress (GPa)
As-received surface	1150 (+50/−50)	
Laser-treated surface	1650 (+50/−50)	−1.8 (+0.06/−0.06)
Laser-treated and mud-removed	1730 (+50/−50)	−2.0 (+0.06/−0.06)

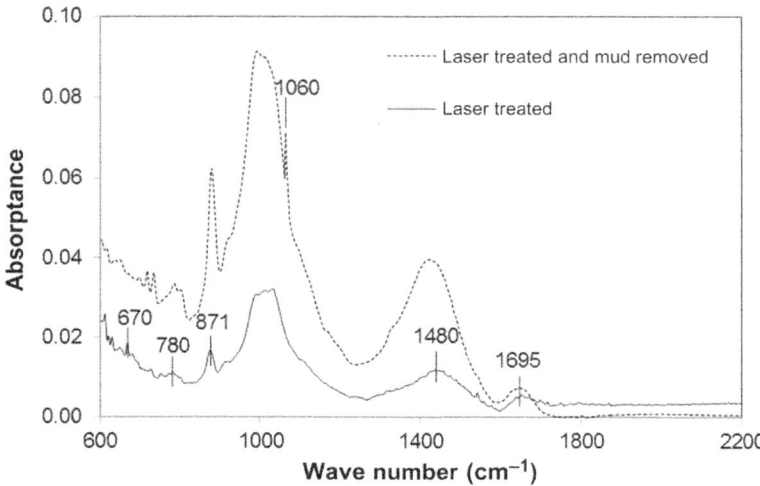

FIGURE 6.44   FTIR spectrums for laser-treated and laser-treated and mud-removed surfaces.

takes place at peak 1060 cm^{-1}, which is observed for laser-treated and dust-removed surface in line with the XRD peaks occurring due to aluminum oxyhydroxide for the laser-treated surface after the mud removal.

Friction coefficients of the as-received and laser-treated surfaces without mud and after the mud removal are shown in Fig. 6.45A and B shows the adhesion force required for the mud removal from the laser-treated surface. The friction coefficient reduces for the laser-treated surfaces, which is associated with the surface hardness improvement after the laser treatment process. The friction coefficient increases for the laser-treated surface after the dust removal, which is related to the mud residue at the surface. The adhesion force required for the mud removal from the laser-treated surface changes across the surface; in which case, the adhesion force increases locally. The attainment of locally scattered high adhesion force is associated with the dried liquid mud solution at the interface of the mud and the workpiece surface. The crystals, which are formed at the surface after drying of the liquid mud solution, enhance the adhesion force. Since the liquid mud solution is not uniformly distributed at the surface, the crystal formation becomes local while altering the adhesion force locally.

6.7.4   Wetting State of Surface After Mud Removal

The liquid—surface interface is composed of liquid—solid and liquid—vapor interfaces. In this case, the equation for the contact angle yields [77] $cos\theta_c = f_1 cos\theta_1 + f_2 cos\theta_2$, where θ_c is the apparent contact angle; f_1 is the surface fraction of the liquid—solid interface; f_2 is the surface fraction of the

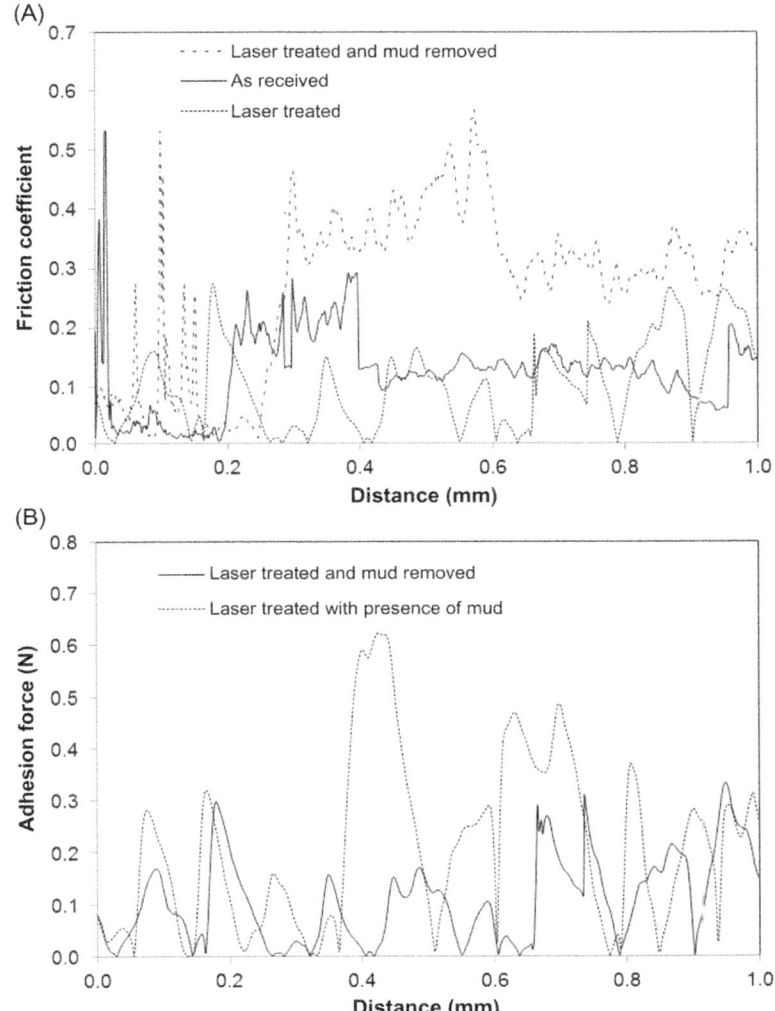

FIGURE 6.45 Scratch test results for: (A) laser-treated with mud-removed and (B) laser-treated with presence of mud surfaces.

liquid−vapor interface; θ_1 is the contact angle for the liquid−solid interface; and θ_1 is the contact angle for the liquid−vapor interface. However, for the air−liquid interface, f_1 can be presented in terms of the solid fraction (f) and air fraction (f_2), which is ($1-f$) and f ranges within 0 to 1. Hence, $f = 0$ corresponds to the liquid droplet, which is not contacting at the surface; however, $f = 1$ represents the surface, which is completely wetted. In reality, depending on the surface energy, the surface texture, and impact of the droplet at the surface, the contact mode changes from the Cassie–Baxter state to the Wenzel state [32]. Two states can coexist on a nanopillared surface [33]; however, when a liquid–air interface can remain pinned at the pillar tops, transition to the Wenzel state is possible.

Fig. 6.46 shows images of the droplets obtained from the contact angle measurements, and the contact angles due to as-received and laser-treated without mud and after mud removal are listed in Table 6.25. Laser treatment results in high contact angles while contact angle reduces significantly for the laser-treated surface after the mud removal. The attainment of high contact angle for the laser-treated surface is associated with: (1) surface texture composed of micro/nanopillars and (2) low surface energy due to the presence of nitride compounds at the surface. It should be noted that the surface free energy of AlN is 38.3 mJ/m^2, and it is 68 mJ/m^2 for Al$_2$O$_3$ [71]. The surface roughness and surface energy alter the wetting properties [77] and the contact angle; consequently, lowering surface energy enhances the contact angle at the surface [77]. In general, the Cassie–Baxter state takes place for the laser-treated surface and the Wenzel state dominates for the as-received and laser-treated after mud-removed surfaces. The dried liquid solution after the mud removal changes the surface energy and partially fills the surface texture of the laser-treated workpiece. This in turn modifies the surface morphology (Fig. 6.42D) and surface hydrophobicity. In addition, the water droplet used in the contact angle measurements diffuses into the mud residue and causes a large area of wetting at the surface. This behavior occurs where the mud residue is highly concentrated at the surface. However, some variations in contact angle data are observed for laser-treated surfaces and the averaged variation is in the order of 5 degrees. This is related to the nonuniform surface texture and nonuniform distribution of nitride compounds at the surface, that is, the presence of the locally isolated

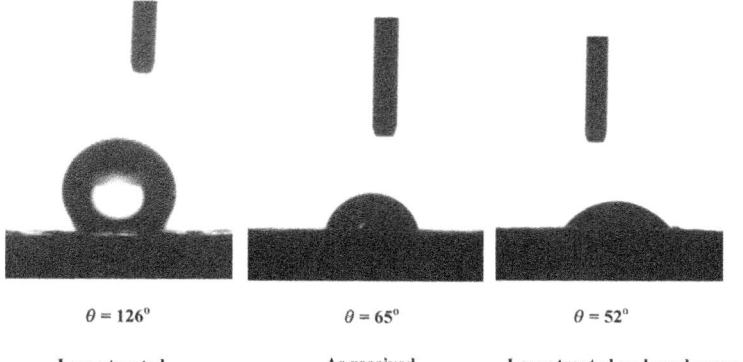

$\theta = 126°$ $\theta = 65°$ $\theta = 52°$

Laser treated   **As received**   **Laser treated and mud removed**

FIGURE 6.46 Images of droplets obtained from the contact angle measurements for laser-treated, as-received, and laser-treated and mud-removed surfaces.

TABLE 6.25 Contact Angles Measurement Results for As-Received, Laser-Treated, and Laser-Treated and Mud-Removed Surfaces

	Contact Angle (degrees)
Untreated surface	65.5 (+5/−5)
Laser-treated surface	125.6 (+5/−5)
Laser-treated and mud-removed	52.4 (+5/−5)

textured alumina surface without forming nitride compounds is responsible for locally reduced contact angles. However, the coverage area of the low contact angle is less than 7% at the surface.

6.8   LASER-TREATED AND SOL-GEL-COATED ALUMINA SURFACE WITH MUD EFFECT

Selective surfaces for thermal harvesting are constructed from materials with high absorptivity of incident solar radiation and low emissivity for thermal radiation emitting from the surface. Various coating processes are adopted to improve characteristics of selective surfaces in terms of absorption and emission. Coatings based on the destructive−interference absorption principle have been developed in the last several years [78]. In general, the coating is composed of multilayer structures such as metal layers [79] double metal-ceramic layers [80], and ceramic layers [81]. The functionality of the coating provides antireflection, solar-spectrum absorption, and transparency characteristics. On the other hand, coating degradation due to (1) temperature increase during absorption of incident solar radiation and (2) formation of excessive thermal stresses under high-temperature gradients are the main concerns in the selection of coating type because coating degradation limits the service life of selective surfaces in solar-thermal receiving systems. The macrostructure of the coatings is usually destroyed due to varying thermal stresses under the transient solar-thermal load created by changing weather conditions. In addition, oxidation is one of the main reasons for coating degradation after long hours of exposure to high-temperature environments [82]. However, one of the solutions to overcome the environmental effects on coating degradation is to introduce optically transparent and environmentally resistive coating on selective surfaces. Sol-gel technology can be used to form this type of coating on selective surfaces [83]. Some sol-gel coatings provide good optical transmittance and high resistance to harsh environments such as sol-gel coating of 3-methacryloxypropyltrimethoxysilane (MAPTMS), zirconium (IV) n-propoxide (ZPO), tantalum ethoxide (TAO), and methacrylic acid (MAAH, $C_4H_6O_2$) [84]. Consequently, sol-gel coating of selective surfaces provides environmental protection without significant loss of optical characteristics of the surfaces.

Selective surfaces can be produced from ceramic nitrides such as zirconium nitride, aluminum nitride, and cermet. One of the methods to form aluminum nitride is to use high-power laser source ablating/melting alumina surfaces under the high-pressure nitrogen environment [75]. Laser texturing through a combination of ablation and melting under the control nitrogen gas environment gives rise to formation of nitride species at the surface and micro/nanotexture, which improve the hydrophobic characteristics at the surface [85]. Although laser texturing provides several advantages over the conventional coating methods, such as fast processing time, precision of operation, local treatment, and low-cost, high-temperature gradients formed in the surface region result in high stress levels in the treated layer. However, the self-annealing effect can be generated to reduce the stress levels in the laser-treated layer through forming closely spaced regular laser-scanning tracks at the treated surface [86]. Considerable research studies were carried out to examine the laser treatment of alumina surfaces. Laser surface modification and characterization of air plasma-sprayed alumina coating was studied by Krishnan et al. [87]. They showed from photoluminescence data that a marginal decline in residual stress and significant increase in density of the laser-treated coatings could be possible to achieve. A study on laser-induced deposition of sol-gel alumina coating on stainless steel under wet conditions was carried out by Adraider et al. [88]. They indicated that the laser-deposited alumina coating had higher mechanical strength than the as-dried xerogel coating, with hardness values four times higher than that of the as-dried xerogel coating. Laser surface treatment of plasma-sprayed alumina-titania coatings was investigated by Ibrahim et al. [89]. The findings revealed that the laser fluence played a major role in modifying the surface morphology of the coating, followed by the pulse repetition rate.

Sol-gel coating of laser-treated alumina surfaces minimizes the chemical activities of the dust solution and lowers the adhesion force between the dry mud and the treated alumina surface. Although laser texturing of alumina surfaces toward forming nitride species were carried out previously [85,86], sol-gel coating of the laser-treated surface and the effect of mud solution formed from the dust particles on sol-gel-coated surface is still needed for the practical applications of selective surfaces. Consequently, in the present section, laser treatment of alumina surface under high-pressure nitrogen environment is considered. The coating method for the laser-treated surfaces is introduced via incorporating sol-gel technique using a solution consisting of MAPTMS, ZPO, TAO, and MAAH, $C_4H_6O_2$. The mud was formed from the dust particles to resemble the condensation of water vapor on to the dust particles in the humid air conditions. The surface characteristics of laser-treated and sol-gel-coated alumina tiles prior and after the mud removal are presented in light of the previous study [7].

6.8.1 Experimental Analysis

Alumina (Al_2O_3) tiles (Ceram Tec-ETEC, 2010) 3 mm thick were used as workpieces. A water-soluble phenolic resin was applied to the surface of alumina tiles to form a coating. The coating was dried in a control chamber at 2 bar pressure and 175°C for 2 hours to obtain a uniform thickness of 40 μm coating at the aluminum tile surface. The tiles were then heated to 400°C in an argon environment at 8 bar pressure for several hours to ensure conversion of the phenolic resin into carbon. This arrangement provided a uniform carbon coating thickness of about 40 μm at the alumina tile surface prior to laser scanning.

The CO_2 laser (LC-ALPHAIII) was used to irradiate the alumina tile surfaces. The nominal output power of the laser was 2 kW and the irradiated spot diameter at the workpiece surface was about 200 μm through focusing. High-pressure nitrogen gas jet emanating from the conical nozzle was utilized during the laser heating of the surfaces. The laser-pulsing frequency was set at 1500 Hz, which in turn gave rise to 70% overlapping ratio for the irradiated spots at the surface. The initial tests were conducted to select the laser treatment parameters so that the laser parameters resulting in crack-free surfaces were selected. The results of the initial tests revealed that increasing laser power by 10% while keeping laser-scanning speed gave rise to large cavity formation at the surface. On the other hand, reducing laser-scanning speed by 10% while keeping the laser output power same resulted in cracks at the surface. Consequently, laser-treated surface properties became high dependent on the proper selection of the laser-processing parameters. Therefore through controlling the laser-power settings, beam intensity distribution, pulse repetition rate, spot size, and scanning speed, crack-free surface texture could be realized. The laser treatment conditions are given in Table 6.26.

Jeol 6460 electron microscopy was used for SEM and EDS examinations and Bruker D8 Advanced with CuKα radiation was used for XRD analysis. AFM (5100 AFM/SPM) in contact mode was used to examine the surface texture characteristics. The tip was made of silicon nitride probes ($r = 20-60$ nm) with a manufacturer-specified force constant, k, of 0.12 N/m.

The sol-gel considered for the coating of laser-treated surface was composed of two hybrid components, silica-based and metal transition-based,

TABLE 6.26 Laser-Processing Parameters [7]

Feed Rate (m/s)	Power (W)	Frequency (Hz)	Nozzle Gap (mm)	Nozzle Diameter (mm)	Focus Diameter (mm)	N_2 Pressure (kPa)
0.1	2000	1500	1.5	1.5	0.3	600

which provided a dense coating with good barrier properties for harsh environmental applications [84]. The sol-gel synthesis was based on the formation of stable and homogeneous solutions obtained from the reaction between photosensitive organically modified precursors MAPTMS, ZPO, and MAAH, which were allowed to react in two different molar ratios. MAPTMS was prehydrolyzed with an aqueous solution (HCl 0.005 N), employing a 1.00:0.75 water to the alkoxide molar ratio. After 20 minutes of stirring, the solution became fully transparent demonstrating the occurrence hydrolysis and condensation reactions leading to the production of alcohol that allows the miscibility of all species present in solution. In parallel, to control the hydrolysis−condensation of ZPO and avoid the formation of any undesired ZrO_2 precipitate, MAAH was used in a 1:1 stoichiometric ratio against ZPO as a chelating agent to bind with the zirconium atom through two oxygen atoms. After 45 minutes of reaction, the prehydrolyzed MAPTMS solution was added dropwise to the zirconium complex. Following another 45 minutes of reaction, in order to improve the homogeneity of both molecular systems, a second hydrolysis employing water (pH 7) was performed leading to a hydrolysis of 50% of the total alkoxide groups. The 5 mol.% of Irgacure 184 against the MAPTMS content was added to the sol to make the materials photoreactive under UV irradiation. The final sol was left stirring for 24 h before use. The laser-textured alumina surfaces were coated by the deposition technique incorporating the spin coating (1000 rpm) for 40 seconds.

The standard test method for Vickers indentation hardness of advanced ceramics (ASTM C1327-99) was adopted to measure the surface microhardness and Microphotonics digital microhardness tester (MP-100TC) was used for this purpose. The measurements were repeated five times at each location for consistency of the results. The XRD technique was used to measure the residual stresses in the surface region because of low penetration depth of $CuK\alpha$ radiation, which was in the order of 5 μm. The residual stress (σ) could be determined from the relationship between the peak shift, which could be written as found in [74]. However, when there is no shear strain in the treated layer, the d spacing changes linearly with $sin^2\psi$. Moreover, the γ-Al_2O_3 peak (ICDD 29−1486) takes place at 37.5 degrees, which corresponds to the (311) plane with the interplaner spacing of 0.239 nm. The linear dependence of interplaner spacing results in the slope of -1.44×10^{-12} m/degrees and the intercept of 0.239 nm. The residual stress determined from the XRD technique at the surface vicinity is in the order of -2.1 ($+0.07/-0.07$) GPa. XRD measurements were repeated three times and the error related to the measurements was in the order of 3%.

The lattice strains were also assessed and the residual stress formed was calculated from the peak shifts of the FTIR data. The equation used to determine the residual stress was [90]: $\sigma = C\varepsilon$, where C is the elastic modulus evaluated from the elastic constants c_{ij} ($[C = c_{11} + c_{12} - (2c_{13}^2/c_{33})]$) [91], and ε is the in-plane invariant strain, which was determined from the FTIR peak shift, that is, $\varepsilon = -(\Delta w/w)$, where Δw was the shift to the higher value as compared to the unstressed value corresponding to the wave numbers (w). The value of C was taken as 308 GPa [92] and ε was determined from the FTIR data as -6.35×10^{-3}. The residual stress determined is in the order of -1.95 GPa, which is comparable to that obtained from the XRD technique (-2.1 GPa).

The contact angle measurements were repeated several times to ensure the repeatability of the contact angle data. The contact angle measurements resulted in the contact angle of as-received alumina surface in the order of 62 degrees with the hysteresis of 42 degrees. The laser-treated surface had a droplet contact angle in the order of 112 degrees with the hysteresis of 21 degrees. The laser-treated and sol-gel-coated gave rise to the water-droplet contact angle of 80.5 degrees with the hysteresis of 16 degrees. The data calculated are given in Table 6.27 for the surface free energy of the laser-treated surfaces. The surface energy from the droplet method resulted in 54.7 mJ/m^2, which was slightly higher than that reported in the previous study (53 mJ/m^2) for AlN [93]. Moreover, the contact angle measurements were repeated for the laser-treated and sol-gel-coated surfaces. The surface free energy for laser-treated and sol-gel-coated surface is in the order of 26.3 mJ/m^2.

To investigate the effect of mud on the surface characteristics of the laser-treated and as-received workpieces, the actual dust accumulation and the mud formation was simulated in laboratory environments. In reality, mud is formed from accumulated dust particles in humid air because of the condensation of water vapor on to the dust particles. To resemble the dust accumulation, the dust thickness was measured in the environments of Saudi Arabia over a period of 2 weeks after a dust storm. The measurements revealed that the thickness of the dust particles layer was in the range of $250-350$ μm. Therefore a layer 300 μm thick, composed of dust particles collected from the local environment, was formed at the laser-treated and laser-treated and sol-gel-coated surfaces. The desalinated water was dispensed gradually on to the dust layer. It should be noted that the water

TABLE 6.27 Surface Energy and Lifshitz–van der Walls Components and Electron-Donor Parameters Determined for Laser-Treated and Laser-Treated and Sol-Gel-Coated Surfaces [7]

	γ_S (mJ/m^2)	γ_S^+ (mJ/m^2)	γ_S^- (mJ/m^2)	γ^p (mJ/m^2)	γ_S^L (mJ/m^2)
Laser-treated and sol-gel-coated	26.3	0.64	11.33	5.38	20.96
Laser-treated	54.7	2.6	17.7	13.5	41.78

dispensed was equal to the water vapor condensate with the same volume of dust in the open environment. To secure the accuracy of amount of the vapor condensation on the dust layer, the initial condensation tests were conducted in the local ambient humid air to determine accurately the amount of condensate accumulated over time. The findings of the initial tests revealed that the amount of condensate had almost the same volume of dust. Moreover, to resemble the water condensation on the dust layer in the humid air conditions, water dispensed was left on the surface of the dust layer without mechanical mixing. This gave rise to formation of mud at the workpiece surfaces the same as that occurring naturally in the open environment. The workpieces were then dried in air ambient at a local room temperature for 3 days. Once the mud dried, the tests were carried out to measure the adhesion force required for removal of the mud from the laser-treated and laser-treated and sol-gel-coated surfaces. In addition, to examine the mud aftereffects on the laser-treated and laser-treated and sol-gel-coated surfaces, the dry mud was removed from the surfaces with a desalinated waterjet of 2 mm diameter and 2 m/s velocity. The cleaning process was continued for 25 minutes for each workpiece surface.

6.8.2 Surface Characteristics of Laser-Treated and Sol-Gel-Coated Alumina Tiles

Fig. 6.47 shows SEM micrographs of laser-treated and laser-treated and sol-gel-coated alumina surfaces. Laser-treated surface composes of closely spaced regular scanning tracks (Fig. 6.47A), which are formed during high-frequency (1500 Hz) laser repetitive pulsing of the surfaces during the scanning. In general, laser-treated surface is free from cracks and large size cavities. Since the heating involves with high-temperature processing, temperature gradient remains high in the close region of the irradiated surface. However, the self-annealing effects created during laser scanning along the closely spaced scanning tracks of 150 µm wide modify the temperature gradients and lower the thermally induced strains in the surface region. This results in crack-free laser-treated surface. Since the laser pulse intensity distribution is Gaussian at the irradiated surface, high intensity occurs at the irradiated surface center, which in turn results in partial evaporation in the close region of the irradiated spot center. However, melting takes place in the region close to the irradiated spot edges. Therefore during the scanning, the melt flow takes place from edges of the recently irradiated spots toward the previously formed cavity. This, in turn, modifies the cavity size and causes development of surface morphology with micro/nanotexture heights (Fig. 6.47B). On the other hand, sol-gel coating covers uniformly the laser-treated surface (Fig. 6.47C) and the laser-coated surface is free from asperities such as voids and large pores. Since the thickness of the sol-gel

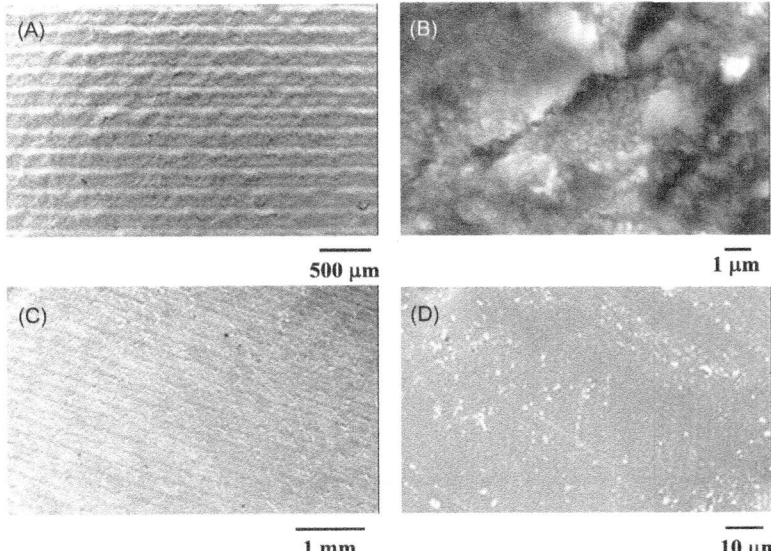

FIGURE 6.47 SEM micrographs of laser-textured, and laser-textured and sol-gel-coated surfaces: (A) regular scanning tracks at the surface; (B) micro/nanotexture at the laser-treated layer; (C) laser-treated and sol-gel-coated surface; and (D) some surface textures are visible for laser-treated and sol-gel-coated surface [7].

coating is small, which is in the order of 6 μm, the original laser-textured surface is visible beneath the coating (Fig. 6.47D).

Fig. 6.48 shows AFM image of laser-treated surface (Fig. 6.48A), and laser-treated and sol-gel-coated surface (Fig. 6.48B) together with the line scan of the surfaces (Fig. 6.48C and D). It is evident that laser treatment gives rise to micro/nanosize textures at the surface. The maximum texture height is in the order of 1.1 μm and the average surface roughness is 0.75 μm for laser-treated surface. The presence of the small peaks around the major peaks indicates the fibrils like formation at the surface. In the case of laser-treated and sol-gel-coated surface (Fig. 6.48D); the texture is rather smooth as compared to the laser-treated surface because of the uniform coverage of coating at the surface. The small peaks disappear from the texture profile because of smoothing of the surface by the sol-gel coating. The surface roughness is in the order of 220 nm for laser-treated and sol-gel-coated surface.

Fig. 6.49 shows SEM micrographs of cross sections of laser-treated layer (Fig. 6.49A), and laser-treated and sol-gel-coated (Fig. 6.49B) workpieces. It should be noted that few cracks are observed across the sample cross-section because of the fracture occurred during the preparation of sample cross-section by a diamond cutter. Unfortunately, cracking could not be avoided

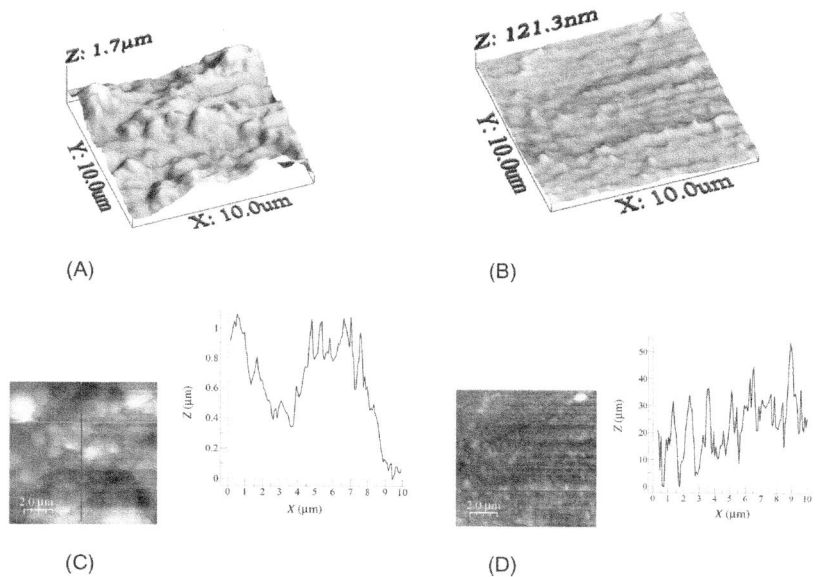

(A)   (B)

(C)   (D)

FIGURE 6.48 AFM micrograph of laser-textured, and laser-textured and sol-gel-coated surfaces: (A) 3D image of laser-treated surface; (B) 3D image of laser-treated and sol-gel-coated surface; (C) line scan at laser-treated surface; and (D) line scan at laser-treated and sol-gel-coated surface [7].

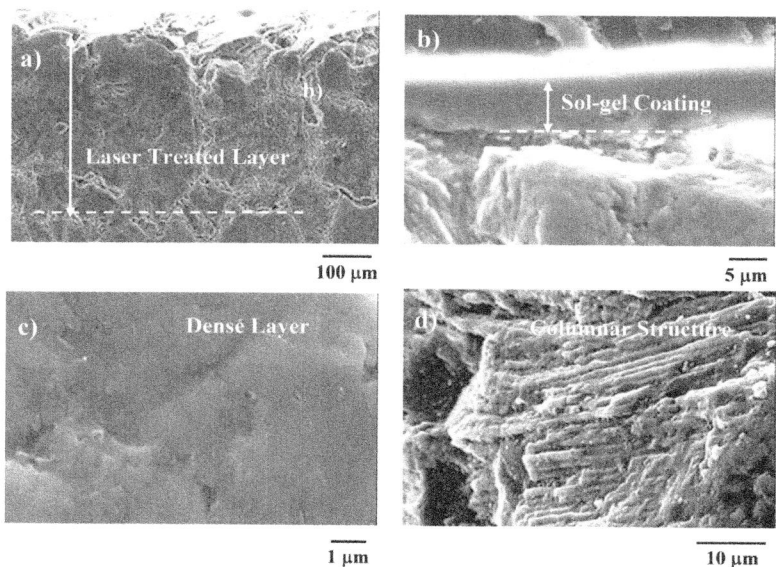

FIGURE 6.49 SEM micrographs of cross section of laser-treated, and laser-treated and sol-gel-coated layer: (A) Uniform depth of laser-treated layer; (B) sol-gel coating on laser-treated layer; (C) dense layer in the surface region of laser-treated layer; and (D) columnar structure below the dense layer [7].

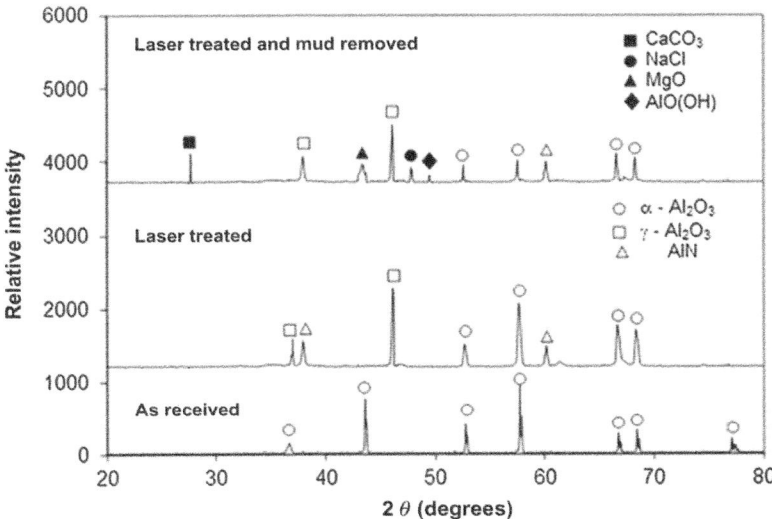

FIGURE 6.50 XRD for as-received, laser-treated, and laser-treated and mud-removed surfaces [7].

despite lowering the cuter speed significantly and using the cooling fluid during the sample cross sectioning. In general, laser-treated layer consists of a dense layer in the surface region (Fig. 6.49C) and columnar structure below the surface is observed (Fig. 6.49D). The close examination of the dense layer reveals that the layer is free from pores and cracks. The columnar structure appears to be small and its size increases as the depth below the surface increases. This is associated with the heat diffusion in alumina during the processing; in which case, temperature gradient reduces with increasing depth [94]. The heat affected zone is not observed below the melted layer, which is attributed to the low thermal diffusivity of alumina.

Fig. 6.50 shows an XRD of the laser-treated workpiece surface. The following peaks appear in the diffractogram: Al_2O_3 (ICSD Collection Codes 010426 and 073076), AlN (ICSD Collection Codes 041358 and 082790), and AlN (ICSD Collection Codes 070032 and 070033). Since the carbon film is formed at alumina tile surface prior to the laser treatment, formation of AlN can be explained through the two-step reactions. In the first step, alumina and carbon form a single alumina oxide and carbon monoxide at the treated surface through a reaction $Al_2O_3 + 2C \rightarrow Al_2O + 2CO$. In the second step, AlN is formed at the surface through the reaction $Al_2O + CO + N_2 \rightarrow 2AlN + CO_2$ under high-pressure nitrogen environment during laser treatment. CO_2 gas leaves the treated surface during the laser treatment process. The EDS data prior and after the laser treatment are given in Table 6.28. Laser treatment does not alter considerably the elemental composition at the surface. However, nitrogen appearing in EDS data suggests the formation of nitride compounds at

TABLE 6.28 EDS Results in the Surface Region of the Laser-Treated Alumina [7]

Spectrum	N	O	Al	
Spectrum 1	4.7	41.1	Balance	
Spectrum 4	4.5	42.5	Balance	
Spectrum 5	0.0	44.9	Balance	
Spectrum 2	0.0	45.8	Balance	
Spectrum 3	0.0	44.1	Balance	

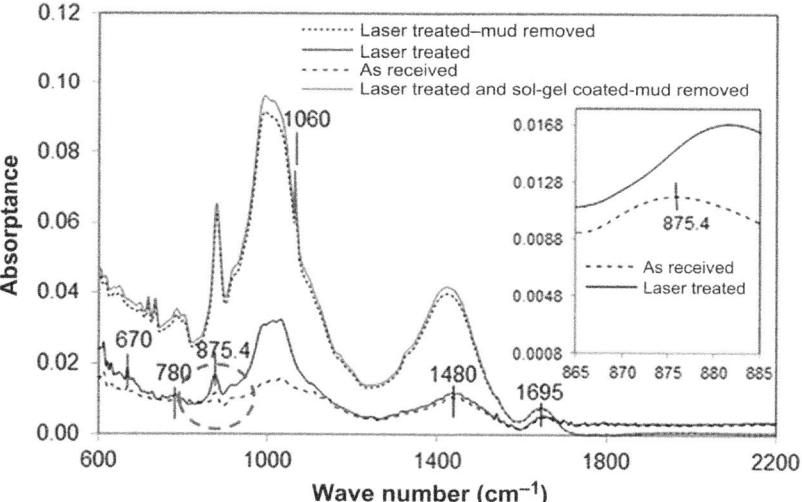

FIGURE 6.51 FTIR data for as-received, laser-treated, laser-treated and mud-removed, and laser-treated and sol-gel-coated and mud-removed surfaces. The small plot in the figure represents magnified view of absorptance with wave number (as shown in a circle in the figure) [7].

the surface despite the fact that the precise quantification of nitrogen is difficult in the EDS system because of being a light element.

Fig. 6.51 shows FTIR data for laser-treated, and laser-treated and sol-gel-coated surfaces. In addition, FTIR data for bare (as-received) alumina is also included for the comparison reason. As-received alumina, FTIR peaks reveal that Al-O vibration due to absorption broadening band takes place in-between 1000 and 400 cm^{-1}. In the case of laser-treated surface, Al-O vibration peaks moves slightly to a higher wave number (as shown in the small

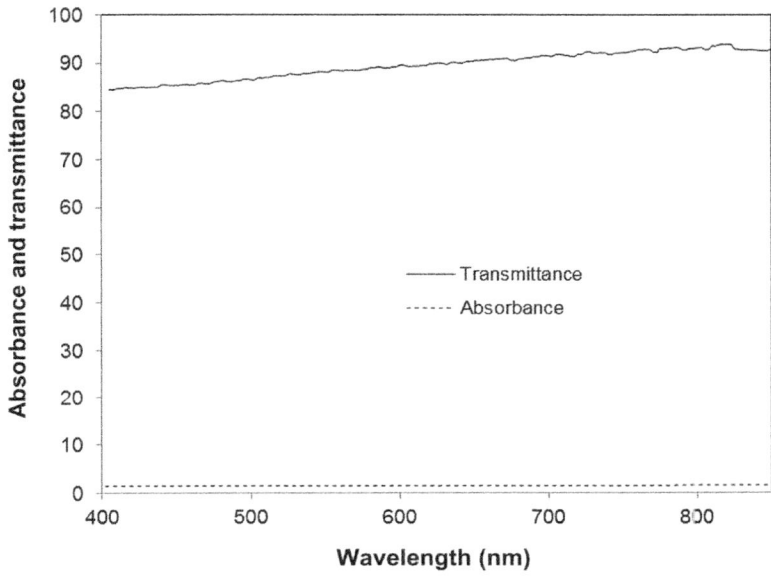

FIGURE 6.52 Absorbance ad reflectance of sol-gel coating [7].

TABLE 6.29 Microhardness and Residual Stress for Laser-Treated Surface [7]

	Hardness HV	Residual Stress (GPa) (XRD Data)	Residual Stress (GPa) (FTIR Data)
As-received surface	1150 ± 50	–	–
Laser-treated surface	1650 ± 50	−2.1	−1.95

plot in the figure). Moreover, the peaks 1695 and 670 cm^{-1} at the absorption bands correspond to AlN vibrations for the laser-treated surface. The peak locations do not change for the laser-treated and sol-gel-coated samples; however, slight increase in absorption occurs in the peak heights. The change in Al-O peak positions for laser-treated and as-received surfaces are associated with residual stress formed in the treated layer.

Fig. 6.52 shows the absorption and reflection characteristics of the sol-gel coating incorporated in the present study [84]. The absorption of the incident radiation and reflection from the coating surface is considerably low for all the wavelengths. Therefore the sol-gel coating does not have significant effect on the FTIR data presented in Fig. 6.51, except some small changes in the peaks.

Table 6.29 gives the surface microhardness and residual stress determined from the XRD technique and the FTIR data. Laser treatment improves

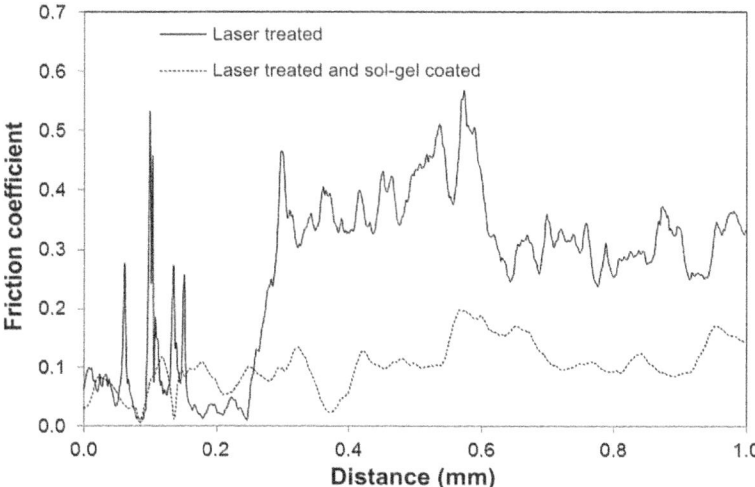

FIGURE 6.53 Friction coefficient of laser-treated, and laser-treated and sol-gel-coated surface [7].

slightly the microhardness of alumina surfaces because of the surface remelting and solidification under the high cooling rates. In addition, the volume shrinkage in the dense layer because of aluminum nitride formation in the surface region contributes to the densification in this region, which is also observed from Fig. 6.49C. The residual stress determined is compressive and is in the order of -2.1 GPa. When comparing the residual stress values obtained from the FTIR data and XRD technique, we find them to be in in good agreement. The difference in the values of the residual stress is small, that is, in the order of 5%.

Fig. 6.53 shows the friction coefficient of as-received, laser-treated, and laser-treated and sol-gel-coated surfaces. The friction coefficient attains the lowest value for the laser-treated and sol-gel-coated surface, and then follows laser-treated and as-received surfaces. The wavy appearance in the friction coefficient curve for the laser-treated surface is associated with the surface texture; however, friction coefficient reduction is because of the enhancement of the surface hardness for the laser-treated workpiece. In the case of sol-gel-coated surface, the friction coefficient reduces significantly.

6.8.3   Laser-Treated/Sol-Gel-Coated Surfaces and Dry Mud Removal

In order to study the effects of the mud, which is formed from the dust particles resembling the humid air environments, on laser-treated, and laser-treated and sol-gel-coated surfaces, the mud is formed from the dust collected at the surfaces without mechanical mixing while simulating the humid environmental conditions. The mud formed at the workpiece surfaces is dried in the control environment for 36 hours prior to the adhesion tests. The adhesion work required to remove the dry mud from the laser-treated, and the laser-treated and sol-gel-coated surfaces is determined from the tangential force recorded by the microtribometer. Once the adhesion tests are completed, the dry mud is removed from the workpiece surfaces using the pressurized desalinated waterjet of 2 mm in diameter and 2 m/s velocity. The desalinated waterjet impinges on to the workpiece surface with an angle of 60 degrees for 20 minutes. Morphological and optical changes of the workpiece surfaces after the dry mud removal are examined using the analytical tools.

Fig. 6.54 shows SEM micrographs of laser-treated (Fig. 6.54A and B), and laser-treated and sol-gel-coated (Fig. 6.54C and D) surfaces after the dry mud removed from the surface. The texture of the laser-treated surface is locally covered by the dry mud residue, which is left over at the surface after pressurized waterjet cleaning (Fig. 6.54A). In general, the mud residue is locally scattered and forms a few small islands at the laser-treated surface; in which case, the strong adhesion force between the dry mud and the laser-textured surface is responsible for the presence of the mud residue at the surface. In the case of laser-treated on sol-gel-coated

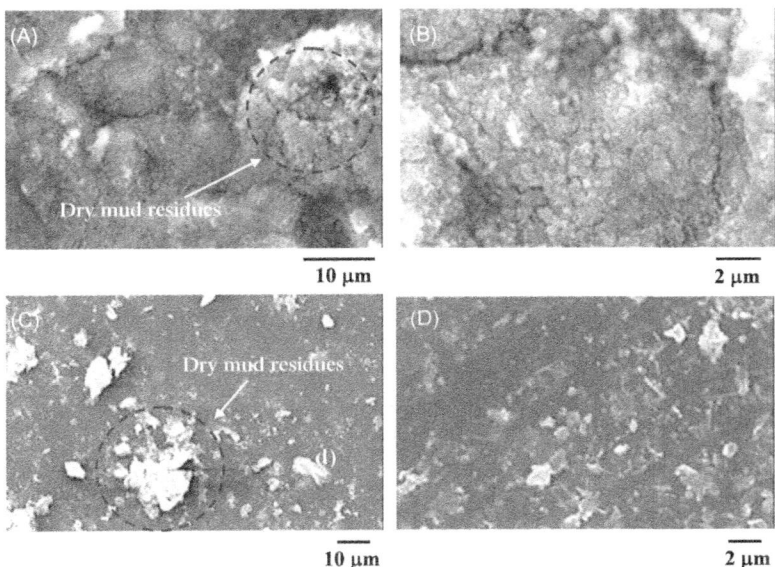

FIGURE 6.54 SEM micrographs of laser-treated, and laser-treated and sol-gel-coated surfaces after dry mud removal: (A) laser-treated surface after dry mud removed; (B) laser-treated surface and small mud residue; (C) laser-treated and sol-gel-coated surface after dry mud removed; and (D) laser-treated and sol-gel-coated surface and dry mud residue [7].

surfaces (Fig. 6.54C and D), the amount of mud residue remains at the surface is smaller than that of the laser-treated and uncoated surfaces. In some regions of the laser-treated and sol-gel-coated surfaces, amount of dust residue becomes almost negligible. This indicates that the adhesion between the dry mud and the laser-treated and sol-gel-coated surface is considerably smaller than that of the laser-treated surface.

The residue of dry mud contains $CaCO_3$ and MgO, which can be observed from Fig. 6.50. In addition, alkaline (Na, K) and alkaline earth metals (Ca) dissolves in the water during the mud formation and the mud solution sediments at the interface between the wet mud and the workpiece surface under the gravitational force. Upon drying of the mud, the dried solution forms an interlayer between the dry mud and the workpiece surface. This situation can be observed from Fig. 6.55A and B, in which the SEM micrograph of the cross section of the dry mud is shown. In order to examine the morphological structure of the dried mud solution at the surface, the study is extended and the mud solution is extracted from the mixture of the dust particles and desalinated water. pH of the mud solution is in the order of 7.8, which shows strong alkaline due to the dissolution of alkaline and alkaline earth metals. The mud solution is, then, dispensed onto the glass surface and left for drying for 36 hours in the control environment.

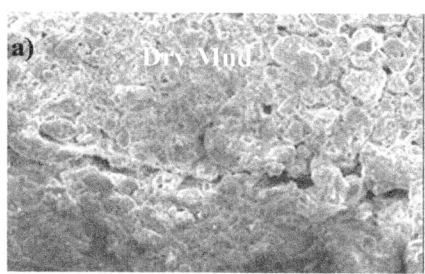

50 μm

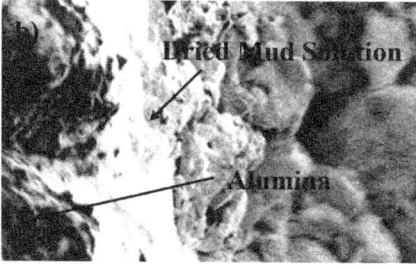

10 μm

FIGURE 6.55 SEM micrograph of dry mud cross section at laser-treated alumina surface: (A) dry mud cross section and (B) interface of dry mud solution indicating dry mud, dried mud solution, and laser-treated alumina surface [7].

Fig. 6.56A and B shows SEM micrograph of the crystallized dry mud solution. It is evident that the dried mud solution forms crystals at the surface indicating the strong bonding at the glass surface. Hence, the dry mud solution at the interface between the dry mud and the workpiece surface alters the adhesion characteristics of the dry mud at the surface. Consequently, the mud residue observed at the surface is associated with the strong ionic bonding of the mud because of the dry mud solution despite the pressurized waterjet cleaning at the workpiece surface.

Fig. 6.57 shows AFM micrographs of laser-treated (Fig. 6.57A), and laser-treated and sol-gel-coated surfaces (Fig. 6.57B) after the dry mud removal. The mud residue at the surface covers a large area with large texture height for the laser-treated surface (Fig. 6.57C), which can also be seen from the line profile of the surface. In this case, the texture height varies in-between 500 and 20 nm. However, the sol-gel-coated surface demonstrates a smooth appearance. The mud residue forms local textures at the surface and, in some cases; texture height remains high as observed from the line scan over the dust residue (Fig. 6.57D). The texture height of the dry mud residue

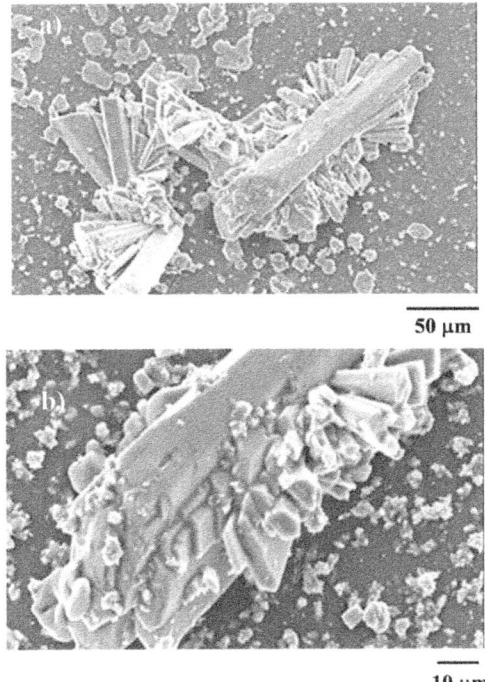

FIGURE 6.56 SEM micrograph of crystals formed from mud solution at the surface: (A) two large crystals and (B) growth of crystal and mud residue around the crystal [7].

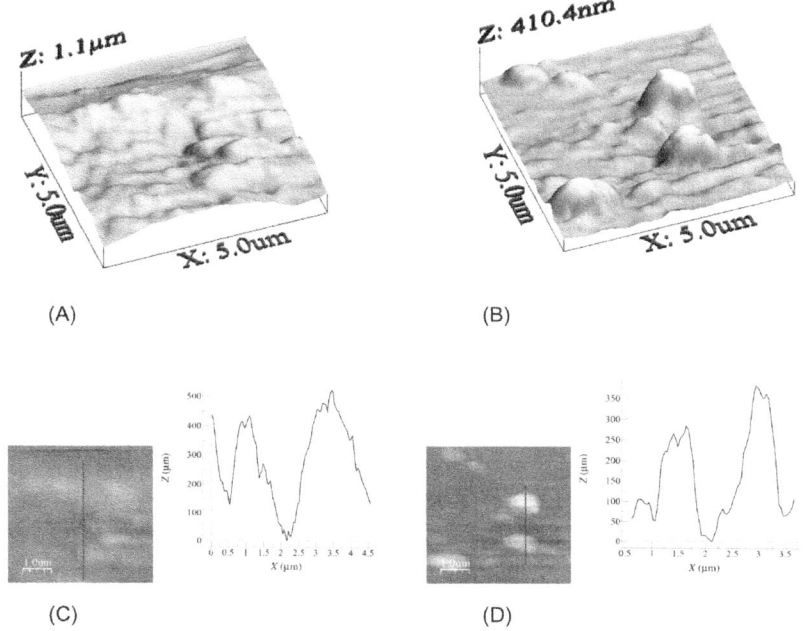

FIGURE 6.57 AFM micrographs of laser-treated, and laser-treated and sol-gel-coated surfaces after dry mud removal: (A) laser-textured after dry mud-removed surface; (B) laser-treated and sol-coated after dry mud-removed surface; (C) line scan for laser-textured surface after dry mud removed; and (D) line scan for laser-textured and sol-gel-coated after dry mud-removed surface [7].

ranges between 300 and 10 nm at the laser-treated and sol-gel-coated surface. This behavior is associated with the formation of a layer of dry mud solution in-between the mud and the sol-gel coating surface; in which case, it increases the adhesion of the dry mud residue at the sol-gel coating surface despite the pressurized desalinated waterjet is used to clean the coating surface. When comparing the texture size and texture height of the dry mud residue at the laser-treated (Fig. 6.57C), and laser-treated and sol-gel-coated (Fig. 6.57D) surfaces, we can observe that the texture size and texture height remains low for the laser-treated and sol-gel-coated surface.

After the mud removal, FTIR data reveals that the absorption band of Al-O vibration is also observed in-between 1000 and 400 cm^{-1} for laser-treated and sol-gel-coated surfaces after dry mud removal (Fig. 6.51). In addition, AlN vibration takes at peaks 1695 and 670 cm^{-1}, similar to that corresponding to laser-treated surface. However, stretching vibration of Al-OH occurs at peak 1060 cm^{-1}, which takes place only for the surface of laser-treated workpieces, which agrees with the XRD peak (Fig. 6.50) corresponding to aluminum oxyhydroxide for the laser-treated surface after the dry mud removed.

Fig. 6.57 shows the tangential force to remove the dry mud from the laser-treated, and laser-treated and sol-gel-coated surfaces. The tangential force required to remove the dry mud is significantly higher for laser-treated surface than that corresponding to laser-treated and sol-gel-coated surface. This is associated with the adhesion between the dry mud and the workpiece surface, which reduces significantly for the laser-treated and sol-gel-coated surface. However, the friction coefficient at the workpiece surface also contributes to the tangential force (Fig. 6.53), in which case the friction coefficient for the laser-treated and sol-gel-coated surface is considerably smaller than that of the laser-treated surface. Since the adhesion between the dry mud and the surface depends on the surface energy of the workpieces, the surface energy of the laser-treated and sol-gel-coated surface is lower than that of the laser-treated surface (Table 6.27). The adhesion work determined from the tangential force is in the order of 0.214 mJ/m for the laser-treated and sol-gel-coated surface while it is in the order of 0.121 mJ/m for the laser-treated surface. The sol-gel coating reduces significantly the adhesion between the dry mud and the workpiece surface, which enables the dry mud removal with relatively smaller efforts as compared to that corresponding to the laser-treated surface. Consequently, coating the laser-treated surface with sol-gel provides environmental protection and easiness for the surface cleaning.

REFERENCES

[1] B.S. Yilbas, H. Ali, M.M. Khaled, N. Al-Aqeeli, N. Abu-Dheir, K.K. Varanasi, Influence of dust and mud on the optical, chemical, and mechanical properties of a PV protective glass, Sci. Rep. 5 (2015) 15833.

[2] B.S. Yilbas, H. Ali, N. Al-Aqeeli, M.M. Khaled, S. Said, N. Abu-Dheir, et al., Characterization of environmental dust in the Dammam area and mud after-effects on bisphenol-A polycarbonate sheets, Sci. Rep. 6 (2016) 24308.

[3] B. Yilbas, H. Ali, C. Karatas, A. Al-Sharafi, Laser texturing of Inconel 718 alloy surface: influence of environmental dust in humid air ambient, Opt. Laser Technol. 108 (2018) 346–354.

[4] B. Yilbas, H. Ali, A. Al-Sharafi, N. Al-Aqeeli, N. Abu-Dheir, F. Al-Sulaiman, et al., Characteristics of a solar selective absorber surface subjected to environmental dust in humid air ambient, Solar Energy Mater. Solar Cells 172 (2017) 186–194.

[5] G. Hassan, B. Yilbas, M.A. Samad, H. Ali, F. Al-Sulaiman, N. Al-Aqeeli, Analysis of environmental dust and mud adhesion on aluminum surface in relation to solar energy harvesting, Solar Energy 153 (2017) 590–599.

[6] B. Yilbas, H. Ali, M. Khaled, N. Al-Aqeeli, F. Al-Sulaiman, Laser gas assisted texturing of alumina surfaces and effects of environmental dry mud solution on surface characteristics, Ceram. Int. 42 (1) (2016) 396–404.

[7] B. Yilbas, H. Ali, N. Al-Aqeeli, M. Oubaha, M. Khaled, N. Abu-Dheir, Laser gas assisted nitriding and sol–gel coating of alumina surfaces: effect of environmental dust on surfaces, Surface Coatings Technol. 289 (2016) 11–22.

[8] M. García-Heras, N. Carmona, A. Ruiz-Conde, P. Sánchez-Soto, J. Benítez, Application of atomic force microscopy to the study of glass decay, Mater. Charact. 55 (4−5) (2005) 272−280.

[9] E. Khalil, F. ElBatal, Y. Hamdy, H. Zidan, M. Aziz, A. Abdelghany, Infrared absorption spectra of transition metals-doped soda lime silica glasses, Phys. B Condens. Matter 405 (5) (2010) 1294−1300.

[10] J. Xiong, S.N. Das, J.P. Kar, J.-H. Choi, J.-M. Myoung, A multifunctional nanoporous layer created on glass through a simple alkali corrosion process, J. Mater. Chem. 20 (45) (2010) 10246−10252.

[11] J.A. von Fraunhofer, Adhesion and cohesion, Int. J. Dent. 2012 (2012) 1−8.

[12] J. Wang, Y. Li, X. Liang, Y. Liu, Research of adhesion force between dust particles and insulator surface using atomic force microscope, High Voltage Eng. 39 (6) (2013) 1352−1359.

[13] S.L. Mkoma, W. Maenhaut, X. Chi, W. Wang, N. Raes, Characterisation of PM10 atmospheric aerosols for the wet season 2005 at two sites in East Africa, Atmos. Environ. 43 (3) (2009) 631−639.

[14] W. Maenhaut, N. Raes, W. Wang, Analysis of atmospheric aerosols by particle-induced X-ray emission, instrumental neutron activation analysis, and ion chromatography, Nucl. Inst. Methods Phys. Res. Sect. B Beam Interact. Mater. Atoms 269 (22) (2011) 2693−2698.

[15] A. Baidourela, K. Zhayimu, Patterns of dust retention by urban trees in oasis cities, Nat. Environ. Pollut. Technol. 14 (1) (2015) 53−57.

[16] S.K. Mishra, R. Agnihotri, P.K. Yadav, S. Singh, M. Prasad, P.S. Praveen, et al., Morphology of atmospheric particles over semi-arid region (Jaipur, Rajasthan) of India: implications for optical properties, Aerosol Air Qual. Res. 15 (3) (2015) 974−984.

[17] W. Petruk, H. Skinner, Characterizing particles in airborne dust by image analysis, JOM 49 (4) (1997) 58−61.

[18] H. Jiang, L. Lu, K. Sun, Experimental investigation of the impact of airborne dust deposition on the performance of solar photovoltaic (PV) modules, Atmos. Environ. 45 (25) (2011) 4299−4304.

[19] M. Gupta, S. Viswanath, Role of metal oxides in the thermal degradation of bisphenol a polycarbonate, J. Thermal Anal. 46 (6) (1996) 1671−1679.

[20] M.J. Allen, H. Salzberg, Organic electrode processes, J. Electrochem. Soc. 105 (7) (1958) 136C.

[21] E. Denisov, L.-P. Oxidation, Comprehensive Chemical Kinetics, Vol. 16, Elsevier, 1980, p. 182.

[22] R.M. Silverstein, G.C. Bassler, Spectrometric identification of organic compounds, J. Chem. Educ. 39 (11) (1962) 546.

[23] T. Ko, H. Chu, Spectroscopic study on sorption of hydrogen sulfide by means of red soil, Spectrochim. Acta Part A Mol. Biomol. Spectrosc. 61 (9) (2005) 2253−2259.

[24] P.S. Kumar, G. Parthasarathy, S. Sharma, R. Srinivasan, P. Krishnamurthy, Mineralogical and geochemical study on carbonate veins of the Salem-Attur Fault Zone, Southern India: evidence for carbonatitic affinity, J. Geol. Soc. India 58 (1) (2001) 15−26.

[25] B. Bhushan, Adhesion and stiction: mechanisms, measurement techniques, and methods for reduction, J. Vacuum Sci. Technol. B Microelectr. Nanometer Struct. Process. Meas. Phenom. 21 (6) (2003) 2262−2296.

[26] B. Jańczuk, W. Wójcik, A. Zdziennicka, Determination of the components of the surface tension of some liquids from interfacial liquid-liquid tension measurements, J. Colloid. Interface. Sci. 157 (2) (1993) 384–393.

[27] E. Chibowski, K. Terpilowski, Surface free energy of polypropylene and polycarbonate solidifying at different solid surfaces, Appl. Surf. Sci. 256 (5) (2009) 1573–1581.

[28] C. Van Oss, R. Good, M. Chaudhury, Mechanism of DNA (Southern) and protein (Western) blotting on cellulose nitrate and other membranes, J. Chromatogr. A. 391 (1987) 53–65.

[29] C. Dayong, L. Wenchang, L. Rongbin, Z. Weihong, Y. Mei, On the accuracy of the X-ray diffraction quantitative phases analysis method in Inconel 718, J. Mater. Sci. 39 (2) (2004) 719–721.

[30] G.J. Ram, A.V. Reddy, K.P. Rao, G.M. Reddy, Improvement in stress rupture properties of Inconel 718 gas tungsten arc welds using current pulsing, J. Mater. Sci. 40 (6) (2005) 1497–1500.

[31] B. Bhushan, Y.C. Jung, Micro- and nanoscale characterization of hydrophobic and hydrophilic leaf surfaces, Nanotechnology. 17 (11) (2006) 2758.

[32] H.-M. Kwon, A.T. Paxson, K.K. Varanasi, N.A. Patankar, Rapid deceleration-driven wetting transition during pendant drop deposition on superhydrophobic surfaces, Phys. Rev. Lett. 106 (3) (2011) 036102.

[33] B. Bhushan, Y.C. Jung, K. Koch, Self-cleaning efficiency of artificial superhydrophobic surfaces, Langmuir 25 (5) (2009) 3240–3248.

[34] D. Murakami, H. Jinnai, A. Takahara, Wetting transition from the Cassie–Baxter state to the Wenzel state on textured polymer surfaces, Langmuir 30 (8) (2014) 2061–2067.

[35] C.V. Oss, R.J. Good, R. Busscher, Estimation of the polar surface tension parameters of glycerol and formamide, for use in contact angle measurements on polar solids, J. Dispersion Sci. Technol. 11 (1) (1990) 75–81.

[36] https://www.twi-global.com/technical-knowledge/faqs/faq-what-are-the-typical-values-of-surface-energy-for-materials-and-adhesives, 2018.

[37] A. Al-Sharafi, B.S. Yilbas, H. Ali, A.Z. Sahin, Internal fluidity of a sessile droplet with the presence of particles on a hydrophobic surface, Numer. Heat Transfer Part A Appl. 70 (10) (2016) 1118–1140.

[38] G. Smith, A. Ben-David, P. Swift, A new type of TiN coating combining broad band visible transparency and solar control, Renew. Energy 22 (1–3) (2001) 79–84.

[39] S. Ishii, R.P. Sugavaneshwar, T. Nagao, Titanium nitride nanoparticles as plasmonic solar heat transducers, J. Phys. Chem. C 120 (4) (2016) 2343–2348.

[40] Y. Yin, L. Hang, S. Zhang, X. Bui, Thermal oxidation properties of titanium nitride and titanium–aluminum nitride materials—a perspective for high temperature air-stable solar selective absorber applications, Thin. Solid Films. 515 (5) (2007) 2829–2832.

[41] X.-H. Gao, Z.-M. Guo, Q.-F. Geng, P.-J. Ma, A.-Q. Wang, G. Liu, Enhanced optical properties of TiN-based spectrally selective solar absorbers deposited at a high substrate temperature, Solar Energy Mater. Solar Cells 163 (2017) 91–97.

[42] H.C. Barshilia, Growth, characterization and performance evaluation of Ti/AlTiN/AlTiON/AlTiO high temperature spectrally selective coatings for solar thermal power applications, Solar Energy Mater. Solar Cells 130 (2014) 322–330.

[43] N. Selvakumar, H.C. Barshilia, Review of physical vapor deposited (PVD) spectrally selective coatings for mid- and high-temperature solar thermal applications, Solar Energy Mater. Solar Cells 98 (2012) 1–23.

[44] F. Cao, L. Tang, Y. Li, A.P. Litvinchuk, J. Bao, Z. Ren, A high-temperature stable spectrally-selective solar absorber based on cermet of titanium nitride in SiO_2 deposited on lanthanum aluminate, Solar Energy Mater. Solar Cells 160 (2017) 12−17.

[45] D. Jiang, W. Yang, A. Tang, A refractory selective solar absorber for high performance thermochemical steam reforming, Appl. Energy 170 (2016) 286−292.

[46] G. Wang, S. Liu, Porous titanium nitride microspheres on Ti substrate as a novel counter electrode for dye-sensitized solar cells, Mater. Lett. 161 (2015) 294−296.

[47] J. Feng, S. Zhang, X. Liu, H. Yu, H. Ding, Y. Tian, et al., Solar selective absorbing coatings TiN/TiSiN/SiN prepared on stainless steel substrates, Vacuum 121 (2015) 135−141.

[48] L. An, S.T. Ali, T. Søndergaard, J. Nørgaard, Y.-C. Tsao, K. Pedersen, Optimization of TiAlN/TiAlON/Si_3N_4 solar absorber coatings, Solar Energy 118 (2015) 410−418.

[49] G. Matenoglou, S. Logothetidis, S. Kassavetis, Durable TiN/TiNx metallic contacts for solar cells, Thin. Solid. Films. 511 (2006) 453−456.

[50] B. Yilbas, S. Shuja, Laser treatment and PVD TiN coating of Ti−6Al−4V alloy, Surf. Coat. Technol. 130 (2−3) (2000) 152−157.

[51] S.C. Costa, A.S.A. Diniz, L.L. Kazmerski, Dust and soiling issues and impacts relating to solar energy systems: literature review update for 2012−2015, Renew. Sustain. Energy Rev. 63 (2016) 33−61.

[52] T. Sarver, A. Al-Qaraghuli, L.L. Kazmerski, A comprehensive review of the impact of dust on the use of solar energy: history, investigations, results, literature, and mitigation approaches, Renew. Sustain. Energy Rev. 22 (2013) 698−733.

[53] T. Sueto, Y. Ota, K. Nishioka, Suppression of dust adhesion on a concentrator photovoltaic module using an anti-soiling photocatalytic coating, Solar Energy 97 (2013) 414−417.

[54] A. Ustaoglu, J. Okajima, X.R. Zhang, S. Maruyama, Performance evaluation of a nonimaging solar concentrator in terms of optical and thermal characteristics, Environ. Prog. Sustain. Energy 35 (2) (2016) 553−564.

[55] B. Yilbas, H. Ali, M. Khaled, N. Al-Aqeeli, N. Abu-Dheir, K. Varanasi, Characteristics of laser-textured silicon surface and effect of mud adhesion on hydrophobicity, Appl. Surf. Sci. 351 (2015) 880−888.

[56] Z. Guan, S. Yu, K. Hooman, H. Gurgenci, J. Barry, Dust characterisation for solar collector deposition and cleaning in a concentrating solar thermal power plant, Heat Exch. Fouling Clean. (2015) 301−307.

[57] A.S. Modaihsh, M.O. Mahjou, Falling dust characteristics in Riyadh city, Saudi Arabia during winter months, APCBEE Proc. 5 (2013) 50−58.

[58] G. Hassan, B. Yilbas, S.A. Said, N. Al-Aqeeli, A. Matin, Chemo-mechanical characteristics of mud formed from environmental dust particles in humid ambient air, Sci. Rep. 6 (2016) 30253.

[59] E. McCafferty, The electrode kinetics of pit initiation on aluminum, Corros. Sci. 37 (3) (1995) 481−492.

[60] E. McCafferty, Sequence of steps in the pitting of aluminum by chloride ions, Corros. Sci. 45 (7) (2003) 1421−1438.

[61] J. Van den Brand, O. Blajiev, P. Beentjes, H. Terryn, J. De Wit, Interaction of anhydride and carboxylic acid compounds with aluminum oxide surfaces studied using infrared reflection absorption spectroscopy, Langmuir 20 (15) (2004) 6308−6317.

[62] J. Van den Brand, O. Blajiev, P. Beentjes, H. Terryn, J. De Wit, Interaction of ester functional groups with aluminum oxide surfaces studied using infrared reflection absorption spectroscopy, Langmuir 20 (15) (2004) 6318−6326.

[63] Y.-X. Zhang, Y. Jia, Z. Jin, X.-Y. Yu, W.-H. Xu, T. Luo, et al., Self-assembled, monodispersed, flower-like γ-AlOOH hierarchical superstructures for efficient and fast removal of heavy metal ions from water, CrystEngComm 14 (9) (2012) 3005−3007.

[64] R. Casasola, J.M. Rincón, M. Romero, Glass−ceramic glazes for ceramic tiles: a review, J. Mater. Sci. 47 (2) (2012) 553−582.

[65] N. Kuangwoo, P. Yunkwon, H. Byeoungju, S. Dongha, S. Insang, P. Jaemoon, et al., Piezoelectric properties of aluminum nitride for thin film bulk acoustic wave resonator, J. Korean Phys. Soc. 47 (2005) 309−312.

[66] S. Klimenko, I. Podchernjaeva, V. Beresnev, V. Panashenko, S.A. Klimenko, M.Y. Kopeikina, et al., B 2 ion-plasma coating for cutting tools of cBN-based polycrystalline superhard material, J. Superhard Mater. 36 (3) (2014) 208−216.

[67] C.-Y. Hsieh, C.-N. Lin, S.-L. Chung, J. Cheng, D.K. Agrawal, Microwave sintering of AlN powder synthesized by a SHS method, J. Eur. Ceram. Soc. 27 (1) (2007) 343−350.

[68] C. Ozgit-Akgun, E. Goldenberg, A.K. Okyay, N. Biyikli, Hollow cathode plasma-assisted atomic layer deposition of crystalline AlN, GaN and Al x Ga 1 − x N thin films at low temperatures, J. Mater. Chem. C 2 (12) (2014) 2123−2136.

[69] S.H. Lee, B.J. Kim, H.H. Kim, J.J. Lee, Structural analysis of AlN and (Ti1 − X Al X) N coatings made by plasma enhanced chemical vapor deposition, J. Appl. Phys. 80 (3) (1996) 1469−1473.

[70] B.S. Yilbas, Laser texturing of zirconia surface with presence of TiC and B4C: surface hydrophobicity, metallurgical, and mechanical characteristics, Ceram. Int. 40 (10) (2014) 16159−16167.

[71] H.-W. Zan, K.-H. Yen, P.-K. Liu, K.-H. Ku, C.-H. Chen, J. Hwang, Low-voltage organic thin film transistors with hydrophobic aluminum nitride film as gate insulator, Org. Electr. 8 (4) (2007) 450−454.

[72] B. Yilbas, N. Al-Aqeeli, C. Karatas, Laser control melting of alumina surfaces with presence of B4C particles, J. Alloys Compd. 539 (2012) 12−16.

[73] B. Yilbas, S. Akhtar, C. Karatas, Laser carbonitriding of alumina surface, Opt. Lasers Eng. 49 (3) (2011) 341−350.

[74] R. Khan, A. Yerokhin, T. Pilkington, A. Leyland, A. Matthews, Residual stresses in plasma electrolytic oxidation coatings on Al alloy produced by pulsed unipolar current, Surf. Coat. Technol. 200 (5−6) (2005) 1580−1586.

[75] B. Yilbas, C. Karatas, A. Arif, B.A. Aleem, Laser gas assisted nitriding of alumina surfaces, Surf. Eng. 25 (3) (2009) 235−240.

[76] H. Park, D.U. Choe, Reversible synthesis of colloidal aluminum oxyhydroxide nanoplatelets from aluminum oxides, J. Ceram. Process. Res. 11 (2) (2010) 191−197.

[77] S. He, M. Zheng, L. Yao, X. Yuan, M. Li, L. Ma, et al., Preparation and properties of ZnO nanostructures by electrochemical anodization method, Appl. Surf. Sci. 256 (8) (2010) 2557−2562.

[78] C.E. Kennedy, Review of Mid-to High-Temperature Solar Selective Absorber Materials, National Renewable Energy Laboratory, Golden, CO, 2002.

[79] F. Bensebaa, D. Di Domenicantonio, L. Scoles, D. Kingston, P. Mercier, G. Marshall, Alternative coating technologies for metal−ceramic nanocomposite films: potential application for solar thermal absorber, Int. J. Low-Carbon Technol. 11 (3) (2016) 370−374.

[80] F. Mammadov, Study of selective surface of solar heat receiver, Int. J. Energy Eng. 2 (4) (2012) 138−144.

[81] L. Rebouta, A. Sousa, M. Andritschky, F. Cerqueira, C. Tavares, P. Santilli, et al., Solar selective absorbing coatings based on AlSiN/AlSiON/AlSiOy layers, Appl. Surf. Sci. 356 (2015) 203−212.

[82] X. Wang, L. Jiang, M. Du, L. Hao, X. Liu, Q. Yu, The degradation of solar thermal absorption coatings, Energy Proc. 49 (2014) 1747−1755.

[83] X. Zhang, R. Cai, J.Y. Wang, F. Liu, L.Y. Wu, Easy-to-clean multifunctional coatings by sol−gel processing for polymer substrates, J. Sol-Gel Sci. Technol. 75 (1) (2015) 17−24.

[84] S. Elmaghrum, A. Gorin, R.K. Kribich, B. Corcoran, R. Copperwhite, C. McDonagh, et al., Development of a sol−gel photonic sensor platform for the detection of biofilm formation, Sens. Actuators B Chem. 177 (2013) 357−363.

[85] B. Yilbas, A. Matthews, C. Karatas, A. Leyland, M. Khaled, N. Abu-Dheir, et al., Laser texturing of plasma electrolytically oxidized aluminum 6061 surfaces for improved hydrophobicity, J. Manuf. Sci. Eng. 136 (5) (2014) 054501.

[86] B. Yilbas, Laser treatment of zirconia surface for improved surface hydrophobicity, J. Alloys Compd. 625 (2015) 208−215.

[87] R. Krishnan, S. Dash, R. Kesavamoorthy, C.B. Rao, A. Tyagi, B. Raj, Laser surface modification and characterization of air plasma sprayed alumina coatings, Surf. Coat. Technol. 200 (8) (2006) 2791−2799.

[88] Y. Adraider, Y. Pang, F. Nabhani, S. Hodgson, Z. Zhang, Laser-induced deposition of sol−gel alumina coating on stainless steel under wet condition, Surf. Coat. Technol. 205 (23−24) (2011) 5345−5349.

[89] A. Ibrahim, H. Salem, S. Sedky, Excimer laser surface treatment of plasma sprayed alumina−13% titania coatings, Surf. Coat. Technol. 203 (23) (2009) 3579−3589.

[90] W. Meng, J. Sell, T. Perry, L. Rehn, P. Baldo, Growth of aluminum nitride thin films on Si (111) and Si (001): structural characteristics and development of intrinsic stresses, J. Appl. Phys. 75 (7) (1994) 3446−3455.

[91] A. Wright, Elastic properties of zinc-blende and wurtzite AlN, GaN, and InN, J. Appl. Phys. 82 (6) (1997) 2833−2839.

[92] K. Jagannadham, A. Sharma, Q. Wei, R. Kalyanraman, J. Narayan, Structural characteristics of AlN films deposited by pulsed laser deposition and reactive magnetron sputtering: a comparative study, J. Vacuum Sci. Technol. A Vac. Surf. Films 16 (5) (1998) 2804−2815.

[93] C.-C. Sun, S.-C. Lee, W.-C. Hwang, J.-S. Hwang, I.-T. Tang, Y.-S. Fu, Surface free energy of alloy nitride coatings deposited using closed field unbalanced magnetron sputter ion plating, Mater. Transact. 47 (10) (2006) 2533−2539.

[94] B. Yilbas, S. Akhtar, C. Karatas, Laser cutting of alumina tiles: heating and stress analysis, J. Manuf. Processes 15 (1) (2013) 14−24.

Chapter 7

Application of Water Droplet for Self-Cleaning of Surfaces

Chapter Outline

7.1 INTRODUCTION

The settling of dust on surfaces degrades the performance of the energy-harvesting devices in terms of efficiency and power output and, in some cases, the efforts required to remove the dust particles for restoring the device performance are significantly high. However, several techniques have been proposed to remove dust particles from surfaces, and most of these methods call for sophisticated devices requiring operating power. In addition, the scarcity of clean water limits practical applications of waterjet and water film cleaning of surfaces. Adopting self-cleaning surfaces can minimize external energy usage in energy-harvesting devices and provide an effective cleaning process for removing dust particles. In general, the self-cleaning process utilizes surfaces with hydrophobic characteristics, where the low free energy of the surface and reduced particle contact area at the surface, due to air gaps within the surface texture, allow only weak adhesion of particles on the surface. The low contact angle hysteresis of hydrophobic surfaces enables water-droplet rolling on an inclined hydrophobic surface. This, in turn, facilitates the removal of weakly adhered dust particles from the

Self-Cleaning of Surfaces and Water Droplet Mobility. DOI: https://doi.org/10.1016/B978-0-12-814776-4.00007-0
375

hydrophobic surface by rolling water droplets. The surface hydrophobicity mainly depends on the interfacial energies of the solid and liquid, the surface texture, and the Laplace pressure of the liquid droplet. The surface hydrophobicity of a substrate can be considerably improved by chemical modification and by increasing the surface roughness. On the other hand, a large number of studies have been conducted to understand water-droplet mobility, rolling, and impact on hydrophobic surfaces [1−4]. Some studies considered droplet mobility through thermal excitation of the liquid droplet utilizing the thermocapillary effect [5]. In this case, rotating circulation cells were formed inside the droplet due to the combination of the Marangoni and buoyancy currents. This, in turn, gave rise to the droplet fluid inertia force, which became greater than the adhesion and fluid shear forces during the thermal excitation period [5]. For textured isothermal hydrophobic surfaces, both the static contact angle and rolling angle increased with an increase in the micropillar height. When the micropillar spacing was increased, the rolling angle decreased; however, the change in the static contact angle was irregular. On superhydrophobic surfaces, the spreading coefficient of the droplet was affected by both the static contact angle and the rolling angle, and the rebounding coefficient of the droplet was highly associated with the rolling angle, where the bigger the inclination angle of the surface, the smaller the rebounding coefficient of the droplet [6]. Newer models have tried to overcome this shortcoming of current models by incorporating the adhesion energy of a droplet using the relationship between the solid and liquid contact areas [7]. Although hydrophobic surfaces show significant asymmetry in the advancing-to-receding profiles of droplets, superhydrophobic surfaces demonstrate advancing-to-receding profiles with different behavior [8]. This, in turn, results in larger force requirements for droplet rolling than that on superhydrophobic surfaces. In addition, the lateral retention (adhesion) forces developed on hydrophobic surfaces are much higher than those formed on superhydrophobic surfaces [8]. Consequently, the inclination angle for droplet rolling to achieve self-cleaning is lower for superhydrophobic surfaces than for hydrophobic surfaces. However, sliding can occur during rolling, which influences the dynamic characteristics of rolling droplets [9]. Asymmetrically patterned surfaces can favor preferential droplet transport through sliding [10]. In this case, the unbalanced capillary force developed at the contact line becomes critical for achieving a preferable direction of liquid motion. However, droplet sliding on an asymmetrically structured surface can result in different migration velocities, which depend on the direction of the structure with respect to droplet movement [10]. Therefore, applications of droplet sliding for self-cleaning purposes require proper design of textural features that enable control of the sliding direction.

Environmental dusts have various shapes and sizes and can scatter and absorb incident solar radiation while reducing the amount of solar radiation that reaches the surface. The scattering and absorption of incident radiation can be detrimental to the solar-harvesting device as the thickness of the dust layer increases on the surface over time. Therefore, the removal of dust from the surface becomes essential to sustain the efficient operation of solar-harvesting devices. On the other hand, the dynamic motion of a water droplet remains critical for removing dust particles from hydrophobic surfaces. Although many research studies have examined water-droplet rolling on hydrophobic surfaces [1−3] and environmental dust characteristics [11−13], their main focus was the droplet dynamics during rolling and the assessment of the environmental dust characteristics. Dust removal from inclined hydrophobic surfaces by water-droplet rolling is often left for future study. In addition, droplet addition to hydrophobic surfaces and droplet impingement on the self-cleaning characteristics of surfaces have been studied [14]. The rolling dynamics of droplet and the mechanism of dust removal from hydrophobic surfaces, which involves cloaking of the dust particles by water, were left for future investigations. The state of water droplet and its dynamics on hydrophobic surfaces and formulation of droplet motion on the inclined surfaces were presented previously [15−20]; in which case, the influence of dust particles, located on the droplet path, on the dynamic characteristics of droplet is not the focus in the previous studies. Dust particles can modify the droplet dynamics on the surface while influencing the performance of the self-cleaning capacity of droplet on hydrophobic surfaces. Consequently, in the present section, dust particle removal from hydrophobic surfaces is analyzed, and the droplet dynamics influencing the dust removal process are presented in line with the previous studies [21].

7.2   DUST PARTICLE REMOVAL THROUGH ROTATING DISK

Although many methods for self-cleaning of active surfaces have been suggested, some of these methods are expensive and the durability of such surfaces under solar radiation is the main concern. Improvement of hydrophobicity of active surfaces is one of the promising methods for self-cleaning applications; however, the requirement of water for surface cleaning and thermal management of treated hydrophobic surfaces under high-intensity solar radiation limit the practical applications of such an approach [22]. Recent developments in coating technology also offer solutions for self-cleaning applications of surfaces; however, the process is, in general, expensive and has a tendency to lower the transmittance of ultraviolet and visible radiation reaching on photoactive surfaces [23]. One of the techniques for surface dry cleaning applications involves spinning of the protective cover of the photovoltaic panels. Although the spinning process requires an external electrical energy source for the rotation, the photovoltaic panel can fulfill this energy requirement from the additional energy harvesting from the sun with a dust-free protective cover surface. Since the mechanical motion of the dust particles involves various forces including adhesion, friction, lift, drag, and centrifugal forces during the spinning of the protective cover, scale analysis of the resulting forces acting on the dust particles and experimental study in relation to the dust removal via spinning of protective cover is essential.

Several research studies were carried out to examine the motion of dust particles under the rotational action of flat surfaces. The effect of rotation on a layer of a ferromagnetic fluid permeated with dust particles was studied by Sunil et al. [24]. They introduced the oscillatory modes of rotation for the removal of the dust particles. The self-powered cleaning of air pollution by wind-driven triboelectric nanogenerator was investigated by Chen et al. [25]. They presented a self-powered air cleaning system focusing on dust removal with the use of a rotating triboelectric nanogenerator driven by the wind power. A study on improved dust removal by cyclone was carried out by Far et al. [26]. Their findings revealed that the exhaust dusty flow of a cyclone could be recycled to the jet-impingement chamber after dedusting. The particle removal efficiency from a wafer surface by means of a buffing disk was examined by Fernando et al. [27]. They developed a mathematical model that considered the toppling of a particle as a result of forces due to friction, hydrodynamic drag, adhesion, and capillary phenomena. The analysis provided correlation of the particle removal efficiency with the flow rate, buffing disk pressure, and relative rotational speed. The motion of micrometer-sized spherical particles exposed to a transient radial flow was investigated by Gonzalez-Avila et al. [28]. They indicated that the expansion of a strong shear layer loosened the particles from the surface through particle spinning and secondly an unsteady boundary layer generated an attractive force, thus, collecting the contamination. The particle removal using the noncontact brush scrubbing for postchemical-mechanical-planarization cleaning was studied by Chein and Liao [29]. They showed that the dominant force to achieve particle removal using a rolling mechanism was the drag force and the electrical double layer; the thermophoretic forces had an insignificant effect on the particle removal mechanism.

Although particle removal from surfaces has been studied previously [27,28], research on the chemo-mechanical effects on the removal of dust particles from rotating surfaces is needed. In the present section, environmental dust removal through rotating disk is considered in relation to surface cleaning of plates. The forces acting on the dust particles are formulated and are predicted in line with the experimental conditions. The adhesion force between the dust residue and the polycarbonate (PC) surface after the tests are measured using the atomic force microscope. The analysis and the findings are presented in line with the previous study [21].

7.2.1 Dynamics of Dust Particles on Rotating Disk

The adhesion of dust particles on solid surfaces is governed by the multi-forces including van der Waals, gravitational, ionic, electrostatic, friction, lift, drag, centrifugal forces, etc. In addition, the relative humidity, contact area, surface roughness, agglomeration, time of contact, temperature, etc., affect the adhesion between the particles and the surface [30]. Fig. 7.1 shows the forces on dust particles located on a rotating disk.

In general, under dry and electrodynamically neutral ambient air, the van der Waals force can be considered to be the most dominant force in particle adhesion [31]. Johnson et al. [32] developed an adhesion force model that considers the van der Waals force of spherical particles on flat surfaces. This model is known as the JKR model. In this case, the adhesion force can be written as $F = (3/2)\pi R\gamma$, where R is the radius of the particle and γ is the surface energy between the two surfaces. This equation is valid for large, soft bodies with high surface energy [32]. Derjaguin et al. [33] developed a model similar to the JKR model but with different constants known as the DMT (after Derjaguin, Muller, and Toporov) model. The adhesion force is $F = 2\pi R\gamma$. However, this equation is appropriate for small, hard solid particles with low surface energy [31]. Hamaker [34] developed a model for formulation of the adhesion force; in which case, the adhesion force is $F = aR/12z^2$, where a is the Hamaker constant and z is the separation distance between the particles and the surface, normally 0.3 or 0.4 nm [31]. The Hamaker model omitted the particle contact area to the surface. Moreover, Rabinovich et al. [35] modified the Rump model, which includes van der walls force due to a particle on the rough surface. They incorporated statistically the effect of root mean square of the asperity of the rough surface (RMS) on the force calculations. The adhesion force is written as $F = (aR/6z^2)\left(\left(1/(1 + \frac{R}{1.48RMS})\right) + \left(1/(1 + \frac{(1.48RMS)}{z^2}))^2\right)\right)$ where a is the Hamaker constant, R is the particle radius, z is gap in between particle and the surface, and RMS is the root mean square of the asperity of the rough surface. In the present case, the surface is smooth; however, the dust particles have irregular shape and surface texture. Consequently, the Rump−Rabinovich model is adopted to determine the adhesion force between the dust particles and the surface.

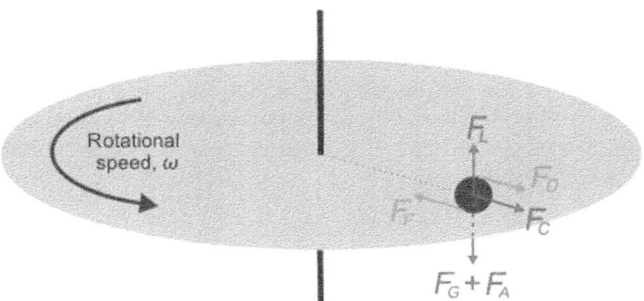

FIGURE 7.1   A schematic view of forces acting on the dust particles during spinning.

One of the important forces acting on a particle is the gravitational force; it can be written for a spherical particle as $F_g = m \times g = (4/3)\pi R^3 \rho g$ where m is the mass of the particle, ρ is the density of the dust particle, and g is the gravitational acceleration. Since the particles are located on a rotating disk, a lift force is generated. The inertial lift causes the force acting on the particle with the point contact between particle the surface. However, the other lift force is also generated because of the flow in the friction layer at the surface. The combination of these forces generates the lifting force acting on the particles located at the rotating surface. Leighton et al. [36] and Cherukat at al. [37,38] derived the lift force from the Navier−Stokes equation for a spherical shaped particle at the surface and low Reynolds number, below unity. Then Zhang et al. [39] simplified this as $F_L = (9.22\mu^2/\rho)(Re^*)^3$, where μ is the dynamic viscosity of the fluid and Re^* is the shear Reynolds number defined by $Re^* = (Ru^*/v)$, where v is the kinetic viscosity and u^* is the friction velocity. $u^* = \sqrt{(2\tau/\rho)}$, where τ is shear stress. Moreover, White [40] derived the shear stress for a rotating disk system incorporating the Navier−Stokes equation. He proposed the equation $\tau_o = \rho r G_o' \sqrt{v\omega^3}$, where τ_o is the shear stress at the wall ($y = 0$), r is the position of the particle from the center of rotation, G_o' is the dimensionless constant at the wall ($G_{z=0}' = -0.61592$), and ω is the angular velocity of the disk.

The particle on the rotating disk suffers from drag force, which can be divided into three types: first, pressure drag force when the area normal to the flow is relatively wide; second, a shear friction drag force caused by a friction force of flow parallel to the particle; and lastly, the combination of pressure and shear drag force. The drag force acting on a spherical shape particle can be considered to be the third type, which is a combination of pressure and shear drag force. O'Neill et al. [41] and Goldman et al. [42] formulated the drag force using the Navier−Stokes equation, which is $F_D = 10.2\pi\mu Ru$, where μ is the fluid viscosity and u is the flow velocity.

The centrifugal force acting on the particle located on the rotating disk can be written as $F = m\omega^2 r = (4/3)\pi R^3 \rho \omega^2 r$. The forces acting on the particles were calculated using Matlab 2014b software. In the simulations, the parameters, which are varied, included the rotational speed of the disk (ω, rad/s), the dust particle radius (R, μm), and the radial distance, where the particle is located, from the center of rotation (r, cm). All the forces including adhesion, gravitational, lift, drag, friction, and centrifugal forces were calculated from the Matlab code. The parameters were varied in the range 0−100 rad/s, 0−10 μm, and 0−10 cm for ω, R, and r, respectively, in line with the experimental conditions.

7.2.2   Experimental Analysis

Elemental analysis and morphological characteristics of the dust particles were analyzed using analytical tools. Scanning electron microscopy (SEM) and energy dispersive spectroscopy (EDS) examinations were performed using a Jeol 6460 electron microscope and X-ray diffraction (XRD) analysis was carried out using a Bruker D8 Advanced with CuKα radiation. A typical setting of XRD was 40 kV and 30 mA and scanning angle (2θ) ranged from 20 to 90 degrees. The atomic force microscope (AFM) tip was made of silicon nitride probes ($r = 20-60$ nm) with a manufacturer-specified force constant, k, of 0.12 N/m.

A disk shape of the polycarbonate wafer of 3 mm thickness and 10 cm diameter were used as workpieces. PC wafer has excellent optical clarity with high toughness and is derived from p-hydroxyphenyl. The PC disks were cleaned ultrasonically prior to tests. The PC disk surface was analyzed using SEM/AFM.

The experiment is designed to simulate the dynamic motion of the dust particles on the PC disk in line with the cleaning applications through dust removal. The equipment used is manufactured by Sparkfun, US [43]. The components included Arduino due and motor shield as controller, DC motor to rotate the PC disk, and direct current (DC) adapter. The source code was made using processing software as interface in the computer to set the parameter and start the rotation. The weight of the PC disk prior and after the dust addition to the surface was recorded. The weight of the disk and dust residues after the tests was also recorded.

7.2.3   Dynamic Analysis of Dust Particles on Rotating Disk

The forces related to the dust particles located on the rotating disk are predicted from the sets equations (Eqs. 7.1–7.11) and the experiments are carried out to identify the amount of dust particles removed from the PC disk surface after completing the rotation of the disk at various rotational speeds. The force simulations were carried out in line with the experimental conditions incorporating the dust particle size, density, and dust locations on the disk surface. It should be noted that the dust density is measured and the averaged dust particle density is in the order of 2600 kg/m^3. The forces acting on the dust particles include van der Waals, electrostatic, capillary, lift, drag, and gravitational forces. However, the adhesion force was modified by using the Rump–Rabinovich model, which incorporates the van der Waals force and the roughness of the surface [35]. The lift force acts in the z-direction, which is opposite to the Rump–Rabinovich and the gravitational forces (Fig. 7.1). The drag and centrifugal forces act on the lateral direction toward the edge of the disk. The friction force acts opposite to the particle motion and is related to the gravitational force. In the analysis, three mechanisms were considered for the removal of the dust particles from the disk surface during the rotation. The first mechanism is associated with lifting; in which case, if the lift force is larger than the combination of the Rump–Rabinovich and the gravitational forces, the particle can be lifted from the surface. The second one is the sliding, which takes place for the case when the centrifugal force is larger than the frictional force. The third mechanism is related to the rolling, which occurs when the centrifugal force is small, but the drag force is higher than the frictional force. In the simulations, the rotational speed of the disk varied from 0 to 100 rad/s and the dust position was altered from the rotational center toward the disk edge within the range of 0−10 cm in line with the experimental conditions. Here, 0 represents the rotational center and 10 cm corresponds to the disk edge. In the simulations, the dust particle size was considered as 10 μm, which is the averaged size of the dust particles in terms of the volume distribution of the dust particles. The tribology tests were carried out to measure the friction coefficient of the PC surface. Fig. 7.2 shows the friction coefficient variation along the scan length on the PC surface. The average friction coefficient was found to be 0.23, which is set in the simulations.

Figs. 7.3−7.5 show the counterplots of ratios of the centrifugal, frictional, drag, lift, and gravitational forces over the adhesion force. To determine the adhesion force between the particle and the disk surface due to the van der Waals force, the Rump−Rabinovich model is considered [35]. Since the adhesion force (F_a) is critical for the removal of dust particles from the disk surface, the forces generated on the dust particles are normalized with the adhesion force. In general, increasing the distance along the disk radius toward the disk edge enhances the centrifugal, drag, and lift forces acting on the dust particles. However, the adhesion force remains almost constant along the disk radius. In addition, increasing the rotational speed of the disk and dust particle size generates similar behavior for the forces. The adhesion force remains higher than the lift, gravitational, drag, and friction forces. At some radial locations on the disk, the centrifugal force becomes larger than the adhesion force, depending on the dust particle size. In this case, the

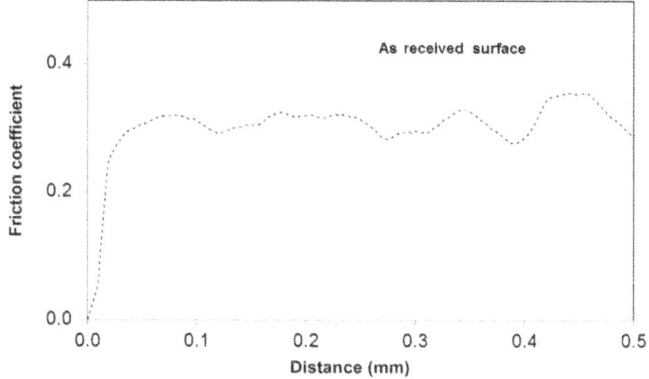

FIGURE 7.2 Friction coefficient for polycarbonate (PC) plane sheet without dust on the surface.

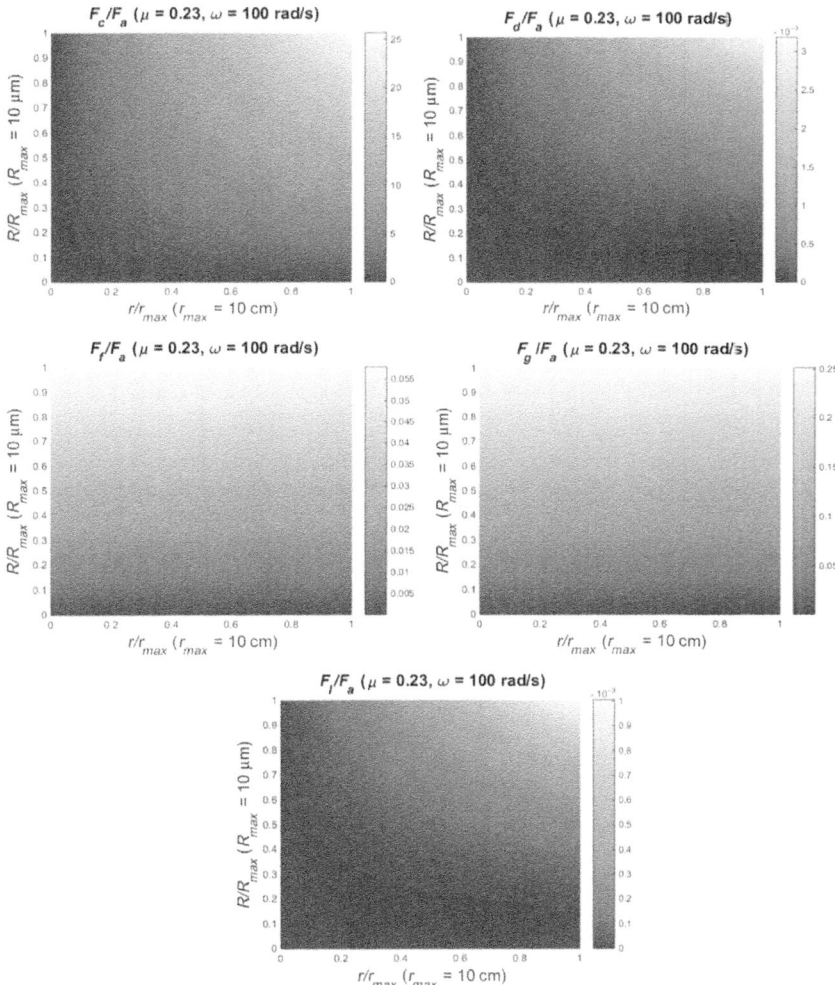

FIGURE 7.3 Contour plots of force ratios along the radial distance (*r*) for different dust particle radius (*R*). The rotational speed is 100 rad/s.

adhesion force remains higher than the centrifugal force for the small particles at locations close to the center of rotation of the disk. Therefore, dust removal from the surface may be possible for large dust particles ($\geq 3 \ \mu m$) located on the disk surface, which is more pronounced for the locations close to the disk edges.

Figs. 7.6—7.8 show the variation of the forces including centrifugal (F_c), gravitational (F_g), adhesion (F_a), frictional (F_f), and drag (F_d) forces. The centrifugal, lift, and drag forces increase with distance along the disk radius;

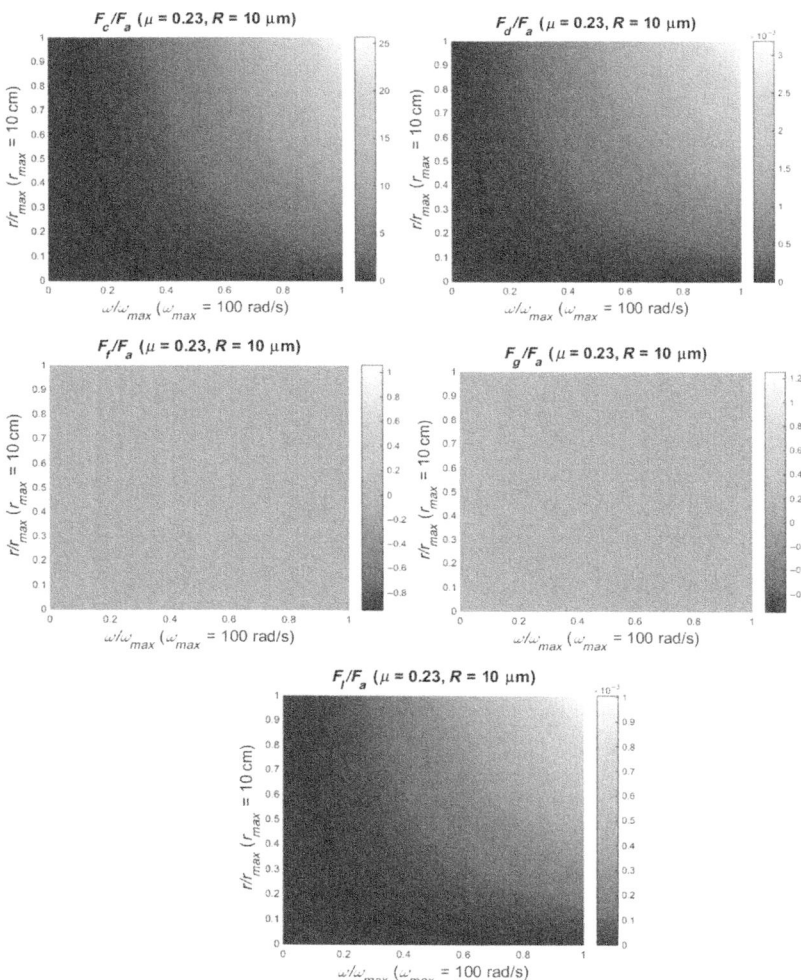

FIGURE 7.4 Contour plots of force ratios along the radial distance (r) for different rotational speed (w). The dust particle radius is 10 μm.

however, the other forces remain almost the same. This behavior is associated with the assumption of constant friction coefficient along the disk radius at the surface and constant mass of the dust particle. The adhesion and gravitational forces remain almost constant along the radius at the surface of the disk. The centrifugal force remains high along the radius and is higher than the other forces. However, the adhesion force becomes higher than the other forces for the particle location close to the rotational center, that is, 2 cm away from the disk center. Therefore, in the close region of the rotating disk, the adhesion force is dominant. In addition, the drag and lift forces become

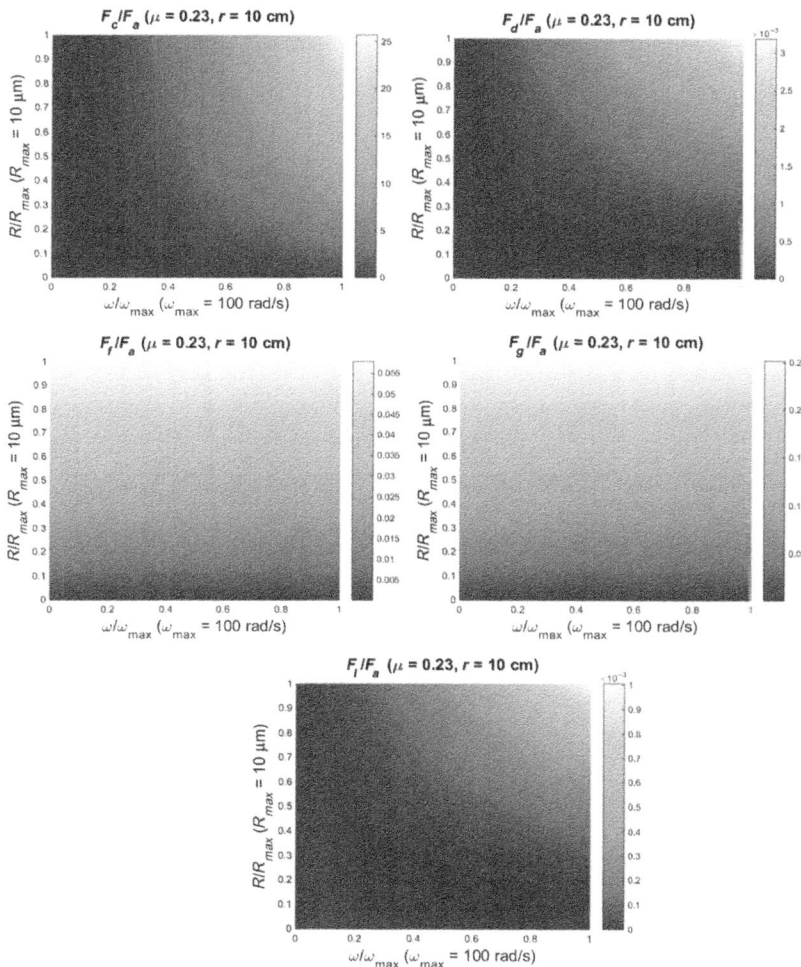

FIGURE 7.5 Contour plots of force ratios for different rotational speed (w) and dust particle radius (R). The radial distance is 10 cm.

extremely small compared to that of the gravitational, friction, centrifugal, and adhesion forces. Similar arguments are also true for small dust particles (5 μm), as can be seen in Fig. 7.6. In the case of force variation with the dust particle size (Fig. 7.7), the lift, drag, gravitational, and centrifugal forces increases with increasing dust particle size. However, this increase is sharp for the dust particles ≤ 3 μm. The adhesion force remains higher than all the other forces for these size particles (≤3 μm). This is true for dust particles located at the edge of the disk. As the dust particle location gets close to the rotational center (5 cm), the adhesion force remains higher than the other

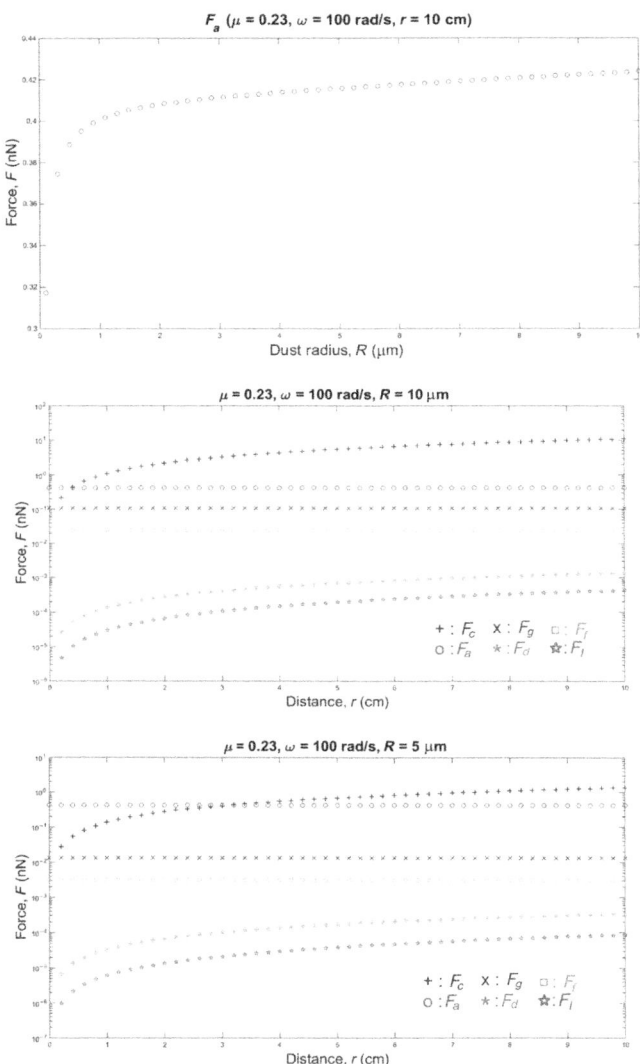

FIGURE 7.6 Semilog plot of forces along radial distance along disk surface for two dust particle sizes. Adhesion force variation along dust particle radius at disk edge and for $\omega = 100$ rad/s.

forces for dust particles $\leq 4\,\mu$m. Consequently, the location of the dust particles at the disk surface is very critical to overcome the adhesion force; in which case, the centrifugal force generated becomes larger than the adhesion force while giving rise to removal of the dust particles with sizes $\geq 3\,\mu$m from the edge of the disk. It should be noted that the adhesion force is not constant along the radial distance on the disk surface; however, the variation

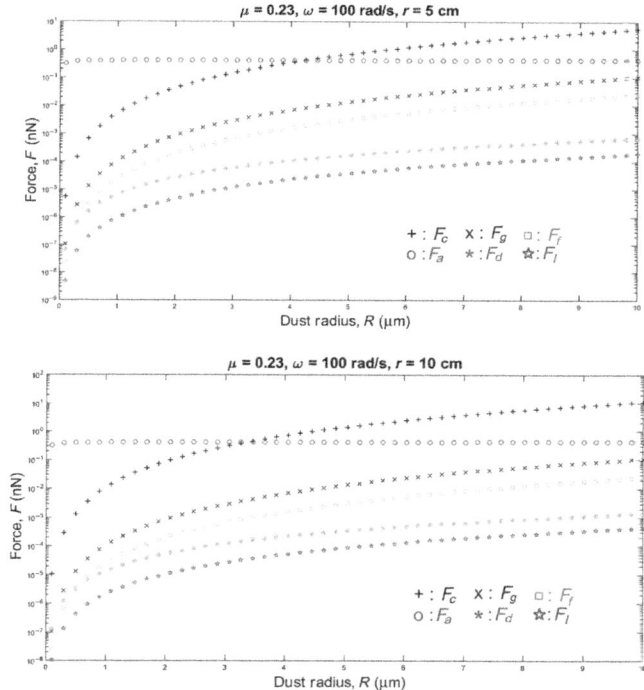

FIGURE 7.7 Semilog plot of forces with dust particle radius (R) for two radial locations on disk.

is not observed along the distance because of the large values of the adhesion force. Nevertheless, the small plot in Fig. 7.6 shows the adhesion force variation with distance. The effect of rotational speed on the forces generated on the dust particles is shown in Fig. 7.8. The centrifugal force is higher than the adhesion force for the rotational speed in the order of 20 Hz for the case where the dust particles are located at the edge of the disk (10 cm). The rotational speed when the centrifugal force becomes higher than the adhesion force occurs almost at 30 Hz for the dust particles located at mid-distance on the disk surface (5 cm). The rotational speed should be large enough to overcome the adhesion force for the particle removal. The dust particle removal from the disk surface is not possible for a certain range of rotational speeds, but this situation changes with particle location at the surface.

In order to investigate the dust removal from the rotating disk, an experiment was carried out and the weight percentage of the dust particles removed from the disk surface recorded. Table 7.1 gives the percentage of dust removed from the surface for various rotational speeds of the disk. It is evident that the percentage of the dust particles removed increases significantly with increasing rotational speed. This behavior is mainly attributed to the

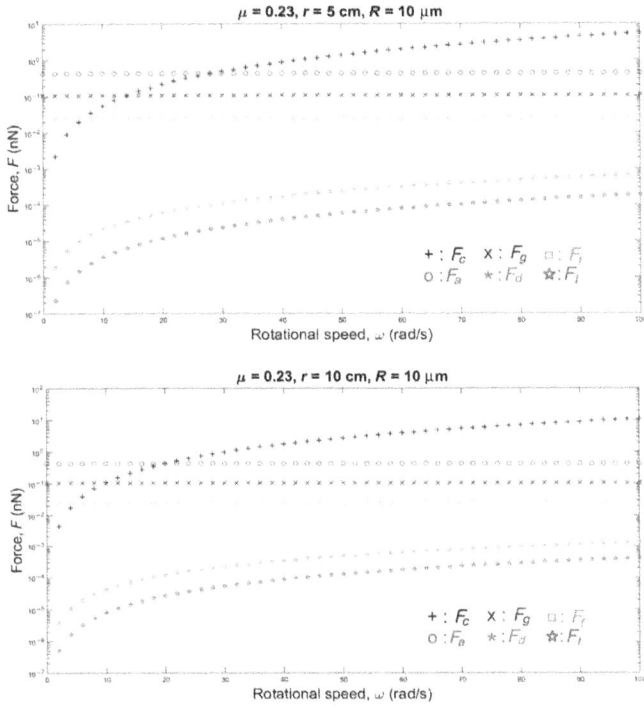

FIGURE 7.8 Semilog plot of forces with rotational speed (ω) for two radial locations on disk.

TABLE 7.1 Dust Removed From Polycarbonate (PC) Disk Surface by Rotation

Speed (rpm)	100	175	250	375
Disk weight (g)	21.998	21.998	21.999	22.000
Disk/dust weight(g)	22.065	22.075	22.068	22.071
Dust weight (g)	0.066	0.076	0.069	0.071
Disk/dust weight after rotation (g)	22.063	22.055	22.026	22.016
Dust loss (g)	0.002	0.020	0.043	0.056
Dust loss percentage (%)	3.167	25.916	61.383	78.230

increase in the centrifugal force at high rotational speeds, which is in agreement with the predictions (Fig. 7.8). To examine the dust residuals on the disk surface after the tests, the geometric features and the elemental composition of the dust residues were analyzed using SEM and EDS.

Fig. 7.9 shows the SEM micrographs of the dust residues on the disk surface. The dust residues comprise fine-size small particles, which attach together at the disk surface. The bright appearance of the dust particles indicates that the electron charge is high at the particle surface. Therefore, the dust residues have a charge field that enhances the adhesion of these particles at the surface. Elemental analysis (Table 7.2) reveals that the dust residues contain alkaline (Na, K) and alkaline earth (Ca) metal compounds. These contribute to the ionic bonding at the surface under the influence of humidity [13], which in turn enhances the adhesion between the dust particles and the disk surface. Therefore, the equation used for the adhesion may not be applicable for this case. This is because the formulation of adhesion based on the Rump—Rabinovich model relies on the van der Waals forces only and the contribution of the ionic compounds and electrostatic forces to

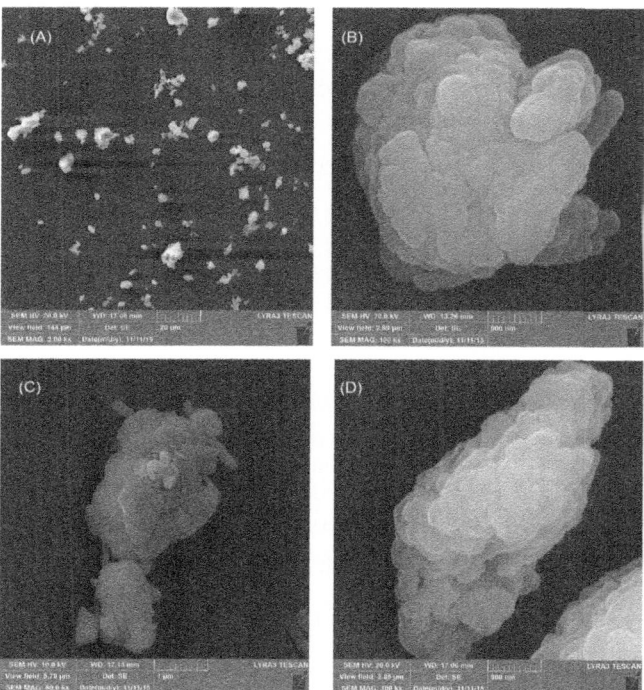

FIGURE 7.9 SEM micrographs of dust residues on polycarbonate (PC) disk: (A) small and large dust residues, (B) dust residues comprise fine-size dust particles, (C) combined dust particles, and (D) elongated dust particles comprise fine-size dust particles.

TABLE 7.2 Elemental Analysis of Dust Residues on the Disk (EDS Data wt.%)

	Weight%	Atomic%
O K	57.93	74.43
Mg K	1.79	1.51
Al K	2.48	1.89
Si K	5.79	4.24
S K	12.92	8.28
K K	0.43	0.23
Ca K	17.63	9.04
Fe K	1.02	0.38

the adhesion force is excluded in the formulation. Consequently, the presence of ionic bonding and electrostatic charge forces in between the dust particles and the disk surface modifies the adhesion force. Nevertheless, the amount of dust residue remaining at the surface is small; therefore, rotation of the disk enables removing the large amount of dust particles from the disk surface.

To assess the adhesion between the dust residuals and the PC surface, an AFM force measurement is used. Fig. 7.10A shows the AFM image of the dust residue on the PC surface while Fig. 7.10B shows the adhesion force obtained from the AFM tip. It should be noted that the sensitivity of the AFM cantilever tip is proportional to the slope of the deflection of the tip while the tip is in contact with the surface. From the deflection relation, the adhesion force can be written as $F = k\sigma\Delta V$, where k is the spring constant of the cantilever tip (N/m), σ is the slope of the displacement over the probe voltage recorded ($\Delta z/\Delta V$, m/V), and ΔV is the voltage recorded during the surface scanning by the tip in the contact mode. In the measurements, the following data was adopted: $k\sigma = 0.960908$ nN/mV. The adhesion force obtained from the AFM measurement is in the order of 3.73 nN while the calculated adhesion force from the Rump−Rabinovich model is in the order of 0.4 nN (Fig. 7.6) for the same size of the dust (1.1 μm) used in the measurements (Fig. 7.10A).

7.3 DROPLET DYNAMICS AND DUST REMOVAL

The dynamics of droplet is important in terms of self-cleaning applications. Droplet rolling and sliding can occur simultaneously depending on the

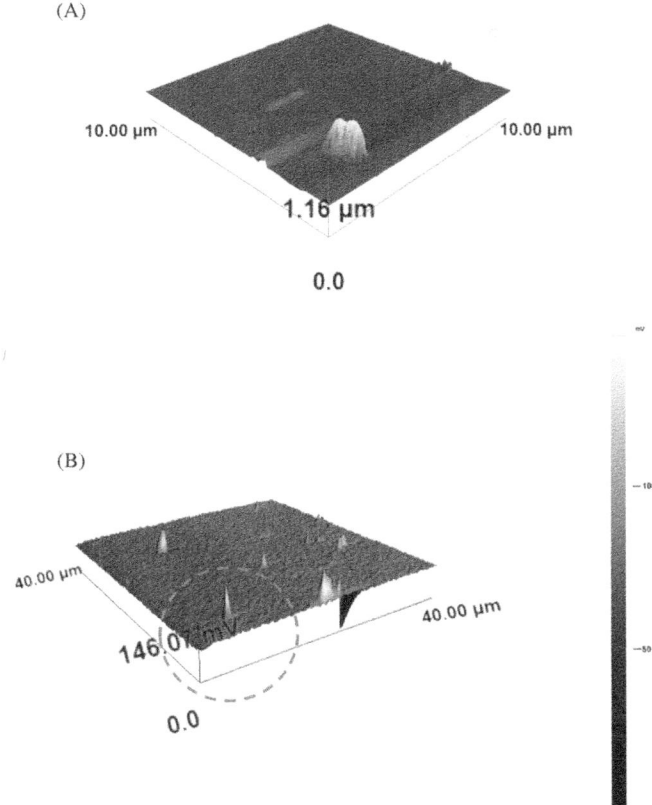

FIGURE 7.10 AFM microimages of dust particle and adhesion force: (A) dust particle on polycarbonate (PC) surface, and (B) tangential force map recorded from AFM. The peak in the *red circle* (gray in print version) represents the tangential force for the dust particle shown above.

surface texture characteristics and wetting state of the surface. In order to generate droplet rolling on surfaces, two basic conditions should be met: the hydrophobic wetting state and low contact angle hysteresis. Droplet takes various shapes during rolling. However, these shapes depend not only on the surface texture characteristics and wetting states, but also on the droplet volume. In this case, large droplets undergo wobbling due to bulging under the gravitational potential while small droplets roll like marbles. Consequently, the force and energy balance on the wetted surface play a key role in the droplet rolling. In the present section, droplet rolling on inclined hydrophobic surfaces and dust particle removal from the surfaces are presented in light of the previous study [44].

7.3.1 Experimental and Texture Characteristics

A PC sheet with dimensions of 30 mm × 200 mm × 3 mm (width × length × thickness) was used as the base. PC was derived from *p*-hydroxyphenyl, which was initially solution crystallized using acetone to create hierarchical textures on the surface. PC wafers were ultrasonically cleaned prior to immersion in acetone for 2 minutes. Several tests were carried out to select the appropriate concertation of acetone and the appropriate immersion duration for crystallization of the PC surface. An acetone concentration of 60% (by volume in water) and immersion duration of 3 minutes were selected in line with a previous study [45]. This arrangement resulted in hierarchical crystal structures with hydrophobic characteristics on the surface.

To generate the lotus effect at the surface to reduce the contact angle hysteresis, functionalized nanosize silica particles were deposited on the solution-crystallized PC surface. The silica nanoparticles were synthesized using a procedure similar to that reported in a previous study [46]. In this case, tetraethyl orthosilicate (TEOS), isobutyltrimethoxysilane (OTES), ethanol, and ammonium hydroxide were used in the synthesis. Prior to the deposition of functionalized silica particles, the crystallized PC surface was washed with piranha solution followed by distilled water to further clean the surface. Solvent casting was applied to deposit the solution onto a glass surface. After all of the solvent was evaporated by vacuum drying, the resulting surface was characterized. The dynamic contact angle of the resulting surface was measured in line with a previous study [47], and a contact angle on the order of 158 degrees with contact angle hysteresis on the order of 2−3 degrees was measured. Consequently, the deposition of functionalized silica particles on the crystallized PC surface resulted in a superhydrophobic surface with significantly reduced contact angle hysteresis.

7.3.2 Droplet Dynamics on Dusty Hydrophobic Surface

The droplet velocity can be formulated after incorporating the energy balance for moving droplet on inclined surfaces. In this case, after considering the conservation of energy, an equation for the droplet velocity on the hydrophobic surface can be formulated. In this case, the change in the potential energy of the droplet prior to rolling remains the same as the summation of the change in the dissipated energy and the kinetic energy of the droplet regardless of location on the inclined surface, that is, $\Delta E_{Tot} - \Delta E_{loss} = \Delta E_{kinetic}$, where ΔE_{tot} ($\Delta E_{tot} = mg\Delta h$, where m is the mass of the droplet, g is the gravitational acceleration, and Δh is the elevation between the droplet location and the reference level) represents the potential energy change of the droplet along the inclined hydrophobic surface. ΔE_{loss} is the dissipation energy, which can be written in the form of $\Delta E_{loss} = \Delta E_{friction} + \Delta E_{deformation} + \Delta E_{retention} + \Delta E_{shear} + \Delta E_{air-drag}$, where

$\Delta E_{friction}$, $\Delta E_{deformation}$, $\Delta E_{retention}$, and $\Delta E_{air-drag}$ are the dissipation due to (1) friction (frictional energy dissipation between the droplet and surface during rolling); (2) elastic deformation of the droplet during wobbling; (3) work done against droplet retention due to dynamic contact angle hysteresis; and (4) air drag during rolling, respectively. Here, the energy dissipation due to friction is $\Delta E_{friction} = \mu \Delta L F_n$, where μ is the dynamic friction coefficient between the water droplet and the hydrophobic surface; ΔL is the distance along the inclined hydrophobic surface; and F_n is the normal force due to the droplet weight. The energy dissipated during deformation of the droplet due to wobbling can be described as $\Delta E_{deformation} \sim \forall_p \gamma_L((D_{h_1} - D_{h_2})/(D_{h_1} D_{h_2}))$, where $\forall p$ is the droplet volume, γ_L is the surface tension of the droplet fluid, D_{h_1} is the instant hydraulic diameter of the droplet at a location on the inclined hydrophobic surface, and D_{h_1} and D_{h_2} are the changes in the hydraulic diameter of the water droplet along the distance ΔL on the inclined hydrophobic surface due to wobbling. The energy dissipation due to the retention force can be described as $\Delta E_{adhesion} \sim (24/\pi^3)\gamma_L Df \Delta L(\cos\theta_R - \cos\theta_A)$, where θ_A is the dynamic advancing angle and θ_R is the dynamic receding angle, which change along the distance ΔL during rolling of the droplet. The energy dissipation due to fluid friction because of the rate of fluid strain can be described as $\Delta E_{shear} \sim A_w(\mu_t(dV_f/dy))\Delta L$, where A_w is the contact area ($A_w = \pi r^2$, where r is the contact area radius); μ_t is the droplet fluid viscosity; V_f is the flow velocity of the droplet; and y is the distance normal to the contact surface. It is assumed that the maximum fluid velocity in the droplet is of the same order as the tangential velocity of the droplet. In the case of air drag loss, $\Delta E_{air-drag} = (1/2)K_L m U_T^2$, where K_L is the loss coefficient due to air drag. On the other hand, the rotation of the droplet can be formulated via incorporating the force balance. The droplet puddling and wobbling modify the line of action of the net force inside the droplet and also alter the dynamic hysteresis of the droplet ($\theta_R - \theta_A$), where θ_R is the receding angle and θ_A is the advancing angle of the droplet during rolling, while changing the droplet retention force on the hydrophobic surface during rolling. The force balance for a steadily rolling droplet around the center of mass when rolling on an inclined surface yields $mg\sin\delta - F_{ad} - F_\tau - F_f - D_a = (2/5)mR\omega^2$, where m is the droplet mass; δ is the inclination angle of the hydrophobic surface; F_{ad}, F_τ, and F_f are the retention, shear, and frictional force between the surface and droplet during rolling, respectively; D_a is the air drag force; R is the droplet radius; and ω is the angle of rotation [44]. The retention force was formulated previously by approximating a three-phase contact line with a single ellipse [48]. Later, a polynomial function was developed using experimental data for the dependence of the contact angle on the position along the three-phase contact line [49]. In this case, the retention force equation yields $F_{ad} = (24/\pi^3)\gamma_{LV}D(\cos\theta_R - \cos\theta_A)$, where γ_{LV} is the surface tension of the liquid on the solid surface; D is the droplet diameter prior to deformation (the same area as the ellipse); θ_R is the receding angle; and θ_A is the advancing angle. Because the hydrophobic surface has texture, the roughness parameter can be introduced in the adhesion force, in line with the Young–Dupre Equation [50]. Hence, it becomes $F_{ad} = (24/\pi^3)\gamma_{LV}Df(cos\theta_R - cos\theta_A)$, where f is the solid-surface fraction (solid–liquid contact fraction).

A shear force is generated when the droplet rolls/slides on the hydrophobic surface due to the rate of fluid strain developed along the contact surface between the water droplet and the hydrophobic surface. Therefore, the shear stress can be expressed as $F_\tau = A_w(\mu(dV/dy))$, where A_w is the contact area ($A_w = \pi r^2$, where r is the contact area radius); μ is the droplet fluid viscosity; V is the flow velocity; and y is the distance normal to the contact surface. The frictional force associated with the droplet and the hydrophobic surface can be presented in terms of the normal force and the friction coefficient of the hydrophobic surface: $F_f = \mu_f F_n$, where μ_f is the friction coefficient of the hydrophobic surface and F_n is the normal force, which is of the same order of the droplet weight (mg, where m is the droplet mass). To obtain the friction coefficient of the hydrophobic surface, AFM friction measurements were utilized and the average friction coefficient was determined to be 0.03. The drag force due to air resistance as a droplet rolls on a surface is related to the pressure drag and frictional drag. However, the simplified form of the drag force for a spherical body due to air resistance is a function of the flow Reynolds number and shape factor. However, the drag force can be related to $D \cong 1/2 C_d \rho_a A_c U_T^2$, where C_d is the drag coefficient [51], and U_T is the air velocity opposing the droplet during droplet movement on the hydrophobic surface. However, it can be considered to be on the same order as the translation velocity of the droplet; in which case, $U_T \cong V$, where V is the translational velocity of the droplet. The droplet rotational speed can be obtained as:

$$\omega = \sqrt{\left(\frac{\frac{5}{2mR}\left(mgsin\delta - \frac{24}{\pi^3}\sigma f(cos\theta_R - cos\theta_A) - \mu_t A_w \frac{\partial u}{\partial y} - \mu_f mg\right)}{1 + \frac{5}{4m}C_d\rho_a A_c R}\right)} \quad (7.1)$$

The tangential velocity (U_T) is determined from the angular rotation of the droplet after introducing the hydraulic radius ($D_H/2$ is the instant hydraulic diameter). The energy balance of the droplet ($\Delta E_{Tot} - \Delta E_{loss} = \Delta E_{kinetic}$) allows formulation of the droplet velocity on the inclined hydrophobic surface, that is:

$$V = \sqrt{\begin{array}{l} 2g[\Delta Lsin\alpha - \mu_f \Delta L - \frac{1}{mg}\frac{24}{\pi^3}\gamma_L Df\Delta L(cos\theta_R - cos\theta_A) \\ -\frac{4\gamma_L}{\rho g\Delta L}\left(\frac{D_{h_1}-D_{h_2}}{D_{h_1}D_{h_2}}\right) - \frac{1}{mg}A_w\left(\mu_t\frac{dV_f}{dy}\right)\Delta L - \frac{1}{2g}K_L U_T^2] \end{array}} \quad (7.2)$$

Fig. 7.11 shows the droplet translational velocity along the hydrophobic surface in the presence of dust particles for three cases incorporating different length scales for droplet acceleration. It should be noted that the location at which droplet rolling is initiated is changed during the experiments. This arrangement provides varying localized acceleration of the droplet prior to reaching the dust particles on the inclined hydrophobic surface. The droplet translational velocity sharply increases toward the location of the dust on the surface. Once the droplet reaches the dust region, the droplet translational velocity increases gradually over the length of the dusty region. The droplet translational velocity increases sharply upon leaving the dusty region. Therefore, droplet acceleration is suppressed by the dust particles on the surface. This may be attributed to one or all of the following: (1) the adhesion of the dust particles on the hydrophobic surface, despite the fact that the adhesion for individual dust particles is small; and (2) the presence of the dust particles increases the friction between the droplet and the surface due to the enhancement in the surface roughness by the dust particles; in which case, the dust particles act as an additional texture with a large texture height on the hydrophobic surface. Fig. 7.12 shows high-speed camera images of the top and side views of the droplet at different intervals when the droplet is in the dusty region. The dust particles are picked up by the droplet as the droplet moves along the dusty region. The droplet picks up almost all of the dust particles along its path.

To assess the dust particle residue left on the hydrophobic surface after droplet motion, a microscopic 3D image of the droplet path was taken and is shown in Fig. 7.13. In general, the droplet picks up almost all of the dust particles on its path; however, a few dust particles remain on the hydrophobic surface along the droplet path (Fig. 7.13). Further examination was carried out to determine the cause of the dust residue remaining on the droplet path. SEM micrographs of typical dust residue are also shown in Fig. 7.13, and Table 7.3 gives the elemental composition of the dust residue. The size of the typical dust residue is in the order of 2 μm, and it has a shape similar to the other dust particles. In addition, the elemental composition (Table 7.3) suggests that silica is the main compound in the dust particles. Therefore, the adhesion of dust residue on the hydrophobic surface is expected to be in the same order as those of the other dusts. Consequently, the possible explanation for the presence of dust residue along the droplet path is neither the shape effect nor the elemental composition of the dust particles, but the surface energy of the dust particle residue. Dust particle(s) can randomly obtain a low surface energy, which may be associated with the prolonged residence of some of the dust particles in air in the region close to the Gulf Sea.

To assess the influence of the surface energy on the dust particles that are picked up from the hydrophobic surface by the water droplet, further experiments were carried out. In this case, the dust particles were functionalized by a surface coating of trichloro($1H,1H,2H,2H$-perfluorooctyl) (PFOTS) via chemical vapor deposition, in line with a previous study [52].

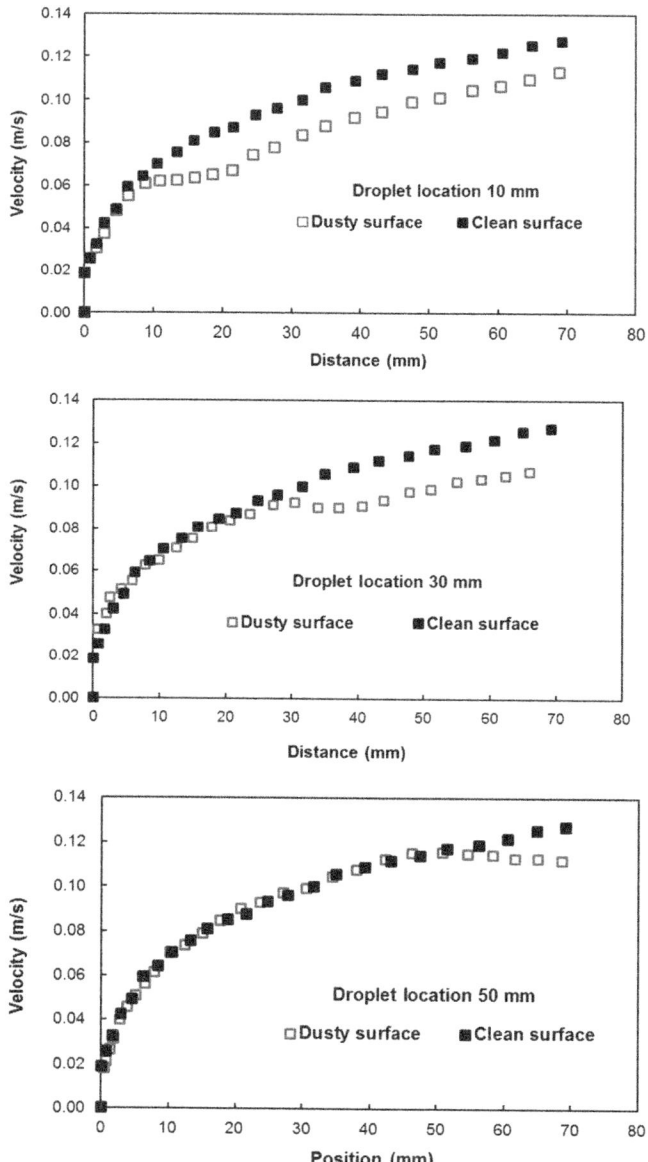

FIGURE 7.11 Translational velocity of the droplet on clean and dusty surfaces for various droplet locations. Droplet location is the distance between droplet initiation and the start of dusty region on the hydrophobic surface.

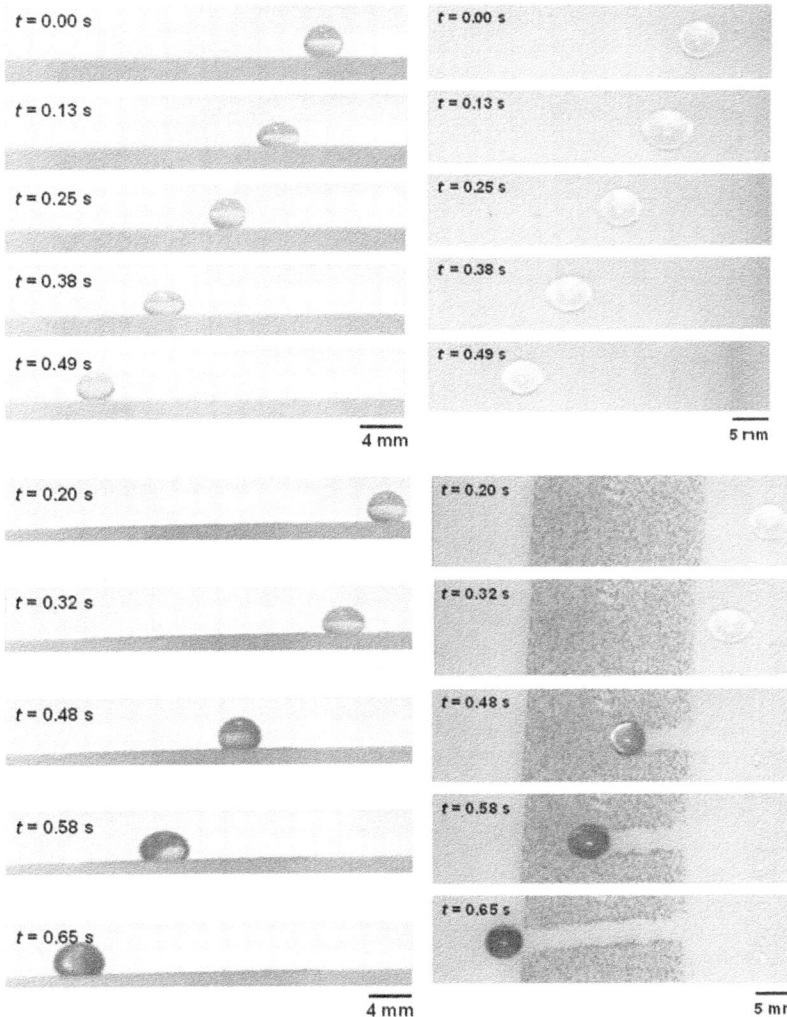

FIGURE 7.12 Side and top images of the droplet obtained from a high-speed camera recording for clean and hydrophobic surfaces at various durations.

The water clocking of the dust particles is important for the water droplet picking up the dust particles from the hydrophobic surface. The cloaking velocity first increases rapidly and then reduces as time progresses. Because water film cloaking occurs opposite to gravity, as the weight of the water film cloaking the dust particles increases, the net driving force opposing gravity for cloaking decreases. Therefore, the water-cloaking experiment was extended to the functionalized dust particles for comparison. Water does not

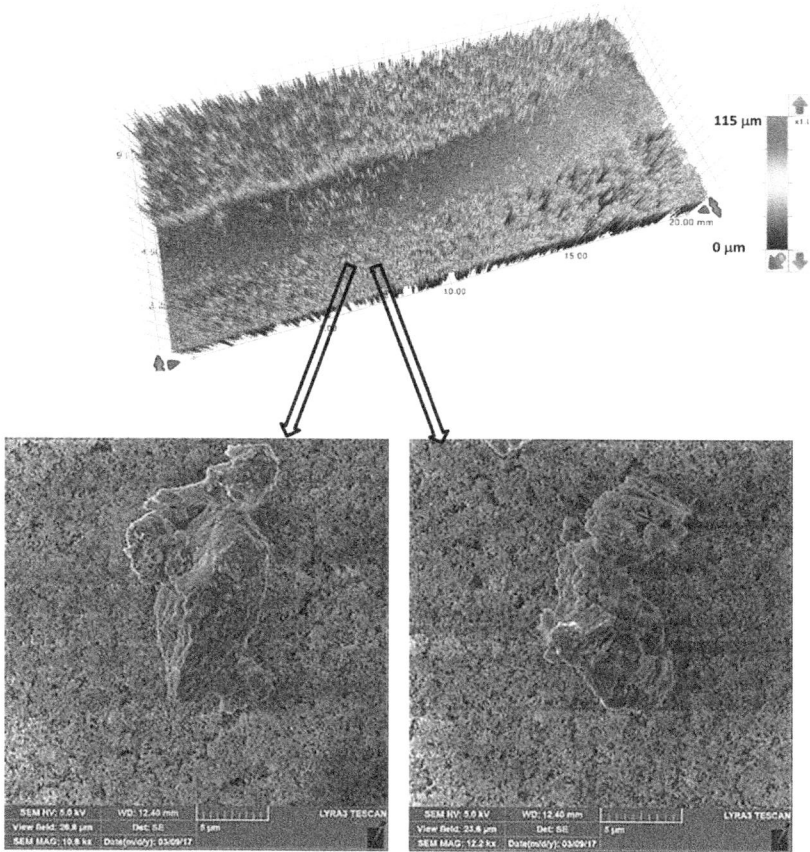

FIGURE 7.13   3D optical image of the droplet path and SEM micrographs of dust particle residue left on the droplet path. The *red circle* (gray in print version) depicts the dust particle residue along the droplet path.

TABLE 7.3 Elemental Composition of Dust (wt.%) Determined by Energy Dispersive Spectroscopy (EDS)

	Si	Ca	Na	S	Mg	K	Fe	Cl	O
Size ≥ 1.2 µm	12.3	8.2	2.1	1.4	2.6	0.7	1.1	0.3	Balance
Size <1.2 µm	10.1	7.1	3.1	2.4	1.1	1.8	1.2	1.2	Balance
Dust residues	12.5	8.1	2.1	1.3	2.4	0.9	1.0	0.4	Balance

cloak the functionalized dust particles. This behavior is associated with the spreading rate of the water film at the dust particle−air interface. The spreading coefficient ($S_{op(a)} = \gamma_{pa} - \gamma_{pw} - \gamma_{wa}$, where γ_{pa} is the interfacial energy at the dust particle−air interface; γ_{pw} is the interfacial energy at dust particle−water interface; and γ_{wa} is the interfacial energy at the water−air interface) must be greater than zero for water to cloak the outer surface of the dust particles. Although the interfacial energy between the dust particle and air is unknown, the condition $\gamma_{pa} > (\gamma_{pw} + \gamma_{wa})$ should be satisfied for water cloaking. Note that the dust particles were compacted into a pellet, and the surface energy of the dust pellet can be determined through experiments incorporating the Owens−Wendt method [53]. The surface energy of the pellet was determined to be in the order of 750 mJ/m², which is between the surface energy of calcite (347 mJ/m²), [54] and silica (1500 mJ/m²) [55]. The actual surface energy of the dust particles may slightly differ from that of the measured value; however, it should remain within a similar order of magnitude because of the major constituting elements, silica and calcite. Because $\gamma_{wa} = 72 \text{mJ/m}^2$, in any case, γ_{pw} should be less than 678 mJ/m². On the other hand, water spreading on the dust particles occurs into two stages. In the first stage, the balance between the surface tension gradient and the shear stress at the water−dust interface results in a monolayer of water spread on the dust particle. In the second stage, the location of water spreading follows Joos' law [56], and the spreading velocity can be related to $V_s \propto (3S_{ow(a)}/4\sqrt{\mu_o\rho_o})^{1/2}t^{-1/4}$, where μ_o is the dynamic viscosity of water, ρ_o is the density of water, and $S_{ow(a)}$ is the spreading coefficient of water on the dust particles [57]. The dissipating force during water spreading around a dust particle can be approximated by the Ohnesorge number ($Oh = \mu_o/\sqrt{\rho_o a \gamma_{oa}}$), where a is the characteristic size of the dust particle [57] that can be considered to be the equivalent diameter [58]. For an average dust particle size of 1.2 μm, Oh well exceeds unity ($Oh > 1$), which implies a large dissipation force for water cloaking of the dust particle. The cloaking rate is associated with cloaking time in the form of $\sim k_m t^{1/4}$, where k_m is the cloaking factor [58], and cloaking is not possible if $k_m t^{-/4} < 1$. In the present case, $k_m t^{1/4}$ was determined to be greater than unity for normal dust particles. Moreover, the cloaking velocity was determined from high-speed camera data and its average value is on the order of 0.3×10^{-3} m/s. From the average cloaking velocity, the duration of complete cloaking of the dust particle within the range of 1.2−10 μm is on the order of 0.0315 seconds. Note that the cloaking velocity is inversely related to cloaking time; in which case, the relation for the cloaking velocity can be approximated in the form of $\sim Ct^{-0.5}$, where C is a constant that varies with the shape of the dust and t is the cloaking time. Moreover, the distance corresponding to the cloaking time and traveled by the droplet on the hydrophobic surface is in the order of 9 μm, which is much less than the contact length of the 40 μL liquid droplet on the solid surface ($l \cong 0.002$ m). Therefore, the dust particles are picked up from the hydrophobic surface by the water droplet cloaking the dust particles. The dust particles picked up by the droplet remain in the droplet fluid and mix with the droplet liquid. On the other hand, the functionalized dust particles do not penetrate into the droplet liquid and instead remain at the droplet surface. Consequently, the droplet fluid and the functionalized dust particles do not mix. The adhesion of the functionalized dust particles on the water-droplet surface can be attributed to the electrostatic attraction developed within the deposited surface of the functionalized dust particles [59,60].

The translational velocity of the droplet is composed of the rolling and slip velocities of the droplet. The slip velocities during droplet movement on the hydrophobic surface with and without dust particles are shown in Fig. 7.14A. Note that the data presented in Fig. 7.14A were obtained from experiments incorporating the high-speed camera. In addition, a comparison

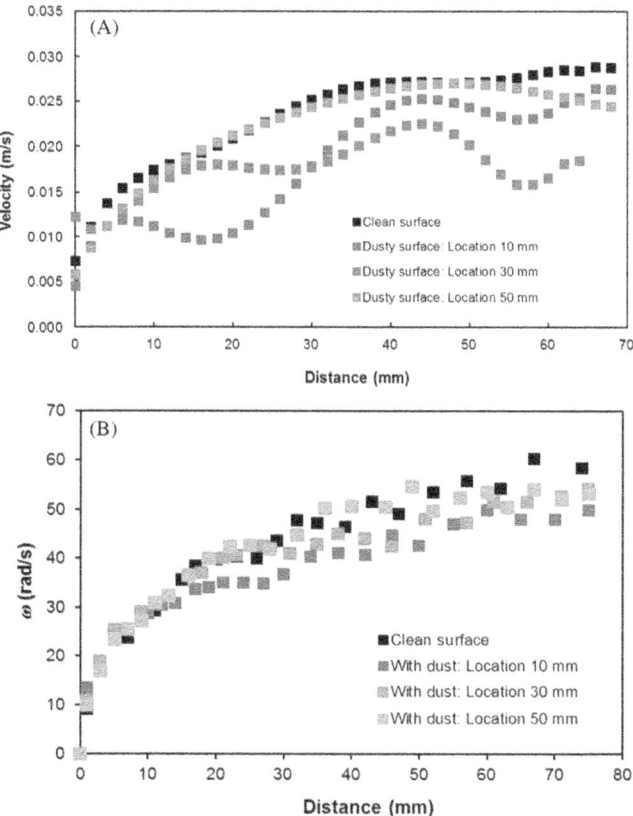

FIGURE 7.14 Slip velocity (Fig. 7.11A) and rotational speed (Fig. 7.11B) of the droplet on clean and dusty hydrophobic surfaces for different droplet locations. The droplet location represents the distance between the droplet initiation and the start of the dusty region.

of the rotational speed of the droplet on the hydrophobic surface with and without dust particles is shown in Fig. 7.14B. The rolling velocity (angular speed) reduces along the region at which the dust particles are located. This behavior is associated with an increased retention force between the droplet and the dusty surface (Fig. 7.15). The sliding velocity of the droplet is small, on the order of 0.02 m/s, which is smaller than the translational velocity. As the droplet progresses along the hydrophobic surface, the sliding velocity slightly increases. The attainment of low sliding velocity is associated with the retention of the droplet on the hydrophobic surface, which is larger for the dusty surface than the clean surface (Fig. 7.15). For the dusty surface, the retention force, resulting from the difference between the advancing and receding angles, is in the order of 80 μN, which is slightly higher than that on the clean surface (50 μN).

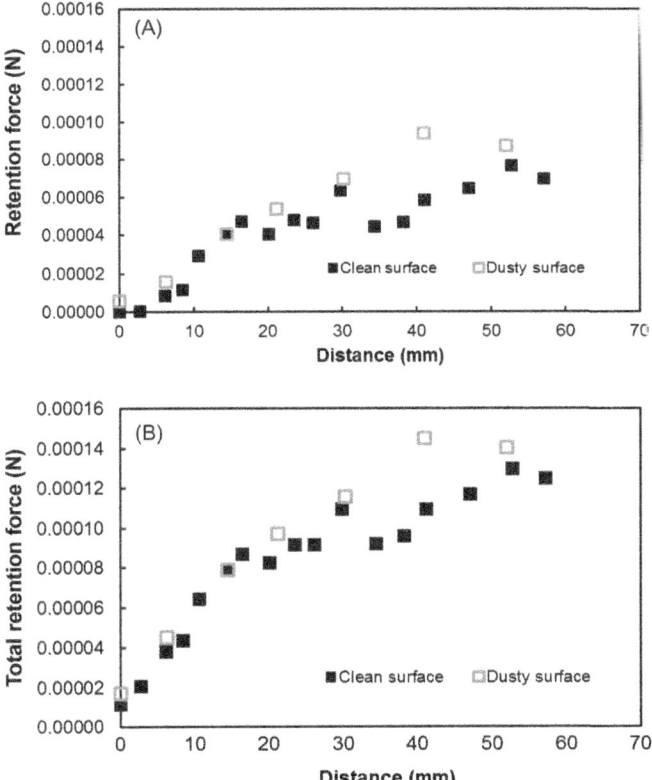

FIGURE 7.15 Forces acting on the hydrophobic surface during droplet motion on clean and dusty surfaces: (A) retention force and (B) total retention force. The forces were calculated analytically after determining the droplet advancing and receding angles and the droplet diameter at each location on the hydrophobic surface.

7.4 THERMOCAPILLARY EFFECT ON DUST REMOVAL FROM HYDROPHOBIC SURFACE

In general, the temperature of the droplet fluid and the substrate surface are different; therefore, the heat transfer from the surface toward the droplet or vice versa can result in thermocapillary flow in the droplet. On the other hand, dust particles have various elemental compositions, shapes, and sizes that influence the resulting forces including buoyancy, drag, Brownian, and Marangoni in the flow field. Consequently, the combination of the geometric features and the elemental composition of dust particles significantly alters the convection forces generated in the droplet. In addition, the dust particles contain soluble alkaline and earth alkaline components [13] that alter the composition of the droplet fluid while modifying the fluid density, thermal properties, and surface tension. Although internal fluidity of a sessile droplet with fine-size dust particles involves a complicated flow field, it can be tailored for dust removal from hydrophobic surfaces. Consequently, in the present section, the movement of the dust particles inside the sessile droplet under the thermocapillary-induced forces is demonstrated, in line with the previous study on achieving the possible application of the self-cleaning of hydrophobic surfaces. The effect of size and number density of the dust particles on the internal fluidity of the droplet is predicted numerically. The model study and findings are presented in line with the previous study [61].

7.4.1 Formulation Numerical Analysis of Thermal and Flow Fields

Internal fluidity of a sessile droplet with microsize particles are simulated in line with the experimental conditions. The actual geometric feature of the droplet and the dust particles are shown in Fig. 7.16. An incompressible flow field is considered to formulate the governing equations of flow and heat transfer. In the case of energy equation, it is assumed that the droplet possesses a slurry single fluid that resembles the water and the dust particle mixture. Since the microsize particle concentration is low (<5%) in the carrier fluid, effective thermal properties are incorporated in the energy equation. The coupled flow and thermal fields are considered simultaneously in the simulations. Details of the formulation of the flow and heat transfer are found in [61].

Since the homogeneous model (slurry) is considered for the energy equation, mass-based averaged thermal properties of the mixture of water and dust particles are incorporated in the simulations. It should be noted that density, thermal conductivity, and specific heat capacity of the dust particles are measured and compared to the data presented in the open literature [62]. However, the density of the homogeneous bulk liquid (mixture of water and particles) can be calculated using mass balance as [63]:

$$\rho_{eff} = c\rho_p + (1 - c)\rho_W \tag{7.3}$$

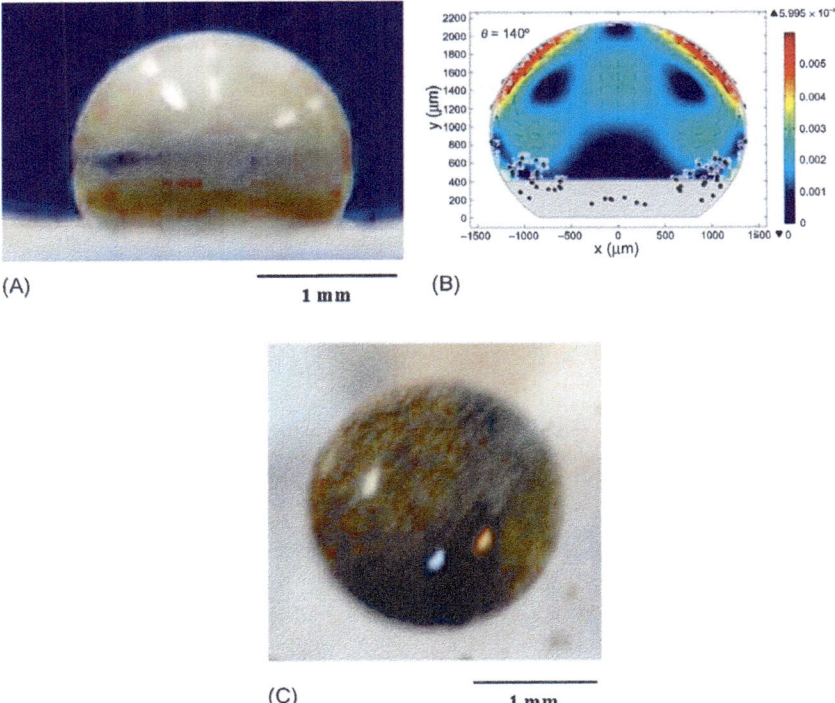

(A)   1 mm   (B)

(C)   1 mm

FIGURE 7.16   Optical image and droplet geometry used in the simulations: (A) side view of optical image on the hydrophobic surface ($\theta = 140$ degrees), (B) geometric configuration used in the simulations and resembling the actual optical image, and (C) top view of optical image of the droplet.

where c is the volume concentration of particles in water and ρ_{eff} is the effective density of the water-particle mixture (slurry). Subscripts: p is for the particles and W represents the water.

Effective thermal conductivity of the slurry is determined using the Maxwell model [64]; therefore, thermal conductivity of the homogeneous mixture can be written as:

$$k_{eff} = k_W \frac{2 + \left(\frac{k_p}{k_W}\right) + 2c\left(\frac{k_p}{k_W} - 1\right)}{2 + \left(\frac{k_p}{k_W}\right) - c\left(\frac{k_p}{k_W} - 1\right)} \qquad (7.4)$$

where k_{eff} is the effective thermal conductivity of the slurry.

The specific heat of the homogeneous mixture of slurry can be calculated using energy balance as [64]:

$$c_{p_{eff}} = \frac{c\left(\rho\, c_p\right)_p + (1 - c)\left(\rho\, c_p\right)_W}{\rho_{eff}} \qquad (7.5)$$

where $c_{p_{eff}}$ is the specific heat of the nanofluid mixture.

The bulk viscosity is calculated as [65]:

$$\mu_{eff} = \left(1 - c - 1.16\ c^2\right)^{-2.5} \mu_w \tag{7.6}$$

The above correlation for bulk viscosity was found to agree well with experimental data presented in Ref. [66] for nanoparticles volume concentrations below 11%.

The relative contribution between natural convection and Marangoni convection can be written in terms of the Bond number (*Bo*), which is proportional to ratio of buoyancy force and the surface tension force, that is:

$$Bo = \frac{\beta g \rho L_c^2}{\left| d\sigma/dT \right|} \tag{7.7}$$

where $\partial \sigma / \partial T$ is the surface tension thermal gradient temperature derivative of the surface tension and (L_c) is the characteristic length, which is considered as:

$$L_c = \frac{\forall}{\pi R^2} \tag{7.8}$$

where $\forall$ is the volume of the droplet and R is the wetting radius [67]. It should be noted that when $Bo < 1$, the internal flow is dominated mainly by the Marangoni convection [68].

Natural convection is driven by the buoyancy force, which overcomes viscous resistance in the flow. To characterize the strength of natural convection, the Grasshoff number (*Gr*) can be introduced, which is given by:

$$Gr = \frac{\beta g \Delta T L^3}{\nu^2} \tag{7.9}$$

where (ΔT) is the temperature difference between the droplet and its ambient and (ν) is the kinematic viscosity of water. It should be noted that the effect of natural convection can be ignored when $Gr < 2400$ [67].

Marangoni convection is driven by the surface tension gradient and the Marangoni number (*Ma*) can be used to describe the intensity of Marangoni convection, which is:

$$Ma = \frac{\left| d\sigma/dT \right| \Delta T L}{\mu \alpha} \tag{7.10}$$

In the simulations, the Newtonian formulation is considered, which adds forces to the particles to influence the particle motion. The particle momentum is determined using Newton's second law, that is:

$$\frac{d\left(m_p\ V\right)}{dt} = F \tag{7.11}$$

where m_p is the particle mass and F is the force exerted on the particle, which is defined as:

$$F = F_D + F_g + F_{ext} \qquad (7.12)$$

F_D is the drag force, which is defined as:

$$F_D = \left(\frac{1}{\tau_p}\right) m_p (V - u) \qquad (7.13)$$

where u is the velocity of the particle, V is the fluid velocity, and τ_p is the particle velocity response, which can be determined from Stokes drag law as:

$$\tau_p = \frac{\rho_p d_p^2}{18\mu} \qquad (7.14)$$

where ρ_p is the particle density and d_p is the particle diameter.

The gravitational force vector can be expressed as:

$$F_g = m_p g \frac{\left(\rho_p - \rho\right)}{\rho_p} \qquad (7.15)$$

F_{ext} is any other external force. COMSOL Multiphysics finite element code [69] is used to solve the governing equations of flow and heat transfer for a stationary droplet incorporating the experimental conditions.

The numerical simulation of the flow and temperature fields in a droplet consisting of water and microsize particles (1.2, 10, and 20 μm diameters) is carried out in line with the experimental conditions. The data reported in the previous study [62] are utilized for thermal conductivity, which is 0.0036 W/mK and specific heat capacity is obtained from the elemental composition of the dust particles, which are listed in EDS data (Table 7.3). However, thermal conductivity of the dust particles is also measured using a C-Therm TCi thermal conductivity analyzer. The thermal conductivity measured is in the order of 0.003 W/mK. The mass-based calculations are conducted to determine the effective heat capacity of the dust particles. The findings of specific heat capacity of the dust particles as determined from the EDS data is 867.4 J/kgK, which is close to the sand particle (800 J/kgK). Initially, it is assumed that the solution domain is at an initial equilibrium temperature (T_{in}), that is, $T = T_{in} = 300K$, and stagnant water is assumed in the droplet. A uniform heating with a constant surface temperature source at 308K is considered along the interface of droplet bottom and substrate surface. The natural convection boundary condition is adopted at the droplet surface with surrounding air temperature of 300K. No slip flow boundary is incorporated at the droplet bottom.

Optical image of the sessile droplet with dust particles inside is shown in Fig. 7.16A together with the droplet geometric configuration, which is used in the simulations (Fig. 7.16B), and the top view of the actual sessile droplet is shown in Fig. 7.16C. The dust particles are located at the droplet bottom resembling the dust layer in the actual droplet (Fig. 7.16A). The droplet cross-section constructed is identical to the optical image captured from the experiment. Although 3D simulation of the flow and temperature field is attempted, the computation power required for the simulation is significantly expensive in terms of memory size and runtime. Hence, simulations were carried out for 3D droplet geometry. Moreover, the time taken is short for the experiment and the predictions, that is, the heating duration is in the order of 30 seconds; therefore, the evaporation from the droplet surface is neglected in the simulations. As indicated earlier, the boundary conditions for the energy balance equation are considered to be in line with the experimental conditions. It is considered that the density of the fluid outside the droplet domain is negligibly small as compared to that inside of the droplet domain. At the droplet-free surface, a no-penetration condition is introduced $(u \cdot n = 0)$. This implies that there is no viscous effect normal to the droplet-free surface (liquid-air interface) and, hence, no boundary layer is developed at the droplet-free surface. This is a reasonable approximation, since the fluid from the droplet surface does not leave from the droplet domain to its surroundings. This condition can be formulated as $u \cdot n = 0$, and $(-pI + \mu(\nabla u + (\nabla u)^T)n = 0$, where n is the unit vector normal to the droplet-free surface. The top surface of the droplet is subjected to a natural convective cooling with $h = 10$ W/m^{2}K. In the simulations, the nonisothermal flow model was used for the flow and heat transfer. The model incorporates conjugate heat transfer and internal flow inside the droplet. This model solves continuity, momentum, and energy equations simultaneously to predict the flow field and the temperature at each node of the computational domain. To incorporate the Marangoni effect, a weak contribution is specified at the outer surface of the droplet. Since the microsize particles possess alkaline (Na, K) and alkaline earth metal (Ca) compounds, they dissolve in the water and modify the surface tension. Table 7.4 gives the data obtained from inductively coupled plasma mass spectrometry. It is evident that the dissolved dust compounds alter the elemental concentration of alkaline and

TABLE 7.4 IPC-MS Data Obtained for Two Concentrations of Water–Dust Mixtures

% of Dust	23Na	24Mg	39K	44Ca	52Cr	56Fe
	ppb	ppb	ppb	ppb	ppb	ppb
1%	71,200	6840	21,400	24,5360	1235	6440
5%	39,420	13,560	92,140	688,520	4320	8710

alkaline earth metals in the desalinated water. Consequently, the surface tension of water–dust particle mixture is measured and incorporated in the simulations.

The mesh is generated in such a way that finer grids are located in the region where the fluxes are high. Mesh independence tests are conducted for each contact angle considered in this study to ensure the mesh-selected results in mesh-independent solutions. Hence, the mesh size that comprises 19,111 cells is selected to accomplish further simulations (for droplet contact angle of 120 degrees). The accuracy of the scheme is governed by the size of the time step, which is considered to be 10^{-4} seconds. The residuals of flow parameters are set as $\left| \psi^k - \psi^{k-1} \right| \leq 10^{-8}$.

7.4.2   Experimental and Validation of Predictions

A solution-crystalized PC sheet was used as a hydrophobic surface in the experiments. A bare PC wafer was rinsed in an ultrasonic shaker for 15 minutes for cleaning purposes. The wafer was immersed in acetone for 4 minutes in line with the previous study [70]. This process results in surface texture of PC wafer with the hieratical structures comprising micro/nanospherulites and fibrils. To examine the surface topology and surface texture of the crystallized PC surface, Jeol 6460 SEM and AFM in contact mode were used. The surface roughness of crystallized PC is in the order of 1.75 μm. The droplet static contact angle was measured incorporating the sessile drop method using Kyowa (model—DM 501) contact angle goniometer. The contact angle measurements revealed that the droplet contact angle varied within 100–150 degrees.

To monitor and track the dust particles in the droplet, a test rig was developed and the textured and hydrophobized PC surface was placed in a control chamber where ambient temperature and humidity were controlled. Water was kept at 300K prior to forming a droplet on the substrate surface. The substrate with the presence of dust particles at the surface was kept at 308K prior to the formation of droplet. In order to assure a sufficient number of dust particles at the surface, a mount of dust particles located at the hydrophobic surface was measured and their distribution examined under the optical microscope. Once the droplet was formed, the flow field and movement of dust particles inside the droplet were recorded under the optical microscope during the first 50 seconds immediately after the droplet formation. The tracking method was used to monitor the particle motion in the droplet from the data recorded under the microscope. The data were then used to determine the particle velocity in line with the previous study [71]. The distance traveled by each particle was small because of low particle velocities in the droplet; therefore, the distance over the time difference was considered to be the velocity of the particles tracked. Several microimages of tracked particles in the droplet were analyzed to minimize the error related to the measurements. Fig. 7.17 shows the images of particles recorded. The curvature effect of the droplet on the images recorded was corrected optically. The experiment was repeated 12 times to monitor the motion of different particles, ensuring the repeatability of the velocity measurements in the droplet. Based on the distribution of the data recorded during the experimental repeats, a confidence level of 93% resulted; in which case, the mean (μ) of the data distribution was within ± 1.95 of the standard deviation of the distribution of a single measurement from that distribution. The experimental uncertainty analysis revealed that the uncertainty was less than 7%.

Table 7.5 gives the velocity data obtained from the simulations and determined from the particle-tracking method. It is evident that both results agreed well and the differences between both results are associated with the experimental error, which is in the order of 7%.

FIGURE 7.17   Image of particles monitored using the optical microscopic system. Particle initial and final locations are shown on the image.

TABLE 7.5 Particle Velocity Measured and Predicted From Simulations

Particle no.	V_{Num} (m/s)	$V_{Exp.}$ (m/s)
1	0.00048579	0.0005224
2	0.00048298	0.0005031
3	0.00048709	0.0004729
4	0.00025225	0.0002574
5	0.00020331	0.0002013
6	0.00017626	0.0001728
7	0.00014298	0.0001505
8	0.00013624	0.000131
9	0.00011831	0.0001195
10	0.0000000	0.0000083

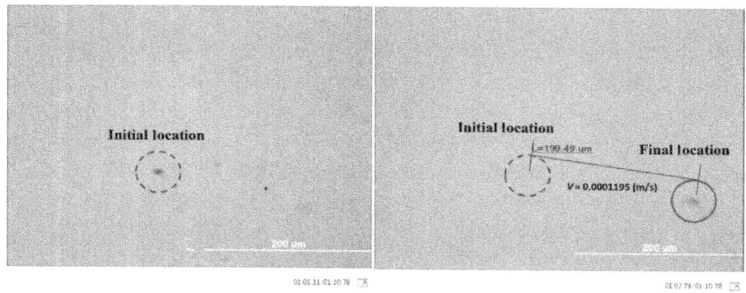

7.4.3 Internal Fluidity of Sessile Droplet and Dust Particles

Since the dust particles have various sizes, three different diameters of dust particles were incorporated in the simulations to assess the effect of dust particle size on the flow field and dust mechanics in the sessile droplet. The diameter of the dust particles includes 1.2, 10, and 20 μm. The density of the dust particles was measured and found to be in the order of 1600 kg/m^3. Moreover, the dust particles comprise alkaline (Na, K) and alkaline earth compounds (Ca) dissolve in the water. This, in turn, modifies the surface tension of the droplet, which is incorporated in the simulations.

Since temperature difference is present in between the hydrophobic surface (308K) and the droplet fluid (300K), surface tension and density variations result in a fluid motion in the droplet under the thermocapillary forces. However, the droplet stability depends on the Rayleigh number, which is related to buoyant convection [72]. The Marangoni force influences the conditions for the droplet instability. In the case of small diameter droplets ($a < 0.3$ mm, where a is the characteristic droplet diameter), rolling off of the droplet can take place because of the buoyant convection-induced instability by the resulting lateral force [72]. The droplet pinning and droplet mobility are also related to the contact-line morphology and droplet contact angle [73]. In this case, the inclination angle of the surface, where the droplet is located, becomes important for the droplet mobility and is related to the contact line. The ratio of contact-line length over the fraction of the projected area of the droplet on the surface is associated with the droplet instability for the rolling off [74]. Therefore, the lateral force balance becomes critical to avoid rolling off of the stationary droplet from the surface. On the other hand, the flow stability in the droplet is related to the characteristic time for thermal diffusion ($\tau_d = L_c^2/\alpha$ where τ_d is the characteristic time for thermal diffusion, L_c is the characteristic droplet diameter, and α is the thermal diffusivity); in which case, the Marangoni velocity over the natural convection velocity is the same order of the thermal diffusion time [74]. Therefore, the velocity ratio is the prime concern for controlling the flow instability in the droplet. The velocity ratio yields $(\partial\sigma/\partial T)\Delta T/\beta\rho g L_c^2$, that is, $Ma = (\partial\sigma/\partial T)(L_c\Delta T/\mu\alpha)$, where L_c represents the characteristics diameter of the droplet; σ is the surface tension; T is the temperature; μ is the dynamic viscosity; and $Ra = \beta g L_c^3\Delta T/v\alpha$, where β is the thermal expansion coefficient, v is the kinematic viscosity, and g is the gravity [74]. In the present case, the forces generated under thermocapillary forces for the droplet diameter (a), which is in the order of 2 mm, removes the possibility of droplet rolling off from the hydrophobic surface. The possible droplet rolling off from the hydrophobic surface is tested experimentally under various contact angles (110 degrees $\leq \theta \leq$ 160 degrees, where θ is the droplet contact angle) and the findings yield that the droplet does not roll off from the surface under the heat transfer conditions defined in the numerical simulations. The velocity ratio defining the characteristic time for stable flow inside the droplet is estimated as 28.4 seconds, that is, the ratio a^2/α (where droplet diameter $a \approx 2 \times 10^{-3}$ m and thermal diffusivity $\alpha = 1.41 \times 10^{-7}$ m^2/s), which is 28.4 seconds for the present situation. Fig. 7.18 shows variation of temperature and velocity along the vertical rake inside the droplet for different heating periods. Flow velocity and temperature along the rake becomes identical for the periods of 20 and 30 seconds. Consequently, the time period of 30 seconds is selected to present the flow and temperature fields as well as the positions of the particles in the droplet.

Figs. 7.19 and 7.20 show velocity and temperature contours inside the droplet with dust particles for two droplet contact angles and after 30 seconds of heating period. Since heat transfer takes place from the hydrophobic

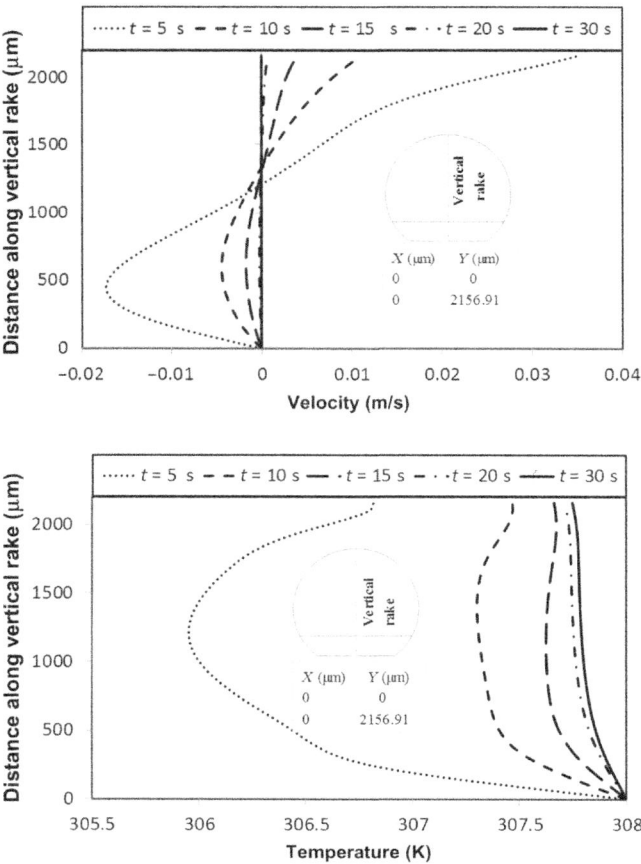

FIGURE 7.18 Velocity and temperature variation along the y-axis (vertical rake) in the droplet for various heating durations.

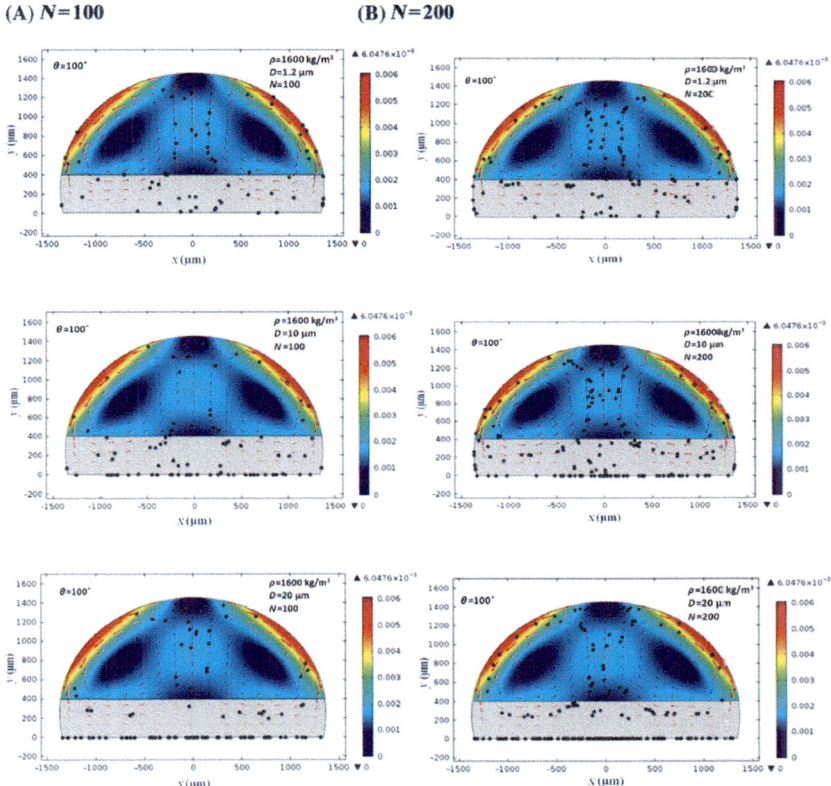

FIGURE 7.19 Velocity contours inside the droplet for 100-degree contact angle, two particle number densities, and three particle diameters.

surface toward the droplet, buoyancy and Marangoni currents are developed in the droplet and flow becomes stabile after 30 seconds of the heating duration. In this case, two counter-rotating circulation cells are formed in the lower part of the droplet for a contact angle of 100 degrees (Fig. 7.19). This is mainly associated with the combination of the Marangoni and buoyancy currents, which generates complex flow structures in the droplet. Since the velocity ratio due to Marangoni current over the buoyancy current is related to $(\partial\sigma/\partial T)\Delta T/\beta\rho gL_c^2$ [74], the ratio becomes larger than unity for the droplet diameter $a = 2 \times 10^{-3}$ m. Therefore, the Marangoni current dominates over the flow field inside the droplet. The buoyancy current generated in the close region of the droplet bottom influences the orientation and size of the circulation cells in the droplet. In the case of large droplet contact angle ($\theta = 140$ degrees), the droplet height along the vertical rake increases while resulting in reduced temperature gradient along the vertical rake (Fig. 7.20). This alters the surface tension gradient and the velocity ratio, due to

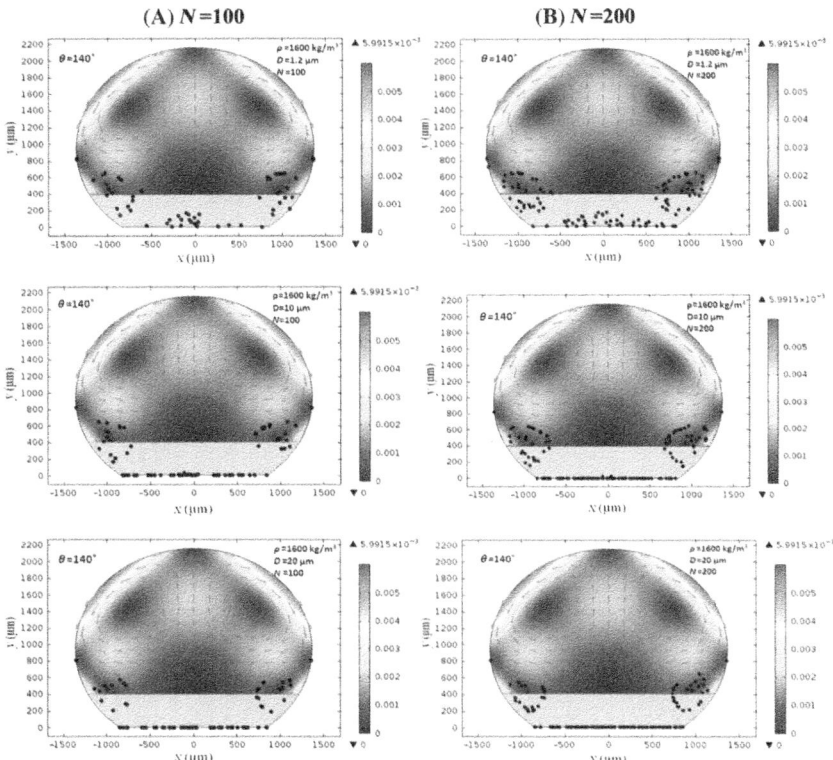

FIGURE 7.20 Velocity contours inside the droplet for 140-degree contact angle, two particle number densities, and three particle diameters after 30 s of heating duration.

Marangoni current over the buoyancy current, reduces. In addition, thickness of the high temperature region in the droplet reduces (Fig. 7.21) while thinning the thermal boundary layer in the droplet bottom. This, in turn, increases the density variation and buoyancy current in this region. Therefore, the combination of the Marangoni and buoyancy currents generates a complex flow structure. In this case, counter rotating four circulation cells are formed in the droplet. The presence of the dust particles in the droplet influences the flow field due to drag, body, and Brownian forces; however, this influence is not substantial.

Figs. 7.22 and 7.23 show the particle distribution inside the droplet for two contact angles after 30 seconds of heating duration. It should be noted that the density of particles is measured (1600 kg/m^3) and used in the simulations. Three particle diameters and two numbers of particles are selected to demonstrate the influence of particle size and number density on the particle distribution in the droplet. The particle distribution almost follows the

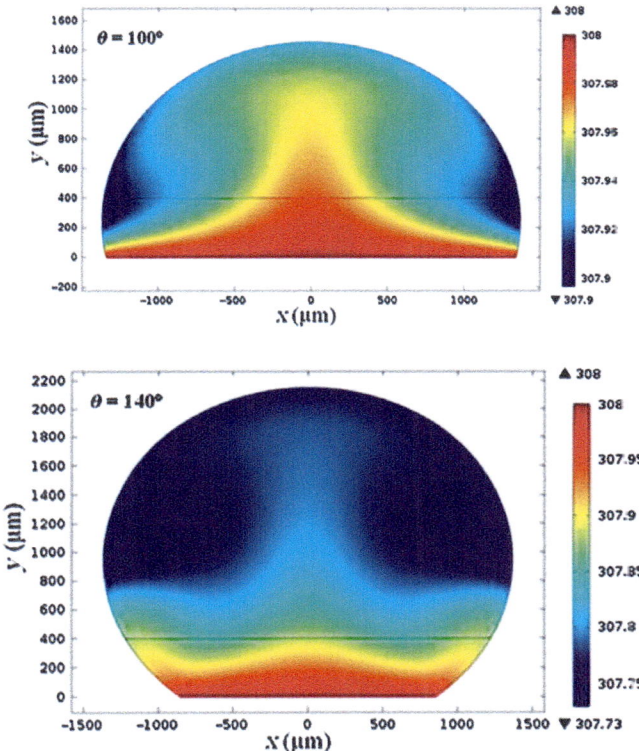

FIGURE 7.21   Temperature contours inside the droplet for 140-degree contact angle after 30 s of heating duration.

streamlines in the circulation cells in the droplet, which is more pronounced for the contact angle of $\theta = 100$ degrees (Fig. 7.22). This is true for all particle number densities and particle sizes. In this case, the forces generated due to gravitational, buoyancy, and Brownian motions are not significantly larger as compared to the forces generated by the combinations of Marangoni and natural convection. Since three dust particle diameters are considered (1.2, 10, and 20 μm) and the averaged density of the particles is in the order of 1600 kg/m^3, the weight of the dust particles varies with in the order of 1.14×10^{-14} N to 5.26×10^{-10} N. The buoyancy force $(F_g = m_p g \left((\rho_p - \rho)/\rho_p \right))$, where m_p is the mass of the particle, ρ_p is the density of the dust particles, and ρ is the density of water in the droplet) is in the order of 4.26×10^{-14} N to 1.97×10^{-10} N. The drag force $F_D = (1/\tau_p) m_p (V - u)$, where u is the velocity of the particle, V is the fluid velocity, and τ_p is the particle velocity response time, can be determined from Stokes drag law ($\tau_p = \rho_p d_p^2 / 18\mu$, where ρ_p is the particle density, d_p is the particle diameter, and μ is the viscosity of water) [75]. The data obtained

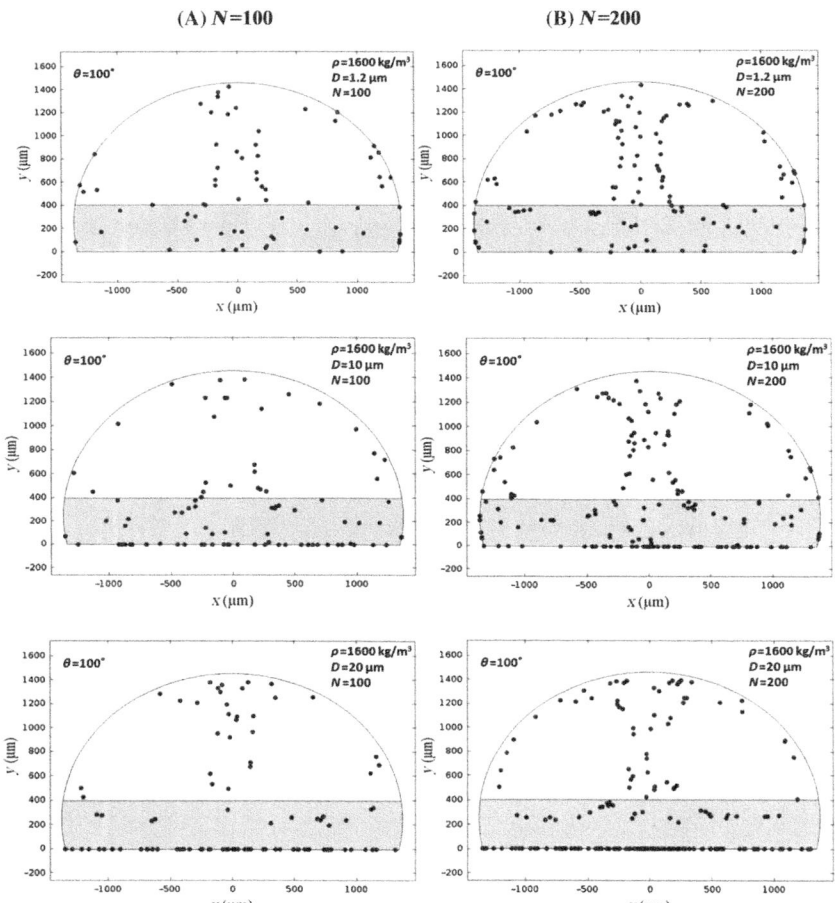

FIGURE 7.22 Particle distribution inside the droplet for 100-degree contact angle, two particle number densities, and three particle diameters after 30 s of heating duration.

from the simulations indicate that the velocity difference between flow and particle $(V - u)$ is in the order of 0.2×10^{-6} m/s; therefore, the drag force $(F_D = (1/\tau_p)m_p(V - u))$ is in the order of 3.10×10^{-14} N to 4.59×10^{-11} N. The hydrodynamic force due to Brownian motion is $F_{ext} = \zeta \sqrt{12\pi k_B \mu T_p r_P / \Delta t}$, where k_B is the Boltzmann constant, μ is fluid viscosity, T_p is particle temperature, r_p is particle radius, ζ is a normally distributed random number with a mean of zero and unit standard variation, and Δt is time step in between iterations. The hydrodynamic force due to Brownian motion is in the order of 9.13×10^{-14} N to 3.73×10^{-13} N. The hydrodynamic force due to Brownian motion of the dust particles is lower than the other forces. The buoyancy force is larger than the drag force, but

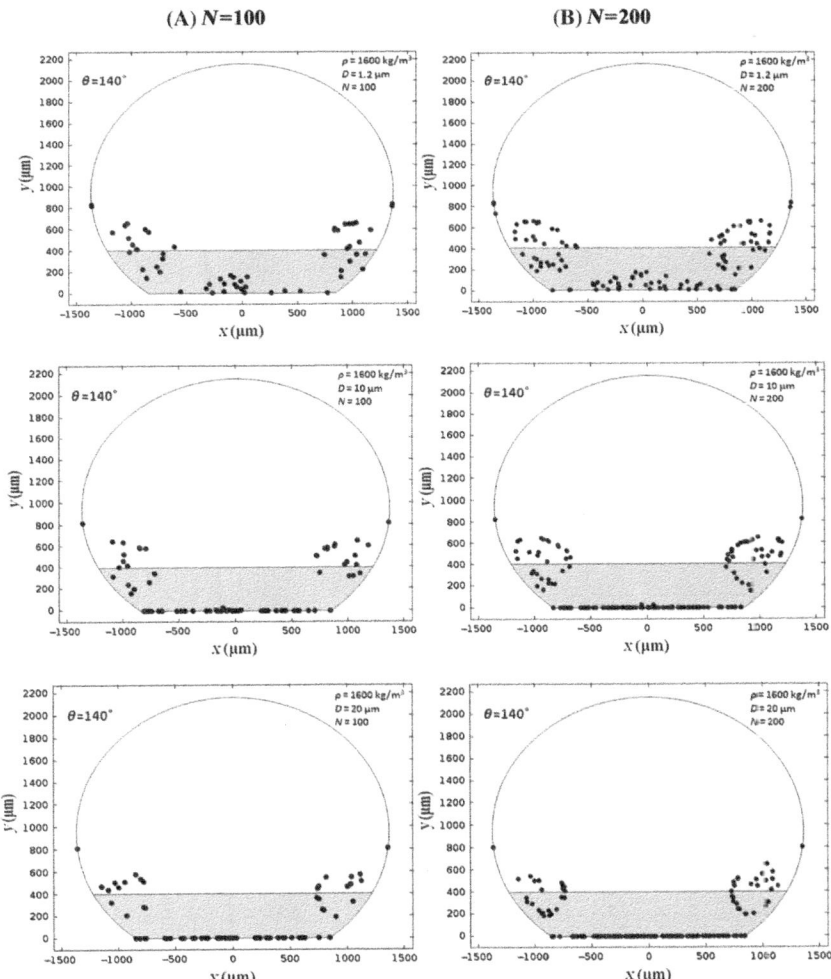

FIGURE 7.23 Particle distribution inside the droplet for 140-degree contact angle, two particle number densities, and three particle diameters after 30 s of heating duration.

less than the gravitational force. Therefore, the generation of the natural convection current results in particle movement in the droplet bottom, which is more pronounced for the small diameter particles for which the weight force is small. Once the particles reach the location where the flow velocity in the circulation cells is strong, the particles follow the stream line in the droplet. Since the particle drag force is considerably small (3.10×10^{-14} N to 4.59×10^{-11} N), the particles do not resist or reorient in the circulation cells because of the drag forces, particularly for small diameter dust particles. In the case of large droplet contact angle (Fig. 7.23), the particles are mainly located in the lower part of the circulation cell. This is attributed to the small buoyancy force because of the density of the dust particles, which is larger than water. In addition, the flow velocity in between the bottom and top circulation cells is significantly small because of the shear layer developed due to counter rotation of the circulation cells. This lowers the inertia force for particles to transfer from bottom circulation cells to the top circulation cells. Consequently, all the particles in the droplet are located at the bottom and in the first circulation cells close to the bottom.

Figs. 7.24 and 7.25 shows the ratio of the number of particles crossing over the horizontal rake in the droplet for two droplet contact angles, respectively. The ratio corresponds to the number of particles crossing over the horizontal rake over the total number of particles in the droplet. It should be noted that the horizontal rake is located at the same height of the sediment dust particles in the droplet (Fig. 7.16A). The particles are first accelerated from the droplet bottom by the natural convection current and move into the droplet upper region above the horizontal rake. This is more pronounced for

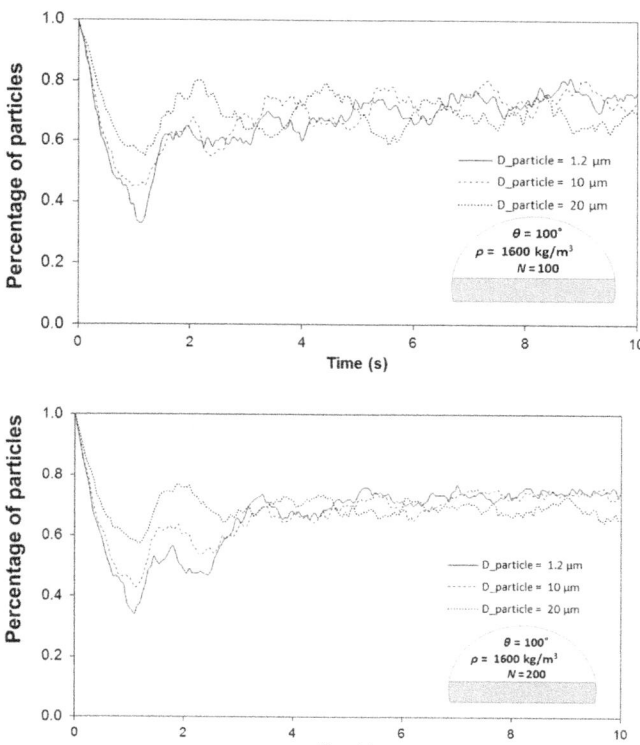

FIGURE 7.24 Temporal variation of particles crossing over the horizontal rake for the droplet contact angle 100 degrees, two particle number densities, and three different particle diameters.

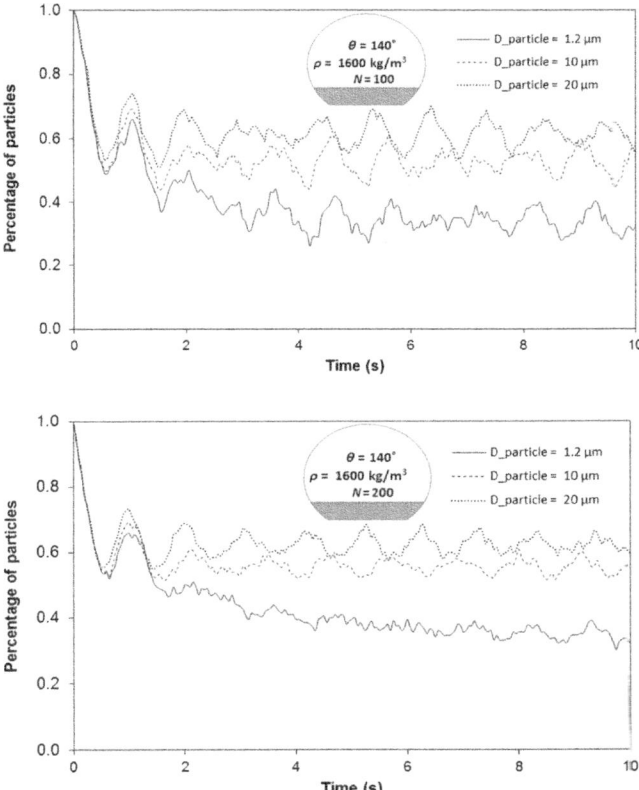

FIGURE 7.25   Temporal variation of particles crossing over the horizontal rake for the droplet contact angle 140 degrees, two particle number densities, and three different particle diameters.

the large particles, despite their high gravitational force. The buoyancy force of the large particles is higher than for the small particles due to the large amount of volume of liquid displaced. In addition, the other forces including drag and Brownian forces are higher for the large particles than the small particles. Since the density of the particles is higher than the water, the particle movement in the droplet requires additional inertia force. Consequently, the convection current in the close region of the droplet bottom initiates the particle movement. Although the large particle displacement requires relatively larger forces, their inertia remains higher than those of the small particles once they move in the droplet. This contributes to percentage of particles crossing over the horizontal rake. Therefore, the percentage of particles crossing over the horizontal rake is higher for the large size particles (20 μm) than those small size particles (10 and 1.2 μm). As the heating period progresses, the percentage of particles crossing over the horizontal rake remains almost constant with time. The particle number ratio crossing over the horizontal rake is larger for low droplet contact angle ($\theta = 100$ degrees) than that of the large droplet contact angle ($\theta = 140$ degrees). This behavior is attributed to the flow field developed around the counter rotating circulation cells; in which case, the centers of the circulation cells take place in the lower part of the droplet for the droplet contact angle $\theta = 100$ degrees (Fig. 7.24) unlike the case for large droplet contact angle (Fig. 7.25). In addition, increasing the number density of the particles lowers slightly the particle ratio crossing over the horizontal rake. This may be because of the requirement of the large inertia force to move the particles from droplet bottom towards the droplet inside. Nevertheless, the particles located on the hydrophobic surface can be removed through thermocapillarity induced forces in a droplet. However, the rate of removal is dependent on the particle size; in which case, large diameter particles can be removed from the droplet bottom into a droplet interior at a high percentage. In addition, increasing particle density slightly lowers the percentage of particles removed from the droplet bottom.

REFERENCES

[1] L. Mahadevan, Y. Pomeau, Rolling droplets, Phys. Fluids 11 (9) (1999) 2449–2453.
[2] J.D. Smith, R. Dhiman, S. Anand, E. Reza-Garduno, R.E. Cohen, G.H. McKinley, et al., Droplet mobility on lubricant-impregnated surfaces, Soft Matter 9 (6) (2013) 1772–1780.
[3] B. Balu, A.D. Berry, K.T. Patel, V. Breedveld, D.W. Hess, Directional mobility and adhesion of water drops on patterned superhydrophobic surfaces, J. Adhesion Sci. Technol. 25 (6–7) (2011) 627–642.
[4] S. Dorbolo, D. Terwagne, N. Vandewalle, T. Gilet, Resonant and rolling droplet, New J. Phys. 10 (11) (2008) 113021–113030.
[5] B.S. Yilbas, A. Al-Sharafi, H. Ali, N. Al-Aqeeli, Dynamics of a water droplet on a hydrophobic inclined surface: influence of droplet size and surface inclination angle on droplet rolling, RSC Adv. 7 (77) (2017) 48806–48818.
[6] Y.H. Yeong, J. Burton, E. Loth, I.S. Bayer, Drop impact and rebound dynamics on an inclined superhydrophobic surface, Langmuir 30 (40) (2014) 12027–12038.
[7] X. Yang, X. Liu, Y. Lu, J. Song, S. Huang, S. Zhou, et al., Controllable water adhesion and anisotropic sliding on patterned superhydrophobic surface for droplet manipulation, J. Phys. Chem. C 120 (13) (2016) 7233–7240.
[8] G. McHale, N. Shirtcliffe, M. Newton, Contact-angle hysteresis on super-hydrophobic surfaces, Langmuir 20 (23) (2004) 10146–10149.
[9] M. Miwa, A. Nakajima, A. Fujishima, K. Hashimoto, T. Watanabe, Effects of the surface roughness on sliding angles of water droplets on superhydrophobic surfaces, Langmuir 16 (13) (2000) 5754–5760.
[10] J.M. Lee, S.-H. Lee, J.S. Ko, Dynamic lateral adhesion force of water droplets on microstructured hydrophobic surfaces, Sensors Actuators B: Chem. 213 (2015) 360–367.
[11] B.S. Yilbas, H. Ali, N. Al-Aqeeli, M.M. Khaled, S. Said, N. Abu-Dheir, et al., Characterization of environmental dust in the Dammam area and mud after-effects on bisphenol-A polycarbonate sheets, Sci. Rep. 6 (2016) 24308.

[12] G. Hassan, B. Yilbas, S.A. Said, N. Al-Aqeeli, A. Matin, Chemo-mechanical characteristics of mud formed from environmental dust particles in humid ambient air, Sci. Rep. 6 (2016) 30253.

[13] B.S. Yilbas, H. Ali, M.M. Khaled, N. Al-Aqeeli, N. Abu-Dheir, K.K. Varanasi, Influence of dust and mud on the optical, chemical, and mechanical properties of a PV protective glass, Sci. Rep. 5 (2015) 15833.

[14] Y.-Y. Quan, L.-Z. Zhang, R.-H. Qi, R.-R. Cai, Self-cleaning of surfaces: the role of surface wettability and dust types, Sci. Rep. 6 (2016) 38239.

[15] C. Extrand, Y. Kumagai, Contact angles and hysteresis on soft surfaces, J. Colloid Interface Sci. 184 (1) (1996) 191–200.

[16] J. Bico, C. Marzolin, D. Quéré, Pearl drops, Europhys. Lett. 47 (2) (1999) 220.

[17] S. Varagnolo, D. Ferraro, P. Fantinel, M. Pierno, G. Mistura, G. Amati, et al., Stick-slip sliding of water drops on chemically heterogeneous surfaces, Phys. Rev. Lett. 111 (6) (2013) 066101.

[18] T. Podgorski, J.-M. Flesselles, L. Limat, Corners, cusps, and pearls in running drops, Phys. Rev. Lett. 87 (3) (2001) 036102.

[19] L. Gao, T.J. McCarthy, Contact angle hysteresis explained, Langmuir 22 (14) (2006) 6234–6237.

[20] A. Gupta, M. Sbragaglia, Deformation and breakup of viscoelastic droplets in confined shear flow, Phys. Rev. E 90 (2) (2014) 023305.

[21] A. Rifai, N.A. Dheir, B.S. Yilbas, M. Khaled, Mechanics of dust removal from rotating disk in relation to self-cleaning applications of PV protective cover, Solar Energy 130 (2016) 193–206.

[22] G. He, C. Zhou, Z. Li, Review of self-cleaning method for solar cell array, Proc. Eng. 16 (2011) 640–645.

[23] G. Wang, X.-F. Cheng, P. Hu, Z.-S. Chen, Y. Liu, L. Jia, Theoretical analysis of spectral selective transmission coatings for solar energy PV system, Int. J. Thermophys. 34 (12) (2013) 2322–2333.

[24] A. Sharma, R. Shandil, Effect of rotation on a ferromagnetic fluid heated and soluted from below in the presence dust particles, Appl. Math. Comput. 177 (2) (2006) 614–628.

[25] S. Chen, C. Gao, W. Tang, H. Zhu, Y. Han, Q. Jiang, et al., Self-powered cleaning of air pollution by wind driven triboelectric nanogenerator, Nano Energy 14 (2015) 217–225.

[26] S.A. Far, S. Zahedi, M. Shirvani, S.P. Far, A new process for improved dust removal by cyclone, in: Conference Proceedings, ISBN-13: 97808169107002011, AIChE Annual Meeting, 2011.

[27] W. Fernando, Y. Lok, M.M. Don, V. Madhaven, W. Tay, Experimental and modeling studies of particle removal in post silicon chemical mechanical planarization cleaning process, Thin Solid Films 519 (10) (2011) 3242–3248.

[28] S.R. Gonzalez-Avila, X. Huang, P.A. Quinto-Su, T. Wu, C.-D. Ohl, Motion of micrometer sized spherical particles exposed to a transient radial flow: attraction, repulsion, and rotation, Phys. Rev. Lett. 107 (7) (2011) 074503.

[29] R. Chein, W. Liao, Modeling of particle removal using non-contact brush scrubbing in post-CMP cleaning processes, J. Adhesion 82 (6) (2006) 555–575.

[30] M. Corn, The adhesion of solid particles to solid surfaces, I. A review, J. Air Pollut. Control Assoc. 11 (11) (1961) 523–528.

[31] Q. Li, V. Rudolph, W. Peukert, London-van der Waals adhesiveness of rough particles, Powder Technol. 161 (3) (2006) 248–255.

[32] K.L. Johnson, K. Kendall, A. Roberts, Surface energy and the contact of elastic solids, Proc. R. Soc. Lond. A 324 (1558) (1971) 301–313.

[33] B. Derjaguin, V. Muller, Y.P. Toporov, Effect of contact deformations on the adhesion of particles, Prog. Surf. Sci. 45 (1–4) (1994) 131–143.

[34] H. Hamaker, The London—van der Waals attraction between spherical particles, Physica 4 (10) (1937) 1058–1072.

[35] Y.I. Rabinovich, J.J. Adler, A. Ata, R.K. Singh, B.M. Moudgil, Adhesion between nanoscale rough surfaces: I. Role of asperity geometry, J. Colloid Interface Sci. 232 (1) (2000) 10–16.

[36] D. Leighton, A. Acrivos, The lift on a small sphere touching a plane in the presence of a simple shear flow, Z. Angew. Math. Phys. ZAMP 36 (1) (1985) 174–178.

[37] P. Cherukat, J.B. McLaughlin, The inertial lift on a rigid sphere in a linear shear flow field near a flat wall, J. Fluid Mech. 263 (1994) 1–18.

[38] P. Cherukat, J. McLaughlin, The inertial lift on a rigid sphere in a linear shear flow field near a flat wall, J. Fluid Mech. 285 (1995) 407.

[39] F. Zhang, A.A. Busnaina, M.A. Fury, S.-Q. Wang, The removal of deformed submicron particles from silicon wafers by spin rinse and megasonics, J. Electr. Mater. 29 (2) (2000) 199.

[40] F.M. White, I. Corfield, Viscous Fluid Flow, McGraw-Hill, New York, 2006.

[41] M. O'neill, A sphere in contact with a plane wall in a slow linear shear flow, Chem. Eng. Sci. 23 (11) (1968) 1293–1298.

[42] A. Goldman, R. Cox, H. Brenner, Slow viscous motion of a sphere parallel to a plane wall—II Couette flow, Chem. Eng. Sci. 22 (4) (1967) 653–660.

[43] https://www.sparkfun.com, 2018.

[44] B.S. Yilbas, G. Hassan, A. Al-Sharafi, H. Ali, N. Al-Aqeeli, A. Al-Sarkhi, Water droplet dynamics on a hydrophobic surface in relation to the self-cleaning of environmental dust, Sci. Rep. 8 (1) (2018) 2984.

[45] B. Yilbas, H. Ali, N. Al-Aqeeli, M. Khaled, N. Abu-Dheir, K. Varanasi, Solvent-induced crystallization of a polycarbonate surface and texture copying by polydimethylsiloxane for improved surface hydrophobicity, J. Appl. Polym. Sci. 133 (22) (2016) 43467.

[46] W.Y.D. Yong, Z. Zhang, G. Cristobal, W.S. Chin, One-pot synthesis of surface functionalized spherical silica particles, Colloids Surf. A: Physicochem. Eng. Aspects 460 (2014) 151–157.

[47] F. Heib, M. Schmitt, Statistical contact angle analyses with the high-precision drop shape analysis (HPDSA) approach: basic principles and applications, Coatings 6 (4) (2016) 57–74.

[48] A. ElSherbini, A. Jacobi, Retention forces and contact angles for critical liquid drops on non-horizontal surfaces, J. Colloid Interface Sci. 299 (2) (2006) 841–849.

[49] D. Pilat, P. Papadopoulos, D. Schaffel, D. Vollmer, R. Berger, H.-J. Butt, Dynamic measurement of the force required to move a liquid drop on a solid surface, Langmuir 28 (49) (2012) 16812–16820.

[50] A.H. Ayyad, Thermodynamic derivation of the Young–Dupré form equations for the case of two immiscible liquid drops resting on a solid substrate, J. Colloid Interface Sci. 346 (2) (2010) 483–485.

[51] B.W. McCormick, Aerodynamics, Aeronautics, and Flight Mechanics, Wiley, New York, 1995.

[52] B.S. Yilbas, M.R. Yousaf, H. Ali, N. Al-Aqeeli, Replication of laser-textured alumina surfaces by polydimethylsiloxane: improvement of surface hydrophobicity, J. Appl. Polym. Sci. 133 (41) (2016) 1–13.

[53] D.K. Owens, R. Wendt, Estimation of the surface free energy of polymers, J. Appl. Polym. Sci. 13 (8) (1969) 1741–1747.

[54] A.T. Santhanam, Y. Gupta, Cleavage surface energy of calcite, International Journal of Rock Mechanics and Mining Sciences & Geomechanics Abstracts, Elsevier, 1968, pp. 253–259.

[55] Y.K. Shchipalov, Surface energy of crystalline and vitreous silica, Glass Ceramics 57 (11–12) (2000) 374–377.

[56] V. Bergeron, D. Langevin, Monolayer spreading of polydimethylsiloxane oil on surfactant solutions, Phys. Rev. Lett. 76 (17) (1996) 3152–3155.

[57] A. Carlson, P. Kim, G. Amberg, H.A. Stone, Short and long time drop dynamics on lubricated substrates, EPL (Europhys. Lett.) 104 (3) (2013) 34008.

[58] S. Anand, K. Rykaczewski, S.B. Subramanyam, D. Beysens, K.K. Varanasi, How droplets nucleate and grow on liquids and liquid impregnated surfaces, Soft Matter 11 (1) (2015) 69–80.

[59] D. Ingber, D.C. Leslie, M. Super, A.L. Watters, A. Waterhouse, Modification of Surfaces for Fluid and Solid Repellency. Google Patents, 2016.

[60] A. Laukkanen, J.-E. Teirfolk, O. Ikkala, R. Ras, H. Mertaniemi, Hydrophobic Coating and a Method for Producing Hydrophobic Surface. Google Patents, 2014.

[61] A. Al-Sharafi, B.S. Yilbas, H. Ali, A.Z. Sahin, Internal fluidity of a sessile droplet with the presence of particles on a hydrophobic surface, Numer. Heat Transfer, Part A: Applicat. 70 (10) (2016) 1118–1140.

[62] M. Krause, J. Blum, Y.V. Skorov, M. Trieloff, Thermal conductivity measurements of porous dust aggregates: I. Technique, model and first results, Icarus 214 (1) (2011) 286–296.

[63] G. Lu, Y.-Y. Duan, X.-D. Wang, D.-J. Lee, Internal flow in evaporating droplet on heated solid surface, Int. J. Heat Mass Transfer 54 (19–20) (2011) 4437–4447.

[64] S. Morsi, A. Alexander, An investigation of particle trajectories in two-phase flow systems, J. Fluid. Mech. 55 (2) (1972) 193–208.

[65] V. Vand, Theory of viscosity of concentrated suspensions, Nature 155 (3934) (1945) 364–365.

[66] Y. Fang, S. Kuang, X. Gao, Z. Zhang, Preparation of nanoencapsulated phase change material as latent functionally thermal fluid, J. Phys. D. Appl. Phys. 42 (3) (2008) 035407.

[67] G. Yali, W. Lan, S. Shengqiang, C. Guiying, Simulation of dynamic characteristics of droplet impact on liquid film, Int. J. Low-Carbon Technol. 9 (2) (2014) 150–156.

[68] J. Maroto, V. Pérez-Munuzuri, M. Romero-Cano, Introductory analysis of Bénard–Marangoni convection, Eur. J. Phys. 28 (2) (2007) 311–320.

[69] http://www.comsol.com/comsol-multiphysics, (2018).

[70] Y. Cui, A.T. Paxson, K.M. Smyth, K.K. Varanasi, Hierarchical polymeric textures via solvent-induced phase transformation: a single-step production of large-area superhydrophobic surfaces, Colloids Surf. A: Physicochem. Eng. Aspects 394 (2012) 8–13.

[71] A. Al-Sharafi, A.Z. Sahin, B.S. Yilbas, S. Shuja, Marangoni convection flow and heat transfer characteristics of water–CNT nanofluid droplets, Numer. Heat Transfer, Part A: Applicat. 69 (7) (2016) 763–780.

[72] S. Chandrasekhar, Hydrodynamics and Hydrodynamic Stability, Clarendon, Oxford, 1961.

[73] A. Nakajima, Design of hydrophobic surfaces for liquid droplet control, NPG Asia, Materials 3 (5) (2011) 49–56.

[74] D. Tam, V. von ARNIM, G. McKinley, A. Hosoi, Marangoni convection in droplets on superhydrophobic surfaces, J. Fluid. Mech. 624 (2009) 101–123.

[75] H. Gelderblom, O. Bloemen, J.H. Snoeijer, Stokes flow near the contact line of an evaporating drop, J. Fluid Mech. 709 (2012) 69–84.

Chapter 8

Concluding Remarks

Self-cleaning of surfaces has several advantages over other conventional cleaning techniques such as mechanical brushing, electrostatic repulsion, air blowing and suction, etc. Self-cleaning of surfaces involves surface texture characteristics and surface free energy of the substrate material; in which case, texture consisting of micro/nano pillars on the surface with low surface free energy are the main requirements for the cleaning process. Many challenges need to be addressed to create texture characteristics with low surface energy, which is particularly important for cost-effective and single-step processing. While surface texturing alters the surface wetting toward the hydrophobic state, it also modifies the optical characteristics on the surface. In this case, scattering and absorption of the incident optical radiation from the surface reduces the optical transmittance on the surface. The use of the optical correction fluid with similar refractive index of the substrate material can improve the optical transmittance; however, maintaining this type of oil-impregnated surface in outdoor environments is challenging. This is because micro/nanosize particles, such as dust particles, are cloaked by the impregnated oil while forming a dense particle layer at the oil and substrate surface interface. In addition, oil impregnation covers the surface texture and changes the wetting state at the surface from hydrophobic to hydrophilic. While research into self-cleaning of surfaces is aimed at meeting the conditions of harsh environments, recent changes in climate cause regular dust storms around the globe, particularly in the Middle East. The dust particles accumulate and and modify its optical and chemical characteristics. The dust effects on surfaces magnify in ambient humid air; in which case, some of the chemical compounds in the dust particles dissolve in water condensate while forming a chemically active solution. The liquid solution gradually forms a film on the solid surface under the gravitational potential energy. Since the liquid solution is chemically active, in some cases, it causes erosion and corrosion on the surface while permanently damaging the surface. Recent research studies also demonstrate that the liquid solution upon drying forms an adhesion layer between the dust particles and the solid surface. This, in turn, significantly increases the efforts required to remove the dust particles on the surface because of the excessive large pinning forces.

Self-Cleaning of Surfaces and Water Droplet Mobility. DOI: https://doi.org/10.1016/B978-0-12-814776-4.00008-2

Water-droplet mobility on hydrophobic surfaces can be utilized to enhance the self-cleaning characteristics of the surface. Research studies demonstrate that several factors affect the droplet mobility on the surface. In general, the inertia force created via surface tilting should overcome retarding forces including the pinning force due to droplet adhesion, frictional force, and drag force. In addition, the size of the droplet remains critical for the droplet motion on the inclined hydrophobic surface because of the fact that the large droplets undergo sliding rather than rolling on the surface because of the large droplet puddling during its motion. The droplet contact hysteresis enhances the droplet adhesion along the three-phase contact line while suppressing the droplet mobility. On the other hand, the thermocapillary effects can be generated while altering the internal fluidity of the droplet. Hence, the Marangoni and buoyancy forces generated inside the droplet, under temperature variation between the droplet liquid and the hydrophobic surface, contribute to the droplet mobility and lower the pinning force. Research into these effects toward achieving high mobility droplets on the hydrophobic surface is in progress. The concluding remarks related to the self-cleaning of surfaces in terms of the dust particles, the mud formation from the dust particles in humid ambient, the droplet mobility, and the thermocapillary effects are given separately in this section.

The dust particles collected from the local area of Dammam in Saudi Arabia comprised of a nonuniform distribution of alkali and alkaline earth metals, oxygen, silicon, sulfur, iron, etc. The average size of the particles is in the order of $1.2\ \mu m$. Morphological examination of the dust particles revealed that the dust shape factor approaches unity for the smaller particles, while for the large particles, the median shape factor reaches almost 3. The small dust particles ($\leq 0.5\ \mu m$) were attached to the surfaces of the large dust particles ($10 \leq 20\ \mu m$) due to the presence of electrostatic charges. The mud formed from the dust on the glass surface significantly influences the glass properties, including the absorption, transmittance, microhardness, and surface texture characteristics. The dissolution of the alkali and alkaline earth compounds in the mud forms a chemically active solution, and the active solution accumulates at the interface between the mud and the glass due to gravity. Because the mud solution contains alkali and alkaline earth hydroxides, it has a pH that reaches approximately 8.4 (basic). The mud solution attacks the glass surface while altering the surface texture and the molecular vibrational states in the surface region. In addition, the diffusion of potassium into the surface region causes toughening of the surface, while increasing the surface microhardness of the glass. The optical transmittance of the glass decreases after the removal of the mud; this reduction is associated with (1) mud residues that remain after cleaning the glass surface and (2) chemical changes in the glass surface due to the alkali and alkaline earth hydroxide attacks. The adhesion and cohesion work required to remove the mud from the glass is higher than the frictional work performed against the glass surface. When the mud is formed from the dust particles on the poly-carbonate (PC) sheet, the dissolved alkaline and alkaline earth metal hydro-xyls (OH^-) sediments precipitated at the interface of the mud and PC surface, forming a layer in this region after drying. Alkaline hydroxyls attacked the PC surface, thereby altering the vibrational state of the macro-molecules at the surface region, and increasing the microhardness and lower-ing the UV−visible transmittance of the resulting PC surface. In addition, the bonding of calcite and the formation of hydroxyls compounds at the PC surface increased the adhesion of the dry mud on the surface. The adhesion work required to remove the dry mud from the PC surface increased signifi-cantly because of the presence of the dried mud solution at the interface of the dry mud and the PC sheet surface. The influence of the chemo-mechanical behavior of the mud formed from the dust particles on the PC surface is novel and significantly alters the properties of the PC. The findings provide broad insight into the performance of PC for solar-energy systems when subjected to environmental dust and mud.

The lateral adhesion force of the droplet overcomes the gravitational force in the direction of surface inclination, which is true for all the inclination angles and droplet volumes considered. The shear stress developed at the droplet bottom due to the rate of fluid strain is considerably smaller than that of the lateral adhesion force; hence, its contribution to the droplet pinning is negligibly small. The normal force generated on the hydrophobic surface due to droplet surface tension is larger than that of the gravitational force for the surface inclination angles up to 115 degrees. As the inclination angle of the hydrophobic surface increases further, the normal component of the adhesion force remains less than the gravitational force while resembling the compo-nent of the weight in the normal direction. Since the droplet pins on the inclined surface beyond 115 degrees, the contribution of the Magdeburg-like forces to the droplet pinning becomes critical. These forces are associated with the change of pressure in the air trapped within the texture gaps when the meniscus geometry of the droplet bottom changes during the inclination. Although some of the textures are fully connected and the air trap in these textures is exposed to atmospheric conditions, some of the textures are not connected and appear closed packed while causing the air pressure to be dif-ferent than the atmospheric pressure. The air trapped in these textures is responsible for the generation of the Magdeburg-like forces at the interface of the droplet meniscus and the surface texture. Heating of the droplet causes formation of two contour-rotating circulation cells inside the droplet for the surface inclination angles within the range of 0 degrees $< \delta < 25$ degrees and 135 degrees $< \delta < 180$ degrees. However, increasing the inclination angle of the surface further, a single circulation cell extends inside the droplet and occupies a large droplet volume, which is more pronounced for the inclina-tion angle and large circulation cell is developed for the inclination angles of 45 degrees $\leq \delta \leq 135$ degrees. This behavior is attributed to the buoyancy and the Marangoni currents, which influence the flow field inside the droplet. The Nusselt number remains high for inclination angles 45 degrees $\leq \delta \leq 135$ degrees, which is attributed to the convection heat transfer inside the droplet; in which case, a single circulation cell is formed inside the droplet, and the enhancement of heat diffusion in the droplet fluid is due to the temperature gradient. The present study provides insight into the droplet adhesion on the inclined hydrophobic surface, and the forces acting on the lateral and normal directions on the hydrophobic surface. Moreover, solvent-induced crystallization of the PC wafer gives rise to surface texture with a surface roughness of 2.8 μm and surface solid fraction within $0.4 \leq f \leq 0.6$. The solid fraction is defined as the ratio of poles area over the total area on the projected surface. The absence of nanofibrils at the surface suppresses the lotus effect and causes strong adhesion of the droplet at surface. This gives rise to a hydrophobic surface with high hysteresis of the water-droplet contact angle; in which case, the water droplet attaches to the surface for various inclination angles of the surface. The flow field generated inside the droplet, due to the inclination of the surface, results in flow acceleration inside the droplet. In this case, a circulation cell is formed inside the droplet. The orientation of the circulation cell center changes with increasing droplet volume and the inclination angle. This is particularly true for large volume droplet (45 μL); in which case, bulging causes excessive body deformation of the large size droplet at high inclination angle of the surface. The moment generated about the locus of the droplet meniscus, due to the droplet inclination, increases with increasing inclination angle; however, further increase in inclination angle reduces the value of moment because of the change of the line of action of the resulting force under the excessive droplet bulging. The adhesion force between the droplet and the surface is higher than the gravitational force component of the droplet in the inclination direction. The shear force developed, due to the flow circulation in the vicinity of the wetted surface, is significantly lower ($\sim 5 \times 10^{-12}$ N) and its contribution to the adhesion force is negligible.

The rotational speed of the rolling droplet is influenced by the inclination angle of the hydrophobic surface and the droplet size. In this case, increasing inclination angle and reducing droplet size enhances the droplet rotational speed. Increasing droplet size gives rise to increased adhesion force along the contact line, drag force, and the shear force. However, the pinning force, comprised of adhesion, drag, and shear forces, increases slightly with increasing droplet volume; however, the increase of the pinning force per unit inclination angle remains slightly larger than that of the increase of the droplet inertia force due to droplet volume increase during the droplet rolling. This, in turn, reduces the rotational speed with increasing droplet radius. The change of the force balance at the droplet–solid interface results in the droplet wobbling on the inclined surface during the rolling. In this case, the height of the droplet changes and becomes almost 24% of the averaged

height of the droplet during the early rolling period. As the rolling progresses, the change in the droplet height remains small. The small diameter
droplets with the same range of capillarity length behave like a quasisolid
sphere and the difference between the maximum and the minimum height of
the droplets becomes small during the rolling. In addition, the difference
between the maximum and the minimum heights remains small for small
values of the rotational Bond number ($Bo_R < 0.02$). The ratio of rotational
speed over the translational speed is critical for the relative magnitude of the
dynamic pressure generated between the droplet center and the droplet ambient pressure. Since the ratio of rotational speed over the translational speed
is in the order of unity, the droplet height does not change considerably
under the influence of relative dynamic pressure during the rolling. A single
circulation cell is formed inside the droplet for various sizes during the rolling and the circulation cell center almost coincides with the droplet mass
center. The rate of fluid strain in the close region of the first and second contact points of the droplet on the inclined surface remains high. This gives
rise to the slightly high shear stresses formation along the contact line in this
region. Nevertheless, the shear stress and shear force become small due to
the low rate of fluid strain and its localized effect, i.e., the shear force acting
on the contact surface is in the order of 5×10^{-8} N. The heat transfer from
the substrate surface to the droplet gives rise to the formation of two
counter-rotating circulation cells. As the droplet contact angle increases, the
center of the circulation cells moves toward the interior of the droplet. In
this case, the combination of the buoyancy and the Marangoni forces mainly
governs the internal fluidity of the droplet. The Bond number remains below
unity for contact angle < 150 degrees and the Marangoni force remains
higher than that of the buoyant force. As the contact angle increases further,
the wetted area reduces significantly while increasing the droplet characteristic length ($L_c = \frac{\forall}{\pi R^2}$). This increases the Bond number significantly. The
Nusselt number also increases with increasing droplet contact angle. In this
case, thinning of the thermal boundary layer takes place in the near region of
the droplet bottom and temperature gradient increases in this region. This, in
turn, accelerates the heat transfer rates and enhances the Nusselt number.
The number of particles above the horizontal rake increases significantly for
low-density dust particles, which is more pronounced during the early heating period. The dust particles almost follow the streamlines in the circulation
cells for the low-density particles. The contribution of the hydrostatic force,
due to Brownian motion, and the combination of the buoyancy and the
Marangoni forces are responsible for the motion of the dust particles in the
droplet. As the density of the dust particles increases, the body force
becomes critical and the number of particles crossing over the horizontal
rake and situated in the central region of the droplet reduces significantly.

   The solution crystallized polycarbonate surface has hydrophobic characteristics with water-droplet contact angle of 130 degrees and contact angle

hysteresis of 36 degrees. The deposition of functionalized silica particles on crystallized polycarbonate surface improves the water-droplet contact angle to 160 degrees and lowers the contact angle hysteresis to 2 degrees. The coating of functionalized silica particle-deposited surface by the continuous film of liquid n-octadecane with the thickness of 1.5 μm changes the wetting state of the surface to hydrophilic state with a water-droplet contact angle of 86 degrees and contact angle hysteresis of 1 degree. Once the continuous film of liquid n-octadecane solidifies by reducing temperature, flakes of n-octadecane are formed on the functionalized silica particle-deposited surface. This gives rise to exposure of functionalized silica particles to the free surface via emerging beside the flakes of solid n-octadecane. This alters the wetting state from hydrophilic to hydrophobic with a droplet contact angle of 140 degrees and contact angle hysteresis of 8 degrees. Once the flakes of the solid n-octadecane are melted on the functionalized silica particle-deposited surface with increasing temperature to liquidus temperature, a continuous film of liquid n-octadecane film is formed on the surface. In this case, the wetting state of the surface changes to hydrophilic. Consequently, the reversible exchange of wetting state occurs once the n-octadecane coating undergoes phase change on the functionalized silica particle-deposited crystallized polycarbonate surface. The liquid n-octadecane forms a ridge around the water droplet due to cloaking. The water droplet moves on the surface onset of the liquefaction of n-octadecane film. This is mainly because of the flow current developed in the liquid phase of the n-octadecane onset of melting. In this case, the n-octadecane ridge formed around the water droplet anchors and moves the droplet on the surface of the liquid n-octadecane film. In order to achieve reversible exchange of wetting state of the functionalized silica particle-deposited surface, n-octadecane coating with 1.5 μm thickness is introduced on the surface. The resulting surface remains hydrophobic with a water-droplet contact angle of 140 degrees and a contact angle hysteresis of 8 degrees when n-octadecane is in a solid phase and it becomes hydrophilic with a water-droplet contact angle of 85 degrees when the n-octadecane is in a liquid phase. The wetting state is reversibly exchanged through melting and resolidification of n-ocadecane coating. The solid-phase n-octadecane coating forms flakes like structures at the surface, which are distributed randomly at the surface, and functionalized silica particles emerge to the free surface in the near region of the flakes. This arrangement gives rise to hydrophobic characteristics at the surface. The liquid phase totally wets the surface and encapsulates (covers entirely) the functionalized silica particle-deposited surface while forming a thin layer of a liquid film with 1.5 μm thickness, which has the hydrophilic characteristics. The translational velocity of the water droplet increases sharply, reaching the terminal velocity along the inclined surface. Increasing droplet volume reduces the translational velocity on the surface; in which case, the droplet wobbling increases while enhancing energy dissipation due to the elastic deformation of the droplet during its movement on the inclined hydrophobic surface. The rolling of the droplet terminates randomly on the inclined surface depending on the surface texture conditions. In this case, closely spaced solid n-octadecane flakes, formed during solidification of the liquid film, change the wetting state from hydrophobic to hydrophilic on the surface. Consequently, the wetting state of the surface changes arbitrarily depending on the solidification pattern of the n-octadecane liquid film. Sliding of the droplet occurs once the rolling ceases on the inclined surface. This behavior is more pronounced for large droplets; in which case, a small liquid tail is formed behind the droplet on the surface.

Droplet heat transfer influences the internal fluidity of the droplet fluid. In this case, the correlation among the variation of the Nusselt, the Bond numbers, and the droplet contact angle is examined thoroughly and a new number, the Ayse number, is proposed to combine the effects of the Bond number and the contact angle. In general, the flow field is comprised of two counter-rotating cells inside the droplet for small contact angles, which correspond to the hydrophilic surface. The number of circulation cells increases to four inside the droplet as the contact angle increases, which resembles the hydrophobic characteristics at the substrate surface. In this case, the neighboring circulation cells demonstrate counter-rotational features. This occurs for the hydrophilic and hydrophobic surfaces. The cell rotation depends on the direction of the heat transfer; in which case, the heat transfer from droplet to the surface (case 1) results in the clockwise rotation of the cell in the right lobe of the droplet. However, the rotation becomes counterclockwise for the case of heat transfer taking place from the substrate surface toward the droplet. This behavior is associated with the combination of the Marangoni and buoyancy forces; in which case, the direction of the Marangoni flow changes in the droplet because of the change of the surface tension gradient ($d\sigma/dT$) sign. The Nusselt and the Bond numbers increase with the contact angle. This is attributed to the increase of Marangoni force in the droplet with increasing droplet contact angle. Since the droplet height increases with increasing contact angle, it in turn enhances the temperature gradient and heat diffusion in the droplet. Consequently, enhancement of the Marangoni current and heat diffusion increases the Nusselt number. The rise of Nusselt number with the contact angle is sharper for the hydrophobic surfaces in both cases of the heat transfer (case 1 and case 2). This argument is also true for the Bond number variation. This demonstrates the strong dependence of the heat transfer rates on the contact angle. Since the Nusselt number and the Bond number variations with the contact angle are similar, a new correlation, the *Ayse* number, is introduced between the Bond number and the droplet contact angle. The *Ayse* number combines the effects of the Marangoni and the buoyancy forces, and the interfacial energies of the fluids on the heat transfer characteristics. In this case, a linear relation is developed between the Nusselt number and the Ayse number. The values of the coefficients of the linear relation change for the hydrophilic and the hydrophobic surfaces as well as the direction of the heat transfer. The inclusion of 1% (weight percent) of CNT into the water leads to the enhancement of the thermal conductivity by 2% and reduces the heat capacity by 0.89%. The Grasshoff, Marangoni, and Bond numbers are incorporated to analyze the influence of buoyancy, viscous, and surface tension forces on the flow field inside the droplet. The flow-velocity predictions are validated with the experimental data incorporating the identical conditions. The predictions of the simulations revealed that Marangoni flow has a dominant contribution to the flow field in the droplet as compared to that of the natural convection. The effect of the droplet size on the flow characteristics is also investigated. This is accomplished through increasing the droplet volume and its contact angle while monitoring the flow characteristics inside the droplet. The findings revealed that the Bond number enhances with increasing the contact angle and the internal flow is dominated by the Marangoni convection for high droplet contact angles. Using the particle tracing model, the transient analysis can be achieved to determine flow characteristics and visualize the motion of CNT particles in the solution domain.

The solvent crystallization of the polycarbonate surface results in hierarchical micro/nanotextures characteristics while achieving hydrophobicity at the surface. The functionalized silica particle deposition on the crystalized surface lowers the contact angle hysteresis to as low as 1 degrees. A new dimensionless number, called the Merve number, introduced to incorporate the effect of the droplet weight and surface tension force, provides useful information between the ratio of the droplet weight over the surface tension force and the Nusselt number. The droplet side surface heating results in two counter rotating circulation cells inside the droplet because of the combination of the Marangoni and the buoyancy currents. The Bond number remains low (less than unity); therefore, the Marangoni current governs the flow field inside the droplet. The heat transfer inside the droplet does not generate a notable temperature gradient in the close region of the droplet–hydrophobic surface interface. Therefore, the heat loss from the droplet fluid toward the hydrophobic surface, due to conjugation of the droplet–solid interface, remains negligibly small. The combination of the Marangoni and the buoyancy currents generates fluid acceleration inside the droplet. The total fluid acceleration increases with reducing droplet volume, which in turn enhances the inertia force of the fluid. The adhesion force increases with the droplet volume and the shear force developed at the hydrophobic surface since the rate of fluid strain in the wall vicinity is negligibly small. The difference between the droplet fluid inertia force and the adhesion force remains greater than zero for all the droplet volumes considered in the present study. This finding reveals that the droplet rolls off on the hydrophobic surface under the net resulting force between the fluid inertial and droplet adhesion forces. The Nusselt and the Bond numbers increase with increasing droplet volume. The Bond number attains values less than unity for all droplet sizes.

This indicates that the Marangoni current dominates over the buoyancy current inside the droplet fluid. Two distinct regions are identified for the Nusselt number variation with the Merve number. The Nusselt number remains at low values in the first region, which extends to the Merve number ≤ 2.5. The Nusselt number increases linearly with increasing the Merve number in the second region, which corresponds to a Merve number ≥ 2.5. This indicates that the droplet size has a considerable effect on the heat transfer rates from the radiative source to the water droplet, i.e., the ratio of gravitational force over the surface tension force becomes important in terms of the heat transfer rates.

Silicon wafer is textured via a lithographic technique to generate micro-post arrays on the surface, which gives rise to hydrophobic wetting state on the surface. The textured surface is replicated by polydimethylsiloxane (PDMS) to enhance optical transparency and reduce surface energy. The water-droplet mobility on the textured surface is examined. The effect of droplet size on the heat transfer rates is incorporated and the new number, the Merve number, is introduced to account for the droplet size. The variation of the Nusselt and the Bond numbers with the Merve number is assessed. The findings reveal that the water-droplet adhesion on PDMS repli-cated surface with micropost arrays depends on the texture structure includ-ing micropost cross-sectional area, micropost pitch, and postheight. The buoyancy and the Marangoni forces generated in the droplet result in forma-tion of two counter-rotating circulation cells. The circulation cell centers move away from the droplet bottom as the droplet size changes. The buoy-ancy current and thermal diffusion at the droplet bottom gives rise to attain-ment of a high-temperature region along the central rake of the droplet. The shear layer developed at the outer region of the circulation cells prevents mixing of the heated fluid in this region; in which case, temperature remains low within the circulation cells. The maximum flow velocity increases with the droplet size, which occurs in the vicinity of the droplet surface in the upper part of the droplet. The Nusselt and Bond numbers increase with Merve number; however, the Bond number remains less than unity for all droplet sizes considered. This indicates that the Marangoni current dominates over the buoyancy current in the droplet. The Nusselt number attains higher values for the micropost-textured surface as compared to the plain surface. This behavior is attributed to the convection current developed at the droplet bottom because of temperature and velocity oscillations along the droplet contact line due to the pitch of the micropost arrays. The present study enhances the understanding of the droplet heat transfer associated with the hydrophobic surfaces comprised of micropost arrays. It also provides useful information on the droplet adhesion on textured surfaces toward self-cleaning applications. In the case of the ferro fluid, the droplet diameter slightly increases at the three-phase contact line as the droplet volume increases beyond 45 µL; in this case, the droplet height increases more than its diameter. This behavior is associated with force balance between the droplet interior pressure and the surface tension forces. The horizontal component of the surface tension gives rise to adhesion of the droplet on the water surface; in which case, the droplet remains stationary on the water surface during the heating period. Two counter-rotating circulation cells are formed separately inside the water and the droplet fluid. The Marangoni and buoyancy forces developed inside the droplet are responsible for the formation of the circulation cells. The maximum flow velocity inside the water remains almost the same for all droplet volumes incorporated. However, the maximum flow velocity inside the droplet increases with increasing droplet volume. In addition, the size of the circulation cells increases with increasing droplet volume. The Bond number attains values larger than unity. Since the Bond number is associated with the buoyancy over the Marangoni forces, the attainment of a large Bond number indicates that the influence of the buoyancy current on the formation of the circulation cells is significant. The temperature in the region close to the cell centers remains low and the temperature increases in the outer region of the circulation cells, which is associated with the convection current and the heat diffusion, i.e., the heated fluid is carried from the droplet bottom toward the droplet interior and heat conduction inside the droplet fluid. The Nusselt number increases with increasing droplet volume. As the droplet volume increases beyond 45 μL, the increase in the Nusselt number becomes gradual. This is associated with the contribution of the buoyancy and Marangoni currents to the heat transfer inside the droplet, which behaves similar to the Bond number variation.

The dynamics of a water droplet on an inclined hydrophobic surface are considered, and the removal of environmental dust particles from the hydrophobic surface by water droplets is examined. The translational velocity of the droplet is comprised of the rotational and slip velocities, and the rotational velocity dominates over the slip velocity along the hydrophobic surface, which is true for the inclined hydrophobic surfaces both with and without dust. The prediction of the rotational speed agrees well with that obtained from experiments. The initial location of the droplet on the hydrophobic surface (standoff distance) is important in terms of the droplet translational velocity; in which case, the translation velocity reaches an almost stable value as the distance along the hydrophobic surface increases. The initial location of the dust particles on the hydrophobic surface is also important in terms of the droplet translational velocity along the dusty region. In this case, the droplet velocity along the dusty region is lower at short distances (10 mm) than at large distances (50 mm). Consequently, the droplet velocity is not affected by the dust particles along its path when the droplet velocity attains high values prior to reaching the dusty region. The retention forces of the clean and dusty surfaces are almost similar; however, the retention force for the droplet on the dusty region is higher than that on the clean surface. On the other hand, droplet cloaking is responsible for the removal of dust particles from the hydrophobic surface by the water droplet; in which case, the time required for cloaking the dust particles is less than the transition time during which the droplet is wet along the length of the hydrophobic surface. Few dust residues are left on the droplet pathway after the droplet passes over the dusty region. The elemental composition of the dust residue is similar to that of the ordinary dust particles. The dust residues have low surface energy while giving rise to a negative spreading rate; thus, the water droplet could not cloak these particles during its transition along the pathway. The hydrophobic surface cleaned by the rolling water droplet significantly improved the optical transmittance of the surfaces over that of the dusty surface. In the case of the rotating disk for the dust particle removal, the shape and the size of the dust particles located on the disk were assessed. The average dust particles measured, based on the volumetric consideration, is in the order of 10 μm. The adhesion force predicted, using the Rump—Rabinovich model, is higher than the gravitational, friction, lift, and drag forces at all locations on the disk surface. During the disk rotation, the centrifugal force attains a higher value than the adhesion force in the region away from the disk rotational center. The effective distance along the disk radius, where the centrifugal force remains higher than the adhesion force, depends on the rotational speed of the disk and the dust particle size. In this case, the effective distance reduces (≤ 100 rad/s) as the rotational speed reduces and dust particle size becomes small (≤ 3 μm). The dust residues after the rotational tests indicate that the dust particles possess alkaline and alkaline earth metals, which causes possible ionic bonding on the disk surface in the humid environment. In addition, dust particles have electrostatic charges that enhance the adhesion between the dust residues and the disk surface. The adhesion force measurements for the dust residues on the disk surface by using AFM reveal that the adhesion force measured is significantly larger than that calculated from the Rump—Rabinovich model. This is attributed to the influence of the electrostatic charges of the dust particles, which contributes to the adhesion force. This is due to the fact that the Rump—Rabinovich model only considers the van der Waals forces between the dust particles and the disk surface. The thermocapillary effect in the water droplets is considered and the motion of microsize particles in a sessile droplet under the thermocapillary-induced forces is investigated for dust particle removal from hydrophobic surfaces. The flow field inside the droplet is predicted numerically in line with the experimental conditions in which temperature difference is created between the droplet fluid and the hydrophobic surface. Various sizes of particles and droplet contact angles are incorporated in the numerical simulations. The number of particles on the hydrophobic surface is also included in the simulations to examine the particle number density on the motion of the particles inside the droplet. The findings reveal that the dust particles collected have various sizes and the average dust particle size is in the order of 1.2 μm and the density of the dust particles is in the order of 1600 kg/m^3. Since the dust particles are heavier than water, they initially sediment at the droplet bottom. However, combination of buoyancy and Marangoni currents results in motion of the particles inside the droplet. In this case, two counter-rotating circulation cells are formed inside the droplet for the droplet contact angle $\theta = 100$ degrees. The number of counter-rotating circulation cells increases to four for the droplet contact angle $\theta = 140$ degrees. The presence of dust particles inside the droplet does not alter significantly the flow field inside the droplet. The forces developed, due to buoyancy, drag, and hydrodynamic Brownian motion, are smaller than the gravitational force. The buoyancy current generated in the close region of the droplet bottom enables removing the particles from the surface into the droplet interior. The particles follow the streamlines in the circulation cells because of a combination of buoyancy and Marangoni currents. The percentage of particles passing over the horizontal rake, which resemble the height of the settled particles in the droplet, remains high for the large particles, and as time progresses, the rate of dust removed from the droplet bottom into the droplet interior remains steady. The particle number density does not significantly influence the percentage of particles removed from the droplet bottom to the droplet interior. The particles, resembling the environment dust particles, can be removed from the hydrophobic surface through thermocapillary-induced forces inside the droplet. However, the particle size is critical for the rate of particle removal from the hydrophobic surface. In addition, particle number density slightly lowers the percentage of particles removed from the droplet bottom.

Index

Note: Page numbers followed by "*f*" and "*t*" refer to figures and tables, respectively.

CPI Antony Rowe
Eastbourne, UK
March 20, 2023